Advanced and Intelligent Manufacturing in China

Series Editor

Jie Chen, Tongji University, Shanghai, China

This is a set of high-level and original academic monographs. This series focuses on the two fields of intelligent manufacturing and equipment, control and information technology, covering a range of core technologies such as Internet of Things, 3D printing, robotics, intelligent equipment, and epitomizing the achievements of technological development in China's manufacturing sector. With Prof. Jie Chen, a member of the Chinese Academy of Engineering and a control engineering expert in China, as the Editorial in Chief, this series is organized and written by more than 30 young experts and scholars from more than 10 universities and institutes. It typically embodies the technological development achievements of China's manufacturing industry. It will promote the research and development and innovation of advanced intelligent manufacturing technologies, and promote the technological transformation and upgrading of the equipment manufacturing industry.

Min Liu · Ling Li · Feng Yan

Intelligent Predictive Maintenance

Chemical Industry Press Co., Ltd.

Springer

Min Liu
College of Electronic and Information Engineering
Tongji University
Shanghai, China

Ling Li
School of Statistics and Mathematics
Shanghai Lixin University of Accounting and Finance
Shanghai, China

Feng Yan
CINF Engineering Corporation Limited
Changsha, Hunan, China

ISSN 2731-5983 ISSN 2731-5991 (electronic)
Advanced and Intelligent Manufacturing in China
ISBN 978-981-97-2676-9 ISBN 978-981-97-2677-6 (eBook)
https://doi.org/10.1007/978-981-97-2677-6

Jointly published with Chemical Industry Press
The print edition is not for sale in China (Mainland). Customers from China (Mainland) please order the print book from: Chemical Industry Press.

The original submitted manuscript has been translated into English. The translation was done using artificial intelligence. A subsequent revision was performed by the author(s) to further refine the work and to ensure that the translation is appropriate concerning content and scientific correctness. It may, however, read stylistically different from a conventional translation.

Translation from the Chinese Simplified language edition: "智能预测性维护" by Min Liu et al., © Chemistry Industry Press 2020. Published by Chemistry Industry Press. All Rights Reserved.

This Springer imprint is published by the registered company Springer Nature Singapore Pte Ltd.
The registered company address is: 152 Beach Road, #21-01/04 Gateway East, Singapore 189721, Singapore

Preface I

This book mainly introduces fault diagnosis and prediction methods, intelligent predictive maintenance technology, systems and frameworks, IoT-based perception resource management frameworks and models, wireless routing models and algorithms for complex manufacturing environments, protocol integration and cases for data collection, data-driven fault diagnosis methods, data-driven fault prediction models and methods, maintenance optimization scheduling and decision-making in intelligent factories, large-scale maintenance service prediction and optimization configuration, and operation process control based on cyber-physical systems.

This book can serve as an important reference in the fields of automatic testing, reliability, and predictive maintenance.

Manufacturing is the mainstay of the national economy, the foundation of a country, the tool of prosperity, and the basis of strength. In the past twenty years, China's manufacturing industry has continued to develop rapidly, its comprehensive strength has been continuously enhanced, and its international status has been greatly improved, has become the country with the largest scale of manufacturing in the world from 2010. However, China is still in the process of industrialization, and the problem of being big but not strong is prominent, and there is still a large gap compared with advanced countries. To solve the problems of being big but not strong in manufacturing, weak independent innovation capability, high dependence on foreign countries for key core technologies and high-end equipment, the State Council issued the "Made in China 2025" national plan on May, 2015. Subsequently, the Ministry of Industry and Information Technology issued the "Made in China 2025" plan, which proposes China's "three-step" development strategy for manufacturing and the struggle goals, guides principles and strategic routes for 2025, and formulated 9 major strategic tasks and 10 key development areas. On August, 2016, the Ministry of Industry and Information Technology, the National Development and Reform Commission, the Ministry of Science and Technology, and the Ministry of Finance jointly issued the implementation guidelines for the five major projects of "Made in China 2025" manufacturing innovation center, industrial strong foundation, green manufacturing, intelligent manufacturing, and high-end equipment innovation.

In response to the major strategic deployment of building a manufacturing powerhouse made by the Party Central Committee and the State Council, local governments, enterprises, and research departments are actively exploring and deploying. Accelerating the integration of new generation information technology and manufacturing technology, promoting the transformation of China's manufacturing mode from "Made in China" to "Intelligent Manufacturing in China", and accelerating the transformation of China's manufacturing industry from large to strong, is becoming our new historical mission. At present, the information revolution process continues to evolve rapidly, and technologies such as the Internet of Things, cloud computing, big data, and artificial intelligence are widely infiltrating all areas of the economy and society. The prosperity of the digital economy has become an important symbol of national strength. Additive manufacturing, robots and intelligent manufacturing, control and information technology, artificial intelligence, and other fields continue to make major breakthroughs, promoting the differentiation and transformation of the traditional industrial system, and will reshape the international division of labor in the manufacturing industry. The integration of manufacturing technology and information technology such as the Industrial Internet has become a major trend and main feature of the new round of technological revolution and industrial transformation. Under the background of the great development and transformation of China's manufacturing industry, the Chemical Industry Press actively complies with the trend of technology and industry development, and organizes the publication of the "Made in China 2025" publishing project series, which can be described as brave to lead and timely.

The "Made in China 2025" publishing project series is a set of high-level, highly original academic monographs closely revolving around the State Council's implementation of the first ten-year action program of the manufacturing powerhouse strategy-"Made in China 2025". The series is based on intelligent manufacturing and equipment, control and information technology, covering a series of core technologies such as the Internet of Things, big data, 3D printing, robots, intelligent equipment, industrial network security, knowledge automation, artificial intelligence, etc. The selection and planning of the series closely combines the "Made in China 2025" plan and 11 supporting implementation guidelines, action plans or special plans, and each volume organizes content for some core technologies in each field, which reflects the technical development achievements in the domestic manufacturing industry. The aim is to strengthen the research and development, promotion and application of advanced technologies, and provide targeted guidance and systematic technical references for the rooting of the "Made in China 2025" action program.

This set of books embodies the following major features:

First, the content of the series strives to be original, with networked and intelligent technologies as the core, gathering many cutting-edge technologies, reflecting some of the latest technological achievements at home and abroad, especially reflecting the relevant original technological achievements in China. These books include many new technologies that have won many national and provincial and ministerial level scientific and technological awards, so the publication of the books is very helpful for the promotion and application of new technologies! These contents not only solve practical problems for technicians, but also provide new directions and expand new ideas for research.

Secondly, while introducing new technologies, new theories, and new methods in the corresponding professional fields, each volume of the series prioritizes the introduction of new technologies with application prospects and their promotion and application examples to promote the transformation of excellent scientific research results into industries.

The series is led by Sun Youxian, a Chinese control engineering expert and academician, and is planned and organized by many academicians such as Wu Cheng, Wang Tianran, and Zheng Nanning. Many Changjiang Scholars, outstanding young people, and young scholars participate in the specific writing work, which has a high academic level and writing quality.

It is believed that the publication of this series of books has a positive significance for promoting the implementation of the "Made in China 2025" national important strategic plan, can effectively promote the research and development and innovation of China's intelligent manufacturing technology, promote the technological transformation and upgrading of the equipment manufacturing industry, improve the design ability and technical level of products, and thus enhance the core competitiveness of China's manufacturing industry from multiple angles.

Shanghai, China Min Liu
Shanghai, China Ling Li
Changsha, China Feng Yan

Preface II

With the development of information, mechanical, system engineering, management, and other disciplines, especially the new generation information technology represented by mobile Internet, big data, artificial intelligence, etc., the rapid development of intelligent manufacturing has been promoted.

In the field of equipment operation and maintenance, in order to solve the problems of remote, real-time, and timely operation and maintenance services of equipment, equipment manufacturing or operating enterprises collect equipment operation data and upload it to the enterprise data center or enterprise cloud, so that system software can monitor and control the equipment in real time online, and perform predictive maintenance through data analysis, which gradually forms an internet based Maintenance, Repair and Operations (MRO) service network and corresponding data support environment. The supporting data includes product status, equipment status, environmental status, business operation status, personnel status, social network data, and customer feedback data, etc., and the MRO services can have the characteristics of globality (Internet-based connection), real time (industrial Internet-supported status detection), and timeliness (cloud-based mobile terminal scheduling). In MRO service network, the knowledge of equipment lifecycle process can be efficiently and spontaneously generated and utilized.

However, due to the high randomness of equipment failures and the high uncertainty of maintenance requirements in the MRO service network, there exist some challenges to face the multiple uncertain collaborative entities and high randomness of equipment failures in the large-scale MRO service network. These challenges are: (1) How to reliably acquire, model, infer, and mine heterogeneous data information from product status, environmental status, equipment operation status, personnel status, business operation data, social network data, and customer product feedback data in the equipment lifecycle process? (2) How to use the above information to accurately predict equipment failures and maintenance requirements, and formulate appropriate maintenance and optimization strategies? (3) How to make full use of the widely the existing, limited and shared service resources in the Internet environment, respond to network maintenance requirements to the greatest extent, and establish long-term comprehensive cooperative relationships with customers? (4)

Furthermore, how to feedback maintenance decisions to the manufacturing system through the Cyber Physical System (CPS), to achieve error compensation, adjustment, and feedback control? These issues have become the common demands of all participants in the MRO service network in the intelligent manufacturing environment, and it is necessary to study the theories and methods of high-quality acquisition and fusion of multi-source heterogeneous data for intelligent operation environment, data-driven equipment and component failure prediction theories and methods, large-scale maintenance decision theories for equipment operation networks, feedback control methods based on CPS, and theories, technologies and methods of intelligent Predictive Maintenance (SPdM) for MRO service networks in the big data environment.

This book systematically introduces the impact of the new generation information technology on manufacturing, the equipment maintenance problems in the intelligent manufacturing, and various maintenance strategies, such as preventive maintenance, predictive maintenance, and intelligent predictive maintenance; it details the method system of fault prediction, including fault prediction based on physical models, fault prediction based on reliability models, data-driven fault prediction, and fusion model-driven fault prediction; combined with the reference architecture of intelligent manufacturing, it systematically introduces the technical system and framework of intelligent predictive maintenance for intelligent factories and intelligent factory operation networks; it focuses on the key theories, methods, and technologies of intelligent predictive maintenance, such as IoT-based perception resource management and workshop wireless routing technology, big data-driven fault prediction theories and methods, maintenance decision and optimization methods, network decision and optimization methods, and CPS-based virtual control technology, etc., it comprehensively and systematically expounds the emerging theories, technologies, and methods in the field of equipment maintenance, with a wide coverage and rich content; it strives to reflect the strategic deployment and implementation of the new generation intelligent manufacturing and the new generation artificial intelligence in China.

The specific content of this book is arranged as follows: Chap. 1 reviews the intelligent predictive maintenance, equipment maintenance strategies and development trends under the intelligent manufacturing mode, and paves the way for the unfolding of the content of the whole book; Chap. 2 introduces fault diagnosis and prediction methods; Chap. 3, combined with the reference system architecture of intelligent manufacturing, systematically introduces the technical system and framework of intelligent predictive maintenance; Chap. 4 introduces the perception resource management framework and model based on IoT; Chap. 5 introduces the wireless routing model and algorithm for complex manufacturing environments; Chap. 6 presents the protocol integration and design case of data collection; Chap. 7 discusses data-driven fault diagnosis methods; Chap. 8 introduces data-driven fault prediction models and methods; Chap. 9 discusses the maintenance optimization scheduling and decision-making of intelligent factories; Chap. 10 introduces the prediction and optimization configuration of large-scale maintenance services; Chap. 11 discusses the operation process control based on the Cyber Physical System.

This book is supported by the National Natural Science Foundation Project "Feature Extraction and Intelligent Prediction of Multi-source weakly Labeled Signals based on Transferable Representation Learning (No. 62273261)" and the National Key R&D Program Project "Research and Development of Key Technologies and Platforms for Network Collaborative Manufacturing for Fine Control of Non-ferrous Metal Smelting Process (2019YFB1704700)". The book is written by Liu Min, Li Ling, and Yan Feng, and special thanks to Ma Yumin for her contribution to Chap. 11. The book was assisted by Zuo Liangqing and Meng Keao, postgraduate students in the System Engineering major of the College of Electronics and Information Engineering, Tongji University, and we would like to express our heartfelt thanks to them.

As this book covers a wide range and discusses relatively new and complex issues, there may inevitably be shortcomings in the book, and we sincerely welcome readers to criticize and correct.

Shanghai, China | Min Liu

Contents

Chapter 1
Introduction

The deep integration of new generation information technologies such as mobile internet, big data, and artificial intelligence with manufacturing processes has driven profound changes in manufacturing systems, production methods, industrial forms, and business models, propelling the transformation and upgrade from traditional manufacturing to intelligent manufacturing. In the field of equipment maintenance, traditional reliability-based preventive maintenance strategies are gradually shifting to data-driven intelligent predictive maintenance strategies. Intelligent predictive maintenance models and analyzes MRO data to realize intelligent services such as fault prediction, maintenance optimization, and decision-making for equipment operation and maintenance networks. This chapter mainly introduces intelligent predictive maintenance, maintenance strategies, and the content of this book.

1.1 Predictive Maintenance

1.1.1 Manufacturing Industry in the Era of Intelligence

A new round of scientific and technological revolution, especially the popularization of the new generation information technology such as mobile Internet, big data, artificial intelligence, Internet of Things (Fig. 1.1), has led to a group technological revolution characterized by green, intelligent, service-oriented and networked in almost all fields. The deep integration of the new generation information technology and manufacturing industry has bred intelligent manufacturing.

Since 2012, with the introduction of intelligent manufacturing strategies such as the US Industrial Internet, German Industry 4.0, and Made in China 2025, the industrial society has entered an era of intelligent interconnection based on big data perception, analysis and application. The interconnection between devices and devices in people's work and life, the interconnection between people and devices based on the

M. Liu et al., *Intelligent Predictive Maintenance*, Advanced and Intelligent Manufacturing in China, https://doi.org/10.1007/978-981-97-2677-6_1

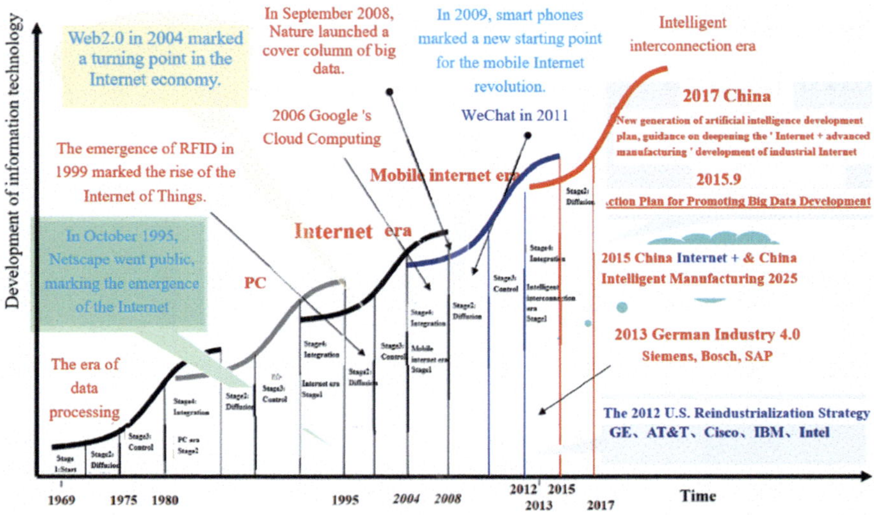

Fig. 1.1 The development process of information technology

Internet of Things and the mobile Internet, and the intelligence based on data analysis and prediction, mark the arrival of the fourth industrial revolution characterized by intelligent interconnection (Fig. 1.2).

Intelligent interconnection is an intelligent network based on Internet of Things technology, with platform-based intelligent hardware as the carrier, in accordance with the agreed communication protocol and data interaction standards, combined with cloud computing, big data and artificial intelligence applications, information

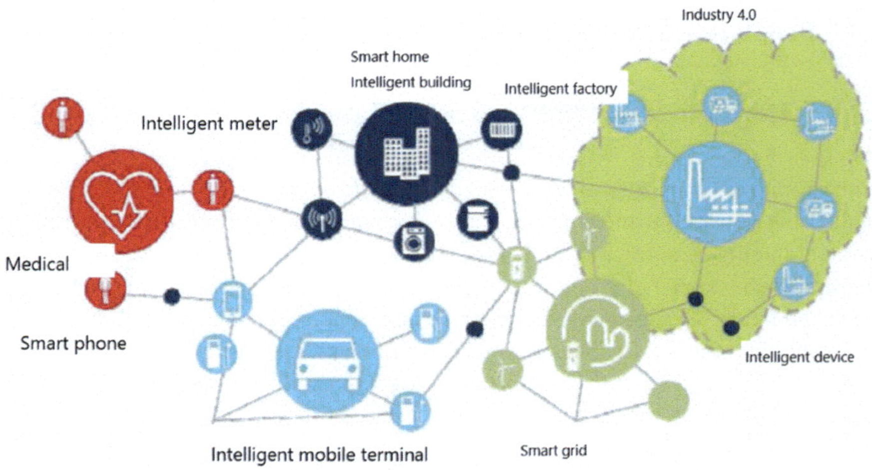

Fig. 1.2 Intelligent interconnected society (Adapted from Industry 4.0)

collection, processing, analysis and application between intelligent terminals, people and cloud services. It has the comprehensive ability of high-speed mobile, big data analysis and mining, intelligent induction and application. It can penetrate and integrate into traditional industries, improve the service ability of traditional industries, connect hundreds of industries, and conduct online and offline cross-border full marketing. In the era of intelligent interconnection, great changes have taken place in the structure, business, business model and basic concept, and its ability is far from the traditional Internet can imagine.

Intelligent interconnection subverts the traditional Internet and reconstructs the relationship between people and people, people and things, including the high reconstruction between people and services, services and services, people and media, virtual and reality, people and the Internet, etc. It is a comprehensive digital reconstruction of traditional enterprises from the product level and marketing level. Intelligent hardware involves cloud computing, big data, sensors, communications and other fields. It is the terminal entrance of the era of intelligent interconnection. It can interact with users intelligently according to the actual scene, so that intelligent hardware becomes a part of the user's work, life, travel and other scenes, and it is just needed in high frequency.

Intelligent interconnection is not to create and form a new industry. It must penetrate into traditional industries, enhance the service and marketing capabilities of traditional industries, change the promotion and dissemination capabilities of traditional industries, and make traditional industries more powerful. In the era of intelligent interconnection, the new generation of information technologies such as mobile Internet, big data and artificial intelligence are deeply integrated with manufacturing industry, which has given birth to a new manufacturing concept-Smart Manufacturing (SM), thus forming new production methods, industrial forms, business models and economic growth points: ① Promote new breakthroughs in 3D (three dimensional) printing, mobile Internet, big data, artificial intelligence, cloud computing, bioengineering, new energy, new materials and other fields; ② Intelligent manufacturing such as intelligent equipment and intelligent factory based on cyber-physical system (CPS) is leading the transformation of manufacturing methods; ③ Network crowdsourcing, collaborative design, mass customization, precision supply chain management, intelligent predictive maintenance, life cycle management and e-commerce are reshaping the industrial value chain system; ④ Wearable smart products, smart home appliances, smart cars and other intelligent terminal products continue to expand new areas of manufacturing.

Intelligent manufacturing is oriented to the whole life cycle of products. It is based on the new generation of information technology and takes the manufacturing system as the carrier. These key links and processes have certain autonomous perception, learning, analysis, decision-making, communication and coordinated control capabilities, and can dynamically adapt to changes in the manufacturing environment, so as to achieve certain optimization goals [2]. The relevant explanation is as follows:

(1) Intelligent manufacturing is oriented to the whole life cycle of products rather than chivalrous processing and production links, including intelligent design

and research and development of products, intelligent production (processing and assembly), intelligent services and other links, and products are the target objects of intelligent manufacturing.

(2) Intelligent manufacturing is based on a new generation of information technology, including the Internet, big data, cloud computing, etc., which is an information-based manufacturing under ubiquitous sensing conditions.
(3) Intelligent manufacturing system (Fig. 1.3) has different levels from micro to macro, such as manufacturing equipment, manufacturing units, manufacturing workshops, manufacturing enterprises and enterprise production systems. The composition of the manufacturing system includes products, manufacturing resources (machines, production lines, people, etc.), various process activities (design, manufacturing, management, services, etc.), and operation and management models;
(4) The application of intelligent manufacturing technology is aimed at the key link and process of manufacturing system, not necessarily all.
(5) Intelligent manufacturing system must have a certain degree of autonomy of perception, learning, analysis, decision-making, communication and coordination control capabilities, which is different from automated manufacturing systems and digital manufacturing systems. At the same time, it is very important to dynamically adapt to changes in the manufacturing environment. A system with only optimized computing power is different from an intelligent system.

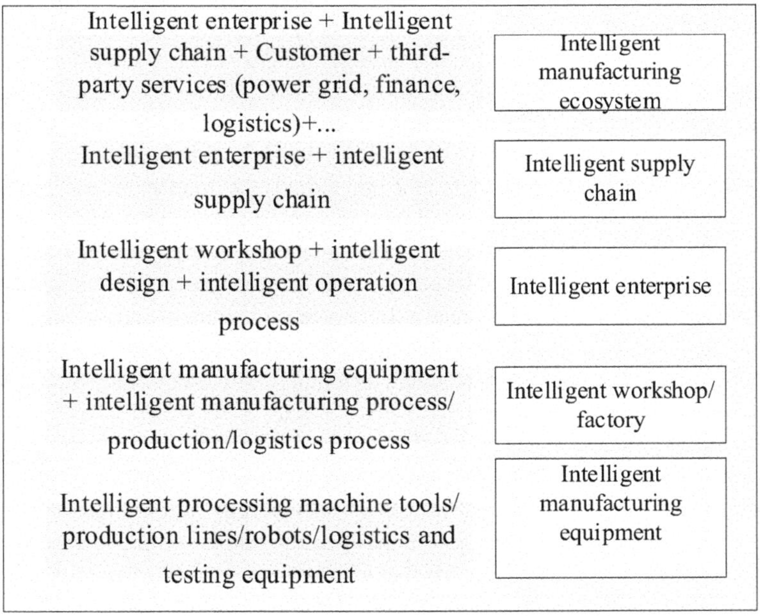

Fig. 1.3 Levels of intelligent manufacturing systems

(6) The construction of intelligent manufacturing system must be to achieve some optimization objectives. There are many optimization objectives, such as enhancing user experience friendliness, improving equipment operation reliability, improving design and manufacturing efficiency, improving product quality, shortening product manufacturing cycle, and expanding value chain space. It should be noted that different levels of manufacturing systems, different links and processes of manufacturing systems, different industries and enterprises, their optimization objectives and importance are different, must be specific analysis.

In the intelligent manufacturing mode, the original basic materials, basic processes and basic parts and other manufacturing bases will be transformed into sensors, big data and software. Manufacturing enterprises have rich big data information such as real-time operating status of equipment, operating environment status, business operation status, personnel status, social network data and customer feedback data. Through the analysis and mining of these data, we can understand the process of the problem, the impact and the way to solve it. These information is abstracted and modeled into knowledge, and then use knowledge to understand, solve and avoid problems, so as to realize intelligent production and decision-making. In this way, the intelligent manufacturing model will change the 'traditional manufacturing operation management theory based on inventory management' to 'intelligent service-oriented manufacturing based on big data'. Under this condition, Smart Factory provides a carrier for the implementation of intelligent manufacturing, and knowledge becomes the basis for the operation of service-oriented manufacturing enterprises. Data, as the main way to generate knowledge, will become the core asset of enterprises.

1. Intelligent Factory

ARC Advisory Group proposed the conceptual model of the early smart factory in 2006. As shown in Fig. 1.4, the three dimensions of engineering (design and technical support for product life cycle), manufacturing (production and operation) and supply chain are used to describe all collaborative manufacturing and management activities of the smart factory.

Based on the concept of manufacturing nation and manufacturing power, German experts and professors applied Cyber Physical System to manufacturing and put forward Cyber Physical Production System (CPPS). CPPS is used as a model to build intelligent factory or digital factory, which points out a realistic and feasible way for the realization of intelligent factory. As a result, Germany's large number of manufacturing-related enterprises, from multinational super large enterprises (Siemens, SAP, etc.) to all kinds of automation products of small and medium-sized enterprises, are considering and brewing how to deal with this development trend.

CPS deeply integrates 3C (computing, communication and control) capabilities to build a safe, reliable, efficient and real-time engineering system based on deep perception of physical facilities. Through the real-time feedback loop between the computational process and the physical process, the complete integration of the information world and the physical world is realized, thus changing the way human beings construct the engineering physics system.

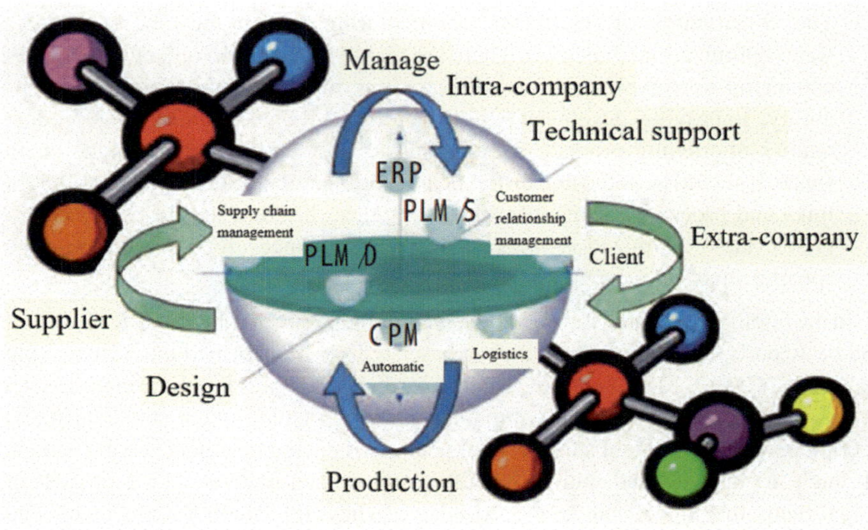

Fig. 1.4 The conceptual model of ARC early smart factory

In 2012, Siemens, Schneider Electric, Rockwell Automation, Yokogawa and other international automation companies, as well as China's mechanical industry instrumentation integrated technical and economic research institutes and other research institutions, participated in the development of IEC / TR 62,794: 2012 digital factory standard, given the definition of digital factory shown in Fig. 1.5: Digital factory is a comprehensive network of digital models, methods and tools (including simulation and 3D virtual reality visualization), integrated together through continuous and uninterrupted data management. It is based on the relevant data of the product life cycle, in the computer virtual environment, the whole production process simulation, evaluation and optimization, and further extended to the whole product life cycle of the new production organization.

The conceptual model of IEC digital factory is divided into three levels. The bottom layer is the physical layer including product components (such as automotive lights, engines, tires, etc.) and factory production resources (such as sensors, controllers and actuators, etc.). The second layer is the virtual layer, which semantically describes the physical entities of the physical layer, converts them into 'mirror' data that can be parsed by the computer, and establishes the connection between the digital product repository and the digital factory repository. The third layer is the tool and application layer related to the product life cycle process, including design, simulation, engineering application, asset management, logistics and other aspects. The greatest contribution of the digital factory concept is to achieve virtual (design and simulation) to reality (resource allocation and production). By connecting product components and production systems, user requirements and product design are input into the resource library through semantic description, and then passed to the resource library of production factors. Manufacturing information can also be

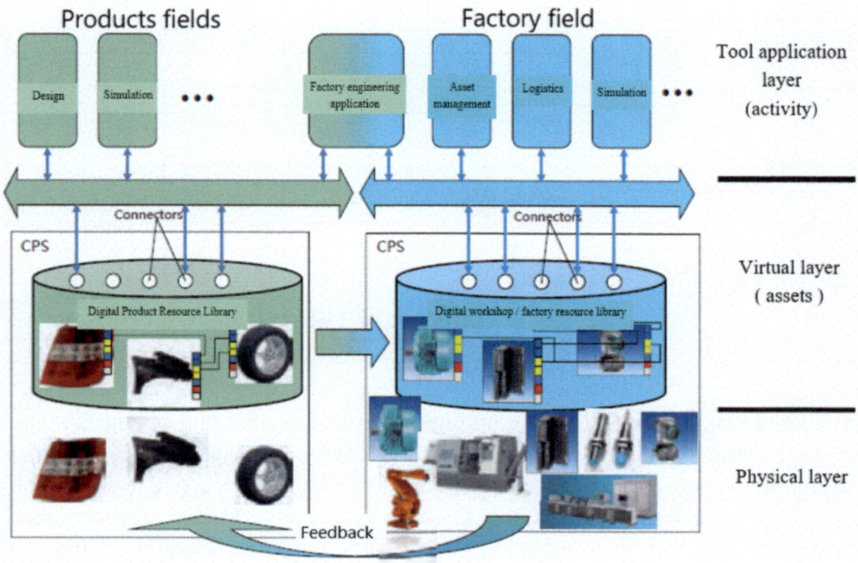

Fig. 1.5 Schematic diagram of the concept of digital factory

fed back to the product resource library, thus bridging the 'gap' between product design and product manufacturing. Furthermore, the resources of the whole network to optimize the production process are realized, the design time is reduced while the quality is improved, and the product development cycle is accelerated.

In the Industry 4.0 framework (Fig. 1.6) proposed in 2013, the Internet and the Service Internet are located at the bottom and top layers of the three-tier information technology infrastructure of the smart factory, respectively. The top level includes ERP, SCM, CRM and so on related to production planning, logistics, energy consumption and management. PLM related to product design and technology is also at the top level, which is closely related to the service Internet. The middle layer realizes the connection with production equipment and production line control, scheduling and other related functions through the information processing system, from the supply of intelligent materials to the output of intelligent products, through the whole product life cycle management. The bottom layer realizes control, execution and sensing through the Internet of Things, and finally realizes intelligent production.

In February 2015, GE's brilliant factory in Pune, India was unveiled. Different from traditional large-scale manufacturing plants, this factory has super flexibility to process and produce seemingly unrelated products such as aircraft engines, fans, water treatment equipment, and diesel locomotive components in the same plant according to GE's needs in different regions of the world.

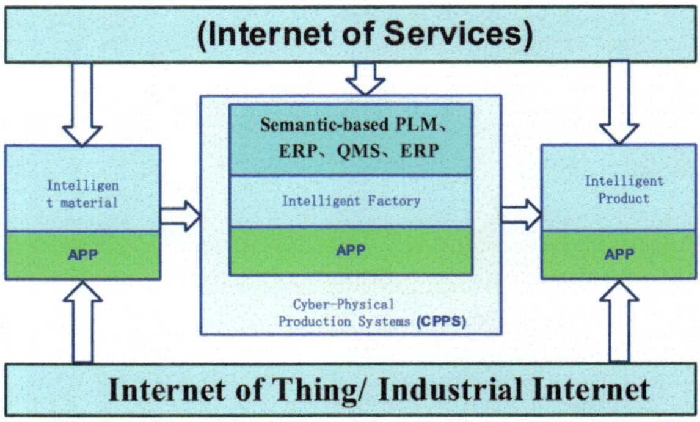

Fig. 1.6 Intelligent factory in the framework of Industry 4.0

GE's brilliant factory (Fig. 1.7) combines industrial Internet and advanced manufacturing technology, and uses digital thread to connect design, process, manufacturing, supply chain, distribution channels, and after-sales service, and forms a cohesive and coherent intelligent system. The factory employs 1500 workers to share the use of production lines, including 3D printers and laser detection equipment. The equipment and computer of the factory communicate and share information with each other, and take measures to ensure quality and prevent equipment failure. Moreover, the factory's production line is connected to suppliers, service providers, and logistics systems through Digital Twin to optimize production.

In March 2017, Accenture proposed the concept of smart factory based on physical factory + virtual factory (Fig. 1.8) [3]. In this model, the physical factory built by automated production equipment is the basis of intelligent manufacturing, and it is also the focus and key of the transformation and upgrading of most Chinese manufacturing factories at this stage, that is, to realize the automation and digitization of the production process. On the basis of digitization, intelligent analysis and

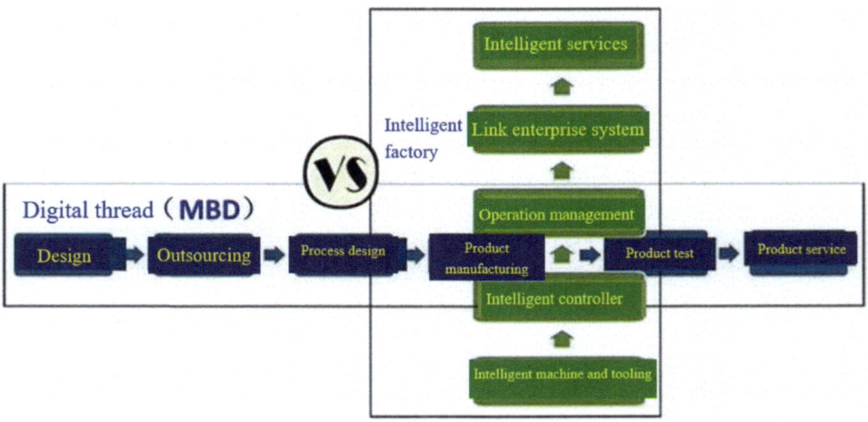

Fig. 1.7 Digital thread driven brilliant factory

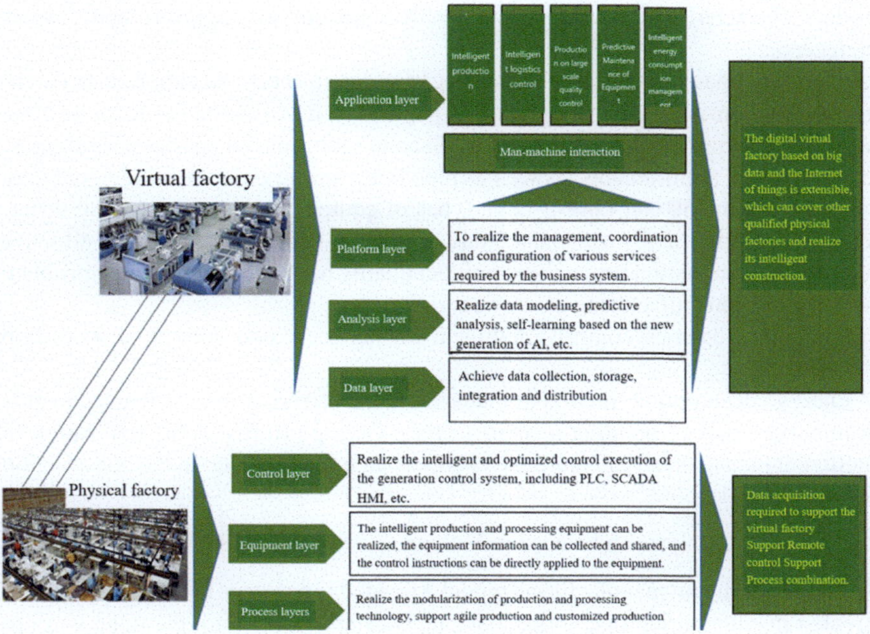

Fig. 1.8 Physical-virtual factory based smart factory framework

application of factories are realized through big data analysis and new generation of artificial intelligence technology.

Intelligent production: Today's market is changing rapidly. In order to cope with the increasing market competition pressure and better meet the market demand, the production mode of enterprises has also changed from previous mass production to mass customization and more advanced personalized customization, which requires the factory's production line to have module combination ability. In the module combination production, with the help of the Internet of Things technology, the real-time tracking, combination configuration and scheduling of personalized customized semi-finished products and finished products produced by mixed lines in the production module are realized, so as to realize the flexible production of the factory and solve the problems of uncertainty, diversity and complexity brought by the transformation from large-scale production to customized production.

Product quality control: In order to better respond to market demand and pursue cost optimization, most companies will set up multiple own factories or use OEMs. Therefore, although it is the same manufacturing process and production process, there are differences in raw materials among different suppliers, and the production process is susceptible to many factors such as climate, temperature and humidity. In this case, how to ensure that products manufactured in the same production process at any factory and at any time in the same factory maintain the same and high-quality product characteristics has become a problem for quality engineers and process engineers.

Predictive maintenance of equipment: In order to meet the different needs of customers, more and more manufacturing plants now have continuous assembly line process equipment for mass production, discrete multi-machine process equipment for customized requirements, as well as pipe linear power equipment, regional duty logistics equipment, etc. For different types of equipment, how to establish differentiated operation and maintenance and guarantee strategies to achieve effective guarantee of production process and product quality has become a common problem faced by large manufacturing enterprises.

Intelligent logistics control: intelligent management and control of workshop logistics.

Energy consumption intelligent management: Enterprise energy management is an important part of building a smart factory. Giving priority to the realization of intelligent management of enterprise energy is an important step in the construction of smart factories.

2. Predictive Analysis and Decision Making

Under the intelligent manufacturing model, knowledge becomes the basis for service-oriented manufacturing operations, data analysis and mining are the main ways to generate knowledge, and data becomes a core asset of the enterprise. The occurrence and resolution of problems in the manufacturing system generate a lot of data. By analyzing and mining these data, we can understand the process of problem occurrence, the impact, and the solution [1] (Fig. 1.9a), this information is abstracted and modeled into knowledge, and then knowledge is used to understand, solve, and avoid problems. The core is to shift from relying on human experience to relying on mining hidden clues in data, making manufacturing knowledge more efficiently and spontaneously generated, utilized, and inherited, bringing a rare historical opportunity for the intelligent development of the manufacturing system.

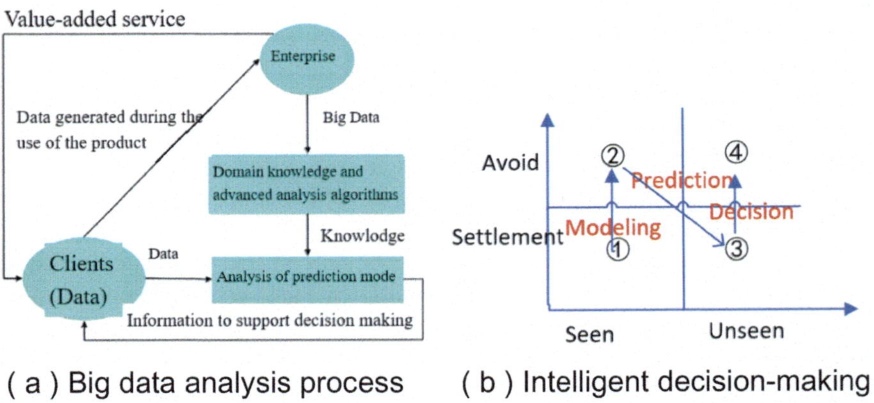

Fig. 1.9 Predictive analysis and decision making based on big data

However, traditional data analysis methods can no longer meet the analysis and processing needs of massive operational maintenance information, making it difficult to effectively use these data. When companies start using big data, they will find that treating big data analysis as a completely independent function is unworkable. The power of data and the scalability of processing will ultimately drive the company's data analysis methods from descriptive analysis (summarizing and describing what has happened in the past) to predictive analysis (predicting what will happen in the future) and prescriptive analysis (deciding what actions to take to make future events happen), and embed data analysis in the company's business processes, directly based on big data predictive algorithms to make plans and decisions and cause actions to occur.

Predictive analysis based on big data can accelerate product innovation, predictive management of production system quality, product health management and predictive maintenance, energy management, environmental protection and safety, supply chain optimization, precision marketing of products, self-inspection and reconstruction capabilities of intelligent equipment and production systems [1], its implementation can be reflected from the following three aspects (Fig. 1.9b).

① Modeling: Turn problems into data, and transform experience into sustainable value through data modeling.
② Prediction: Turn data (such as statistical features of time-domain signals, frequency-domain features of waveform signals, energy spectrum features, and signal readings under specific working conditions, etc.) into knowledge, analyze visible problems, and predict invisible problems.
③ Decision-making: Turn knowledge into data, driving product, process, production, operation, decision-making innovation.

1.1.2 Intelligent Predictive Maintenance of Equipment

In the field of equipment management, to solve the problems of remote, real-time, and timely operation and maintenance of equipment, equipment operation enterprises use the Internet of Things or industrial internet technology shown in Fig. 1.10 to collect equipment operation status data (perception intelligence), and upload it to the enterprise operation center (or enterprise cloud) through the (mobile) internet; system software monitors and controls the real-time status of the equipment online, and predicts maintenance needs and schedules maintenance through data analysis (equipment intelligence). In the process of equipment status monitoring and prediction, enterprises have gradually formed an intelligent predictive maintenance service platform based on the (mobile) internet and equipment maintenance, repair, and operation MRO (Maintenance Repair & Operation) data environment (equipment structure and performance, equipment operation status, environmental status, business operation status, personnel status, maintenance social network data, and customer feedback data, etc. big data information), providing remote personalized equipment intelligent maintenance, repair, and operation services, and reflecting global (internet

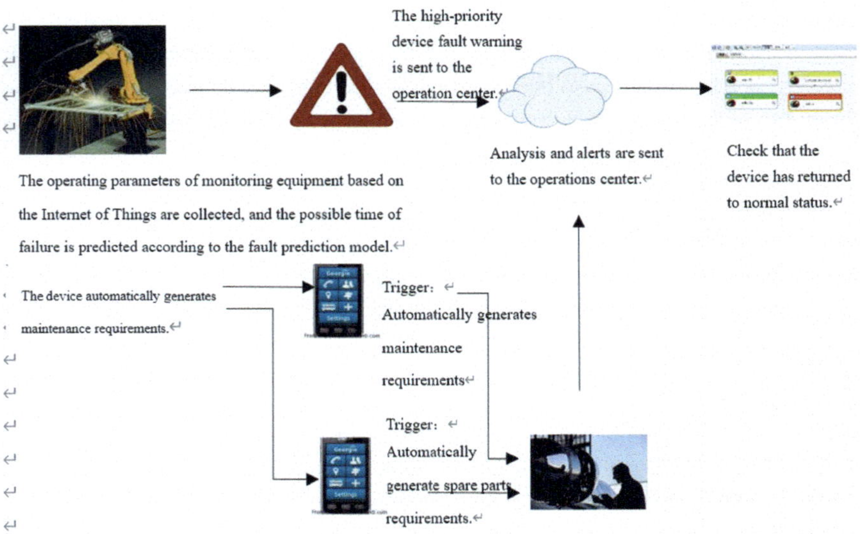

Fig. 1.10 Predictive maintenance of equipment

connection), real-time (industrial internet-supported status detection), and timely (mobile terminal scheduling) characteristics.

In the era of intelligent manufacturing, by analyzing and mining data information such as equipment operation status and historical maintenance information, we can understand the process of equipment failure, the impact, and the solution. This information is abstracted and modeled into knowledge, and then knowledge is used to predict faults and execute maintenance decisions, realizing the self-inspection and intelligence of the equipment, making the knowledge of the entire life cycle of the equipment efficiently and spontaneously generated and used, and also completing the transformation of equipment maintenance strategies from traditional preventive maintenance (Preventive Maintenance, PM) strategies to intelligent predictive maintenance (Intelligent Predictive Maintenance, IPdM) strategies [2, 3].

Vestas of Denmark (a leader in technological development in the world wind power industry) has installed sensors, storage devices, processors, and supervisory control and data The company uses systems such as Supervisory Control and Data Acquisition (SCADA) to implement real-time monitoring of wind turbine operation through intelligent predictive maintenance technology. The company also embeds microcontrollers in wind power turbines, which can control the angle of the blades in each rotation, thereby maximizing the capture of wind energy. It can also control each turbine to minimize the impact on adjacent turbines while maximizing energy efficiency. The company also processes real-time data to predict potential faults in wind turbine components, reducing possible instability in wind turbines, and uses different tools to optimize these data to achieve optimal wind turbine performance. Under the premise that airlines are unwilling to spend money on large-scale regular maintenance or overhaul of engines, the General Electric Aviation Group in the

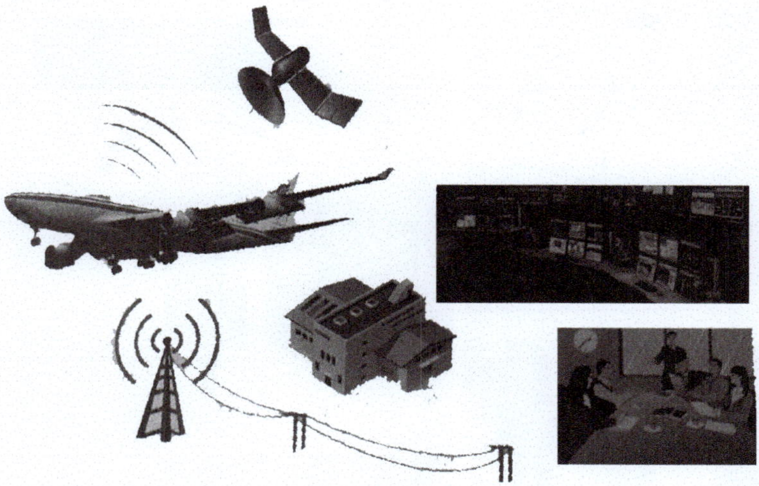

Fig. 1.11 Maintenance service platform of general electric aviation group

United States has launched a service innovation model of "providing airlines with flight usage time services for aviation engines based on intelligent predictive maintenance" by providing engine health monitoring, engine maintenance scheduling/pre-scheduling, and flight scheduling management services as shown in Fig. 1.11. The company undertakes the purchase, maintenance, debugging, and upgrading of aviation engines, rents aviation engines to airlines, and sells flight usage time of engines, without the need for airlines to pay additional maintenance and repair costs. This model has been widely used in the CF6-80 engines of Boeing 767, the CFE738 engines of Falcon 2000 jet aircraft, the TF34 engines of A-10 attack aircraft, and the TF39 engines of C-5 transport aircraft.

The complexity of equipment operating conditions and the diversity of equipment are the biggest challenges faced by equipment health assessment. Since 2010, Nissan in Japan has introduced a health management service platform for industrial robots, as shown in Fig. 1.12, to avoid production losses caused by failures. The main strategy consists of four aspects. First, for a large number of industrial robots, health analysis is carried out using the monitoring parameters in their controllers: fixed signal statistical features such as RMS, variance, extreme values, kurtosis values, and specific position load values are extracted according to the action cycle of each mechanical arm, and the method of similar comparison is used to eliminate the modeling difficulty caused by diverse working conditions. Second, the predictive analysis process includes: selection of key components (such as mechanical arm drive motor), data collection (such as load, torque, position, cycle time, robot model parameters), signal processing and feature extraction, health modeling (such as similarity clustering, positioning), fault diagnosis. Third, the clustering process includes: the first round of clustering based on equipment model and usage time, the second round of clustering based on equipment tasks, environment, and working conditions. Fourth, generate health

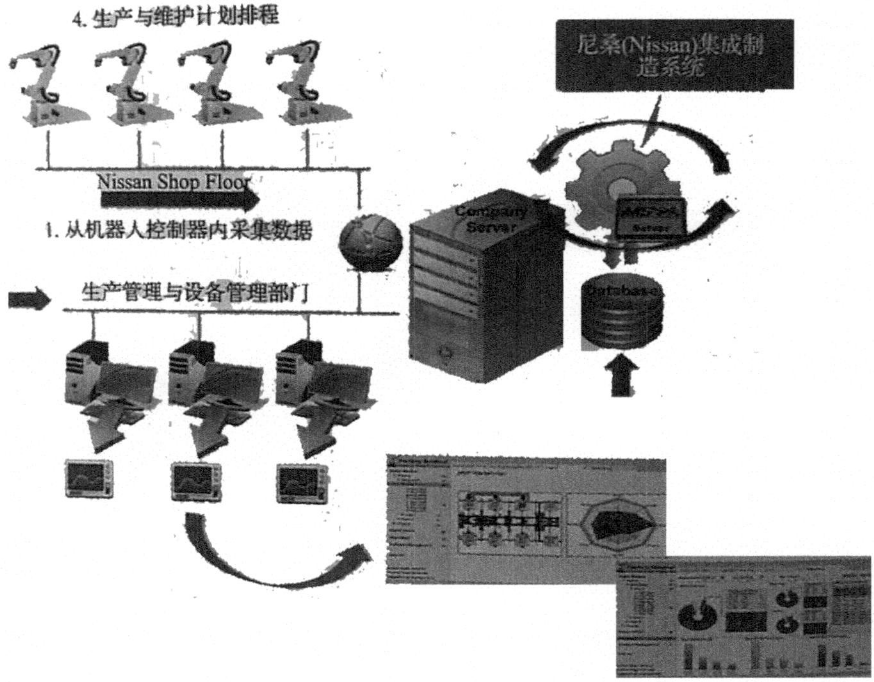

Fig. 1.12 Health management service platform for industrial robots of Nissan company

reports every day, and schedule maintenance plans and production plans based on the real-time status of the equipment.

Since 2008, Sany Heavy Industry Co., Ltd. in China has also carried out the practice of IoT (collection, communication, data collection to big data platform), and currently has 200,000 devices with more than 5000 parameters connected to the Enterprise Control Center (ECC) system, which can monitor the operation data of the equipment in real time, and carry out fault alarm, fault prediction, parts prediction, intelligent service, auxiliary R&D, and credit management.

1.1.3 Intelligent Predictive Maintenance for Equipment Operation Network

With the evolution of enterprise business models and value chains, equipment users outsource the operation and maintenance of internal equipment to professional MRO technical service companies, gradually forming an MRO (service) network covering spare parts suppliers and maintenance providers, equipment manufacturers and maintenance providers, equipment users, domain experts (experts and scholars from the fields of whole machine and component operation, maintenance, fault diagnosis and repair from inside and outside the enterprise), and cloud platform providers. These networks include MRO data acquisition networks, business support networks, social networks, data analysis and support networks, and other value chain networks.

Based on (mobile) internet and value chain networks, MRO service networks can automatically connect to the cloud platform to search for suitable domain experts or solutions to handle problems, and domain experts will also determine the optimal maintenance strategy and maintenance service more effectively through integrated predictive models, knowledge platforms, and mobile devices. These constitute the Smart Predictive Maintenance (SPdM) service platform for equipment MRO networks based on equipment IPdM strategies.

Due to the high randomness of equipment system failures and the high uncertainty of maintenance needs in the MRO service network, how to reliably acquire, model, and mine heterogeneous data information from product status, environmental status, equipment operation status, personnel status, business operation data, social network data, and customer product feedback data in the entire life cycle of equipment? How to use the above information to accurately predict equipment failures and maintenance needs, formulate appropriate maintenance and optimization strategies? How to make full use of the widely available shared service resources in the internet environment under limited service resources, respond to network maintenance needs to the greatest extent, and establish a long-term comprehensive cooperative relationship with customers? Furthermore, how to feed back maintenance decisions to the manufacturing system through the cyber-physical system to achieve error compensation, adjustment, and feedback control? These issues have become the common demands of all participants in the MRO service network in the intelligent manufacturing environment, and it is necessary to deeply study the theories and methods of high-quality acquisition and processing of multi-source heterogeneous data for the MRO network environment, data-driven equipment and component fault diagnosis and prediction theories and methods, maintenance optimization scheduling and decision-making for intelligent factories, large-scale maintenance service prediction and optimization configuration theories and models for equipment operation networks, operation process control based on cyber-physical systems, and construct the SPdM theoretical system, framework, methods, and models for equipment MRO networks.

1.2 Maintenance Strategies

The equipment maintenance strategy is a maintenance policy formulated for product degradation, including the basis for degradation assessment, maintenance decision actions, and planning, etc. [5, 6]. The maintenance strategy of the equipment has evolved from the early (1952) Corrective Maintenance (CM), through widely used Time-based Maintenance (TBM), Proactive Maintenance (PaM), and Reliability-centered Maintenance (RCM) and other Proactive Maintenance (PM) strategies, to the Predictive Maintenance (PdM) strategies represented by Condition-based Maintenance (CBM) and CBM + , forming a relatively complete theoretical and methodological system for maintenance task prediction, planning, design, and management.

With the advent of the era of intelligent manufacturing, the rapid development of new generation information technologies such as (mobile) internet, industrial big data, artificial intelligence, Cyber Physical System (CPS), and the Internet of Things, on the basis of predictive maintenance strategies, intelligent predictive maintenance strategies such as IPdM for equipment and SPdM for equipment MRO networks have been born based on big data analysis and the application of CPS technology.

1.2.1 Corrective Maintenance

According to the national standard GB/T 2900.99—2016 "Electrical Terminology—Reliability", failure is defined as the event where a product ceases/loses the ability to perform the specified function, and a fault is the state where a product cannot perform the specified function. Faults usually occur after the product itself fails, but they may also exist before failure. This book does not strictly distinguish between failure and fault, which can be collectively referred to as faults.

Corrective Maintenance (CM) refers to the identification, isolation, and repair of faults, enabling faulty equipment, machines, or systems to return to the allowable error range of normal operation [7]. The European 2017 Maintenance-Maintenance terminology standard defines it as: "Maintenance activities aimed at restoring the equipment to normally perform its intended function after fault detection [EN 13,306—2017 (E/F/D)]".

According to the degree of recovery of the operating status of the maintained equipment, CM is divided into the following categories [8].

① Perfect Repair/Maintenance: Maintenance that restores the system or system's operating status to as if it were brand new. In other words, under perfect maintenance, a repaired system has the same life cycle distribution function and failure rate function as a brand new system. Generally speaking, replacing a faulty system or device with a brand new system or device is a perfect repair, and a thorough overhaul of an engine with a broken connecting rod is also an example.

② Minimal Repair/Maintenance: A maintenance action that restores the system to the failure rate at the time of failure. Minimal repair was first proposed by Barlow and Proshan in 1965, and after minimal repair or maintenance, the system's operating status is usually referred to as an old operating status. Replacing tires on a car or replacing a damaged fan belt on an engine are examples of minimal repair, as the overall failure rate of the car has basically not changed.

③ Imperfect Repair/Maintenance: Maintenance activities that restore the system to a non-brand new state but the system's operating status does indeed improve.

Imperfect maintenance usually restores the system's operating status to somewhere between brand new and old. Imperfect maintenance is a common maintenance activity, including two extreme cases: minimal maintenance and perfect maintenance. Engine tuning is an example of imperfect maintenance: engine tuning may not make the engine run like new, but its performance will indeed greatly improve.

④ Worse Repair/Maintenance: Maintenance activities that increase the system's failure rate or actual service life but the system does not collapse, the operating condition of the repaired system is worse than before the failure.

⑤ Worst Repair/Maintenance: Maintenance activities that unintentionally cause the system or device to fail or collapse.

Corrective Maintenance, also known as Break-down Maintenance, is maintenance based on whether the equipment is intact or usable, adopting a fault maintenance strategy that only maintains when there is a fault, that is, only after the equipment is partially or completely faulty, it is restored to its original state, which is to repair after it is broken, belonging to unplanned maintenance. Corrective Maintenance can be subdivided into Immediate Corrective Maintenance (maintenance begins immediately after a fault) and Deferred Corrective Maintenance (maintenance begins after a delay consistent with given maintenance rules).

1.2.2 Preventive Maintenance

Preventive maintenance is to keep the equipment or system in a satisfactory operating state, before a major failure occurs, according to the production plan and maintenance experience, to carry out shutdown tests, inspections, disassembly, replacement of parts at specified times or intervals, to prevent damage, secondary destruction, and production losses. Preventive maintenance, also known as timed maintenance, is time-based maintenance, including planned maintenance, proactive maintenance, and Condition based Maintenance, etc., and is currently the commonly used maintenance method.

The preventive maintenance decision process of complex equipment systems is shown in Fig. 1.13. First, the key failure modes of the system are identified using the system's structural and functional characteristics; second, an appropriate remaining life prediction method is selected based on the obtained degradation or failure data samples; third, the preventive maintenance and post-failure repair cost functions of each failure mode are constructed using the predicted remaining life distribution; finally, the optimal preventive maintenance plan is obtained by implementing the maintenance decision that minimizes the cost rate using the optimization algorithm.

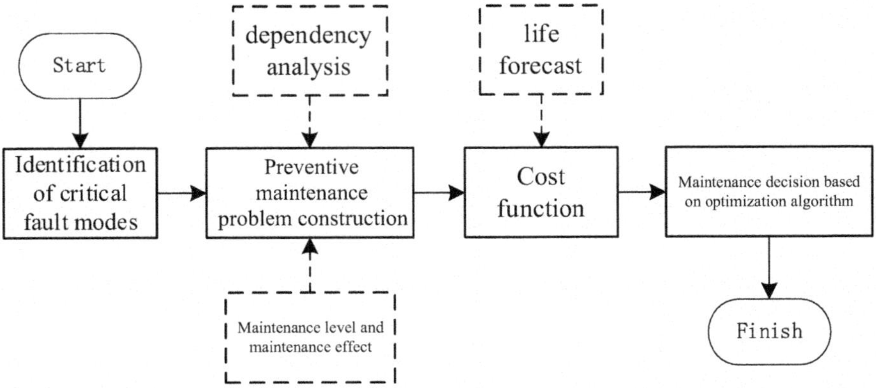

Fig. 1.13 Preventive maintenance decision process of complex equipment systems

1.2.3 Condition Based Maintenance

State monitoring identifies an emerging fault by monitoring the state parameters (such as vibration, temperature, etc.) of the machinery. State monitoring technology can be divided into state detection technology (detecting the impact of potential faults, such as changes in vibration characteristics, temperature changes, lubricating oil particles, and chemical effects, etc.), product quality detection technology (defects in the products manufactured by the equipment often have a direct relationship with the defects of the equipment itself), main parameter detection technology (such as the speed, flow rate, pressure, power, current of the equipment) and human sensory-based inspection technology (such as seeing, hearing, touching, and smelling). State monitoring is commonly used in rotating equipment, auxiliary systems, and other machinery, such as compressors, pumps, motors, internal combustion engines, presses, etc. Some static equipment, such as steam boilers, pipelines, and heat exchangers, often use non-destructive testing techniques and applicability assessments for regular inspections.

Condition based Maintenance is a preventive maintenance strategy based on state monitoring technology, attempting to maintain the right equipment at the right time. Condition based Maintenance is based on the fact that a large number of faults do not occur instantaneously, in fact, faults take time to develop. If this fault development process is detected, measures can be taken to prevent faults or avoid serious consequences. During the fault development process, it can be detected that a fault is occurring or about to occur, this point is called the potential fault point P, as shown in Fig. 1.14. A potential fault is a recognizable actual state that can show that a functional fault is about to occur or is occurring, for example, showing the vibration of a bearing near failure, showing the cracks of metal fatigue, showing the particles in the gearbox lubricating oil of a gear near failure, excessive tire tread wear, etc.

The performance of the equipment exists a gradual process from the beginning of deterioration and entering the potential fault period, the interval from potential

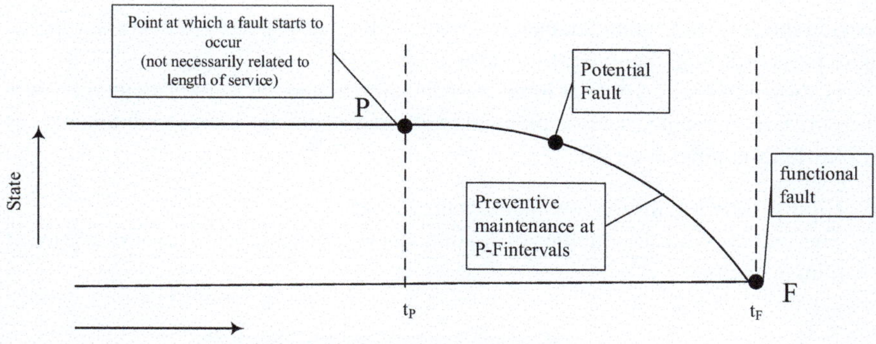

Fig. 1.14 P-F curve of equipment performance

fault to functional fault is the P-F interval. The P-F curve shown in Fig. 1.14 reveals the general process from the start of the fault, deterioration to the potential fault point P that can be detected by the fault, if not detected and not corrected, the performance continues to deteriorate until it reaches the functional fault point (F) of the equipment. If a potential fault is detected between point P and point F, then functional faults can be prevented or the consequences of faults can be avoided. Condition based Maintenance is to check the potential faults of the equipment in order to take measures to prevent faults or avoid the consequences of functional faults.

The inspection frequency of Condition based Maintenance must be much smaller than the P-F interval (the time required for the fault to be detectable from the potential fault point to the development of the fault into a functional fault point), otherwise, the fault may be completely missed. Before most functional faults appear, there is more than one (often several different) potential faults, each potential fault has a different P-F interval. For example, before the bearing functional fault, vibration characteristics are checked through vibration analysis (P-F interval 1–9 months), wear particles are detected through oil quality analysis (P-F interval 1–6 months), noise can be heard (P-F interval 1–4 weeks), heating (hand touch) (P-F interval 1–5 days), etc.

As shown in Fig. 1.15, Condition based Maintenance technology includes data acquisition, feature extraction, fault detection, fault diagnosis, and fault prediction. Data acquisition may involve various types of data information, such as vibration, temperature, pressure, speed, voltage or current, stress/strain/impact, position, particle count/composition, etc. Feature extraction calculations include fast Fourier transform, wavelet transform, adaptive time–frequency analysis, data filtering/smoothing, temperature/pressure ratio, energy efficiency, etc. Fault detection algorithms alert users to potential problems or unknown failures in the equipment. Fault diagnosis algorithms are used to isolate and identify faults in specific components or subsystems. Fault prediction algorithms estimate the available service life RUL or failure probability of the equipment based on historical and current operating conditions. Supervised reasoning algorithms coordinate conflicting information and provide the following maintenance recommendations: inspection intervals, repairs, spare parts ordering, equipment downtime, etc.

As shown in Fig. 1.16, Condition based Maintenance has some advantages over preventive maintenance such as planned maintenance and proactive maintenance, as well as fault maintenance [9]:

① High system reliability, low maintenance cost;
② The reduction in the number of maintenance operations reduces the impact of human errors.

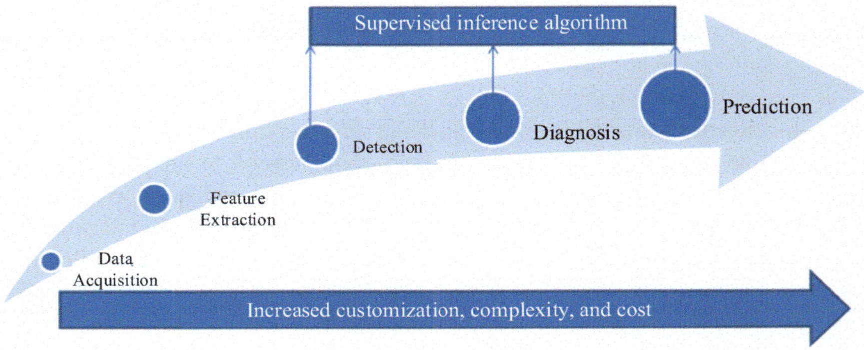

Fig. 1.15 Condition based maintenance technology for complex equipment systems

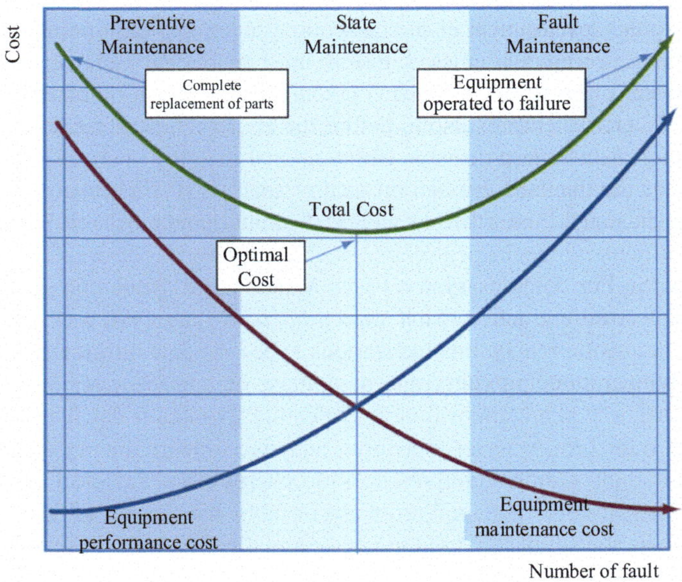

Fig. 1.16 Condition based maintenance optimizes costs

The disadvantages of Condition based Maintenance are:

① Higher installation costs;
② Unpredictable maintenance cycles lead to uneven cost allocation;
③ A large number of parts need to be maintained and inspected (including the installation of the Condition based Maintenance system itself).

1.2.4 Predictive Maintenance

Predictive Maintenance (PdM) is a preventive maintenance developed based on CBM. When the machine or equipment is running, it periodically or continuously (online) monitors the state and predicts faults of its main (or necessary) parts or areas, to determine the state of the equipment, thereby predicting the future development trend of the equipment state, and based on the development trend of the equipment state and possible fault modes, a predictive maintenance plan is formulated in advance, determining the time, content, method, and necessary technical services and material support for the machine to be maintained.

Predictive maintenance integrates equipment status monitoring, fault diagnosis, fault (status) prediction, maintenance decision support, and maintenance activities. It is a type of preventive maintenance. Predictive maintenance has two concepts: narrow and broad. The narrow concept is based on "status monitoring (status maintenance)", emphasizing "fault diagnosis". It refers to irregular or continuous monitoring of equipment status, determining whether there are any abnormal conditions or fault trends based on the results, and then timely arranging maintenance tasks. The narrow concept of predictive maintenance does not have a fixed maintenance cycle, but arranges maintenance plans in a timely manner based on the results of monitoring and diagnosis. It emphasizes the tripartite process of monitoring, diagnosis, and maintenance, which is widely applicable to process industries and large-scale production methods. The broad concept of predictive maintenance combines status monitoring, fault diagnosis, status prediction, and maintenance decision-making. Status monitoring and fault diagnosis are the foundation, status prediction is the focus, and maintenance decision-making provides the final requirements for maintenance activities. The broad concept is a systematic process that incorporates maintenance management into the scope of predictive maintenance, considering the entire maintenance process and content related to maintenance activities. Predictive maintenance is the execution of maintenance tasks at a predetermined time point, set before the equipment loses performance within a threshold, and when maintenance activities are most cost-effective. Up to now, predictive maintenance has basically formed the technical system shown in Fig. 1.17. At the current stage, in addition to preventive maintenance technology, reliability-centered maintenance (RCM) emphasizes the use of predictive maintenance PdM technology.

(1) Status Monitoring Technology Status monitoring technology has developed to the present and has formed its own monitoring methods in various engineering

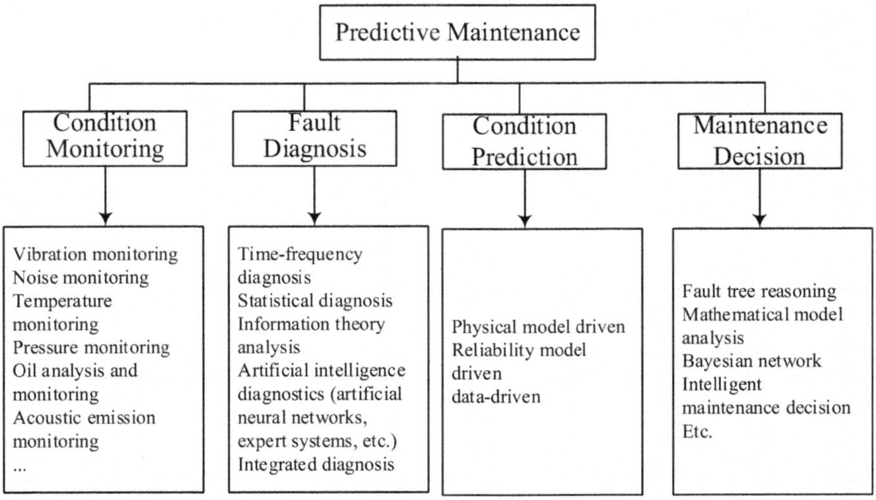

Fig. 1.17 Predictive maintenance technical system

fields. The methods of status monitoring are divided into many types according to different status detection means, commonly used ones include: vibration monitoring, noise monitoring, temperature monitoring, pressure monitoring, oil analysis monitoring, acoustic emission monitoring, etc.

(2) Fault Diagnosis Technology Fault diagnosis is of great significance in continuous production systems. According to the principles of diagnosis methods, fault diagnosis can be divided into time–frequency diagnosis, statistical diagnosis, information theory analysis, and other artificial intelligence diagnosis methods (expert system diagnosis, artificial neural network diagnosis, etc.), fuzzy diagnosis, grey system theory diagnosis, and integrated diagnosis methods (such as fuzzy expert system fault diagnosis, neural network expert system fault diagnosis, fuzzy neural network diagnosis, etc.).

(3) Status Prediction Technology Status prediction is to evaluate the current status of components and predict the future status based on the operation information of the equipment. Common methods include cumulative damage model prediction, random degradation model prediction, time series model prediction, grey model prediction, particle filter prediction, and neural network prediction. From the classification of status prediction methods, there are generally three basic approaches to the development of fault prediction: physical model-driven, reliability model-driven, and data-driven. In practical applications, the three approaches can also be integrated to form a fault analysis and prediction method based on information fusion models, and mixed fault prediction of digital information and symbolic information can be performed, which is more effective for implementing predictive maintenance.

(4) Maintenance Decision Technology Maintenance decision-making is to start from multiple aspects and angles such as personnel, resources, time, cost,

and benefits, according to the results of status monitoring, fault diagnosis, and status prediction, combined with production plans and spare parts inventory to analyze the feasibility of maintenance, formulate maintenance plans, determine maintenance support resources, and give the time, place, personnel, and content of maintenance activities. The methods for making maintenance decisions generally include fault tree reasoning, mathematical model analysis, Bayesian network method (suitable for expressing and analyzing uncertain and probabilistic things), and intelligent maintenance decision-making method.

1.2.5 Intelligent Predictive Maintenance

The rapid development of intelligent manufacturing technology has been driven by new generation information technologies such as mobile internet, big data, artificial intelligence, Internet of Things, and cloud computing. In the technical system built through intelligent manufacturing technology, the intelligent factory monitors physical objects and processes through the Cyber Physical System (CPS), creates virtual replicas of the physical world, and carries out decentralized control and decision-making. CPS communicates and cooperates with people or other CPS systems in real-time through the Internet of Things (IoT), and participants in the value chain provide and utilize services within or across organizations through big data analysis and the Internet of Services (IoS). In the intelligent factory, the CPS system is a multi-dimensional complex system integrating computing, networking, and physical environments. Through the organic integration and deep collaboration of 3C technology, it realizes real-time perception, dynamic control, and information services of large-scale engineering systems. Intelligent predictive maintenance, based on predictive maintenance, transmits equipment status data generated by CPS to cloud storage servers through IoT, and realizes functions such as pre-processing of sensor signals, feature extraction, fault diagnosis, and fault prediction through big data analysis technology; then, it evaluates maintenance performance indicators and makes maintenance optimization decisions based on the results of fault diagnosis and prediction; finally, it corrects, compensates, and controls the manufacturing system through the CPS system according to the results of optimization decisions. Compared with predictive maintenance, intelligent predictive maintenance technology is mainly realized through CPS, IoT, IoS, and big data analysis technology. For the problem of intelligent predictive maintenance of equipment in intelligent workshops or factories, Lee et al. proposed an intelligent predictive maintenance IPdM framework based on 5C, and Wang et al. proposed an intelligent predictive maintenance IPdM framework based on CPS. With the in-depth application of (mobile) internet, big data, and artificial intelligence, equipment MRO operation services extend from a single intelligent workshop or factory to multiple collaborative entities in the MRO network such as spare parts suppliers and maintenance providers, equipment manufacturers and maintenance providers, equipment users, domain experts, and cloud platform providers. Intelligent predictive maintenance SPdM for equipment MRO operation networks has become a key research direction and content in the field of future intelligent manufacturing.

1.2.6 Selection Methods of Maintenance Strategies

The selection of equipment maintenance strategy is related to the frequency of failures and the impact of failures [10], as shown in Fig. 1.18.

① A well-developed device and system should not have a high failure rate and impact.
② For high failure rate and minor impact, choose CBM strategy, collect signals with warnings, and prepare to replace spare parts.
③ For low failure rate and low impact, use inspection, timely replacement or preventive replacement, no need for monitoring and data collection.
④ If the impact of failure is significant, which may cause a long downtime and high maintenance costs, a detailed failure mode and impact analysis is needed to determine the data collection object and maintenance strategy, and carry out corresponding predictive maintenance and risk management.

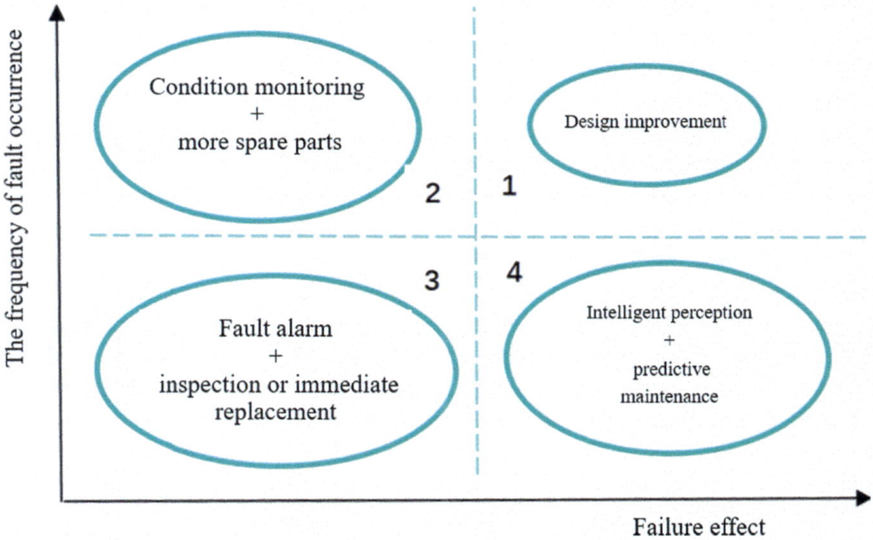

Fig. 1.18 Selection method of maintenance strategy

1.3 Research and Development Trends

1.3.1 Research Status

The academic and industrial circles have carried out research and exploration from the aspects of equipment maintenance strategy and structural health detection, maintenance scheduling optimization and decision-making, and have achieved many valuable research results.

1. Research on equipment maintenance strategy, use technology and fault prediction model

The development process of equipment maintenance strategy is shown in Table 1.1.

(1) The development process of maintenance strategy.

Early Corrective Maintenance (CM) was mainly aimed at the time (1930s–1950s) of industrial mechanization degree is not high, downtime is insignificant, in the minds of most managers, prevent equipment failure is not a top priority. At the same time, most of the equipment is relatively simple, and the design margin is large, which makes the equipment more reliable and easy to repair. In addition to simple daily maintenance work such as cleaning, maintenance and lubrication, there is no need for systematic maintenance, and the skill requirements for workers are relatively low.

Since the Second World War, the increase in demand for various items and the sharp decline in industrial labor have led to an increase in mechanization. The number of equipment is more and more complex. Production has begun to rely on these equipment, and downtime has become a prominent problem (for example, in the just-in-time production system implemented in the 1960s and 1970s, very small faults are likely to cause the entire factory to shut down, making the impact of downtime more severe, which requires equipment to have higher availability). People think that it is possible and should prevent equipment failure, and then form the concept of early preventive maintenance. Early preventive maintenance in the 1960s was mainly manifested in the use of planned maintenance and active maintenance strategies to regularly overhaul equipment. At this stage, the maintenance cost begins to increase sharply, which makes the maintenance plan and control system come into being (the joint optimization of maintenance and inventory produces lower maintenance cost); on the other hand, the sharp increase in the amount and cost of capital occupied by equipment has led people to seek ways to extend the life of equipment (longer service life).

In the early stage of preventive maintenance, the 'MSG (Maintenance Steering Group)' idea of the Federal Aviation Administration's maintenance steering group, which has far-reaching influence, has formed a relatively complete set of methods and theoretical systems for maintenance task planning, design and management [8]. In 1960, the Federal Aviation Administration and airlines formed a special working group to investigate the effectiveness of preventive maintenance and developed a 'Federal Aviation Administration/Civil Aviation Industry Reliability Program'.

Table 1.1 Development process of equipment maintenance strategy and technology

Stage of development	Stage 1 40 s–50 s	Stage2 50 s–70 s	Stage3 70 s–2000	Stage4 2000–2015	Stage5 2015-
Maintenance strategy	Fault maintenance	Planned/ proactive maintenance	Preventive maintenance	Predictive maintenance	Intelligent predictive maintenance
Technology	Repair or replacement if there is a fault	Regular overhaul Work planning and control system A large, slow computer	Condition monitoring Reliability and maintainability design Failure Modes and Effects Analysis Hazard analysis Expert system Fault tree reasoning Fast computer Multiple skills and cooperation	CBM SHM PHM Joint optimization of production / maintenance / inventory High speed computer	IOT/CPS/IOS Big data analysis Real-time fault prediction Large-scale parallel prediction Large-scale machine learning Network maintenance demand forecasting Cloud computing
Classic maintenance thought and birth time		Decision diagram/1965 MSG-1/1967 MSG-2/1969	RCM/1978 RCMII/1991	CBM/1981 SHM/1995 PHM/2000 CBM + / 2002 IMS/e-Main/ 2003	Smart Main./ 2015 CPS&IPdM/ 2016
Maintenance subject	Part	Parts	Equipments system	Equipment users	Equipment manufacturers/ operators

(continued)

Table 1.1 (continued)

Stage of development	Stage 1 40 s–50 s	Stage2 50 s–70 s	Stage3 70 s–2000	Stage4 2000–2015	Stage5 2015-
Maintenance aim	Equipment recovery	Timely repair of equipment	Timely repair of equipment and minimum inventory cost	Equipment users minimize operation and maintenance costs	Minimize operating costs of equipment operation and maintenance network
Fault predictive model	Fault	Physical model-driven prediction models, such as failure physics, cumulative damage model, fatigue life model, etc.; Weibull distribution	Based on the reliability prediction model, the life is estimated by Weibull, exponential, Poisson, logarithmic, binomial, normal (Gaussian), Gama, discrete and other distribution functions The Bayesian method of posterior probability was added in 1985	Data-driven prediction models, such as degradation models, Kalman and particle filter models, machine learning, etc	Big data-driven prediction models, such as deep learning and wide learning

Then, people try to organize what they have learned from various reliability outlines to develop a reasonable and universal method for formulating preventive maintenance outlines. In 1965, an immature decision diagram method was invented, and a report on its application was published at the Civil Aircraft Design and Operation Conference of the American Aeronautics and Astronautics Society in June 1967. The methodology was refined and incorporated into a manual on maintenance assessment and outline development (known as MSG-1 document) drafted by the maintenance steering group established to manage the development of the initial outline of the new Boeing 747 aircraft, thus forming the first scheduled maintenance outline based on reliability (the ability of the product to complete the specified functions under the specified conditions and within the specified time is called the reliability Realiability of the product). The maintenance outline of the Boeing 747 aircraft was successful. The decision diagram method was further improved in use, and two years later (1969) was compiled into the second document, namely 'MSG-2: Airline/Manufacturing Company Maintenance Outline Formulation'. Using MSG-2, the preventive maintenance outline of Lockheed's 1011 aircraft and military aircraft S-3, P-3, McDonald's F4J aircraft and Douglas's DC-10 aircraft was successfully developed. The core purpose of the MSG-1 and MSG-2 files is to develop a predetermined maintenance outline that can ensure the maximum safety and reliability that the equipment can achieve at the lowest cost.

Since the mid-1970s, the process of change in the industrial field has been extremely rapid. With the growth of mechanization and automation, reliability and availability have become key factors in health monitoring, data processing and remote communication industries. A higher degree of automation also means more failures (higher product quality) affecting the ability to maintain satisfactory quality standards; more and more failures have serious safety consequences and environmental consequences, and safety and environmental standards have been rapidly improved (higher safety and environmental friendliness); maintenance costs have been rising (higher cost-effectiveness), both in absolute terms and in proportion to total costs. In order to meet the above requirements of high quality, maintenance technologies such as condition monitoring, reliability and maintainability design, hazard analysis, expert system, fault tree reasoning, failure mode and impact analysis have been adopted to form a reliability-based preventive maintenance strategy.

In the stage of preventive maintenance based on reliability, Nowlan and Heap formally proposed the concept of RCM (Reliability Centered Maintenance) in 1978 [9], and formed the decision diagram of RCM. Based on the RCM decision diagram, the civil aviation industry has formed the MSG-3 decision diagram, and sometimes the RCM of the aviation industry is also known as MSG-3; by 1991, the British scholar John Moubray improved the RCM decision diagram proposed by Nowlan and Heap, forming a new second-generation RCM (RCMII) decision diagram shown in Table 1.2 [11]. RCMII systematically solves seven problems of preventive maintenance, namely: (1) the function and performance standards of facilities; (2) Functional failure; (3) Causes of failure (i.e. Failure Mode); (4) Fault Effect Analysis; (5) Fault consequence-the importance of distinguishing faults; (6) How to prevent-select preventive maintenance strategy; (7) If there is no effective prevention work, change the product structure. The corresponding preventive maintenance strategy has three options: (1) adopt predetermined Condition based Maintenance work, including predictive maintenance, Condition based Maintenance, state detection, etc.; (2) Adopt a scheduled repair work, that is, re-process a component or overhaul a component before a specific length of service, regardless of its state at that time; (3) Adopt a scheduled end-of-life work, that is, scrap parts before a specific length of service, regardless of their status at the time.

Since the MSG-1 document was formulated in 1967, MSG-1, MSG-2, RCM, MSG-3 and RCMII have been widely used in aviation, electricity, petroleum, railways, steel mills, automobile manufacturing and other industries. However, the demand management of preventive maintenance strategies based on MSG and RCM is a maintenance service demand response with equipment as the main body. The maintenance goal is the timely repair of faulty equipment and the minimum operation and maintenance cost of equipment. There is no organic connection between maintenance ideas and maintenance support systems (including material resources, human resources, information resources and management means).

Table 1.2 Decision diagram of RCM

1. Seven basic problems of RCM
2. Equipment registration
3. Function and fault
4. Failure consequences
5. Preventive work
6. Provisional measures;
7. CM decision diagram (1978Nowlan's RCM decision diagram/MSG-3 version/1991RCMII version)
8. The planning, organization and control of the proposed work
9. Fault characteristics and technical experience
10. Implementation of RCM
11. The effect of RCM

With the development of sensing and information technology, predictive maintenance based on condition monitoring has shown more and more advantages in achieving maintenance goals, and structural health monitoring (SHM) and Condition based Maintenance have emerged.

SHM is developed on the basis of condition monitoring technology, which infers the change of structural characteristics by measuring the response of the structure before and after the supernormal load, and then detects and evaluates the damage of the structure, or predicts the long-term degradation and life of the structure through continuous monitoring [14]. Structural health monitoring technology is suitable for all kinds of structures, such as aerospace, mechanical structure, civil engineering and other simple or complex structures. SHM was fully studied in the above fields in the middle and late 1990s.

With the development of CBM technology, in 2001, led by Boeing from industrial manufacturing, military, commercial manufacturing, sensor technology and research institutes and other more than 10 organizations, developed the open system architecture for condition-based main tenance (OSA-CBM) shown in Fig. 1.19.

The main functions of OSA-CBM include: (1) Data acquisition and transmission: use various sensors to collect relevant parameter information of the system, and have data conversion and data transmission functions; (2) Data processing: Accept signals and data from sensors and other data processing modules, and process the data into the format of subsequent processing requirements such as condition monitoring, health assessment and fault prediction. The output results include filtered, compressed and simplified sensor data, spectrum data and other feature data. (3) Condition monitoring: Accept data from sensors, data processing and other condition monitoring modules, compare these data with the predetermined failure criteria to monitor the current state of the system, and provide fault alarm capability according to the predetermined limit/threshold of various parameter indicators. (4) Health assessment: Accept data from different condition monitoring modules and other health assessment modules, evaluate the health status of the monitored system (or subsystems, components, etc.) (such as whether there is parameter degradation, etc.),

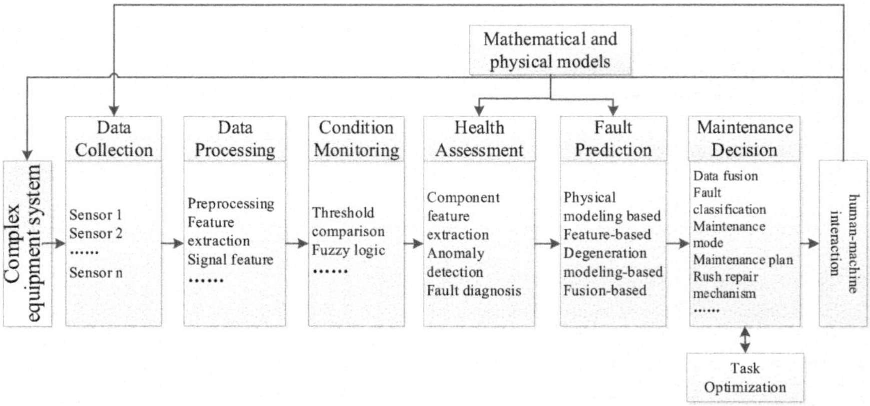

Fig. 1.19 Open architecture for condition based maintenance OSA-CBM [15]

generate fault diagnosis records and determine the possibility of failure. Fault diagnosis should be based on various health status historical data, working status and maintenance historical data; (5) Fault prediction: Comprehensive utilization of the data information of the above parts, evaluate and predict the future health status of the monitored system, including the remaining life; (6) Maintenance decision-making: accept data from condition monitoring, health assessment and fault prediction, generate recommended measures such as replacement and maintenance activities, and take maintenance measures at the appropriate time before the failure of the monitored system; (7) Human–computer interaction includes man–machine interface and machine-machine interface. Man–machine interface includes warning information display of condition monitoring module and data information representation of health assessment, prediction and decision support module. Machine-machine interface makes the data information between the above modules and between the system and other systems can be transmitted. The architecture has been applied and verified in many systems of the U.S.Navy, including ship systems, aircraft, civilian vehicles and other industrial fields.

Based on CBM and SHM, the US Department of Defense proposed the concept and implementation framework of CBM + (CBM plus) in 2002 [16]. CBM + unifies planned maintenance and CBM by adopting RCM idea, and forms supply chain integrated maintenance idea through JTAV (Joint Total Asset Visibility) and ALIS (Automatic Logistics Information System), so as to realize low cost and high efficiency of maintenance. The main contents include PHM (Prognostics and Health Management), CBM, RCM, JTAV, ALIS and other modules based on monitoring and diagnosis. CBM + implementation guidelines are shown in Table 1.3 (from the (DoDI) 4151.22 'Enhanced Condition-Based Maintenance in Equipment Maintenance' document released by the US Department of Defense in December 2007).

In 2003, Iung et al. proposed the concept of e-maintenance [17], and gave Fig. 1.20 Plug and Play distributed maintenance platform for intelligent manufacturing system

Table 1.3 Guidelines for implementation of CBM +

1. Improve the efficiency and effectiveness of maintenance, and achieve the integration of various functional elements of the life cycle management process, such as development, finance, distribution, supply chain management, continuous support and systems engineering
2. Create an integrated, predictive approach to maintenance that minimizes unplanned maintenance, eliminates unnecessary maintenance, and applies the most cost-effective system health management process;
3. Under the integrated life cycle framework, the best combination of various maintenance technologies (such as condition monitoring, diagnosis and prediction), best practices, reliability-centered maintenance (RCM) processes and enabling technologies (such as wholly owned visualization) is achieved;
4. Implement data collection and analysis requirements according to DoD8910.1-M (Department of Defense Procedures for Management of Information Requirements) to measure equipment continuous support performance characteristics and support the degree of continuous support effectiveness throughout the life cycle;
5. Reduce equipment failure during the mission, determine the best time to complete the required maintenance, and then increase the number of combat assets, thereby improving equipment reliability and system life cycle readiness;
6. Promote the widespread use of CBM + enabling technologies through a network-centric approach using an open architecture and open standards, using the Department of Defense architecture framework as defined in the terminology;
7. Improve the reliability of the equipment through rigorous analysis of fault data, thereby improving the design and usage practices, which will ensure that the equipment can meet the target performance standards in the combat context;
8. Providing technicians and support personnel with timely status information, accurate failure pattern recognition and accurate technical data can speed up the repair and support process, optimize the life cycle process and minimize the average downtime;
9. By eliminating unnecessary maintenance activities, the effective support scale required to meet the operational requirements is accurately determined, thereby reducing the cost of use and support capability

(Plug and Play distributed maintenance platform) -IMS-TEMIIS, which realizes tele-maintenance of industrial manufacturing equipment system.

On the basis of CBM + and e-Maintenance, Professor Jay Lee of the University of Cincinnati in the United States proposed the Intelligent Maintenance System (IMS) shown in Fig. 1.21 [18], and combined with dozens of companies such as Ford, Intel and Toshiba in the industry, the IMS system blueprint was developed. IMS is a maintenance system that integrates enterprise e-commerce platforms through Web-driven intelligent agents, and uses performance degradation analysis and prediction methods to achieve near-zero fault performance of products or equipment.

The demand management strategy of CBM + and IMS is based on the core enterprises of the supply chain (equipment users) or large customers. The maintenance goal is to repair the equipment in time and minimize the operation and maintenance cost of the equipment users by means of intelligent maintenance and spare parts supplier stockpiling. However, the result of supplier stockpiling will inevitably lead to high inventory cost operation of suppliers and asymmetric supply and demand of product spare parts. How to accurately predict the maintenance demand, minimize

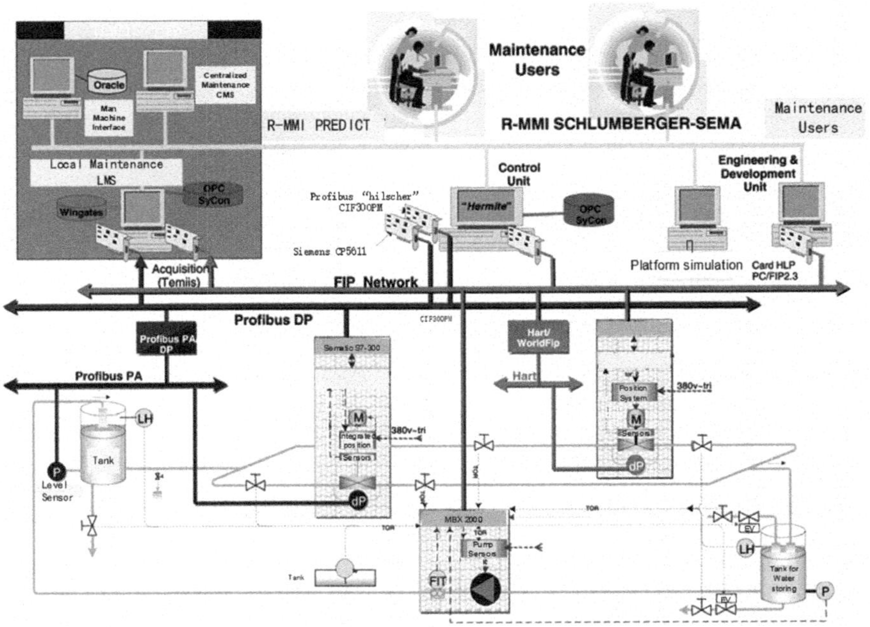

Fig. 1.20 Plug-and-play distributed remote maintenance platform. LMS: Local Maintenance Station; CMS: Centralized Maintenance Station; R-MMI: Romote Maintenance Station (PREDICT/ SCHLUMBERGER-SEMA)

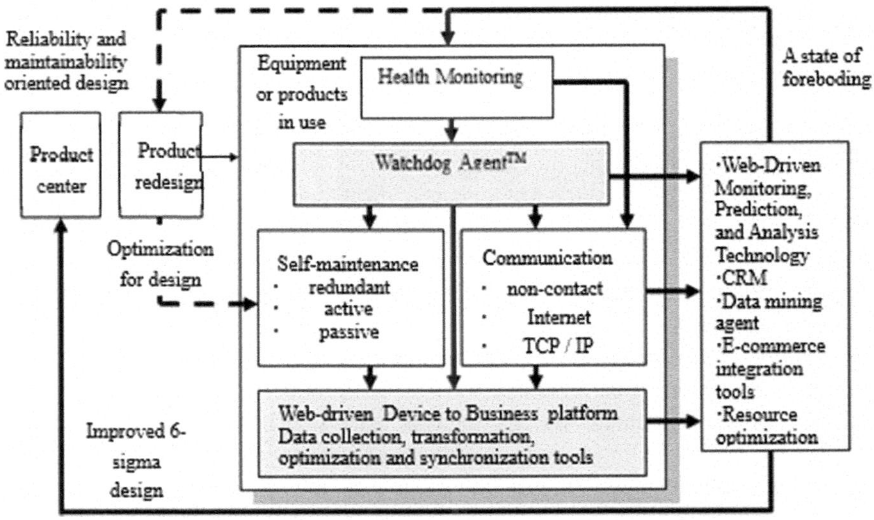

Fig. 1.21 Intelligent maintenance system framework

the safety stock (even 'zero inventory'), and meet the supply demand of spare parts is a common problem for all spare parts suppliers in the MRO maintenance network.

CBM + and IMS also do not consider the sudden and timely response characteristics of large-scale fault maintenance in the equipment maintenance network, the sociality and autonomy of maintenance experts and knowledge (from the social netizen group, according to business expertise, with the ability to participate in the maintenance of the network, as well as the power of innovation or profit). There are problems of difficult matching of service resources and non-sharing of knowledge resources. How to respond to maintenance needs to the greatest extent under limited service resources, and provide matching MRO services, establish long-term and comprehensive cooperation with customers, and how to find expert communication through an information service platform. Finding the answer to the problem is the common appeal of equipment manufacturers, maintenance providers and professional practitioners in the MRO maintenance network.

With the rapid development of information and communication technology (ICT), the emerging information technologies represented by the Internet, big data and artificial intelligence have led to a group technological revolution characterized by green, intelligent, networked and service in almost all fields. Cyber-physical systems and Advanced Analytics of Big Data have also been integrated into the manufacturing, product and service processes. Many industries are facing new opportunities and challenges in maintaining their competitiveness and market demand. In this case, future forecasting and service innovation methods based on CPS and big data analysis and intelligent predictive maintenance based on CPS, IOT and IOS have emerged. CPS, IOT and IOS promote the systematic transformation of big data, make invisible degradation patterns and low efficiency visible, and generate optimal decisions.

(2) Research on structural health monitoring technology

Fault detection and material change belong to the field of diagnosis and are related to the health status of the equipment system. The change of material is related to its surface fatigue, wear and adhesion. The health knowledge of the system can be used as a prerequisite for predicting the time of failure. Monitoring appropriate system operating variables, extracting and selecting appropriate characteristic variables will significantly affect fault diagnosis and prediction.

Among the feature variables extracted from monitoring variables using classical methods (non-big data-based supervision methods), only a part can be used to describe the degradation of components or systems. These feature variables are crucial for fault prediction. Feature extraction and selection are highly dependent on the quality of the monitoring data, which usually contains some redundant, irrelevant, or noisy data. Various methods such as data clustering, filtering, classification and pattern recognition have been mentioned in the past literature [19].

State-Of-Health (SOH) assessment and actual wear state assessment are important features of modern fault detection systems. The goal is to evaluate the failure risk probability (load rate) and the actual remaining useful life (RUL) in the state detection domain, machine monitoring, structural health monitoring, and related fields. The purpose of most methods is to detect changes in system operation and aging. In the traditional method, there are three main problems to be solved: data processing needs specific analysis of specific problems, and at the same time needs to meet the real-time requirements; explain the measurement data according to the machine state (filtering); extract task and application related knowledge (classification) from measurements and data.

In order to directly measure the wear effect, signal-based condition monitoring methods such as filtering and classification are often used. Statistical features are commonly used for feature extraction, such as root mean square, skewness, kurtosis and standard deviation. After filtering (including feature extraction), features must be evaluated and can be used for classification. Fuzzy-based methods usually combine fuzzy logic with fuzzy neural networks or fuzzy c-means. Support vector machine is widely used in condition monitoring and fault diagnosis. Multi-stage classification method can effectively analyze ROC features (detection rate, false alarm rate).

Acoustic emission is considered to be one of the high-fidelity non-destructive testing methods, which allows the observation of physical phenomena of wear mechanisms (cracks, fractures, plastic deformation), which ultimately lead to machine component failure. As the wear phenomenon changes, the material will emit energy in the form of high-frequency mechanical vibration, which propagates in the form of Rayleigh waves with a frequency of 100 kHz to 1MHz throughout the material. The American Nondestructive Testing Academic Committee defines AE as 'transient elastic wave phenomenon caused by the rapid release of energy from local sources in the material' [20]. Since the 1970s, AE has been considered as the main method for detection, microstructure characterization and damage process monitoring. Due to friction, AE is generated by the impact of the friction surface, surface damage (such as surface fatigue) and adhesive bonding. Since the late 1980s, many scholars have studied the relationship between AE behavior and mechanical wear mechanism. The parameter-based method extracts relevant and important AE features from the original AE signal measured in the time domain, and distinguishes different wear stages according to these significant AE features. By examining the AE characteristics during fatigue crack propagation, different wear stages can be clearly distinguished, and robustness is demonstrated using features based on correlation frequencies. Therefore, AE is an ideal indicator for fault detection and health status (aging) analysis.

The gradual deterioration of the system eventually leads to failure, and the time point when this occurs is called service life termination. In order to avoid catastrophic events (life-threatening injuries to humans), the service life can be extended if maintenance and operational actions are carried out in a timely manner at the right time. Therefore, it is necessary to estimate the current information about the system degradation level through appropriate technical or non-technical measurements, and it is very important to establish a life model between operating conditions (model input) and degradation parameters (RUL). Model-based methods try to establish a mathematical description of the physical relationship of the degradation process. Data-based methods usually rely on statistical and artificial intelligence methods, such as hidden Markov models, neural networks, and infinite Kalman filtering (UKF). Experience-based methods take into account the previously accumulated knowledge in the system and integrate it with the system into the model. Palmgren and Miner introduced the cumulative damage model to determine the system life [21]. Under the assumption that the stress remains unchanged, the cumulative damage model needs to know the relationship between the stress and the number of cycles the system is subjected to. Henry [22] and Marco and Starkley [23] increased the nonlinearity of the Palmgren–Miner model to adapt it to specific systems. Recent life model studies have focused on stochastic models that can describe the stochastic nature of degradation processes. [24] introduced a RUL estimation algorithm based on the combination of rain flow counting algorithm and Palmgren–Miner rule, which is used to simulate the thermal mechanical fatigue of semiconductors and applied to wind turbines, speed transmission devices, electric vehicles and aircraft. Singleton et al. [25] compared the performance of Kalman filter and extended Kalman filter algorithm in the process of bearing RUL estimation, calculated the characteristics (variance and entropy) based on time and time frequency, and compared their applicability to prognosis. The results show that entropy is more able to describe the gradual damage progression, especially in the stage of perceptual degradation. [26] proposed the classification of probabilistic and non-probabilistic models, and took Wiener, Gamma and Brownian processes as examples of probabilistic models. [27] gives a detailed introduction to the statistical data-driven method for system lifetime estimation. The statistical data-driven method estimates RUL by fitting the deployed statistical model to the previous (past) data captured by the condition monitoring system. [28] The models considering indirect observable states are based on stochastic processes, such as hidden Markov models and semi-hidden Markov models. [29] developed a method based on state machine, which is trained based on the normalized load usage relationship to apply to random disturbance systems with similar load distribution. This method can guarantee the prediction accuracy of $\pm$ 5% after the service life exceeds 20%. It can be concluded that many different life models have been developed over the years; these models have specific limitations and do not depend on the current SoH; some models require prior knowledge of potential physical degradation processes, some of which depend to a large extent on the quality of environmental monitoring data, showing different model complexity, or facing practical application problems.

(3) Research on PHM technology of fault prediction and health management

In order to meet the requirements of self-protection and self-diagnosis, fault prediction and health management is the development of CBM. PHM emphasizes state perception in equipment management, monitors equipment health status, fault-prone areas and cycles, and predicts the occurrence of faults through data monitoring and analysis. PHM originated in the 1970s. Early applications mainly focused on the field of aero-engines, such as the F404 engine of General Electric Company (GE) and the F117 engine of Pratt & Whitney Group (PW). At that time, there was no fault prediction and no data analysis capability, which belonged to condition monitoring (Fig. 1.22 shows the health monitoring system architecture of a certain type of aircraft propulsion system). In 2000, the intelligent logistics information system

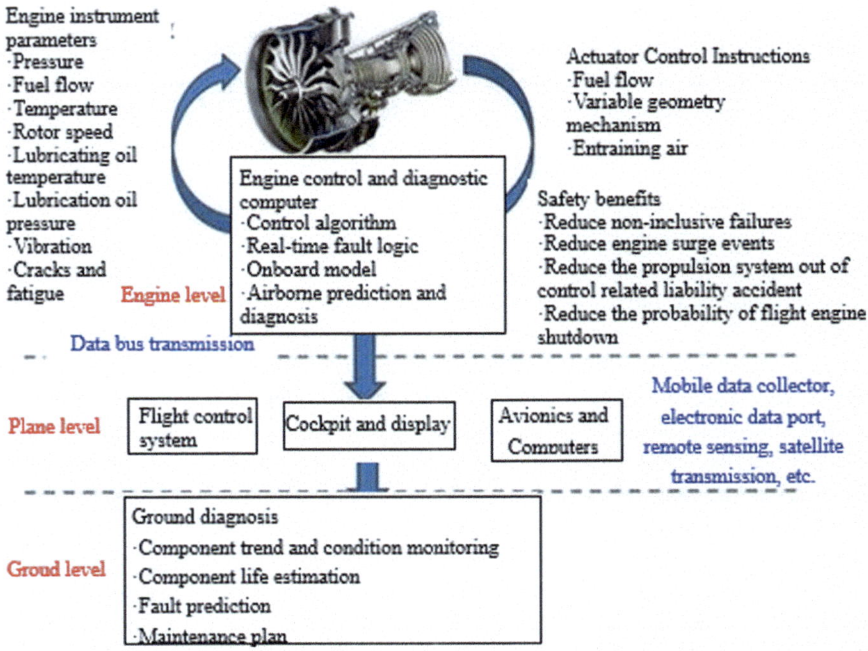

Fig. 1.22 Health monitoring system architecture of a certain type of aircraft propulsion system

of the F35 joint fighter project included the functions of aircraft system status monitoring, health assessment, fault prediction, maintenance plan, logistics support and so on, becoming the first PHM system with real fault prediction function.

The core of equipment maintenance strategy and monitoring management in PHM is equipment failure prediction. There are four types of related methods: physical model-based method, reliability model-based method, data-driven method and information fusion-based method.

1. Equipment health prediction technology based on physical model

Physical model-based equipment health prediction technology is an effective combination of condition monitoring data and equipment-specific mechanical dynamics characteristics to predict life and state. Common models include Physics of Failure (PoF), fatigue life model, crack propagation model, cumulative damage propagation model, random damage propagation model, crack diagnosis and prediction method, etc. The decision diagram of RCMII and the reasoning based on fault tree in expert system are based on the above model and expert experience for maintenance decision-making. Most of the physical model-based methods are applied to mechanical and electromechanical systems such as aircrafts and rotating mechanisms, or focus on the underlying basic units or components of systems such as materials, structures, and mechanical components. For complex systems or processes, due to the relatively complex failure modes or failure mechanisms, the modeling of fault prediction is relatively lagging behind; moreover, various models are difficult to describe a trend of equipment degradation and a detailed modeling process; it is necessary to stop the equipment, which is not allowed in the process of equipment production and operation.

2. Method based on reliability model

The method based on the reliability model calculates the reliability function of the product according to the life and related distribution. In the early 1950s, many military (such as the US military standard MIL-HDBK217) and industrial standards were developed based on the assumption of exponential distribution. Since the 1960s, Weibull distribution (introduced in 39 years) and extreme value distribution have been paid more attention to describe aging, wear and other phenomena, such as fatigue failure, vacuum tube failure and bearing failure. Later, Gamma distribution, log-normal distribution and truncated normal distribution were used to fit the life time and repair time. A large number of discrete life distributions were also used, such as 0–1 distribution used for the shape of the success or failure components such as switches and relays, binomial distribution used to describe the characteristics of several independent distribution of the success or failure components, and other geometric distribution, negative binomial distribution, Poisson distribution and discrete Weilbull distribution. Based on the method of reliability model, the reliability index of the system is divided into non repairable system and repairable system. The unrepairable system is modeled by series, parallel, series–parallel, parallel-series, K-out of n (F), cold standby, hot standby, linear ring repair and other processes. The repairable system can be described by Markov repairable model, renewal process model and Markov renewal repairable model. However, the above classical statistical theory framework does not mention the prior information such as the background information of statistical problems such as the experimental environment and experimental methods of the product. In 1985, with the emergence of Bayesion method, Bayes inference was introduced into the analysis process of reliability model, and Bayes method was formed. Bayes method takes prior information as an important source of information for statistical analysis. Considering the current reliability experimental data information and prior information, the posterior distribution of parameters is obtained, and the problem is evaluated and tested based on the posterior distribution. The maintenance strategy based on the reliability model method is generally selected: (1) the timing replacement strategy with the minimum total cost; (2) Consider the availability of maintenance strategy.

3. Data-driven equipment health prediction method

Degradation process model-driven equipment health prediction methods usually use mathematical models to describe the degradation behavior of equipment, and then predict the health status and residual life of equipment through mathematical models. The method assumes that both operational data and mathematical models are available. For example, empirical methods (Bayesian method, Dempster-Shafer theory, fuzzy logic, regression model, autoregressive model, Gaussian process regression, time series model, etc.), statistical information filtering methods (extended Kalman filter KF, particle filter PF), Markov model/hidden Markov model/hidden semi-Markov model, stochastic process model (Poisson process, Gamma process, Wienner process, Levy process, Inverse Gaussian process, etc.), grey model, etc. The identification and training process of these methods takes a long time. It is also unable to consider complex factors such as operating environment, historical state, and degradation characteristics. It is often used for offline health prediction and is not suitable for online health prediction of equipment. The degradation-based model describes the product failure process (Degradation-based Reliability Modeling) [30] by selecting physical variables that are highly related to product life and reliability (which can be called reliability characteristic quantities) and using quantitative mathematical models to describe their changes over time (performance degradation process model). The method process is shown in Table 1.4. In the product degradation model, the level of state perception includes three layers: Failure driven Reconfigration, Degradation based Adaption and Performance driven Optimization.

The method based on machine learning does not require an accurate analytical model of the fault evolution process or the life degradation process. It directly analyzes all kinds of available data of the object system. Through various data processing and analysis methods (such as multivariate statistical methods, cluster analysis, spectrum analysis, wavelet analysis, etc.), the hidden health status or degradation feature information in the object system data is mined, and the failure time of the equipment is predicted to obtain the health status and residual life of the equipment. The research and application of machine learning-based methods are mainly focused on computational intelligence, machine learning, statistical signal processing and other models and algorithms, which are divided into data preparation, data processing, feature engineering, modeling (Bin-classi/Regression/Multi-classi), simulation and other processes. This kind of method has two relatively large defects, very slow convergence and easy to fall into local optimal solution. Due to data-driven reasons, these models have high computational complexity and are easy to cause computational explosion problems.

Table 1.4 Life prediction process based on degradation model

1. Failure mechanism analysis
2. Determination of degradation amount: (1) physical frequency, vibration mode, eigenvalue, image, spectrum, chromatography, iron content, etc.; (2) the stiffness, damping, crack, size, structural parameters of the structure; (3) Various statistics, eigenvalues and eigenvectors of mathematics
3. Degradation experiment design and analysis
4. Degradation data collection and processing: (1) Preprocessing: remove abnormal data smoothing, feature recognition; (2) Feature extraction
5. Degradation model determination
6. Degradation model identification
7. Determination of failure threshold
8. Calculation of failure probability
9. Calculation of life distribution

The method based on deep learning solves the problem of automatic feature extraction in machine learning. Machine learning originated from pattern recognition and data mining algorithms in the 1970s and 1990s (Fig. 1.23). Pattern recognition emphasizes the establishment of models to characterize existing features. The sample is used to estimate the parameters in the model, and its foothold is perception. For example, after integrating a lot of wisdom and intuition into the recognition of the number '3', people have constructed a program to distinguish '3' and 'B' or '3' and '8'.Optical character recognition is born from the pattern recognition community, and decision tree, heuristic and quadratic discriminant analysis were born in this era. Pattern recognition can be called 'intelligent' signal processing in the 1970s and early 1990s. The concept of machine learning was born in the early 1990s. It is mainly an intelligent program that learns from samples, including (non-) supervised training, feature engineering, modeling algorithms, prediction and classification.

Machine learning is based on the sample training model. For example, the trained neural network is a model for a specific classification problem. The process of training the model is learning, and the foothold of machine learning is thinking. At present, machine learning is currently moving faster in the field of artificial intelligence in the narrow sense, but the breadth is still wide. Pattern recognition is constantly developing in many classic fields, such as signal processing, computer image and computer vision, and natural language analysis. With the emergence of deep learning, a unified big data predictive analysis framework has been formed, which can be verified and evaluated through the test process shown in Table 1.5.

2. Fault prediction model based on information fusion

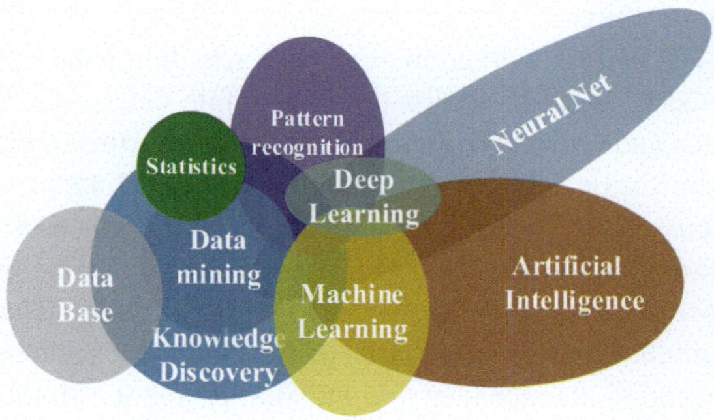

Fig. 1.23 The difference and relationship between pattern recognition and machine learning

Table 1.5 Test bed of verification and valiclation processing

1. Verification and Valiclation system
2. Validation data generation and evaluation techniques based on Monte Carlo simulation (single-parameter data MC simulation, multi-parameter Markov MC simulation (MCMC))
3. Validation and evaluation index system
4. Test bed
5. Platform

Information fusion is a comprehensive intersection of computer science, mathematics, intelligent algorithms and management fields. It means that when facing the same monitoring object, the monitoring information of various sensors and the processing methods of monitoring information can be comprehensively applied to obtain the comprehensive monitoring information of the equipment. With the improvement of the complexity of the equipment, the requirements for the performance of the equipment are gradually increasing, and the requirements for the number and types of sensors are also increasing. The method based on information fusion has received more and more attention. The information fusion method uses the intelligence and fast computing power of the computer to eliminate the difference between data information, which is conducive to the improvement of data processing quality, and also makes up for the shortcomings of different methods, such as multi-sensor fusion (sensor data processing level), data fusion (data feature level), decision fusion (fusion of different data-driven methods at the model level, data-driven and physical model fusion, model-driven and data-driven method fusion, etc.). The monitoring information from multiple sensors in the equipment MRO operation network has the characteristics of diversity, complexity and large information capacity. The processing of such information is different from the processing of single sensor information. The fusion and parallelization of models and data need to be studied by means of big data analysis and prediction methods.

3. Research on maintenance scheduling and optimization decision-making

The research on maintenance scheduling and optimization decision-making can be divided into the early research stage of non-networked maintenance scheduling optimization problem for equipment users and the current research stage of maintenance scheduling optimization problem based on Internet environment for equipment manufacturers.

The non-networked maintenance scheduling and optimization problem is mainly aimed at a single equipment or a single enterprise. Combined with fault maintenance and preventive maintenance strategies, the maintenance and maintenance task optimization scheduling problem is discussed. Domestic and foreign scholars have discussed the problem of equipment maintenance scheduling and optimization of a single equipment or a single enterprise from the aspects of equipment maintenance scheduling, production and equipment maintenance synchronous scheduling. Combined with maintenance strategy, maintenance model and inventory strategy, the preventive maintenance optimization problem of joint optimization of maintenance and inventory is studied, and a two-stage maintenance scheduling is proposed. Predictive maintenance scheduling is based on preventive maintenance. By observing the operation status of the equipment, the health status of the equipment is predicted, the stability of the system is improved, and the maintenance cost is reduced. The main research focuses on four aspects: equipment health prediction based on physical model, equipment health prediction based on data-driven, equipment health prediction based on model-driven, and equipment health prediction based on information processing. The above research has laid a foundation for the preventive maintenance scheduling and optimization theory of a single enterprise or a single equipment system, and most of them have created a theoretical model suitable for the research background. However, due to the development of new technologies, these studies neither consider the networking characteristics of MRO maintenance services, the uncertainty of large-scale random failures and the rapid response of services, nor solve the planning and real-time scheduling requirements of maintenance services in the MRO network environment, nor consider the collaboration characteristics of each participant in the network.

In recent years, many scholars have proposed research on the optimal scheduling of equipment maintenance based on the Internet environment for equipment manufacturers. Houssin and Coulibaly use the information of product use and maintenance stage in the Internet and big data environment to track and record the number and causes of equipment product downtime, and establish a model to evaluate the safety and availability of each part of the product [31], so as to make improvements in the next generation of product design stage. Munirathinam uses a variety of data mining and machine learning to analyze big data, realizes predictive maintenance of semiconductor equipment, reduces fault shutdown consumption, and improves equipment quality [32]; based on the big data analysis method, Tamura used the jump diffusion model with two Wiener processes to analyze the stability and reliability of cloud software [33]. Based on the prediction of multiple faults of equipment by quantum multi-agent neural network model, a win–win service strategy based on fuzzy random programming is proposed. The random behavior of each member in E-MRO mode and the fuzzy factors affecting its behavior are analyzed. Aiming at the selection of maintenance services, a local optimal service selection model of multi-dimensional service quality based on fuzzy analytic hierarchy process is constructed. These studies have preliminarily explored the maintenance scheduling and decision-making for equipment manufacturers in the Internet and big data environment.

1.3.2 Development Trends

With the development of information, mechanical, automation, system engineering, and management disciplines, especially the rapid development of intelligent manufacturing technology driven by new generation information technologies such as mobile internet, big data, artificial intelligence, Internet of Things, cloud computing, research institutions and scholars in the fields of advanced manufacturing, automation, computer, management science have conducted a large amount of research and exploration around equipment failure prediction and its operation and maintenance optimization, achieving a series of important research results, showing research development trends in big data-supported complex data processing, integrated and parallelized failure prediction, networked operation and maintenance, and joint optimization decision-making of production maintenance inventory, especially in the following aspects.

① Pay more attention to the fusion problem of equipment operation and maintenance data under the Internet environment. Under the support of the Internet and big data, the equipment operation and maintenance process has appeared full lifecycle business operation data, various sensor real-time status data, expert dynamic interaction knowledge of the maintenance network, and environmental factor information. How to obtain and use these multi-source heterogeneous big data in the equipment lifecycle and their related fusion analysis methods have attracted the attention of many research institutions and scholars.
② Emphasize more on large-scale failure prediction based on big data analysis. Traditional predictions based on physical models, data-driven predictions, model-driven predictions, predictions based on information fusion, etc., mostly predict the health status and failure status of a single device in a single enterprise, which is difficult to meet all task requirements of failure prediction in complex equipment networked operations. Under the Internet and big data environment, it is necessary to comprehensively consider the historical and status information of similar devices and similar parts in the MRO operation network, effectively integrate and integrate various failure prediction methods, based on big data analysis, enhance large-scale online machine learning capabilities, improve the real-time, intelligence, computational efficiency and predictive performance of the prediction system, and remotely and parallelly predict the large-scale failure status and maintenance needs in the equipment operation network in real time.
③ Pay more attention to the optimization, scheduling, and decision-making problems of equipment operation and maintenance activities based on the value network. Facing uncertain maintenance needs, in the global collaboration network, jointly plan and optimize equipment production operations, maintenance, and inventory activities, optimize shareable network service resources, and make real-time scheduling and decision-making for maintenance services.

1.4 Contents of This Book

The chapters and content arrangement of this book are shown in Fig. 1.24.

Chapter 1 introduces the problem of intelligent predictive maintenance of equipment under the intelligent manufacturing mode, the basic concepts of equipment maintenance strategy and model, and the development trend; Chapter 2 discusses the methods of fault diagnosis and prediction; Chapter 3 introduces the technical system and framework of intelligent predictive maintenance; Chapter 4 presents the perception resource management framework and model based on IoT; Chapter 5 discusses the wireless routing model and algorithm for complex manufacturing environments; Chapter 6 discusses the integration and design case of data collection protocols; Chapter 7 studies the data-driven fault diagnosis framework and method; Chapter 8 studies the data-driven fault prediction model and method; Chapters 9 and 10 discuss the maintenance optimization scheduling and decision-making method for intelligent

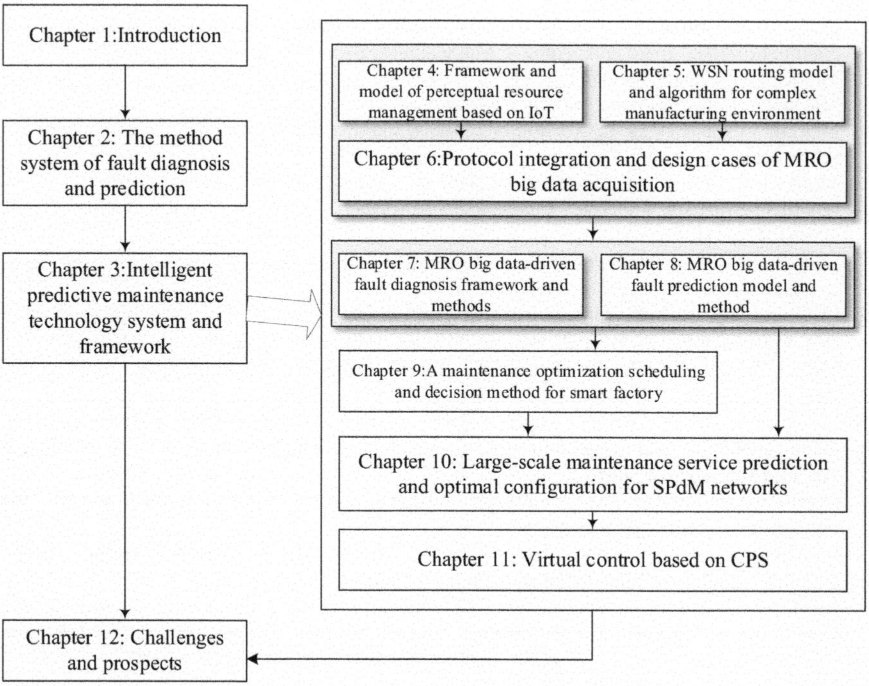

Fig. 1.24 Contents of this book

factories and the large-scale maintenance service prediction and optimization configuration method for SPdM networks, respectively; Chapter 11 explores the operation process control method based on the cyber-physical system.

References

1. Andrew K. Smart manufacturing must embrace big data[J]. Nature, 2017, 544:23–25.
2. National Manufacturing Strong Country Construction Strategy Advisory Committee. Intelligent Manufacturing [M]. Beijing: Electronic Industry Press, 2016 (in Chinese).
3. Accenture China. Outlook for New Manufacturing [March 2017 (Volume 1, 2017)] [M]. Shanghai: Shanghai Jiao Tong University Press, 2017 (in Chinese).
4. Li J., Ni J., Wang A. From big data to intelligent manufacturing [M]. Shanghai: Shanghai Jiao Tong University Press, 2016 (in Chinese).
5. Lee J., Ardakani H D, Yang S, et al. Industrial big data analytics and cyber-physical systems for future maintenance & service innovation. 4th International Conference on Through-Life Engineering Services (Procedia CIRP). Location: Cranfield, ENGLAND Date: NOV 03–04, 2015.
6. Wang K., Dai G., Guo L. Intelligent Predictive Maintenance (IPdM) for Elevator Service Through CPS, IOT& S and Data Mining. Proceedings of the 6th International Workshop of Advanced Manufacturing and Automation. Location: Univ Manchester, ENGLAND Date: NOV 10–11, 2016.

7. Li L., Liu M., Shen W., et al. An improved stochastic programming model for supply chain planning of MRO spare parts,Applied Mathematical Modelling,2017, 47(7), pp 189–207.
8. Mobley.. Translated by Shi L. and Gu N. Reliability centered maintenance [M]. Beijing: Machinery Industry Press, 1995 (in Chinese).
9. Nowlan F S and Heap H, Reliability-centered maintenance, Springfield Virginia: national information service, US Department of Commerce, 1978.
10. Morow L C. Maintenance Engineering Hand Book,Mc Graw Hill,New York,1952.
11. Pham H., Wang H. Imperfect maintenance. European Journal of Operational Research, 94 (1996): 425-438.
12. Larry A., Toms. Machinery Oil Analysis: methods, automation benefits. Society of Tribologists and Lubrication Engineers (3rd edition), 2008.
13. Li J. Industrial Big Data: Industrial Transformation and Value Creation in the Industry 4.0 Era [M]. Translated by Qiu B. et al.. Beijing: Machinery Industry Press, 2015 (in Chinese).
14. Yuan S. Structural health monitoring [M]. Beijing: National Defense Industry Press, April 1, 2007 (in Chinese).
15. Di scenzo F M, Nickerson W, Mitchell C E, et al. Open systems architecture enables health management for next generation system monitoring and maintenance[R] . Development Program White Paper, 2001.
16. Capt. Timothy Smith. USAF Condition-based maintenance plus initiative AFLMA report. Air Force Logistics Management Agency, 2003.
17. Iung B. From remote maintenance to MAS-based e-maintenance of an industrial process. Journal of Intelligent Manufacturing. 2003, 14(1): 59–82.
18. Lee J, Ni J, Djurdjanovic Dragan et al. Intelligent prognostics tools and e-maintenance. SIG2 Conference on Manufacturing Scheduling and Control in the Extended Enterprise. Location: Como, Italy, Date: May 19, 2004. Computers in Industry, 2006, 57(6):476–489.
19. Javed K.,Gouriveau R.,Zerhouni N. and Nectoux P.,Enabling health monitoring approach based on vibration data for accurate prognostics. Industrial Electronics,IEEE Transactions on,2015,62(1): 647–656.
20. American Society for Testing and Materials, Standard Terminology for Nondestructive Examinations, E 1316–90, 1990.
21. Palmgren A.,Lifetime of ball bearing (in German). VDI-Z,1924, 68: 339–341.
22. Henry D. L.,A theory of fatigue-damage accumulation in steel. Trans. of the ASME,1955, 77:913–918.
23. Marco S. M. and Starkley W. L.,A concept of fatigue damage. Trans. of the ASME,1945,76: 627–632.
24. Reddy G. R. L.,Tolbert L. M.,Ozpineci B. and Pinto J. O.,Rainow Algorithm-Based Lifetime Estimation of Power Semiconductors in Utility Applications. Trans. on Industry Applications, 2015,51(4):3368–3375.
25. Singleton R. K.,Strangas E. G. and Aviyente S.,Extended Kalman filtering for remaining-useful-life estimation of bearings. Trans. on Industrial Electronics,2015,62(3): 1781–1790.
26. Le Son K.,Fouladirad M.,Barros A.,et al.,Remaining useful life estimation based on stochastic deterioration models: A comparative study. Reliability Engineering & System Safety,2013,112: 165–175.
27. Si X. S., Wang W., Hu C. H.and Zhou D. H. Remaining useful life estimation - a review on the statistical data driven approaches. European Journal of Operational Research,2011,213(1): 1–14.
28. Beganovic N., Söffker D. Remaining lifetime modeling using State-of-Health estimation. Mechanical Systems and Signal Processing,2017,92: 107–123.
29. Si X, Wang W, Hu C, Zhou D. Remaining useful life estimation – a review on the statistical data driven approaches. European Journal of Operational Research 2011; 213 (1), 1–14.
30. Jin G. Reliability Technology Based on Degradation - Models, Methods, and Applications [M]. Beijing: National Defense Industry Press, August 2014 (in Chinese).
31. Houssin R. and Coulibaly A. Safety-based availability assessment at design stage. Computers & Industrial Engineering,2014,70:107–115.

32. Munirathinam S.,Ramadoss B. Big data predictive analytics for proactive semiconductor equipment maintenance. 2014 IEEE International Conference on Big Data,2014: 893–902.
33. Tamura Y. and Yamada S. Reliability Analysis Based on a Jump Diffusion Model with Two Wiener Processes for Cloud Computing with Big Data. Entropy 2015,17,4533–4546.

Chapter 2
Methods of Fault Diagnosis and Prediction

This chapter first discusses the general classification of fault diagnosis and prediction methods, then introduces fault prediction methods based on physical models, reliability models, data-driven, and fusion model-driven, and finally discusses the selection of fault prediction methods in combination with equipment failure samples and degradation history data.

2.1 Classification of Fault Diagnosis and Prediction Methods

The equipment system is composed of several parts, components, and subsystems. The fault mode (i.e., fault phenomenon) can describe the status of parts, subsystems, and even the entire equipment system, and has a more explicit life definition. Each part may have several fault modes as shown in Fig. 2.1, these fault modes have completely different fault mechanisms and fault rates.

For complex equipment systems, mechanical parts continuously experience wear, aging, fatigue, and other degradation faults during use, gradually manifesting as a decline in some key system performance. Sometimes these degradation faults are difficult to monitor and it is difficult to evaluate the impact of component degradation on the overall function and performance of the system. Non-mechanical parts may also experience sudden faults such as circuit breaks, short circuits, impacts, leading to instantaneous loss of system function. Therefore, the system's key fault modes can be divided into the degradation fault mode set D and the sudden fault mode set F [1] as shown in Fig. 2.2.

According to the equipment performance P-F curve shown in Fig. 1.6, degradation fault modes often have a noticeable detectable degradation process until they reach the specified fault threshold, such as performance degradation, wear, cracks, etc. These fault modes generally do not cause system function loss after degradation

M. Liu et al., *Intelligent Predictive Maintenance*, Advanced and Intelligent Manufacturing in China, https://doi.org/10.1007/978-981-97-2677-6_2

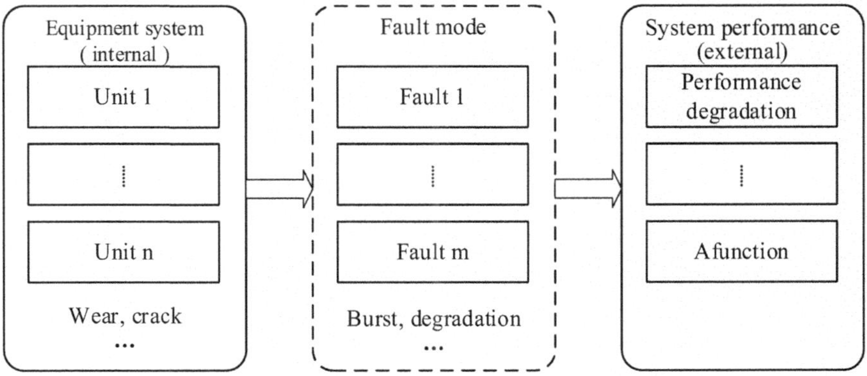

Fig. 2.1 Equipment system, fault modes, and system performance

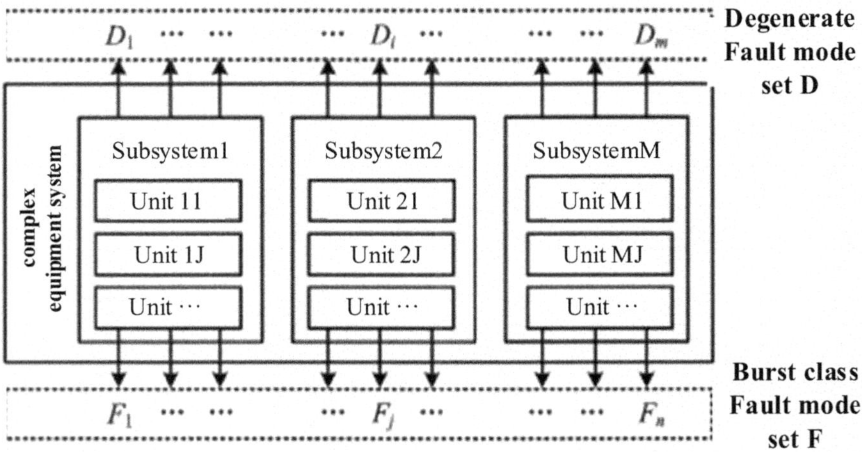

Fig. 2.2 Classification of key fault modes for a complex equipment system

exceeds the threshold, and the system can still operate with faults; sudden fault modes often do not have a noticeable degradation process or the degradation process is undetectable, usually occur instantaneously and cause system function loss.

2.1.1 Fault Diagnosis Methods

When a fault occurs, fault diagnosis involves fault detection, isolation, and identification. Fault detection is to detect whether a fault has occurred in the system, fault isolation is to locate the fault parts, subsystems, systems, and fault identification is to identify the occurred fault mode [2]. That is to say, fault diagnosis is to use the state information of the equipment under operation or relatively static conditions, such as vibration, sound, image, temperature, and wear, debris, etc., combined with the historical status of the diagnostic object, through processing and analysis to quantitatively identify the real-time technical status or fault phenomenon of mechanical equipment and its parts, components, and predict related abnormalities, faults, and predict its future technical status.

Whether it is degradation or sudden faults, the general steps and methods of fault diagnosis are as follows [3].

(1) Collect and master relevant data and information

① The collected data types are divided into three categories.

a. Numerical data: The collected data is a single value, such as oil analysis data, temperature, pressure, and humidity, etc.
b. Waveform data: The data collected in a specific time domain is a time series, usually referred to as a time waveform, such as vibration data and acoustic data are waveform data.
c. Multidimensional data: The collected data is multidimensional, the most common multidimensional data is image data, such as infrared thermal images, X-ray images, visual images, etc.

② Collect and master related data and information.

a. Machine structural performance data: Including the working principle of the machine; its position and role in the production process; important dynamic parameters, such as driving power, flow, pressure, speed change range, current, voltage, temperature, etc.; machine structure composition and parameters, such as bearing form, sealing structure, coupling structure, gear number, blade number, resonance frequency, critical speed, etc.
b. Operation and running conditions: Including load and its changes, lubrication conditions, startup and shutdown conditions, process parameter changes, etc.
c. The impact of the surrounding environment of the machine: Including the impact of factors such as temperature, humidity, association with other machines, foundation settlement, voltage fluctuation on machine performance.
d. Fault and maintenance conditions: Including the time of the last overhaul, what adjustments were made during the overhaul, records and archives of faults and fault handling since operation, weak links of the machine and the expected types and parts of faults that are likely to occur, faults of other machines under the same model and working conditions, etc.

(2) Vibration signal collection

Including instrument configuration, parameter setting, and auxiliary testing, etc.

(3) Fault analysis and diagnosis

Pay attention to the development and changes of signals, analyze the frequency components of vibration signals, the directionality and amplitude stability of vibration signals, the phase of each frequency component, side frequency analysis, waveform change analysis, shaft center trajectory analysis, and holographic spectrum analysis, etc., in this basis, fault diagnosis is carried out through diagnostic methods based on analytical models, based on signal processing (data processing of waveform and multidimensional data is also called signal processing), and based on knowledge.

For traditional diagnostic methods, if a relatively accurate mathematical model of the monitoring object can be established, such as state estimation, parameter estimation, consistency testing, etc., the preferred choice is the method based on the analytical model. The disadvantage of this method is the quality and capacity of the samples, the limitations of the model itself, the presence of noise, and the complexity and poor accuracy of the system. When the input and output signals of the controlled object can be obtained, but it is difficult to establish an analytical mathematical model, methods based on signal processing can be used, such as amplitude analysis, frequency domain analysis, wavelet analysis, adaptive time–frequency analysis, etc. This method is often combined with other methods for pre-processing of monitoring data. When it is difficult to establish a mathematical model of the controlled object, knowledge-based methods can be used, including expert systems, neural networks, fuzzy algorithms, genetic algorithms, rough sets, artificial immune algorithms, fault trees, support vector machines, and other fault diagnosis methods. With the development of new generation information technologies such as mobile internet, big data, and artificial intelligence, fault diagnosis methods based on deep learning and other new generation data-driven methods are receiving more and more attention from the industrial and academic communities [4–6], detailed content can be found in Chap. 7.

(4) Common mechanical fault identification and examples

① Imbalance, such as centrifugal compressor imbalance, compressor imbalance, etc.
② Misalignment, such as poor compressor alignment, motor and generator alignment faults, etc.
③ Mechanical looseness, such as motor imbalance and bearing looseness, generator set turbine bearing looseness, etc.
④ Rotor or shaft cracks.
⑤ Rolling bearing damage, failure.
⑥ Sliding bearing faults.
⑦ Gearbox faults.
⑧ Transmission belt faults.
⑨ Impeller, blade, and rotor fractures.
⑩ Motor faults.

1. Resonance faults, etc.

2.1.2 *Fault Prediction Methods*

Fault prediction is used to determine whether a fault is about to occur and to estimate the speed and likelihood of the fault occurring. Diagnosis is post-analysis, prediction is pre-analysis. In terms of achieving zero downtime performance, prediction is more effective than diagnosis. When a failure occurs due to a prediction failure, fault diagnosis is required [2].

Currently, there are two mainstream types of fault prediction: ① Based on the current equipment condition and past operation, predict how much time is left before a fault (one or more faults) occurs, i.e., predict the remaining life of the equipment; ② In some cases, especially when a fault or failure is catastrophic, such as a nuclear power plant, based on the current equipment condition and past operation, predict the likelihood of a fault or failure occurring at a future time (e.g., the next inspection interval), i.e., predict the probability of a fault occurring within a certain future time period. In fact, in any case, the probability that the equipment will not fail before the next inspection (or status monitoring) interval can be a good reference for maintenance personnel to judge whether the inspection interval is appropriate. In the literature on fault prediction, most papers only discuss the first type of prediction, and only a few papers involve the second type of fault prediction.

In the technical system of preventive maintenance and predictive maintenance, fault prediction is the core method to realize the prediction of the degradation state and remaining life of the object system. The classification methods of different research institutions and organizations are not the same. The most typical classification method comes from Professor Peng Xiyuan of Harbin Institute of Technology, who divides the fault prediction methods into four categories: fault prediction methods based on physical models, fault prediction methods based on reliability models, data-driven fault prediction methods, and fusion model-driven fault prediction methods [7]. This book, combined with the development and application of new generation information technologies such as mobile internet, big data, and artificial intelligence, incorporates the current fault prediction methods based on big data analysis (such as machine learning, deep learning, etc.) into data-driven prediction methods, forming the fault prediction classification method shown in Fig. 2.3.

1. Fault prediction methods based on physical models

Fault prediction methods based on physical models effectively combine state monitoring data and equipment-specific mechanical dynamics features to predict remaining life and state. Common models include failure physical models, fatigue life models, crack expansion models, cumulative damage expansion models, random damage propagation models, crack diagnosis and prediction methods, etc. The failure mode effect analysis in the RCMII decision chart and the reasoning based on the fault tree in the expert system are also based on the above physical models and expert experience for maintenance decision-making.

Fault prediction methods based on physical models are mostly applied to aircraft, rotating mechanisms, and other mechanical and electrical systems, or focus on materials, structures, mechanical components, and other system-level basic units or components. For complex systems or processes, due to the complexity of the failure modes or failure mechanisms, the modeling of their fault prediction is relatively lagging. Moreover, fault prediction methods based on physical models often use mathematical models to describe the degradation behavior of equipment, and

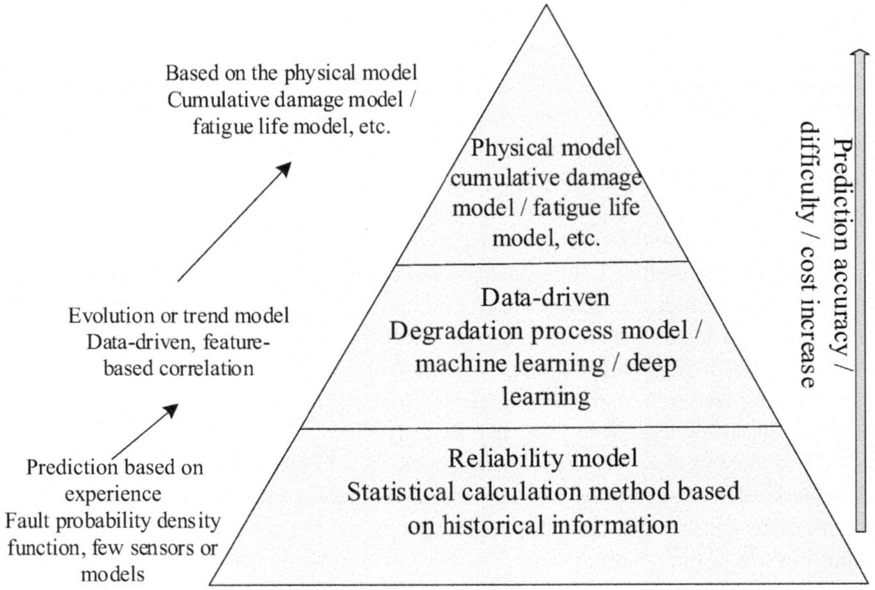

Fig. 2.3 Classification of fault predictive methods

then predict the health status and remaining life of the equipment through mathematical models. The method assumes that operable data and mathematical models are available. However, sometimes various models are difficult to describe the trend of equipment degradation and the detailed modeling process, and it is necessary to stop the equipment for processing, but this is not allowed in the production operation process of the equipment.

2. Fault prediction methods based on reliability models

Fault prediction methods based on reliability models calculate the reliability function of the product according to the density distribution function and life distribution function of historical test data [8]. In the early 1950s, life experiments based on exponential distribution began, and many military standards (such as the American military standard MIL-HDBK217) and industrial standards were formulated based on the assumption of exponential distribution. After the 1960s, Weibull distribution and extreme value distribution were paid attention to, used to describe phenomena such as aging, wear, etc., such as fatigue failure, vacuum tube failure, and bearing failure; later, gamma distribution, log-normal and truncated normal distributions were used to fit life time and repair time, a large number of discrete life distributions were also used, such as 0–1 distribution used to describe the characteristics of success-failure type components such as switches, relays, binomial distribution used to describe the characteristics of several independent success-failure type components, and others include geometric distribution, negative binomial distribution, Poisson distribution and discrete Weibull distribution, etc.

In the methods based on reliability models, the reliability indicators of the system are divided into non-repairable systems and repairable systems. Non-repairable systems are modeled through series, parallel, series–parallel, parallel-series, K -out of n (F), cold standby, hot standby, linear ring repair and other processes; repairable systems can be described using Markov-type repair models, renewal process models, Markov renewal-type repair models. However, the framework of the above classical statistical theory does not mention and consider the background information of statistical problems such as the experimental environment and experimental methods of the product, and other prior information.

In 1985, with the emergence of Bayesian analysis methods [9], the introduction of Bayesian inference into the analysis process of reliability models formed the Bayesian method. The Bayesian method takes prior information as an important source of statistical analysis information, comprehensively considers the current reliability experimental data information and prior information, obtains the posterior distribution of parameters, and evaluates and tests the problem based on the posterior distribution, which can improve prediction accuracy.

The maintenance strategy generally chosen based on the method of reliability model is: ① the timing replacement strategy with the minimum total cost; ② the maintenance strategy considering availability. The failure prediction method based on the reliability model is based on the probability density distribution function and life distribution function of component historical failure data, so it can also be called reliability model prediction based on failure. However, this model does not focus on the evolution of performance status during equipment failure.

(3) Data-driven failure prediction method

The data-driven failure prediction method recognizes the health/unhealthy behavior of the object system from historical data through the state monitoring of the object system, transforms the original monitoring data into related information and behavior models, and judges the probability of future object system failure, estimates the moment when the system fails or reaches the life threshold. Data-driven methods can be divided into methods based on degradation process models, methods based on machine learning, and methods based on deep learning according to different models.

The method based on the degradation process model often uses mathematical models to describe the decline behavior of the equipment and predict the health status and remaining life of the equipment. The method assumes that operable data and mathematical models are available, such as empirical model methods (such as Bayesian methods, D-S evidence theory, fuzzy logic, regression models, autoregressive models, Gaussian process regression, time series models, etc.), statistical filtering methods (extended Kalman filtering, particle filtering), Markov model/hidden Markov model/implicit semi-Markov model, stochastic process model (Poisson process, gamma process, Wiener process, inverse Gaussian process, etc.), gray model, etc.

Corresponding to the reliability model based on failure, the prediction model based on the degradation process model selects physical variables that are highly related to product life and reliability, adopts predictive maintenance strategies, and describes its changes over time with quantitative mathematical models to depict the failure process of products or equipment [10]. The modeling process is as follows: ① failure mechanism analysis; ② determination of degradation quantity (physical, such as frequency, vibration mode, characteristic information, image, spectrum, chromatogram, iron content, etc.; structural, such as stiffness, damping, crack, size, structural parameters, etc.; mathematical, such as various statistics, eigenvalues and eigenvectors); ③ degradation experiment design and analysis; ④ degradation data collection and processing (preprocessing, such as removing abnormal data/smoothing/feature recognition; feature extraction); ⑤ determination of degradation model; ⑥ identification of degradation process model; ⑦ determination of failure threshold; ⑧ calculation of failure probability; ⑨ calculation of life distribution. However, the identification and training process of this kind of method takes a long time, and it cannot consider complex factors such as the corresponding operating environment, historical status, degradation characteristics, etc. It is often used for offline health prediction and is not suitable for online health prediction of equipment.

The method based on machine learning does not require an accurate analytical model of the failure evolution process or life degradation process, and directly analyzes various available data of the object system. Through various data processing and analysis methods (such as multivariate statistical methods, cluster analysis, spectrum analysis, wavelet analysis, etc.), it mines the hidden health status or degradation feature information in the object system data, predicts the failure moment of the equipment, and obtains the health status and remaining life of the equipment. The methods based on machine learning that are widely studied and applied mainly focus on models and algorithms such as computational intelligence, machine learning, statistical signal processing, etc., and are divided into data preparation, data processing, feature engineering, modeling, simulation and other processes. This kind of method has two major defects, that is, very slow convergence and easy to fall into local optimal solutions. Due to the data-driven reason, these models have high computational complexity and are prone to computational explosion problems.

The method based on deep learning solves the problem of automatic feature extraction in machine learning. Machine learning originated from pattern recognition and data mining algorithms in the 1970s and 1980s [11], and the emergence of algorithms such as decision trees, heuristics, and two-class discriminant analysis gave birth to primary intelligent programs. The concept of machine learning was born in the early 1990s, mainly intelligent programs that learn from samples, including (non) supervised training, feature extraction, modeling algorithms, prediction and classification steps. With the emergence of deep learning algorithms, a unified big data prediction analysis framework has been formed, and three different analysis strategies have been produced: deep learning improves algorithm accuracy through data learning; width learning improves algorithm efficiency (up to 1000 times) through model structure diffusion; hybrid learning strategy balances the accuracy and efficiency of learning algorithms, and can realize the adaptive learning process of data and model structure.

(4) Fusion model-driven failure prediction method

Information fusion is an interdisciplinary integration of computer science, mathematics, intelligent algorithms, and management fields, indicating that when facing the same monitoring object, the monitoring information of various sensors and the processing methods of monitoring information can be comprehensively applied to obtain comprehensive monitoring information of the equipment. With the increase in equipment complexity, the requirements for equipment performance are gradually increasing, and the requirements for the number and type of sensors are also increasing. Methods based on information fusion are receiving more and more attention.

The method of information fusion utilizes the intelligence and fast computing power of computers, eliminates the differences between data information, is conducive to the improvement of data processing quality, and also makes up for the shortcomings of different methods, such as multi-sensor fusion (sensor data processing level), data fusion (data feature level), decision fusion (fusion of different data-driven methods at the model level, fusion of data-driven and physical models, fusion of model-driven and data-driven methods, etc.). The monitoring information from multiple sensors in the MRO operation network of the equipment has the characteristics of diversity, complexity, and large information capacity. The processing of this type of information is different from the processing of single sensor information. The fusion and parallelization of models and data need to be studied with big data analysis methods.

2.2 Physical Model Based Fault Prediction Methods

The fault prediction method based on the physical model generally starts from the internal working mechanism of the object system, establishes a mathematical model that can reflect the physical laws of system performance degradation under given load, working and environmental conditions, to predict the degradation development trend of the system, and can deeply understand the essence of the object system and obtain more accurate prediction results. However, in most practical applications, it requires a lot of expert knowledge to establish a performance degradation physical model of the system starting from the internal working mechanism of the system, especially when the degradation process is complex and the degradation mechanism is not fully understood, it is often impossible to construct an effective physical model.

The fault prediction method based on the physical model is closely related to the physical and electrical properties of the object. Different parts or units of the object have large differences in their physical models. Common methods based on physical models include failure physical models, fatigue life models, crack expansion models, cumulative damage expansion models, random damage propagation models, crack diagnosis and prediction methods, etc. The more mature physical model objects include mechanical materials or rotating mechanical parts, lithium-ion batteries, high-power electronic components, electronic mechanical traditional devices, etc.

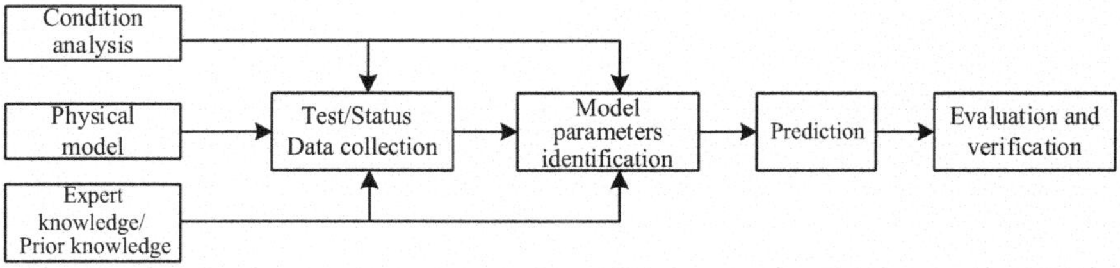

Fig. 2.4 Modeling process of fault prediction method based on physical model

The fault prediction method based on the physical model does not need to collect so much data to express the fault logic of the system, but it needs the support of experts to construct and express the model, and it may also lose some nonlinear and interrelated relationships. The general modeling process is shown in Fig. 2.4. The fault prediction method based on the physical model uses the physical model to predict the degradation rate of the system under given load and usage conditions, and needs to use the knowledge of the specific object's professional field for physical modeling; because the fault mechanism is clear, the prediction accuracy is high; considering the complexity of the prediction model, it is generally only used in specific component-level and part-level fault prediction; it is difficult to establish and abstract physical models for complex electromechanical systems.

2.3 Reliability Model Based Fault Prediction Methods

When the number of sensors in the equipment system is small or it is difficult to obtain the physical model of the system, the fault prediction of the equipment system can be realized by using statistical reliability methods based on a large amount of historical failure data. By fitting the product life distribution with the failure data from reliability experiments or the use of the same type of product, the average remaining life of the equipment system in use can be obtained. Such as early reliability analysis based on Weibull distribution, binomial distribution, exponential distribution, etc., and Bayesian methods based on posterior distribution probability in the mid-to-late 1980s. The use of reliability methods is wide-ranging, especially suitable for commercial products with zero batches and large quantities, but the accuracy is lower.

For general products, statistical analysis can be performed on historical data containing failure faults. The most classic reliability growth model—the Duane model was proposed by J. T. Duane in 1962, which gives the changes in the measured reliability parameters in the way shown in Fig. 2.5, and can also obtain numerical estimates of reliability parameters. In addition, there is the AMSAA method proposed by the US Army Equipment Analysis Agency.

The Duane model assumes that the cumulative test time of repairable products is t, and the cumulative number of product failures in the development period $(0, t)$ is

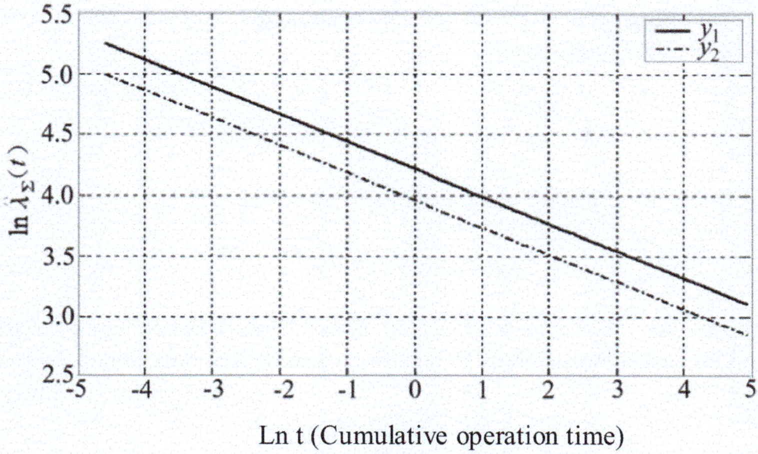

Fig. 2.5 Duane model

$N(t)$. The cumulative failure rate of the product $\lambda_{\Sigma}(t)$ is defined as the ratio of the cumulative number of failures $N(t)$ to the cumulative test time t, that is:

$$\lambda_{\sum}(t) = \frac{N(t)}{t} \tag{2.1}$$

The Duane model points out that in the product development process, as long as the exposed systematic failures are continuously corrected and improved so that they no longer occur, then the cumulative failure rate $\lambda_{\Sigma}(t)$ will decrease.) and the cumulative experimental time t can be approximately described by a straight line on a double logarithmic coordinate paper, and the mathematical expression is:

$$\ln \lambda_{\sum}(t) = \ln \lambda_I - \alpha \ln t \tag{2.2}$$

That is:

$$\lambda_{\sum}(t) = \lambda_I t^{-\alpha} \tag{2.3}$$

where, $\lambda_{\Sigma}(t)$ is the cumulative failure rate observed at time t; λ_{I} is the estimated initial failure rate, that is, the failure rate at $t = 1$, also known as the scale parameter, its geometric meaning is the intercept of the Duane curve on the vertical axis of the double logarithmic coordinate paper; α is the estimated reliability growth rate, its geometric meaning is the slope of the Duane curve on the double logarithmic coordinate paper.

The trend of the cumulative failure rate to steadily decrease is obtained from formula (2.1):

$$N(t) = \lambda_{\sum}(t)t = \lambda_I t^{1-\alpha} \tag{2.4}$$

The relationship between the instantaneous failure rate and the cumulative failure rate can be obtained from formula (2.4):

$$\lambda(t) = \frac{dN(t)}{dt} = (1-\alpha)\lambda_I t^{-\alpha} = (1-\alpha)\lambda_{\sum}(t) \tag{2.5}$$

Thus, the estimated value of the instantaneous failure rate is:

$$\hat{\lambda}(t) = (1-\alpha)\hat{\lambda}_I t^{-\hat{\alpha}} \tag{2.6}$$

Therefore, the cumulative value of the Mean Time Between Failures (MTBF) $\theta_{\Sigma}(t)$ and the instantaneous value $\theta(t)$ can be solved by the following formula:

$$\theta_{\sum}(t) = \frac{1}{\lambda_{\Sigma}(t)} = t^{\alpha}/\lambda_I \tag{2.7}$$

$$\theta(t) = \frac{1}{\lambda(t)} = t^{\alpha}/(1-\alpha)\lambda_I \tag{2.8}$$

Based on the statistical analysis of historical data, given a certain statistical sample, the failure rate can generally be estimated to predict the accuracy and precision of the results. The most typical probability-based prediction method is the Weibull distribution function (also known as the bathtub curve) shown in Fig. 2.6. This curve represents the pattern of equipment failure development within the operating life, manifested as three stages of failure rate changes, corresponding to three basic types of failure distribution, namely, the initial failure period is failure-decreasing, the occasional failure period is failure-constant, and the degradation failure period is failure-increasing.

The first stage is the initial failure period, also known as the early failure period, which refers to the process from the installation and debugging of new equipment (or overhauled equipment) to the handover to production trial use. Due to design and manufacturing defects, component processing quality, and operators not yet fully

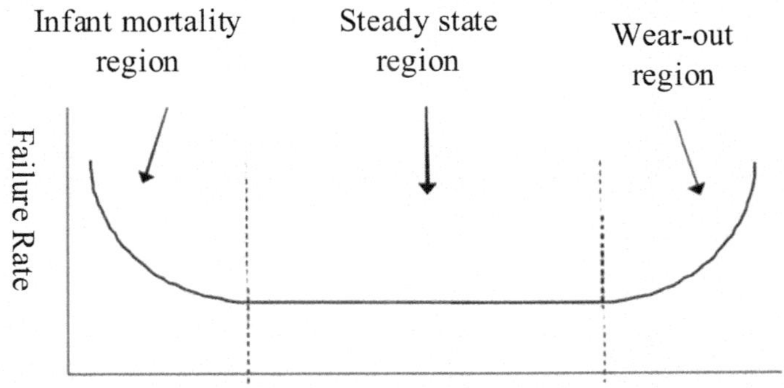

Fig. 2.6 Weibull distribution function

proficient, this stage has more failures and problems are fully exposed. As debugging and troubleshooting progress, the equipment gradually operates normally, and the failure rate gradually decreases.

The second stage is the occasional failure period. At this time, the moving parts of the equipment have entered the normal wear stage, and the operators have gradually mastered the performance, principles, and adjustment characteristics of the equipment, and the failures have significantly reduced, and the equipment has entered the normal operation stage. The failures that occur at this stage are generally due to improper equipment maintenance, improper use, deterioration of working conditions (load, temperature, environment, etc.), or due to material defects, control failures, unreasonable structure, and other design and manufacturing problems.

The third stage is the degradation failure period, also known as the wear failure period. As the use time of the equipment extends, various parts of the machine fail due to increasing wear, corrosion, fatigue, material aging, etc., causing more equipment failures, decreased production efficiency, and the time and difficulty required to eliminate failures gradually increase, and maintenance costs rise. At this time, different forms of overhaul or technical transformation should be adopted to restore production efficiency. If continued use, it may cause accidents.

The Weibull distribution is used to describe the life distribution characteristics of object elements. According to the initialization of historical data samples, a certain distribution parameter model is obtained, and then the Weibull life distribution parameters are dynamically estimated according to the changes in state data during actual use and operation (environmental conditions, load conditions, etc.), to predict the remaining useful life (RUL) of the object elements and their probability distribution function (PDF).

Since the reliability-centered maintenance theory was proposed in the 1960s, for complex equipment failures, in addition to the bathtub curve failure model, there are five other failure models shown in Fig. 2.7 [10]. Among them, the three types of failure probabilities that are stable or slowly rising (as shown in Fig. 2.7d–f) account for 89% of the total failure probability, and the rest account for 11%.

However, the failure prediction method based on the reliability model is based on a large number of sample reliability experiments, which is not applicable to complex and expensive equipment and systems such as spacecraft and aircraft. Especially for the health status assessment of individual samples, this method cannot consider many complex factors, such as operating environment, historical status, degradation characteristics, etc., and the false alarm rate of the forecast is high.

2.4 Data-Driven Failure Prediction Methods

The basic premise of data-driven failure prediction is that the feature values of the product under various failure states are known. When a failure sign is detected, the feature value of the actual operating state of the product is analyzed and compared with the known failure state feature value to judge the current health level, and then

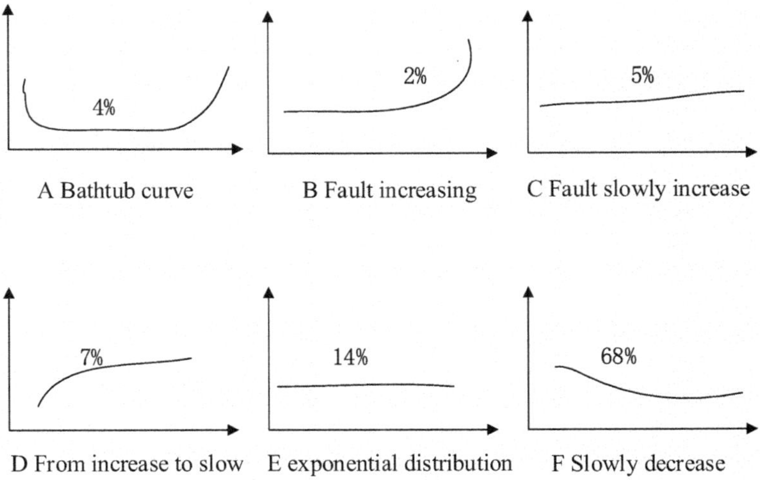

Fig. 2.7 Common failure distribution function models

the future health state is inferred or its remaining life is predicted by comparing with the feature value corresponding to the known failure development law. The process of data-driven failure prediction includes data collection, data storage, data analysis, and data application as shown in Fig. 2.8. Data-driven failure prediction methods are divided into methods based on degradation process models, methods based on machine learning, and methods based on deep learning.

2.4.1 Degradation Process Model Based Methods

Methods for failure prediction based on degradation process models often use mathematical models of the degradation process to describe the decline state or degradation mechanism of the equipment, and then predict the health status and remaining life of the equipment through the degradation model. Degradation models can use empirical models, statistical filtering, Markov chain models, stochastic process models, and gray models for description and modeling.

The Markov process in discrete time, also known as the Markov chain, has a finite or countable state space and is often used to describe degradation processes. When using the Markov chain in discrete time for modeling, the degradation state of the equipment system is divided into finite states. After specifying the probability transition matrix between all states, these models can be used to calculate the failure rate distribution under any state. Markov chain models are suitable for components whose degradation state cannot be accurately measured and must be roughly classified. Hidden Markov models, implicit semi-Markov models (extensions of hidden Markov models), etc., are variants of Markov chain models. Wang [12] constructed a

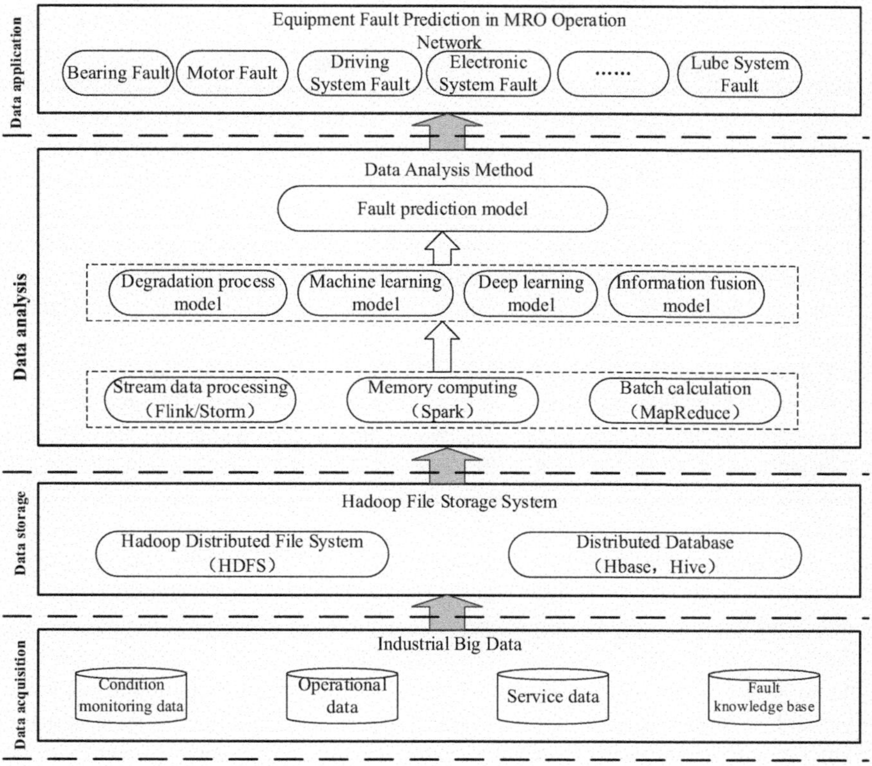

Fig. 2.8 Data-driven failure prediction model and process

state-based spare part replacement and supply strategy for degradation systems with the same components by considering the availability of spare parts and the degradation state of the unit. In recent decades, the rapid development of sensor technology has made it possible to accurately measure the degree of degradation online. In this case, the degradation state is considered to be continuous rather than discrete. Continuous-time Markov processes (stochastic process models) have independent increments, such as Brownian motion with drift (also known as Gaussian or Wiener process), compound Poisson process, and gamma process, etc., are often used to characterize the degradation process of components [13]. Among these stochastic process models, Wiener process, gamma process, inverse Gaussian process and their variants have attracted a lot of research and attention due to their good mathematical properties and clear physical interpretation [14]. These models have independent degradation increments. Brownian motion with drift is a stochastic process with independent, real-valued increments and normal distribution. The compound Poisson process is a Poisson process with independent and identically distributed random processes. The gamma process is a monotonically increasing random process with independent increments that follow the same parameter gamma distribution. Like the compound

Poisson process, the gamma process is an increasing process. The compound Poisson process has a certain number of increments in a finite time interval, and the gamma process has an infinite number of increments in a finite time interval. The former is suitable for simulating damage caused by scattered impacts, and the latter is suitable for describing damage gradually caused by continuous use. Condition based Maintenance strategies based on Wiener and gamma process degradation models have been widely studied [15, 16]. However, since the Wiener process is not always monotonic, it cannot effectively model some degradation processes such as crack growth and wear processes. For such processes with monotonic decreasing properties, gamma process and inverse Gaussian process can be used to model sample paths with monotonic constraints. The gamma process is suitable for damage that gradually accumulates in small increments over a period of time, such as wear, fatigue, corrosion, crack expansion, erosion, consumption, creep, expansion, etc. [17–19]. The inverse Gaussian process was introduced into reliability engineering by Wang and Xu [20], and further studied by Ye [21]. Due to differences in usage and environment, the degradation characteristics of components from the same population often differ. Many literatures have studied degradation models that often exist in actual systems and cause different degradation pattern heterogeneity [22–24]. Generally speaking, there are two main methods to consider heterogeneity. The first is to use random effects models, and the second is to impose prior distributions on one or some parameters in the model parameters. The essence of these two methods is the same, that is, some parameters are component-specific and the rest are shared parameters for the whole. When more degradation observations are collected, the component-specific parameters can be more accurately evaluated, and the degradation process can be better characterized. Although heterogeneity has received widespread attention and has been applied to the modeling process, it is difficult to find the optimal Condition based Maintenance strategy because the degradation process becomes non-stationary and age-dependent, and the inverse Gaussian stochastic process model can improve these problems well. Other models based on the degradation process, such as empirical statistical models, statistical filtering, and gray models, etc., are discussed in literature [6] and some specific algorithm literature, and will not be detailed here.

2.4.2 Machine Learning Methods Based Methods

With the rise of the Internet of Things and its application technology, data collection and processing technology has matured enough to generate, transmit, store, and analyze various data in batches or in real time. Methods based on machine learning are increasingly attracting the attention of the industrial sector. (1) Machine Learning Methods Machine learning is an interdisciplinary field involving probability theory, statistics, approximation theory, convex analysis, algorithm complexity theory, etc., studying how computers simulate or implement human learning behaviors to acquire new knowledge or skills, reorganize existing knowledge structures to continuously

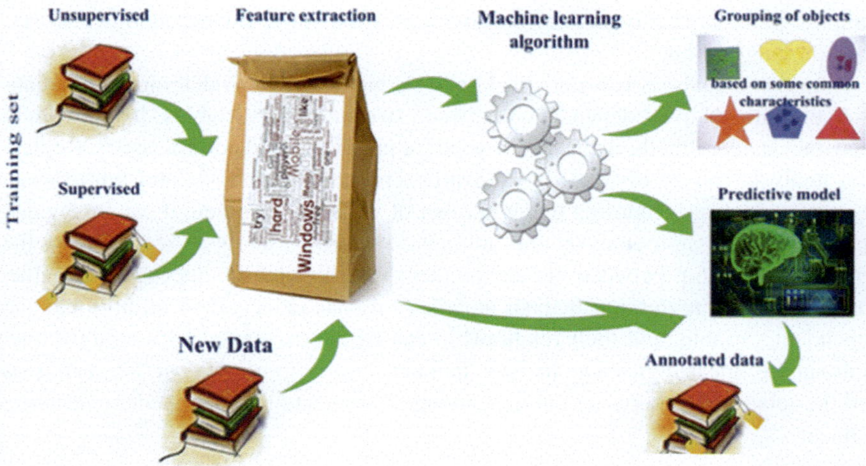

Fig. 2.9 Machine learning process

improve their performance. It is the fundamental way to make computers more intelligent, and its applications are all over the fields of artificial intelligence. The machine learning process is shown in Fig. 2.9. According to the different input data (training set), the methods of machine learning can be mainly divided into supervised learning and unsupervised learning.

① Supervised Learning: Supervised learning refers to deriving a prediction function from labeled training data. Labeled training data means that each training instance includes input and expected output, mainly used for prediction of classification and regression problems. Supervised learning learns a function from a given training data set. When new data arrives, the result can be predicted based on this function. The training set of supervised learning requires input and output, or features and targets. The targets in the training set are labeled by people. Problems labeled with continuous numerical values (such as temperature) are called regression, and problems labeled with discrete values (such as image categories) are called classification. The input data of supervised learning is called training data. Each set of training data has a clear label or result, such as spam and non-spam in the anti-spam email system, 1, 2, 3, 4, etc. in handwritten digit recognition. When building a prediction model, supervised learning establishes a learning process, compares the predicted result with the actual result of the training data, and continuously adjusts the prediction model until the prediction result of the model reaches an expected accuracy.

Classification is the process of learning a classification function or constructing a classification model, i.e., a classifier, based on existing data. This classifier can map data records in the database to a certain category in the given categories, and then be used for data prediction. A classifier is a general term for methods of classifying samples, including decision trees, logistic regression, naive Bayes, neural networks, and other algorithms.

Regression analysis is a statistical analysis method used to determine the quantitative relationship of dependence between two or more variables. It is a widely used data analysis method that can be used for prediction and control issues. Regression analysis can be divided into univariate regression analysis and multivariate regression analysis according to the number of variables involved; it can be divided into linear regression analysis and nonlinear regression analysis according to the type of relationship between the independent variable and the dependent variable. Univariate linear regression analysis includes only one independent variable and one dependent variable, and their relationship can be approximated by a straight line; multivariate linear regression analysis includes two or more independent variables and the relationship between the dependent variable and the independent variables is linear.

② Unsupervised learning aims to obtain the intrinsic structure of the sample data or the association between features from the sample data. The problem of finding similar samples is called clustering, the problem of analyzing the distribution of samples in data space is called density estimation, the problem of finding a more concise data representation is called data dimensionality reduction, and the problem of discovering the association between features is called association rule mining.

Cluster analysis refers to the analytical process of grouping a collection of physical or abstract objects into multiple classes composed of similar objects. Its goal is to collect data based on similarity for classification. Clustering originates from multiple disciplines such as mathematics, statistics, computer science, biology, and economics. Traditional statistical clustering analysis methods include hierarchical clustering, decomposition, addition, dynamic clustering, ordered sample clustering, overlapping clustering, and fuzzy clustering, etc. Clustering analysis tools using k-means, k-medoids, and other algorithms have been added to many famous statistical analysis software packages, such as SPSS, SAS, etc.

From the perspective of machine learning, a cluster is equivalent to a hidden pattern. Clustering is the unsupervised learning process of searching for clusters. Unlike classification, unsupervised learning does not depend on predefined classes or training instances with class labels. It requires the clustering learning algorithm to automatically determine the labels, while the instances or data objects in classification learning have class labels. Clustering is observational learning, not example-based learning.

Cluster analysis is an exploratory analysis. In the process of classification, it is not necessary to give a classification standard in advance. Cluster analysis can start from sample data and automatically classify. Different methods used in cluster analysis often lead to different conclusions. Different researchers may not agree on the number of clusters when performing cluster analysis on the same set of data.

Data dimensionality reduction is the operation of transforming high-dimensional data into low-dimensional data. Dimensionality reduction methods can be divided into linear dimensionality reduction and nonlinear dimensionality reduction. Linear dimensionality reduction includes Principal Component Analysis (PCA), Independent Component Analysis (ICA), Linear Discriminant Analysis (LDA), Logical Framework Analysis (LFA), Locally Preserving Projection (LPP). Nonlinear dimensionality reduction is divided into methods based on kernel functions and methods based on eigenvalues. Methods based on kernel functions include Kernel Principal Component Analysis (KPCA), Kernel Independent Component Analysis (KICA), Kernel Discriminant Analysis (KDA), and methods based on eigenvalues include Isometric Feature Mapping (ISOMAP), Local Linear Embedding (LLE), Laplacian Eigenmap (LE), Local Tangent Space Alignment (LTSA), Maximum Variance Unfolding (MVU), etc.

Association rules were initially proposed for the problem of shopping basket analysis. Suppose a store manager wants to understand more about customers' shopping habits, especially which products customers might buy at the same time during a shopping trip. To answer this question, shopping basket analysis can be conducted. This process analyzes customers' shopping habits by discovering the associations between different products that customers put in their shopping baskets. The discovery of these associations can help retailers understand which products are frequently bought by customers at the same time, thereby helping them formulate better marketing strategies.

In 1993, Agrawal and others first proposed the concept of association rules and provided the corresponding mining algorithm, but the performance of the algorithm was poor. In 1994, they established the theory of itemset lattice space and proposed the famous Apriori algorithm, which is still widely discussed as a classic algorithm for association rule mining.

③. Other machine learning methods include semi-supervised learning, reinforcement learning, transfer learning, and self-learning, which have evolved from the two learning methods mentioned above.

Semi-supervised learning is between supervised learning and unsupervised learning. Part of the input data is labeled, and when predicting, the model first needs to learn the intrinsic structure of the data in order to organize the data reasonably. The algorithms mainly include extensions of some commonly used supervised learning algorithms. These algorithms first try to model the unlabeled data, and then predict the labeled data, such as graph-based inference or Laplacian Support Vector Machines, etc.

Reinforcement learning, also known as reinvigoration learning, evaluative learning, or enhancement learning, is developed from theories such as animal learning and parameter perturbation adaptive control. It is used to describe and solve the problem of an intelligent agent (Agent) achieving maximum reward or achieving specific goals through learning strategies in the process of interacting with the environment. It has many applications in fields such as intelligent control of robots and analysis prediction. The principle is: if an Agent's behavior strategy leads to a positive reward (reinforcement signal) from the environment, then the tendency of the Agent to generate this behavior strategy will be strengthened in the future. The goal of the Agent is to discover the optimal strategy in each discrete state to maximize the expected discounted reward.

Transfer learning can transfer knowledge from existing data to help future learning. Machine learning assumes that training data and test data follow the same data distribution, but in many cases, this same distribution assumption is not satisfied. The usual situation that may occur is that the training data is outdated, that is, the hard-to-label data has to be discarded, and a large amount of new data has to be re-labeled. The purpose of transfer learning is to use the knowledge learned from one environment to help the learning task in a new environment. Currently, there are only a small amount of new labeled data, but a large amount of old labeled data (even valid data of other categories). By selecting the valid data from these old data and adding it to the current training data, a new model can be trained.

Self-learning first extracts a set of features from unlabeled natural images, so that any labeled and unlabeled image can be represented by these features. Since each labeled sample is represented by these features (capturing the high-level structure of the image), the represented labeled samples can be trained into a classifier for classification.

Self-learning, like semi-supervised learning, currently only has a small number of training samples (small samples) at hand, but there are also a large number of unlabeled samples (no labels) around. Take the classic example of separating elephants and rhinos: supervised learning means that there are a large number of labeled samples of elephants and rhinos at hand, and then the classifier is trained for classification; transfer learning means that there are a large number of samples of sheep and horses (large heterospecific samples) and a small number of elephant and rhino samples (small samples) at hand, and then effective samples are selected from the samples of sheep and horses to be added to the labeled samples of elephants and rhinos (weak labels), and then the classifier is trained using supervised learning methods; if there are only a small number of labeled samples of elephants and rhinos (small samples weak labels) at hand, and there is a pile of unlabeled data of elephants and rhinos (the data is either an elephant or a rhino, no other species), semi-supervised learning is to use these samples to train the classifier to achieve classification; unsupervised learning is to obtain the inherent structure or feature association of data from unlabeled samples; self-learning is that there are only a small number of labeled samples of elephants and rhinos (small samples weak labels) at hand, and there is a large pile of natural images (natural images are pictures of various species including pictures of elephants and rhinos). (2) Common algorithms in machine learning Based on the similarity of the functions and forms of the algorithms, the algorithms can be classified, such as tree-based algorithms, neural network-based algorithms, etc. However, the scope of machine learning is very large, and some algorithms are difficult to classify clearly. For some classifications, the same classification algorithms can be used for different types of problems.

① Regression algorithms: Regression algorithms (Fig. 2.10) are a class of algorithms that try to explore the relationship between variables by measuring errors, and are tools for statistical machine learning. In the field of machine learning, regression sometimes refers to a class of problems, and sometimes to a class of algorithms. Common regression algorithms include least squares, logistic regression, stepwise regression, multivariate adaptive regression splines, and local scatter smoothing estimation.

② Instance-based algorithms: Instance-based algorithms (Fig. 2.11) are often used to build models for decision-making problems. A batch of sample data is first selected, and then new data is compared with sample data based on some approximation to find the best match. Instance-based algorithms are often also referred to as memory-based learning. Common instance-based algorithms include k-nearest neighbors, learning vector quantization, self-organizing map algorithms, etc.

③ Regularization algorithms: Regularization algorithms (Fig. 2.12) are extensions of linear regression algorithms and solve many problems of overfitting in regression algorithms due to many input characteristics. Regularization methods prevent overfitting by adding artificial penalties to model coefficients and adjusting the algorithm. Common regularization algorithms include ridge regression, least absolute shrinkage and selection operator regression, elastic network, etc.

④ Decision tree algorithms: Decision tree algorithms (Fig. 2.13) build decision models based on the attributes of the data using a tree structure to solve classification and regression problems (solving the problem that linear regression cannot easily represent the non-linear relationship of features). Common decision tree algorithms include classification and regression trees, ID3, C4.5, chi-square automatic detection method, single-layer decision tree, random forest, multivariate adaptive regression splines, and gradient boosting machine, etc.

⑤ Bayesian algorithms: Bayesian algorithms (Fig. 2.14) are a class of algorithms based on Bayes' theorem, mainly used to solve classification and regression

Fig. 2.10 Regression algorithms

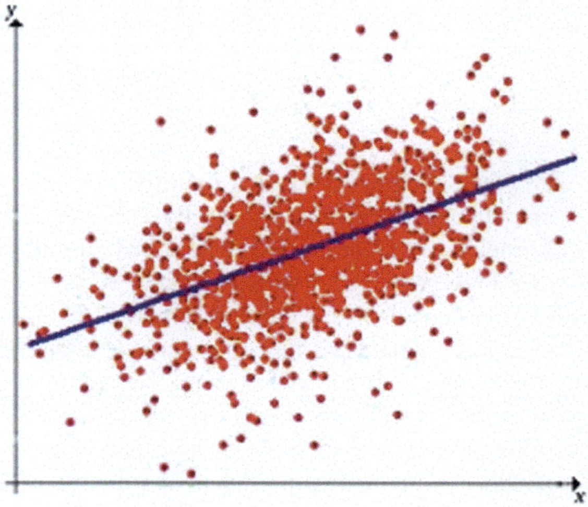

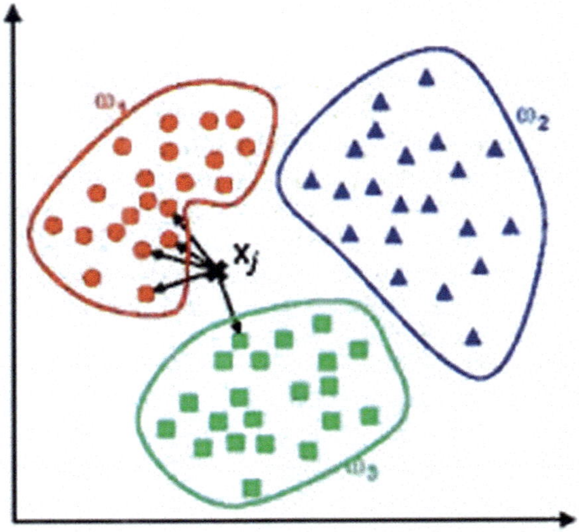

Fig. 2.11 Instance-based algorithms

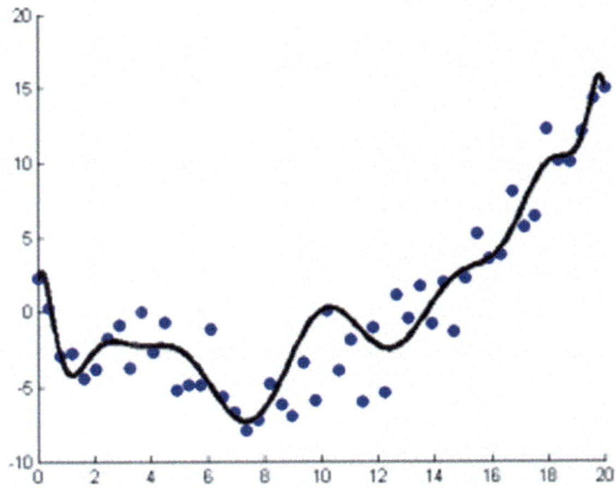

Fig. 2.12 Regularization algorithms

problems. Common Bayesian algorithms include naive Bayes algorithm, average one-dependence estimators, and Bayesian belief networks.

⑥ Kernel-based algorithms: Kernel-based algorithms (Fig. 2.15) map input data to a high-order vector space, making some classification or regression problems easier to solve. Common kernel-based algorithms include support vector machines, radial basis functions, and linear discriminant analysis, etc.

⑦ Clustering algorithms: Clustering, like regression, sometimes describes a class of problems, and sometimes describes a class of algorithms. Clustering algorithms usually merge input data according to the center point or hierarchical method, trying to find the inherent structure of the data, so as to classify the

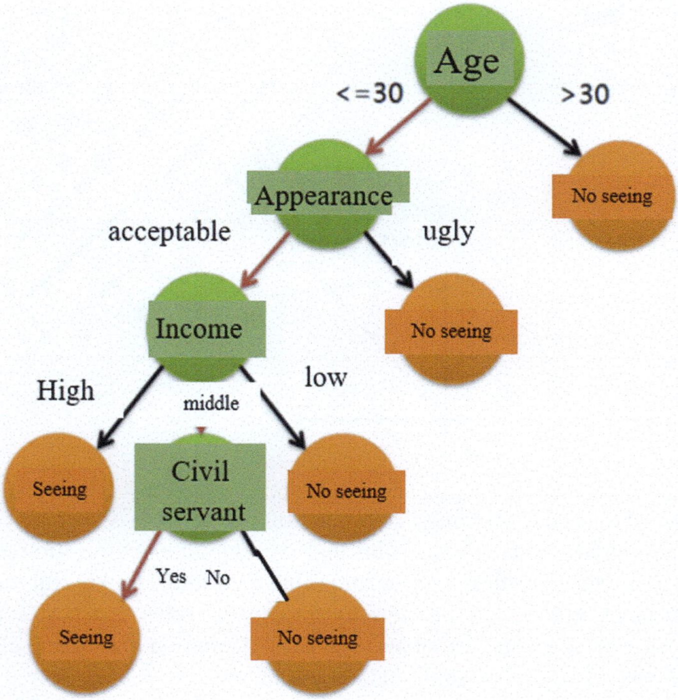

Fig. 2.13 Decision tree algorithms

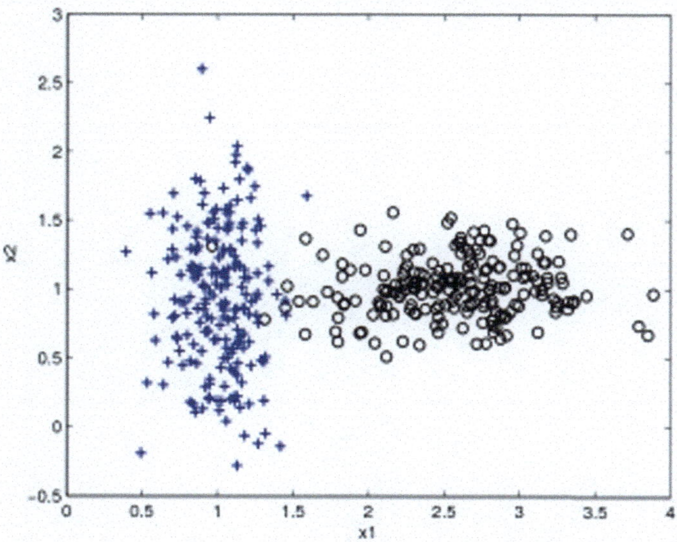

Fig. 2.14 Bayesian algorithms

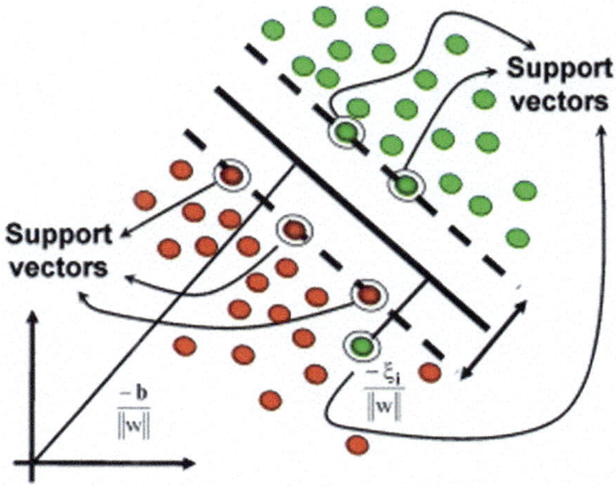

Fig. 2.15 Kernel-based algorithms

data according to the maximum common point (Fig. 2.16). Common clustering algorithms include k-means algorithm and expectation–maximization algorithm.

⑧ Association rule algorithms: Association rule algorithms find useful association rules in large multivariate datasets by finding the rules that best explain the relationship between data variables (Fig. 2.17). Common association algorithms include Apriori algorithm and Eclat algorithm, etc.

⑨ Multilayer artificial neural network algorithms: Multilayer artificial neural network algorithms (Fig. 2.18) simulate biological neural networks, are a class of pattern matching algorithms, and are usually used to solve classification and

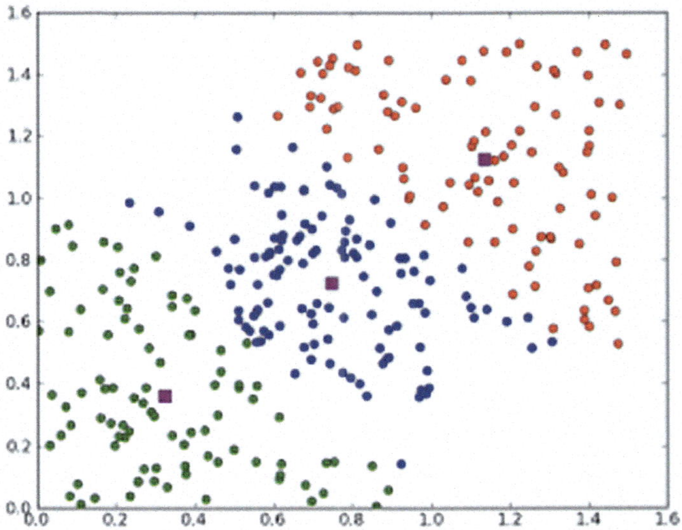

Fig. 2.16 Clustering algorithms

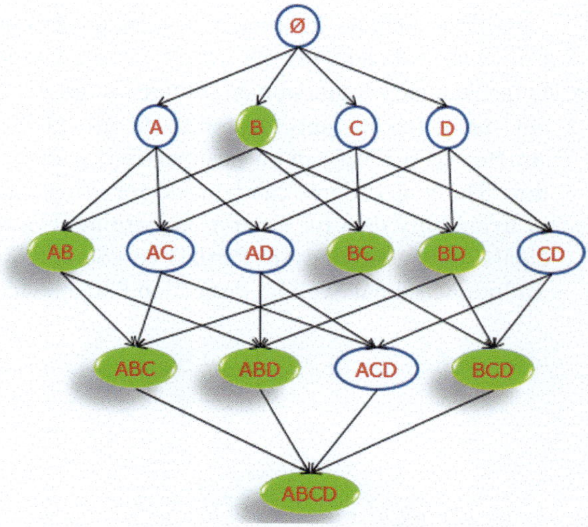

Fig. 2.17 Association rule algorithm

regression problems. Multilayer artificial neural network algorithms are a huge branch of machine learning, with hundreds of different algorithms, and deep learning is one of them. Important multilayer artificial neural network algorithms include perceptron neural networks, backpropagation, Hopfield networks, self-organizing maps, learning vector quantization, etc.

⑩ Deep Learning Algorithm—The deep learning algorithm is a development of the multilayer artificial neural network algorithm and has recently gained a lot of attention. With computing power becoming increasingly affordable, deep learning attempts to build much larger and more complex neural networks. Many deep learning algorithms are semi-supervised learning algorithms used to handle large datasets with a small amount of unlabeled data. Common deep learning

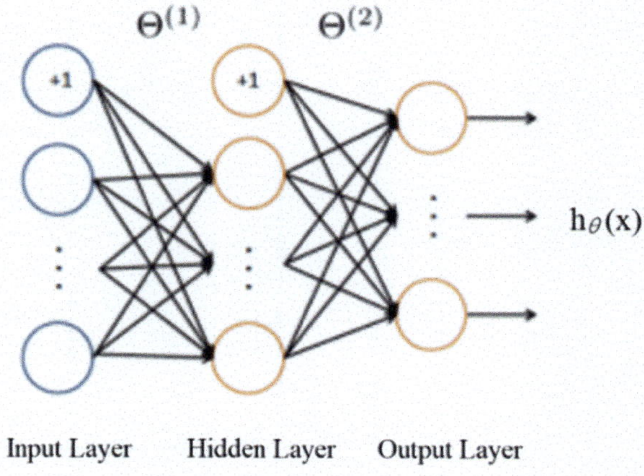

Fig. 2.18 Multilayer artificial neural network algorithm

algorithms include Restricted Boltzmann Machines, Deep Belief Networks, Convolutional Networks (Fig. 2.19), Stacked Autoencoders, etc.

⑪ Dimensionality Reduction Algorithm—Like clustering algorithms, dimensionality reduction attempts to summarize or explain data by analyzing its inherent structure (Fig. 2.20) in an unsupervised learning manner using less information. These algorithms can be used for visualization of high-dimensional data or to simplify data for supervised learning. Common algorithms include Principal Component Analysis, Partial Least Squares Regression, Sammon Mapping, Multidimensional Scaling, Projection Pursuit, etc.

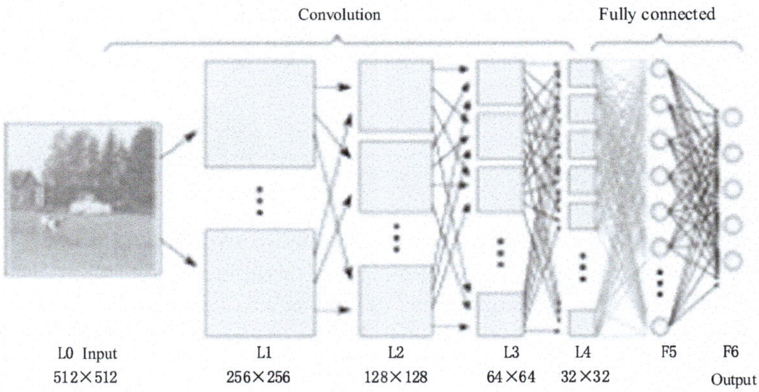

Fig. 2.19 Deep learning algorithm

Fig. 2.20 Dimensionality reduction algorithm

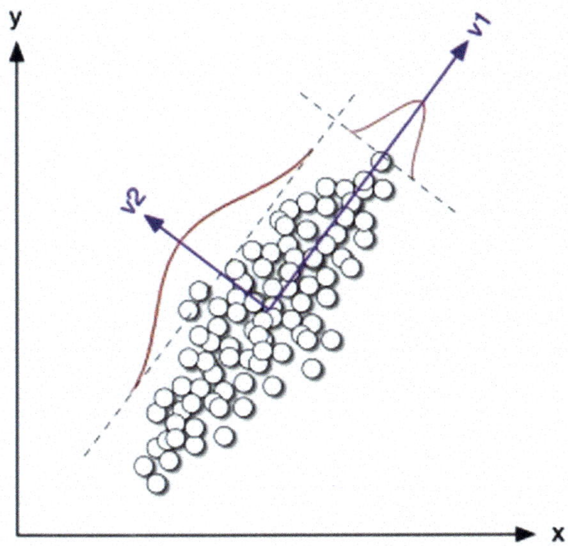

⑫ Ensemble Learning Algorithm—The goal of supervised learning is to learn a stable model that performs well in all aspects, but the reality is often not ideal, sometimes only obtaining multiple biased models (weak supervised models that perform better in some aspects). Ensemble learning uses multiple weak classifiers, such as decision trees, neural networks, Bayesian classifiers, k-nearest neighbors, etc., to form a strong classifier, and then integrates the results for overall prediction. In this case, even if a weak classifier makes a wrong prediction, other weak classifiers can correct the error. The difficulty lies in which independent weaker learning models to integrate and how to integrate the learning results.

Common ensemble learning algorithms include Boosting, Bootstrapped Aggregation (Bagging), AdaBoost, Stacked Generalization, Gradient Boosting Machines, etc. Among them, Bagging uses complex base models, trying to smooth the model's predictions by reducing the overfitting of complex models, while Boosting uses simple base models, trying to increase the overall complexity of the model by improving the flexibility of simple model predictions. When the base model is a decision tree, the ensemble learning algorithms corresponding to Bagging and Boosting are Random Forest and Boosting Tree, respectively (Fig. 2.21).

Ensemble learning has good strategies on datasets of all sizes. For large datasets, they can be divided into multiple small datasets, learning multiple models for combination; for small datasets, the Bootstrap method can be used for sampling, obtaining multiple datasets, training multiple models, and then combining them.

(3) Fault Prediction Based on Machine Learning

The fault prediction and decision-making process based on machine learning is shown in Fig. 2.22, which includes data preparation, data processing, feature engineering, predictive modeling, training, simulation and testing, maintenance decision-making, etc. [25].

① Data Preparation—The common data elements of predictive maintenance problems can be summarized as follows.

a. Fault History: Records of faults in internal parts or components of equipment, such as flight delay dates, aircraft component fault dates and types, ATM withdrawal transaction faults, train door faults, elevator door faults, brake disc replacement dates, fan fault dates, and circuit breaker command faults, etc.
b. Maintenance History: Records of error codes, maintenance activities, or component replacements in the repair and maintenance history of equipment, such as flight error records, ATM transaction error records, train maintenance records, and circuit breaker maintenance records.
c. Equipment Status and Usage: Machine operation status data collected from sensors, such as flight routes and times, sensor data collected from aircraft engines, ATM sensor readings, train event data, sensor readings from wind turbines, real-time data from elevators and connected cars, etc.

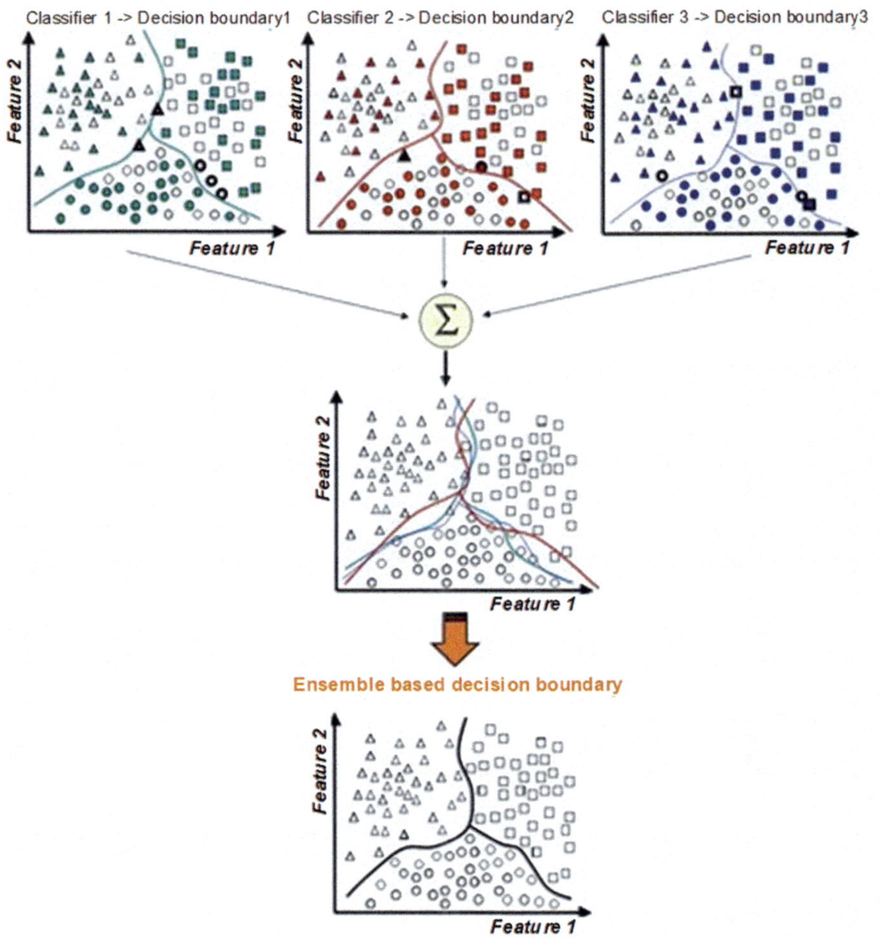

Fig. 2.21 Ensemble learning model

d. Equipment Features: Feature information describing the size of the machine engine, manufacturer and model, location, such as circuit breaker technical specifications, geographical location, car specification descriptions (such as brand, model, engine size, production equipment), etc.
e. Operator Features: Features of the operator, such as gender, past experience, etc.

Usually, the fault history is included in the maintenance history (for example, in the form of special error codes or component order dates). In these cases, data can be extracted from the maintenance data. In addition, different business domains may contain various other data sources that affect fault patterns, which are not exhaustively

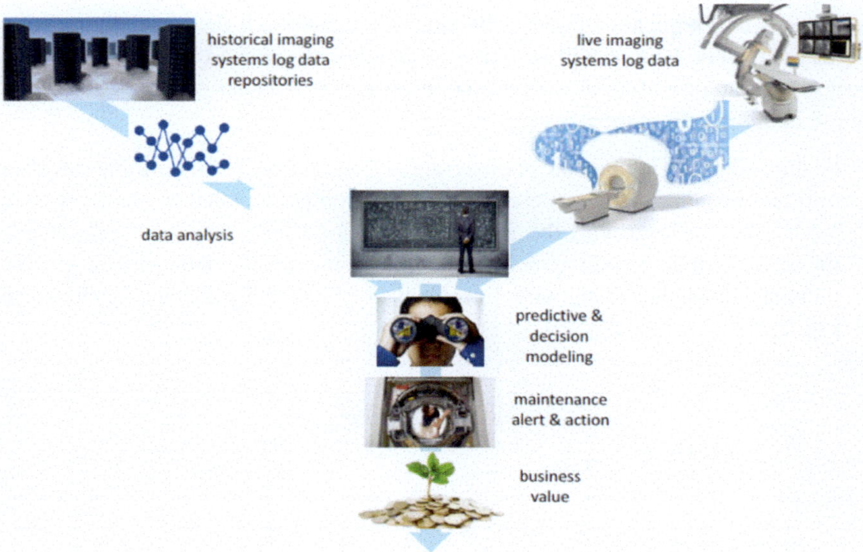

Fig. 2.22 Fault prediction and decision-making process based on machine learning

listed and should be identified by consulting experts in the relevant field when building predictive models.

Given the above data sources, the two main types of data observed in predictive maintenance are temporary data and static data. Fault history records, machine conditions, repair history records, usage history records are almost always accompanied by timestamps indicating the collection time of each data. Machine characteristics and operator characteristics are usually static, typically describing the technical specifications of the machine or the attributes of the operator. These characteristics may change over time, and if so, should be considered as timestamped data sources.

② Data Processing Before entering any type of feature engineering or label engineering, it is necessary to first prepare the data in the form required for creating features. The ultimate goal is to generate a data record for each device or asset in each time unit and input its features and labels into the machine learning algorithm. To prepare a clean final dataset, some preprocessing steps should be taken. The first step is to divide the duration of data collection into time units, where each record belongs to the time unit of the asset. Data collection can also be divided into other units such as operations, but for simplicity, time is chosen as the unit.

The measurement unit of time can be seconds, minutes, hours, days, months, quarters, cycles, etc., the choice of which depends on the efficiency of the data preparation process, or the state change of the device from one time unit to another, or other factors in a specific field. In other words, in many cases, from one unit to another, the data may not show any difference, and the time unit does not have to be the same as the frequency of data collection. For example, if the temperature value is collected every 10 s, it will increase the number of cases in the entire analysis process without providing any additional information. A better strategy could be to choose an hour as the time unit.

③ Feature Engineering Feature engineering is the process of transforming raw data into features that better represent the actual problem handled by the predictive model, thereby improving the accuracy of unknown data. Feature engineering is the construction, extraction, reduction, or combination of features using specific domain knowledge of the target problem or automated methods, as shown in Fig. 2.23.

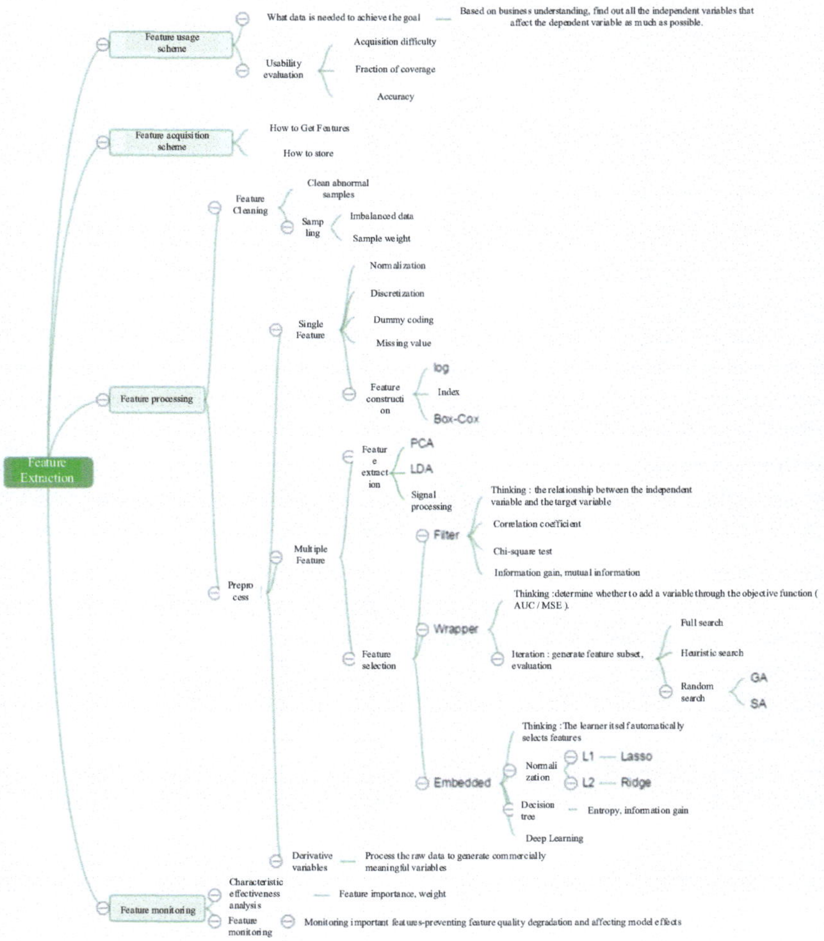

Fig. 2.23 Content of feature engineering

Raw features or data, such as various physiological indicators of the human body (to describe health status), the gray value of each point in a digital image (to describe image content) are directly measured and often not used in machine learning for several reasons.

- Raw data does not reflect the essential features of the object.
- High-dimensional raw data is not conducive to classifier design: the calculation amount is large, for example, for a grayscale image with pixels of 1024 × 768 and a grayscale level of 256 levels, the direct representation requires 786432B, and the space, time, and calculation amount required for training and recognition are unacceptable; redundancy, in the raw data space, a large amount of data are strongly correlated redundant features; sample distribution is sparse, for a limited number of training samples, the distribution in the high-dimensional raw data space is very sparse.
- If a large number of measurement values are used directly as classification features without analysis, it not only consumes time but also affects the classification effect, causing the problem of dimensionality disaster.

In view of the characteristics and deficiencies of the above raw features or data, in order to design a better classifier, it is usually necessary to analyze the set of measurement values of the original data, and after transformation and selection processing, form effective recognition features. The processing methods include:

Reduce the dimensionality of features while ensuring a certain classification accuracy, so that the classifier can achieve fast, accurate, and efficient classification;

- Remove ambiguous features that are not conducive to classification, so that the provided features have better separability, and the classifier is easy to distinguish;
- The provided features should not be repeated, and features with strong correlation but no additional classification information should be removed.

Therefore, the purpose of feature engineering is to discover important features, divided into feature construction, feature extraction, and feature selection.

Feature construction is generally to construct new features by performing arithmetic operations on the original features. For example, if the original features are x_1 and x_2, then $x_1 + x_2$ is a new feature, or when x_1 is greater than a certain number c, a new variable x_3 is generated, and $x_3 = 1$, when x_1 is less than c, $x_3 = 0$, many features can be constructed in this way.

The number of original features may be large, and it is necessary to transform (map) the high-dimensional feature space to a low-dimensional space. These secondary features are generally some combination of the original features. Feature extraction is to transform n features $\{ x_1, x_2, \ldots, x_n\}$ through some transformation, producing m features $\{y_1, y_2, \ldots, y_m\}$ $(m < n)$ as new classification features (or called secondary features). For example, Principal Component Analysis (PCA), Factor Analysis, Linear Discriminant Analysis (LDA) can all perform feature extraction on the original data, each principal component after PCA reduces the original data represents a new feature, and factor analysis can find the latent factors behind the latent variables. The convolution layer of the Convolutional Neural Network is also a feature extraction process. After a picture is continuously scanned by convolution, some features in the original picture will be gradually extracted. In fact, PCA itself is a linear combination of initial variables, and its essence also belongs to feature construction, but general feature construction refers to simple arithmetic operations.

Feature selection is from n measurement value set $\{x_1, x_2,..., x_n\}$. In, a subset is selected according to a certain criterion for classification C_n^m, as a dimension reduction (m dimension, $m < n$) classification feature, the combination number is large, and some algorithms are needed to avoid exhaustive search. Common feature selection methods include filter, wrapper, and embedded methods.

Filter feature selection is to screen features by evaluating the correlation between each feature and the result, leaving the strongest correlated features. The core idea is: first select features for the dataset, and then train the model. The advantage of filter feature selection is its simple idea, often calculating correlation through methods such as Pearson correlation coefficient, chi-square test, mutual information, etc., and then retaining the strongest N features for model training; the disadvantage is that it does not consider the correlation between features, resulting in the final training effect of the model not being so good.

Wrapper feature selection uses the machine learning model to be used and the performance evaluation indicators (such as MSE, AUC, etc.) as important basis for feature selection. Each time, it uses complete search (such as dynamic programming, branch and bound), heuristic search (such as A algorithm, A^* algorithm) or random search (such as genetic algorithm, simulated annealing, taboo search, hill climbing search) algorithms to select several features or exclude several features. Generally, the effect of wrapper feature selection is better than that of filter feature selection, but the training process takes a long time and the system overhead is also greater. The most typical wrapper algorithm is the recursive feature deletion algorithm. Its principle is to use a base model (such as random forest, logistic regression, etc.) for multiple rounds of training. After each round of training, eliminate several features with low weight coefficients, and then train a new round based on the new feature set.

Embedded feature selection is to analyze the importance of features based on machine learning algorithms and models, and then select the most important N features. The biggest difference from wrapper feature selection is that embedded feature selection combines the feature selection process with the model training process, so that the best feature set can be quickly found, which is more efficient and faster. Common embedded feature selection methods include feature selection based on regularization (such as ridge regression, Lasso regression) and feature selection based on decision tree models. Among them, Lasso regression is a regularization method. In regression, it mainly controls the regression coefficient, which cannot be too large. It can not only constrain the coefficient, but also constrain the unimportant coefficient to 0 when the model is optimal, directly achieving feature selection or variable selection, which is very suitable for high-dimensional data analysis; the biggest difference between ridge regression and Lasso regression is that ridge regression introduces L2 norm penalty term, Lasso regression introduces L1 norm penalty term, Lasso regression can make many coefficients in the loss function become

0, which is better than ridge regression (all coefficients exist), and the calculation amount of Lasso regression will be much smaller than that of ridge regression. Decision tree models have strong interpretability. They divide y according to the value of x. The basis of good or bad division is purity. In a division block, the higher the purity, the better the division, which also shows that the division variable is selected well; random deep forest, Bagging, Boosting, Gradient Booting, XGBoost and other algorithms all have feature selection functions, and neural networks, support vector machines, deep learning, etc. also have feature selection functions.

Feature engineering is the basis for building predictive maintenance machine learning algorithms and data label engineering. How to choose the right algorithm will largely depend on the collected data and corresponding business problems. Next, combined with a case, discuss how to apply data sources with timestamps to construct lagged features, and how to use static data sources to construct static features and other feature construction functions.

a. Lagged features. As mentioned earlier, in predictive maintenance, historical data usually comes with timestamps indicating the collection time of each data. There are many ways to create features from timestamped data. Since feature engineering is considered a highly creative field in predictive modeling, there may be many other ways to create features. Here, some general techniques are provided.

- Rolling aggregation features. For each record of a device, select a rolling window of size "W", which is the number of time units for calculating historical aggregation. Then, use the W cycles before the record date to calculate the rolling aggregation features. Some examples of rolling aggregation can be rolling count, average, standard deviation, outliers based on standard deviation, CUSUM measures, minimum and maximum of the window, and also use anomaly detection algorithms in abnormal data to capture trend changes, peaks and level changes.

As shown in Fig. 2.24, the sensor values of each asset per unit time are represented by a blue line. The rolling average characteristics of the records at t_1 and t_2 are calculated and marked as $W = 3$ (indicated by orange and green groups respectively). Taking aircraft component failure as an example, rolling averages, standard deviations, and sums of sensor measurements from the past week, the past three days, and the last day can be created. For ATM failures, original sensor values, rolling averages, medians, ranges, standard deviations, the number of outliers exceeding three standard deviations, consumption upper and lower limits, etc. can all be used as predictive features. For flight delay prediction, features are created using error code counts from last week. For train door failures, lagged features are created using the event count for the last day, the event count for the past two weeks, and the variance of the event count for the past fifteen days. The same counts are used for maintenance-related events. By choosing a very large W (for example, a year), you can view the entire history of the asset, such as all maintenance records, failures, etc. This method is used to count circuit breaker failures in the last three years. Similarly, for train failures, count all maintenance events to create features that capture long-term maintenance effects.

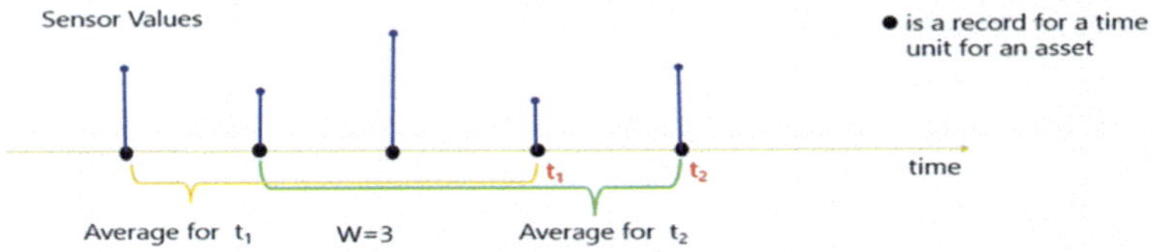

Fig. 2.24 Rolling aggregation features (electronic version 1)

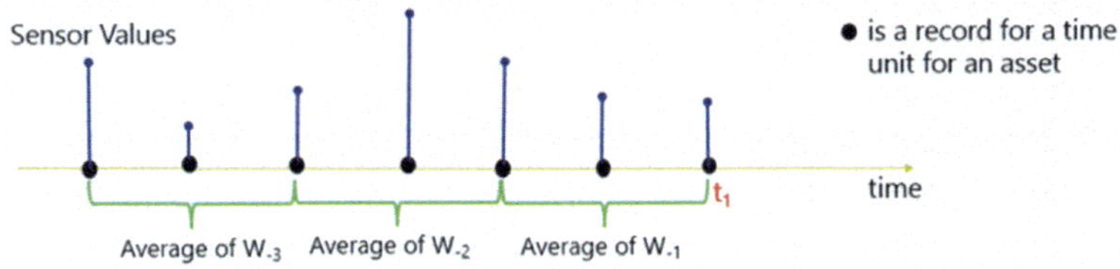

Fig. 2.25 Rolling aggregation features (electronic version)

- Rolling aggregation features. For each device's tag record, select a window of size "W_{-k}", where k is the number or window of size "W" for creating lagged features. k can be chosen as a large number to capture long-term declining patterns, or a small number to capture short-term effects. Using k to flip window W_{-k}, $W_{-(k-1)}$, …, W_{-2}, W_{-1} to create aggregated features (Fig. 2.25).

For example, for each wind turbine, create lagged features using the top and bottom outliers of the data for the first three months, using $W = 1$ and $k = 3$ months.

a. Static features. Static features are the technical specifications of the device, such as the manufacturing date, model, location, etc. While lagged features are mainly numerical, static features often become categorical variables in the model, such as the voltage, current, and power specifications used by the circuit breaker, as well as the type of transformer, power supply, etc. For brake disc failures, the type of tires (for example, whether they are alloy or steel) is used as some static features.

The feature table shown in Table 2.1 is formed through the above feature engineering, where the time unit is days.

④ Predictive modeling, training, simulation, and testing. A large amount of fault data is accumulated during the operation of the equipment. By deeply mining and analyzing these fault big data, valuable knowledge and rules can be extracted. Applying these knowledge and rules to the fault prediction process of the equipment helps the stable and efficient operation of the equipment. Based on the equipment's status data and environmental operation data (from spot checks, status detection data), build a fault prediction model, predict the available life and function loss rate of the equipment and core components, and then give predictive maintenance needs and plans.

Table 2.1 Static features

Device ID	Record time	Feature column	Label
1	Day 1		
1	Day 2		
…	…		
2	Day 1		
2	Day 2		
…	…		

In the fault prediction modeling process, each piece of equipment status data records the operating status of the equipment, so clustering analysis can be performed according to the similarity of status parameters, and then horizontal and vertical comparisons can be made for the equipment in different clusters (Table 2.2). Horizontal comparison refers to the comparison of status parameters under the same time and operating conditions, which can understand the differences within the same cluster and quickly judge which equipment is in an abnormal operating state; another dimension is vertical comparison, that is, the comparison of equipment on the time axis, for the same equipment, quantify its status decline based on the difference between its current status and historical status, and judge whether a fault is about to occur; for the same type of equipment under the same operating environment, vertical comparison can predict whether a fault occurs based on the similarity of one equipment to the historical status of another equipment.

⑤ Maintenance Decision Making: Based on the prediction results and maintenance response time, combined with the spare parts inventory strategy (continuous and periodic inventory ordering strategy), considering the output rate of the production plan and the cost of order delay, decisions are made for predictive maintenance requirements (from predictive maintenance requirements and plans), deterministic maintenance requirements (from preventive maintenance plans), and uncertain requirements (from random failures), providing decision-making strategies for production, maintenance, and spare parts inventory within the enterprise.

Table 2.2 Data modeling methods based on machine learning

Comparison object	Comparison dimension	Modeling purpose
Self	Vertical on time axis	Predict equipment health decline and evaluate differences through supervised learning
Cluster	Horizontal on time axis	Perform difference comparison and anomaly detection through unsupervised learning
Cluster	Vertical on time axis	Predict the life cycle stage and remaining useful life of the equipment through supervised learning

2.4.3 Deep Learning Based Methods

With the rapid development of big data and the new generation of artificial intelligence, the research of fault prediction methods faces the problem of automatic learning and processing of big data. Deep learning automatically learns signal features from raw data and deep nonlinear networks in a supervised, semi-supervised, or unsupervised manner. Some deep learning models, such as Stacked Autoencoder (SAE), Deep Belief Network (DBN), etc., have been applied to the field of fault prediction.

Convolutional Neural Networks (CNN) have powerful data processing capabilities in dealing with image and audio data, and have been widely applied to natural language processing, image recognition, audio recognition, and other fields, and have achieved good results in some fault diagnosis and prediction tasks. The CNN model is not suitable for directly analyzing raw time-domain signals. It needs to use wavelet transform, EMD (Empirical Mode Decomposition), and other time–frequency analysis methods to convert original vibration, acoustic emission, and other non-stationary signals into image data. Although time–frequency analysis fully retains fault characteristics, the algorithm is complex and time-consuming. Some low-complexity methods can be used to improve the conversion efficiency, such as Continuous Wavelet Transform (CWT), etc.

2.5 Fusion Model Driven Fault Prediction Methods

The overall idea of the fault prediction method driven by the fusion model is to play the respective advantages of different types of methods and make up for the shortcomings of different methods, such as the combination and fusion of different data-driven methods, the fusion of data-driven and physical model methods, which can effectively improve the overall performance of fault prediction.

2.5.1 Information Fusion Technology

Information fusion was initially known as data fusion, originating from the sonar signal processing system developed by the U.S. Department of Defense in 1973. In the 1980s, to meet the needs of warfare in the military field, Multi-Sensor Data Fusion (MSDF) technology was born. In the 1990s, with the widespread development of computer technology, information fusion with a more generalized concept was proposed, and the application field quickly expanded from military to civilian use, mainly including robot and intelligent instrument systems, intelligent manufacturing systems, battlefield tasks and unmanned aerial vehicles, space applications, target detection and tracking, image analysis and understanding, inertial navigation, pattern recognition, and other fields.

(1) Information Fusion Model

Various information fusion models have been proposed, the common point is to carry out multi-level processing in the process of information fusion. Existing fusion models can be roughly divided into two categories: ① functional models, mainly built according to the order of nodes; ② data models, mainly built according to data extraction. In the 1980s, typical functional models mainly included the UK Intelligence Cycle and Boyd Control Loop (OODA Loop); typical data models included the JDL model, and in the 1990s, the Waterfall model and Dasarathy model were developed. In 1999, Mark Bedworth integrated several models and proposed a hybrid model.

The UK Intelligence Cycle describes information processing as a circular structure, including four stages: ① collection, including initial intelligence data from sensors and human information sources; ② collation, associating and assembling related intelligence reports, some data merging and compression processing will be carried out at this stage, and the results obtained will be simply packaged for use in the next stage; ③ evaluation, at this stage, intelligence data is fused and analyzed, and the analyst also directly assigns tasks to intelligence collection; ④ distribution, at this stage, the military commander sends the fused intelligence to the user for decision-making action, including the next collection work. The advantage of this model is that it can constantly review the processing process and intelligence collection strategy and correct it at any time. The disadvantage is that its application scope is limited.

In 1984, the U.S. Department of Defense established the Joint Command Data Fusion Laboratory and proposed the JDL model. After gradual improvement and promotion, this model has become a practical standard for the U.S. defense information fusion system. The JDL model divides data fusion into three levels: the first level of processing is for target optimization, positioning, and target recognition; the second level of processing is for situation assessment, building a situation map based on the information provided by the first level of processing; the third level of processing is for threat assessment, interpreting the results of the second level of processing based on possible actions, and analyzing the pros and cons of various actions. Among them, process optimization is a repetitive process, which can be called the fourth level. It monitors system performance throughout the fusion process, identifies potential information sources, and optimizes sensor deployment. Other auxiliary support systems also include data management systems, storage and retrieval, preprocessing data, and human–machine interfaces, etc.

The Boyd Control Loop, or OODA Loop (Observe, Orient, Decide, Act), was proposed by John Richard Boyd, initially applied to military command processing, and is now widely used in information fusion. The Boyd Control Loop makes the feedback iterative characteristics of the problem very obvious, including four processing stages: ① observation, obtaining target information, equivalent to the first level of JDL and the collection stage of the intelligence cycle; ② orientation, determining the direction, equivalent to the second and third levels of JDL, and the collection and collation stages of the intelligence cycle; ③ decision, formulating a response plan, equivalent to the fourth level of JDL process optimization and the distribution behavior of the intelligence cycle, as well as logistics management and plan formulation, etc.; ④ action, executing the plan, unlike the above models, only this stage considers the issue of decision-making efficiency in practice. The advantage of the OODA Loop is that it forms a closed loop at each stage, showing the cyclicity of data fusion. It can be seen that as the fusion stage progresses, the amount of data passed to the next level of fusion stage decreases. The shortcoming of the OODA model is that the decision and action stages lack the ability to influence other stages of the OODA Loop, and each stage is also executed in sequence.

The Extended OODA model is an information fusion system structure developed by Lockheed Martin in Canada, which has been used on the Halifax missile frigate in Canada. This model combines the advantages of the above models and provides a mechanism for concurrent and potentially interactive information fusion processes.

The waterfall model is a project development framework, where the development process unfolds sequentially through a series of stages, starting from system requirements analysis to product release and maintenance. Each stage generates cyclical feedback, so if any information is not covered or a problem is discovered, it is best to "return" to the previous stage and make appropriate modifications. The project development process "flows" from one stage to the next, which is the origin of the name "waterfall model".

The Dasarathy model includes the 5 fusion levels shown in Table 2.3.

In summary, the waterfall model clearly distinguishes the underlying functions, the JDL model clearly divides the middle-level functions, and the Boyd control loop explains the high-level processing in detail. The intelligence cycle covers all processing levels, but does not describe them in detail. The Dasarathy model is built according to fusion tasks or functions, and can effectively describe fusion behaviors at all levels.

The hybrid model combines the cyclical characteristics of the intelligence cycle and the feedback iteration characteristics of the Boyd control loop, while applying the definitions in the waterfall model. Each definition is linked to each level of the JDL and Dasarathy models. In the hybrid model, feedback can be clearly seen. The model retains the structure of the Boyd control loop, thereby clarifying the cyclical characteristics of information fusion processing. The descriptions of the four main

Table 2.3 The 5 fusion levels of the Dasarathy model

Input	Output	Description
Data	Data	Data layer fusion
Data	Features	Feature selection and extraction
Features	Features	Feature layer fusion
Features	Decision	Pattern recognition and processing
Decision	Decision	Decision layer fusion

processing tasks in the model, observation, orientation, decision, and execution, have achieved good reproducibility.

(2) Levels of Information Fusion

The definition of information fusion can be summarized as integrating local data resources provided by multiple similar or different types of sensors distributed in different locations (data information from the same detection source signal), using computer technology to analyze them, eliminating possible redundancy and contradictions between multi-sensor information, complementing each other, reducing their uncertainty, obtaining a consistent interpretation and description of the measured object, thereby improving the speed and correctness of system decision-making, planning, and response, and enabling the system to obtain more comprehensive information. Information fusion appears at different levels, including data layer fusion, feature layer fusion, and decision layer fusion.

① Data Layer Fusion—Data layer fusion is a technique that combines information from multiple sources to form a unified image [26]. The inherent defects of data are the most fundamental challenge for data fusion systems, so most research work is focused on solving this problem. There are many mathematical theories that can be used to represent the imperfection of data, such as probability theory, fuzzy set theory, possibility theory, rough set theory, D-S evidence theory (Dempster-Shafer Evidence Theory, DSET), fuzzy D-S evidence theory and random finite set theory, etc. Most of these methods can represent certain specific aspects of imperfect data, for example, probability theory represents the uncertainty of data, fuzzy set theory can represent the fuzziness of data, and evidence theory can represent uncertainty as well as fuzzy data. Figure 2.26 outlines the mathematical theories for handling data defects mentioned above, and the horizontal axis introduces various aspects of data imperfection.

For the data collected by sensors, fusion of the same type of data is performed according to the type of sensor. The data to be processed in data layer fusion are all collected under the same category of sensors, so data fusion cannot handle heterogeneous data.

② Feature Layer Fusion—Feature layer fusion is the extraction of feature vectors contained in the collected data to reflect the attributes of the physical quantity being monitored. This is fusion oriented towards the features of the monitored object. For example, in the fusion of image data, edge feature information can be used to replace all data information.

③ Decision Layer Fusion—Decision layer fusion refers to making certain judgments, classifications, and simple logical operations based on the data features obtained from feature layer fusion, and making higher-level decisions according to application needs. This is high-level fusion. Decision layer fusion is application-oriented fusion. For example, in the monitoring system of forest fires, the dryness of the forest and the possibility of a fire can be determined by fusing the data features of temperature, humidity, and wind force, etc. In this

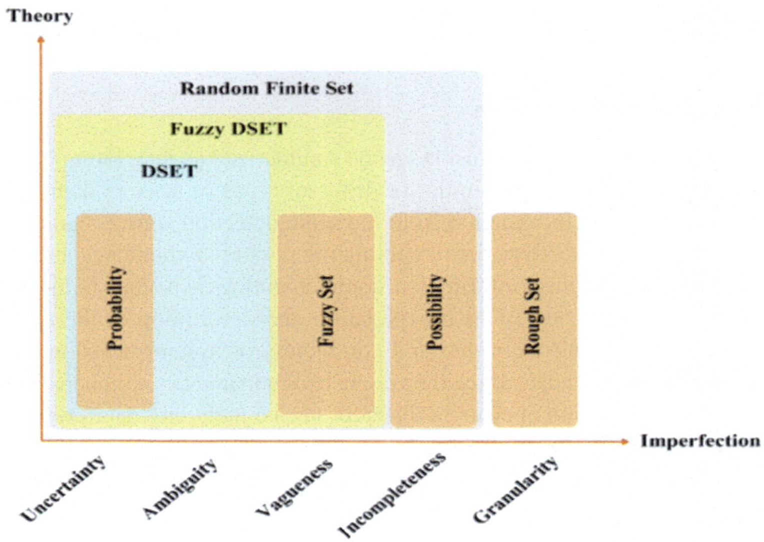

Fig. 2.26 Imperfect data and their corresponding handling methods

case, the data to be sent is not the values of temperature and humidity and the size of the wind force, but only the possibility of a fire and the degree of harm, etc.

Decision layer fusion performs data reduction mapping from multiple inputs to fewer outputs. Decision layer fusion usually does not assume any statistical model of the output parameters of any local data processor [27]. The goal of decision layer fusion methods is to synthesize the results of a set of models used into a single consensus. Three general decision layer fusion methods are linear opinion pool, logarithmic opinion pool, voting or ranking methods.

Decision layer fusion is usually based on the knowledge accumulated in the field of data fusion to make decisions. These techniques aim to make high-level inferences about events and activities generated by detected targets. These techniques usually use symbolic information, and the fusion process needs to reasonably interpret uncertainty and constraints. These methods belong to the 2nd and 4th levels of the JDL data fusion model, such as Bayesian methods, D-S evidence theory, semantic methods, etc.

In the specific data fusion practice of sensor networks, the fusion method can be selected according to the characteristics of the application.

(3) Information Fusion Algorithms

The common methods of multi-sensor information fusion can basically be summarized into two categories: stochastic and artificial intelligence. Stochastic methods include weighted average method, Kalman filtering method, Bayesian estimation method, D-S evidence reasoning, and generative rules, while artificial intelligence methods include fuzzy logic theory, neural networks, rough set theory, expert systems, etc.

The weighted average method is the simplest and most intuitive method of signal-level fusion. This method takes the redundant information provided by a group of sensors and averages it by weight, with the result serving as the fusion value. It is a method that directly operates on the data source.

The Kalman filtering method is mainly used for fusing low-level real-time dynamic multi-sensor redundant data. This method uses the statistical characteristics of the measurement model to recursively determine the optimal fusion and data estimation in a statistical sense. If the system has a linear dynamics model and the errors of the system and sensors conform to the Gaussian white noise model, then the Kalman filtering method will provide the only optimal estimation in a statistical sense for the fused data. The recursive nature of the Kalman filter means that the system does not require a large amount of data storage and computation. When using a single Kalman filter to statistically process data from a multi-sensor combined system, there are many serious problems: in the case of a large amount of redundant combination information, the computation will increase dramatically with the cube of the filter dimension, and real-time performance cannot be satisfied; the increase in sensor subsystems leads to an increase in failures, and when a failure occurs in a system and is not detected in time, the failure will contaminate the entire system, reducing reliability.

The Bayesian estimation method is a common method for fusing high-level information from multiple sensors in a static environment, allowing sensor information to be combined according to the principle of probability. Measurement uncertainty is expressed as conditional probability. When the observation coordinates of the sensor group are consistent, the data from the sensors can be directly fused, but in most cases, the sensor measurement data must be fused indirectly using Bayesian estimation. Multi-Bayesian estimation treats each sensor as a Bayesian estimate, combining the associated probability distributions of each individual object into a joint posterior probability distribution function. By using the likelihood function of the joint distribution function as the minimum, it provides the final fusion value of multi-sensor information. The fused information and a prior model of the environment provide a feature description of the entire environment.

D-S evidence reasoning is an extension of Bayesian estimation. Its three basic points are: basic probability assignment function, trust function, and likelihood function. The structure of D-S evidence reasoning is top-down and divided into three levels. The first level is target synthesis, which combines the observation results from independent sensors into a total output result. The second level is inference, which obtains the observation results of the sensors and makes inferences, extending the sensor observation results into target reports. The basis of this reasoning is: a certain sensor report will produce some credible target reports with a certain degree of credibility. The third level is updating. All kinds of sensors generally have random errors, so a group of continuous reports from the same sensor that are sufficiently independent in time are more reliable than any single report. Therefore, before reasoning and multi-sensor synthesis, the observation data of the sensors should be combined (updated).

Generative rules use symbols to represent the connection between target features and corresponding sensor information. The confidence factor associated with each rule represents its degree of uncertainty. When two or more rules form a joint rule in the same logical reasoning process, fusion can occur. The main problem with using generative rules for fusion is that the definition of the confidence factor of each rule is related to the confidence factors of other rules in the system. If a new sensor is introduced into the system, additional rules need to be added.

Fuzzy logic is multi-valued logic, which specifies a real number between 0 and 1 to represent the degree of truth, equivalent to the premise of an implicit operator, allowing the uncertainty in the process of fusing information from multiple sensors to be directly represented in the reasoning process. If a systematic method is used to model the uncertainty in the fusion process, consistent fuzzy reasoning can be produced. Compared with probabilistic statistical methods, logical reasoning has many advantages. It overcomes the problems faced by probability theory to some extent, and its representation and processing of information are closer to human thinking, generally more suitable for high-level applications (such as decision-making). However, logical reasoning itself is not mature and systematic enough. In addition, due to the large subjective factors in the description of information by logical reasoning, the representation and processing of information lack objectivity. The practical value of fuzzy logic theory for data fusion lies in its extension to fuzzy logic. Fuzzy logic is a kind of multi-valued logic, and the degree of membership can be seen as an imprecise representation of a data truth value. In the MSDF (Multi-Sensor Data Fusion) process, the existing uncertainty can be directly represented by fuzzy logic, and then, using multi-valued logic reasoning, various propositions are merged according to various operations of fuzzy set theory, thereby achieving data fusion.

Neural networks have strong fault tolerance as well as self-learning, self-organization, and self-adaptation capabilities, and can simulate complex nonlinear mappings. These characteristics of neural networks and their powerful nonlinear processing capabilities meet the requirements of multi-sensor data fusion technology. In multi-sensor systems, the environmental information provided by each information source has a certain degree of uncertainty, and the fusion process of these uncertain information is actually an uncertain reasoning process. Neural networks determine classification standards based on the similarity of the samples currently accepted by the system. This determination method is mainly reflected in the weight distribution of the network. At the same time, specific learning algorithms of neural networks can be used to acquire knowledge and obtain an uncertain reasoning mechanism. By using the signal processing capability and automatic reasoning function of neural networks, multi-sensor data fusion is achieved.

Common data fusion methods are shown in Table 2.4. The method used usually depends on the specific application. Due to the complementarity between various methods, two or more methods are often combined for multi-sensor data fusion.

Table 2.4 Common data fusion methods

Fusion method	Operating environment	Information type	Information representation	Uncertainty	Fusion technique	Applicable range
Weighted average	Dynamic	Redundant	Original reading value		Weighted average	Low-level data fusion
Kalman filter	Dynamic	Redundancy	Probability distribution	Gaussian noise	System model filtering	Low-level data fusion
Bayesian estimation	Static	Redundancy	Probability distribution	Gaussian noise	Bayesian estimation	High-level data fusion
Statistical decision theory	Static	Redundancy	Probability distribution	Gaussian noise	Extreme decision	High-level data fusion
D-S evidence reasoning	Static	Redundant complementary	Proposition		Logical reasoning	High-level data fusion
Fuzzy logic	Static	Redundant complementary	Proposition	Membership	Logical reasoning	High-level data fusion
Neural network	Dynamic/ static	Redundant complementary	Neuron Input	Learning error	Neural network	Low/high level
Generative rules	Dynamic/ static	Redundant complementary	Proposition	Confidence factor	Logical reasoning	High-level data fusion

(4) Current Issues

Although information fusion technology has achieved many results after years of research at home and abroad, and has been successfully applied in many fields, there is still no complete theoretical system and effective fusion algorithm. Most of them are for specific problems and specific fields of study, that is, existing research is based on the type of problem, specific objects, and specific levels to establish their own fusion models and inference rules. Some form the so-called optimal solution on this basis. But the design of multi-sensor data fusion system is somewhat blind, it is necessary to establish a complete methodological system to guide the design of data fusion system. The shortcomings are as follows.

① No basic theoretical framework and generalized fusion algorithm have been formed. At present, most of the fusion research is carried out for specific problems in specific application fields (distributed hybrid structure), that is, according to the type of problem, intuitive fusion criteria are established, and the "best" fusion scheme is formed, and a complete theoretical framework and fusion model have not been formed, making the design of the fusion system somewhat blind. The unified data fusion theory must be based on sensor signal and data processing theory, C3I system intelligence processing theory and command decision theory, etc., on the basis of engineering practice, the process of recreating the fusion mechanism of the upper level. The difficulty lies in determining the fusion criteria in a large number of random and uncertain problems, these uncertainties are reflected in inaccurate, incomplete, unreliable, fuzzy measurements, and even information conflicts.

② The ambiguity of association. The ambiguity of association is a major obstacle to data fusion. Before fusion processing, it is necessary to associate the observation results from multiple sensors to ensure that the fused information comes from the same observation target or event, to ensure the consistency of fusion information. The inaccuracy and interference of sensor measurement are factors causing association ambiguity. How to reduce the ambiguity of association is a problem that needs to be solved urgently in the field of data fusion research.

③ The fault tolerance or robustness of the fusion system has not been well resolved.

④ The research on specific methods of data fusion is still in its early stages.

⑤ There are still many practical problems in the design of data fusion systems, such as the establishment of sensor measurement error models, real-time response of systems under complex dynamic environments, etc.

2.5.2 Fusion Modeling Ideas and Methods

In actual applications, there are various ways to implement fault prediction fusion methods. Here are some modeling ideas and methods:

① Dynamic parameter identification of physical models based on data-driven;

② Building mathematical models based on data-driven as physical models for degradation modeling and life prediction;

③ Combination of data-driven feature extraction, degradation pattern matching and model methods;

④ Model methods as initial conditions or auxiliary models for data-driven learning and training;

⑤ Component level, part level, subsystem level, system level, cluster level, factory level, operation network level, etc., can adopt different characteristics of prediction algorithms or models, when extending to a higher level, the methods or models at the lower level can be fused with the algorithms or models at the higher level;

⑥ Direct decision-level fusion or combination of multiple prediction method results.

2.5.3 A Multi-classifier Fusion Model

Pattern recognition and classification is one of the most important application areas of decision layer fusion. Multi-classifier systems are often practical and effective solutions for difficult pattern recognition tasks, which can be referred to as classifier fusion, decision combination, expert mixing, classifier ensemble, consensus ensemble, dynamic classifier selection, hybrid methods, etc.

Classification data fusion is a multidisciplinary field involving multiple domains, making it difficult to establish a clear and rigorous classification. The methods and techniques used can be divided according to the following criteria [27]: ① Focus on the relationships between input data sources proposed by Durrant-Whyte, which can be defined as complementary, redundant, or cooperative data; ② According to the input/output data types and their properties proposed by Dasarathy; ③ The level of abstraction of the data used, such as raw measurements, signals, features, or decisions; ④ According to different data fusion levels defined by JDL.

The main purpose of merging classifiers into a redundant set is to improve their generalization ability. The inherent redundancy within the ensemble can also prevent the failure of individual classifiers. The reason why classifiers may be expected to fail on some inputs is based on the fact that they have only been trained on a limited dataset, and the target function is estimated based on the training data. Unless the function is simple, or the training set is a perfect representation of the data, the data pattern can achieve perfect generalization, otherwise the estimate and expected target will be different.

Once a set of classifiers has been created, an effective method must be found to merge the outputs. Several schemes have been proposed to combine multiple classifiers. Majority voting is by far the most popular method, other voting schemes include minimum, maximum, median, average, and product schemes. The weighted average method attempts to evaluate the best weights for the various classifiers used. The Behaviour Knowledge Space (BKS) selects the best classifier in a certain region of the input space and makes decisions based on its output. Other methods combined with classification include rank-based methods such as Borda count, Bayesian methods, D-S evidence theory, fuzzy theory, probabilistic schemes, and combinations with neural networks. The combiner can be seen as a scheme to assign valuable weights to classifiers.

Figure 2.27 presents a multi-classifier fusion prediction model based on the Markov prediction model, Naive Bayes training model, and neural network prediction model. The key to the Bayesian classification model is to find out the probability of each fault class and the conditional probability of each feature attribute under this fault class. The classification training problem is transformed into counting the number of each fault class in the sample and the count of each fault feature attribute under this class. The Naive Bayes training model is implemented based on the Resilient Distributed Datasets based Naive Bayes Model (RDD-NB) algorithm, and the neural network prediction model is implemented through the Resilient Distributed Datasets based Back Propagation Neural Network (RDD-BPNN) algorithm.

Bayes' theorem is a theorem about the conditional probability of random events A and B (assuming events A and B are unrelated), its expression is:

$$P(A/B) = \frac{P(B/A)P(A)}{P(B)}$$

In the formula, $P(A)$ is the prior probability of event A; $P(B)$ is the prior probability of event B; $P(A/B)$ is the conditional probability of event A occurring after event B;

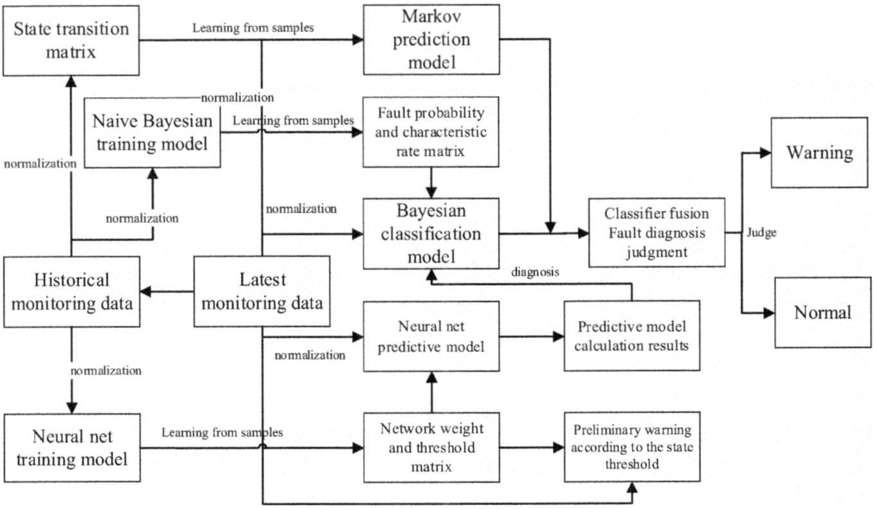

Fig. 2.27 Multi-classifier fusion prediction model based on Markov prediction model, Naive Bayes training model, and Neural network prediction model

$P(B/A)$ is the conditional probability of event B occurring after event A. Knowing Bayes' theorem, we use Naive Bayes to classify equipment failure data.

Let $x = \{a_1, a_2,\ldots, a_n\}$ be a fault record to be diagnosed, a_i is a fault feature attribute. The known fault category set $C = \{c_1, c_2,\ldots, c_n\}$, c_i It is a type of fault. x The probabilities of belonging to each fault type are respectively $P(c_1|x)$, $P(c_2|x)$, …, $P(c_m|x)$, the maximum value is the fault type to which x belongs, then:

$$P(c_k|x) = \max\{P(c_1|x), P(c_2|x), \ldots, P(c_m|x)\}$$

Then the fault type of x is c_k. If the feature attributes are conditionally independent, then according to Bayes' theorem:

$$P(c_k|x) = \frac{P(x|c_k)P(c_k)}{P(x)}$$

For a certain x, $P(x)$ is fixed, $P(x|c_k)$ $P(c_k)$ is the maximum k is the fault type to which x belongs. Therefore:

$$P(x|c_k)P(c_k) = P(c_k)\prod_{i=1}^{n} P(a_i|c_k)$$

The BP neural network is currently the most widely studied and mature neural network. The project realizes the parallel design of the BP neural network algorithm by designing a three-layer neural network model. Define the error $E_\text{p} = 0.5$

$\Sigma(T_{pi}-Y_{pi})^2$, where Y_{pi} is the calculated output of the output node, T_{pi} is the corresponding actual output, the activation function is the Sigmoid function, and the weight correction is:

$$\Delta W_{i,j}(n+1) = h\varphi_i Y_j + \alpha \Delta W_{i,j}(n)$$

In the formula, h is the learning factor; φ_i is the calculation error of output node i; Y_j is the calculated output of output node j; α is the momentum factor. The time window is set to 3, and the fault characteristic value of the next moment is used as the output attribute.

The parallel learning steps of the algorithm are summarized as follows.

① Randomly generate the initial connection weights between each node and the thresholds of the hidden layer and output layer nodes. According to multiple experiments, this book sets them between (−2, 2). Set the number of training times and error threshold.

② Repeat the following two processes until convergence or the maximum number of training times is reached.

a. Forward learning process: For each example, start forward learning from the input layer, calculate the input and output of the hidden layer, and obtain the actual output of the output layer. Calculate the error and total error by comparing with the expected output. If the total error meets the requirements, break the loop, otherwise proceed to the next step.
b. Backward error propagation process: Calculate and adjust the values of the weight and bias matrices based on the error obtained in the previous step.

2.6 Fault Prediction Method Selection Based on Failure Samples

Based on the failure samples and degradation history data of mechanical equipment, fault prediction can be summarized as the method shown in Fig. 2.28.

When there are sufficient failure samples, Weibull distribution, Duane model, and Bayesian statistics and other reliability models based on failure are commonly used methods for remaining life and fault prediction; when there are a certain amount of degradation samples of the same type of equipment, various random process models, empirical models, statistical filtering, Markov chains, gray models and other degradation process models, or machine learning methods such as Artificial Neural Network (ANN), Support Vector Machine (SVM) can be considered; for long-life complex mechanical equipment that are greatly affected by working conditions, it may be impossible to obtain sufficient failure and degradation samples, and small sample modeling tools such as Auto-Regressive Moving Average Model (ARMA), Exponential Smoothing (ES), Relevance Vector Machine (RVM) and deep learning can be tried. That is, appropriate modeling analysis methods can be selected based

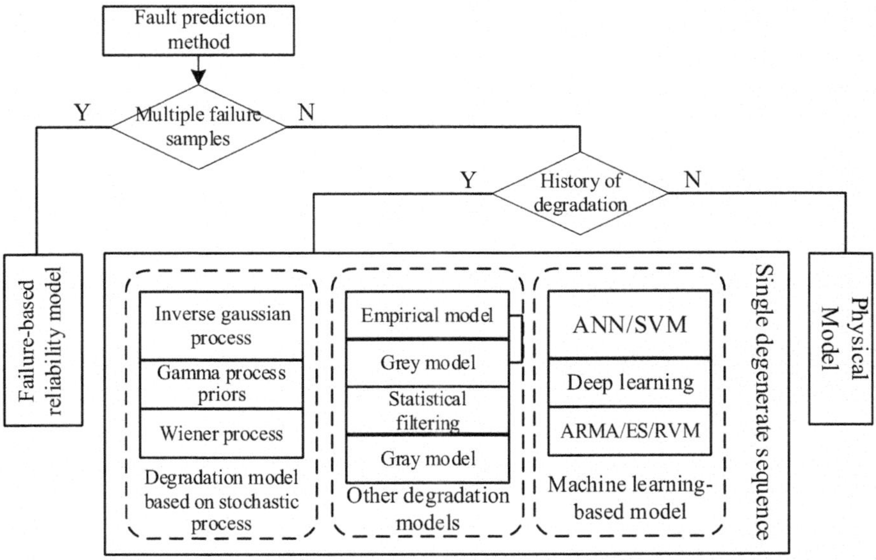

Fig. 2.28 Fault prediction method for a complex device

on the available failure and degradation samples to predict the remaining life or life distribution of the product.

References

1. Wang Yuanhang, Deng Chao, Hu Xianghong, et al. Preventive maintenance decision of complex mechanical equipment based on multiple fault modes. Computer Integrated Manufacturing Systems, 2015, 21(9): 2504–2514.
2. JARDINE A, LIN D, BANJEVIC D. A review on machinery diagnostics and prognostics implementing condition-based maintenance. Mechanical Systems and Signal Processing, 2006, 20: 1483–1510.
3. Niu Mingzhong, Wang Baohua, Wang Guiliang. Vibration identification method and example of equipment failure. Beijing: Metallurgical Industry Press, 1995.
4. CAI B, ZHAO Y, LIU H, et al. A data-driven fault diagnosis methodology in three-phase inverters for PMSM drive systems. IEEE Transactions on Power Electronics, 2016, 32(7): 5590–5600.
5. WEN L, LI X, GAO L, et al. A new convolutional neural network based data-driven fault diagnosis method. IEEE Transactions on Industrial Electronics, 2018, 65(7): 5990–5998.
6. PANDYA D H, UPADHYAY S H, HARSHA S P. Fault diagnosis of rolling element bearing with intrinsic mode function of acoustic emission data using APF-KNN. Expert Systems with Applications, 2013, 40(10): 4137–4145.
7. Peng Xiyuan, Peng Yu, Liu Datong. Data-driven fault prediction. Harbin: Harbin Institute of Technology Press, 2016.
8. Zhang Zhihua. Reliability theory and engineering application. Beijing: Science Press, 2012.
9. BERGER J O. Statistical decision theory and Bayesian analysis. Berlin: Springer-verlag, 1985.

10. Jin Guang. Reliability Technology Based on Degradation - Models, Methods and Applications. Beijing: National Defense Industry Press, 2014.
11. DAVID G S, RICHARD O D, PETER E H. Pattern classification. 2nd ed. New Jersey: Wiley Interscience, 2017.
12. WANG L, CHU J, MAO W. An optimum condition-based replacement and spare provisioning policy based on Markov chains. Journal of Quality in Maintenance Engineering, 2008, 14(4): 387–401.
13. Ye Z S, Xie M. Stochastic modeling and analysis of degradation for highly reliable products. Applied Stochastic Models in Business and Industry. 2015, 31: 16–32.
14. CHEN N, YE Z S, XIANG Y, et al. Condition-based maintenance using the inverse Gaussian degradation model. European Journal of Operational Research, 2015, 243(1): 190–199.
15. ELWANY A H, GEBRAEEL N Z, MAILLART L M. Structured replacement policies for components with complex degradation processes and dedicated sensors. Operations Research, 2011, 59, 684–695.
16. GUO C, WANG W, GUO B, et al. A maintenance optimization model for mission-oriented systems based on Wiener degradation. Reliability Engineering & System Safety, 2013, 111, 183–194.
17. DIEULLE L, BERENGUER C, GRALL A, et al. Sequential condition-based maintenance scheduling for a deteriorating system. European Journal of Operational Research, 2003, 150, 451–461.
18. GRALL A, BERENGUER C, DIEULLE L. A condition-based maintenance policy for stochastically deteriorating systems. Reliability Engineering & System Safety, 2002, 76, 167–180.
19. LIAO H, ELSAYED E A, CHAN L Y. Maintenance of continuously monitored degrading systems. European Journal of Operational Research, 2006, 175, 821–835.
20. WANG X, XU D. An inverse Gaussian process model for degradation data. Technometrics, 2010, 52, 188–197.
21. YE Z S, CHEN L, TANG L, et al. Accelerated degradation test planning using the inverse Gaussian process. IEEE Transactions on Reliability, 2014, 63, 750–763.
22. CHEN N, TSUI K L. Condition monitoring and residual life prediction using degradation signals: Revisited. IIE Transactions, 2013, 45, 939–952.
23. LIAO H, TIAN Z. A framework for predicting the remaining useful life of a single unit under time-varying operating conditions. IIET Transactions, 2013, 45, 964–980.
24. YE Z S, CHEN N, TSUI K L. A Bayesian approach to condition monitoring with imperfect inspections. Quality & Reliability Engineering International, 2015, 31(3): 513–522.
25. Azure AI guide for predictive maintenance. https://docs.microsoft.com/en-us/azure/machine-learning/team-data-science-process/cortana-analytics-playbook-predictive-maintenance, 2018.
26. KHALEGHI B, KHAMIS A, KARRAY F O, et al. Multisensor data fusion: a review of the state-of-the-art. Information Fusion, 2013, 14: 28–44.
27. SINHA A, CHEN H, DANU D G, et al. Estimation and decision fusion: a survey. Neurocomputing, 2008, 71: 2650–2656.

Chapter 3
Intelligent Predictive Maintenance System and Framework

Under the intelligent manufacturing model, manufacturing enterprises have more abundant big data information such as real-time operating status of products, operating environment status, business operation status, personnel status, and social network data. By analyzing and mining these data, we can understand the process of equipment failure, the impact, and the solution. These information are abstracted and modeled into knowledge, and then use knowledge to predict failures and execute intelligent predictive maintenance, so that the knowledge of the entire life cycle of the equipment can be efficiently and spontaneously generated and utilized. This chapter discusses two intelligent predictive maintenance strategies for equipment and equipment operation networks based on the discussion of the reference architecture of intelligent manufacturing, and presents a technical system and framework for intelligent predictive maintenance based on data mining.

3.1 Reference System Architecture of Intelligent Manufacturing

Combining the advantages of various intelligent manufacturing enterprise modeling reference system architectures such as the industrial 4.0 reference architecture RAMI4.0 [1], the industrial internet reference architecture IIRA [2], and the intelligent manufacturing system architecture of Made in China 2025 [3, 4], we establish the reference architecture of the intelligent manufacturing system from the four dimensions of life cycle, view, physical world, and virtual world (Fig. 3.1) to summarize the core features and elements of intelligent manufacturing.

The life cycle dimension is a chain collection of a series of interrelated value creation activities composed of prototype development (R&D/design/use/maintenance), product production (production/sales/use/maintenance), and recycling (recycling/remanufacturing); the view dimension includes four levels of resources,

M. Liu et al., *Intelligent Predictive Maintenance*, Advanced and Intelligent Manufacturing in China, https://doi.org/10.1007/978-981-97-2677-6_3

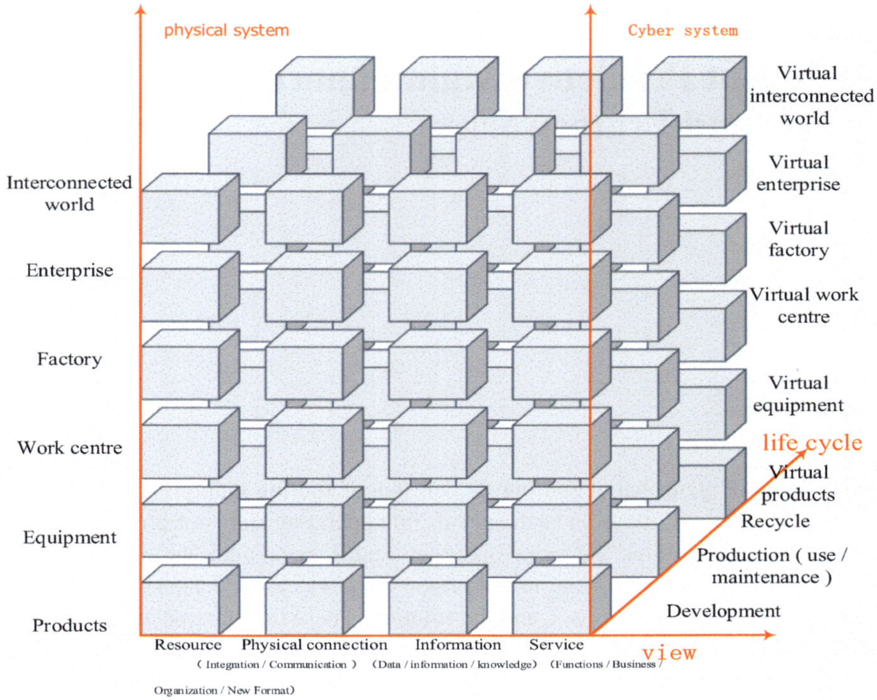

Fig. 3.1 Virtual and real combined intelligent manufacturing reference system architecture

physical connection (integration/communication), information (data/information/knowledge), and services (function/business/organization/new business form); the physical world dimension includes six layers of product layer, equipment layer, work center layer, workshop (factory) layer, enterprise layer, and interconnected world layer, reflecting the interconnection and intelligence of equipment and factories; the virtual world dimension corresponds to the physical world dimension, including six layers of virtual product layer, virtual equipment layer, virtual work center layer, virtual workshop (factory) layer, virtual enterprise layer, and virtual interconnected world layer, reflecting the mapping of equipment and factories from the physical world to the virtual world, as well as the virtualization trend of the manufacturing process.

(1) Life Cycle Dimension

The life cycle dimension provides modeling support for the system life cycle, consisting of three modeling levels in the enterprise model development life cycle, similar to the life cycle dimension of RAMI4.0, including prototype development, product production, and recycling.

Prototype development uses the method of RAMI4.0. According to the goals of the enterprise, the prototype product of the enterprise is described from the initial design to the finalization through a simple and clear graphic system, including various tests and verifications.

Product production takes into account the demand definition of the prototype development layer and the implementability of the implementation tools and user interface, designs the communication methods and forms between the two, carries out the design, simulation, manufacturing and other scale and industrial production of the product, as well as sales, use and maintenance services, each product is an instance of the prototype.

Recycling specifically completes the scrapping and recycling of the product.

(2) View Dimension

The view dimension is a window for observing and controlling different aspects of the enterprise, consisting of four views: resources, physical connection, information, and services.

Resources include design and construction drawings, product process documents, raw materials, manufacturing equipment, production workshops, and factories, among other physical entities, as well as energy sources such as electricity and gas. In addition, personnel can also be considered a component of resources.

The Internet of Things refers to the integration of raw materials, components, energy, equipment, and other manufacturing resources (from small to large, realizing the integration of intelligent equipment to intelligent production units, intelligent production lines, digital workshops, intelligent factories, and even intelligent manufacturing systems) through information technologies such as QR codes, radio frequency identification, and software. This is achieved through wired and wireless communication technologies, enabling interconnection and communication between machines, between machines and control systems, and between enterprises.

Information refers to the collection of all information within an enterprise, describing the transaction information structure used in the operation of the enterprise, and ensuring the security of information, realizing the collaboration and sharing of data, information, and knowledge, based on the Internet of Things using new generation information technologies such as big data, the Internet of Things, and cloud computing.

Services attempt to present the service functions of an enterprise and their relationships through functions, business, organization, and emerging business models. Functions describe all functions and sub-functions of an enterprise and their global and subordinate relationships; business maps and implements related business processes; organization considers issues related to enterprise organization, describing the corresponding relationships and structural relationships between organizations; emerging business models include personalized customization, remote maintenance, and industrial cloud and other service-oriented manufacturing models.

(3) Physical World Dimension

The physical world dimension of the RAMI4.0 reference architecture includes six levels: product layer, equipment layer, work center layer, workshop (factory) layer, enterprise layer, and interconnected world layer.

(4) Virtual World Dimension

The virtual world dimension corresponds to the physical world dimension, realizing the digital modeling, intelligent prediction, and adaptive feedback control of physical objects through virtual objects.

3.2 Equipment-Oriented Intelligent Predictive Maintenance Strategy

With the introduction of the concept of intelligent manufacturing, Cyber Physical Systems (CPS) (complex systems integrating computing, networking, and physical environments) have realized real-time perception, dynamic control, and information services of large-scale engineering systems through the organic integration and deep collaboration of 3C (Computer, Communication, Control) technologies.

In the intelligent factory of Industry 4.0, CPS monitors physical objects and processes, creates virtual replicas of the physical world, and carries out decentralized control and decision-making. The information physical system communicates and cooperates with people and other CPS systems in real time through the Internet of Things, and the participants of the value chain provide and utilize services within or across organizations through big data analysis and the Internet of Services.

In the predictive maintenance process, CPS promotes the systematic transformation of large-scale data, making invisible equipment degradation modes and inefficiencies in the intelligent factory visible, and generating optimal decisions, forming predictive maintenance models based on 5C and intelligent predictive maintenance models based on CPS.

3.2.1 Predictive Maintenance Model Based on 5C

With the rapid development of information and communication technology, advanced analysis technologies have also been integrated into manufacturing, product, and service processes. Many industries are facing new opportunities, as well as challenges to maintain their competitiveness and market demand. This integration (currently known as Cyber Physical Systems CPS) is transforming the industry to the next level: CPS promotes the systematic transformation of large-scale data, making invisible degradation modes and inefficiencies visible, and generating optimal decisions. Professor Jay Lee of the University of Cincinnati analyzed the development trends of industrial big data and CPS, discussed the 5C architecture (intelligent connection, data conversion, information, cognition, configuration) of applying CPS in the production process as shown in Fig. 3.2, and the necessary steps to fully integrate CPS in manufacturing [5], forming the predictive maintenance model based on 5C as shown in Fig. 3.3 [6].

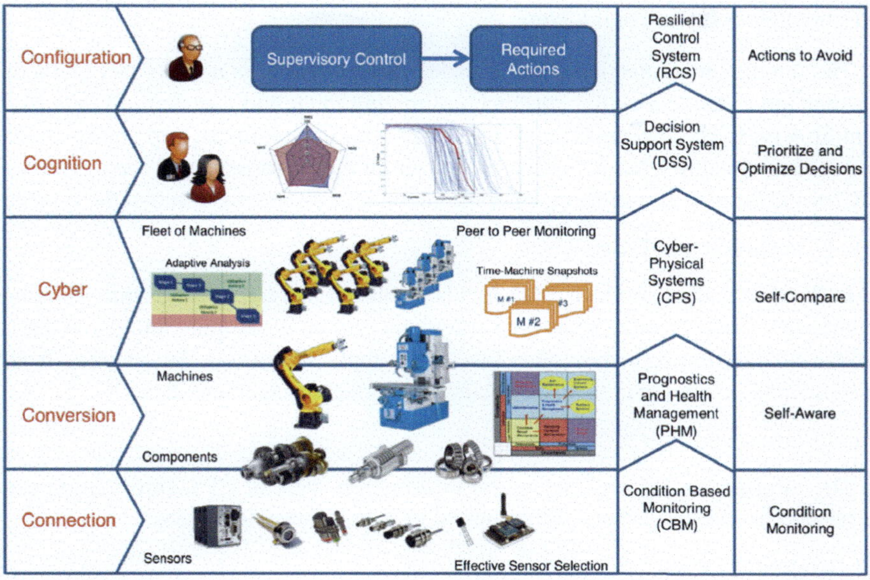

Fig. 3.2 5C-based cyber physical system framework

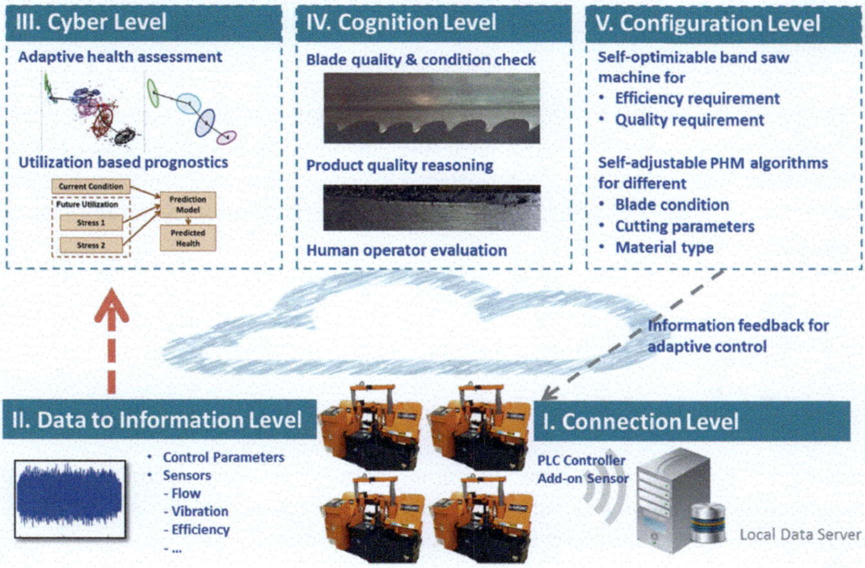

Fig. 3.3 Predictive maintenance model based on 5C

(1) Intelligent Connection Layer

The intelligent connection layer includes various seamless non-contact methods for managing data collection systems, optimizing data, and transmitting data to a central server. The choice of appropriate sensors, data sources, and transmission protocols at this layer has a significant impact on the performance of the next layer of CPS and the quality and accuracy of the knowledge discovered through the system.

(2) Data Conversion Layer

In the data conversion layer, data can be analyzed and transformed into valuable knowledge. Recent research has focused extensively on developing intelligent algorithms and data mining technologies that can be applied to various data sources, from machine and process data to business and enterprise management data.

(3) Information Layer

The information layer is the central information stack of CPS. Information can be pushed from every aspect and compiled into an information space.

(4) Cognitive Layer

This layer generates a comprehensive understanding of the monitored system. The correct representation of knowledge obtained by experts can support decision-making.

(5) Configuration Layer

The configuration layer is the feedback from the information space to the physical space, playing a supervisory control role, enabling equipment to achieve self-configuration and adaptation. At this stage, it can serve as a Resilience Control System (RCS), using the correct preventive decisions for control and monitoring of equipment systems.

3.2.2 Intelligent Predictive Maintenance Model Based on CPS

Industry 4.0 is a collective term for value chain organization technologies and concepts, based on Radio Frequency Identification (RFID), Cyber Physical Systems, Internet of Things (IoT), Internet of Services (IoS), and Data Mining. Data Mining (DM) has promoted the realization of the vision of intelligent factories. In the modular and structured intelligent factories of Industry 4.0, Cyber-Physical Systems (CPS) monitor physical objects and operational processes, create virtual replicas of the physical world, and implement discrete process control and decision-making. Information Physical Systems can communicate and cooperate with other CPS systems and humans in real time through the Internet of Things (IoT). Through data mining and the Internet of Services (IoS), value chain participants can provide and utilize services

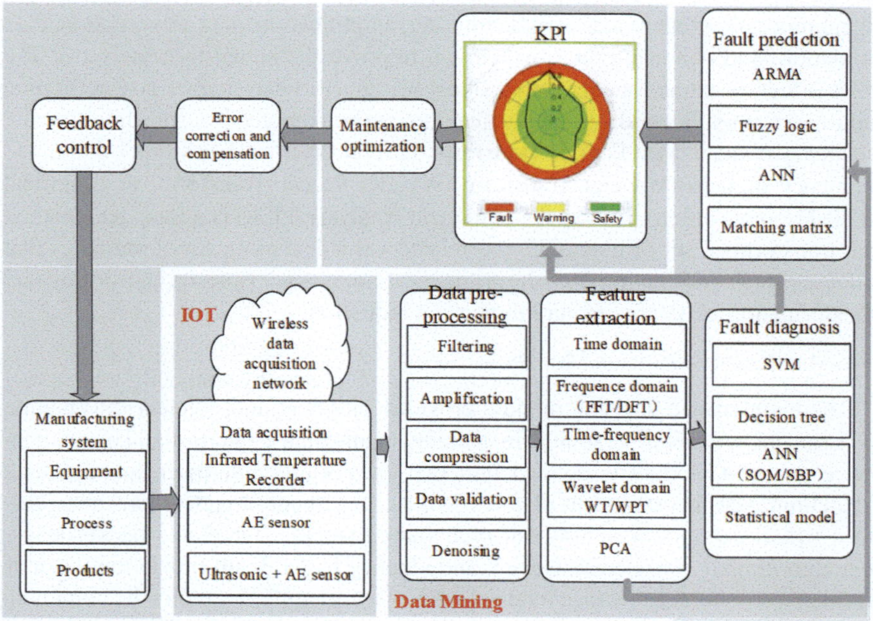

Fig. 3.4 Intelligent predictive maintenance model and system for implementing zero-defect manufacturing

internally and across organizations. Wang et al. from Norway provided a definition of Industry 4.0, identified some key principles for implementation, and demonstrated an intelligent predictive maintenance model and system for implementing Zero-Defect Manufacturing (ZDM) as shown in Fig. 3.4 [7].

The IPdM system is based on many key technologies such as CPS, IoT, IoS, computational intelligence, data mining, swarm intelligence (SI), etc., which need to be researched and developed to meet industry needs. There are six main modules in IPdM: sensor and data acquisition, signal preprocessing and feature extraction, maintenance decision-making, key performance indicators (KPI), maintenance scheduling optimization, error correction, compensation, and feedback control.

(1) Sensor and Data Acquisition Module

This is the first step in implementing equipment diagnosis and predictive IPdM maintenance strategies. The task of this module is to select the appropriate sensors and the best sensor installation strategy. The data acquisition process turns sensor signals into equipment status information, with different sensors used to collect different data, such as temperature sensors, vibration sensors, acoustic emission sensors, etc.

(2) Signal Preprocessing and Feature Extraction Module

Generally speaking, the processing of sensor signals has two steps. One is signal processing, which can enhance the characteristics and quality of the signal. Relevant signal processing techniques include filtering, amplification, data compression, data validation, and denoising, etc., which can improve the signal-to-noise ratio. The other is feature extraction, extracting signal features that can characterize accidental failures or errors. Generally speaking, features can be extracted from the time domain, frequency domain (Fast Fourier Transform, Discrete Fourier Transform), and time–frequency domain (Wavelet Transform, Wavelet Packet Transform, or Empirical Mode Decomposition). Principal Component Analysis (PCA) is a statistical method that transforms a set of potentially correlated variables into a set of variables that reflect as much information from the original variables as possible through orthogonal transformation, achieving feature dimension reduction.

(3) Maintenance Decision Module

The maintenance decision module provides sufficient and efficient information to assist maintenance personnel in making maintenance decisions. The maintenance decision module is divided into two categories: fault diagnosis and fault prediction. The focus of fault diagnosis [including support vector machines, artificial neural networks (such as self-organizing mapping, integrated backpropagation algorithms)] is to detect, isolate, and identify faults. Fault prediction (such as autoregressive moving average models, ANN, etc.) attempts to predict the remaining useful life (RUL) of the equipment before errors or failures occur. Decision support models can be divided into four categories: physical models, statistical models, data-driven models, and hybrid models. Since IPdM strategies mainly rely on signals and data reflecting equipment conditions, data-driven models and hybrid models will dominate.

(4) Key Performance Indicators Module

The Key Performance Indicators (KPI) graph, also known as a spider web graph or health radar graph, is used to display the degree of degradation of components. Each radar line displays the condition of the component from 0 (perfect) to 1 (damaged). The color shows the level, such as safety, warning, alarm, fault, and defect, which can visually assess equipment performance.

(5) Maintenance Scheduling Optimization Module

Maintenance scheduling optimization is an NP problem (Non-deterministic Polynomial problem), and swarm intelligence algorithms are a good way to solve such problems. IPdM can apply genetic algorithms (GA), particle swarm optimization (PSO), ant colony algorithms, and bee swarm algorithms to dynamically find the optimal predictive maintenance scheduling scheme.

(6) Error Correction, Compensation, and Feedback Control Module

This module will use the results of the maintenance decision support module for error correction, compensation, and feedback control.

3.3 Intelligent Predictive Maintenance Strategy for Equipment Operation Network

Based on the 5C predictive maintenance model and the CPS-based intelligent predictive maintenance model, an intelligent predictive maintenance system and method framework for equipment systems have been built from aspects such as data collection and processing, fault diagnosis and prediction, maintenance optimization and decision-making, system compensation, and control. The research on equipment fault prediction and health management mainly focuses on equipment health prediction based on physical models, equipment health prediction based on data-driven, equipment health prediction based on model-driven, and equipment health prediction based on information fusion. These four fault prediction models and predictive maintenance strategies have laid the foundation for preventive maintenance scheduling and optimization theory for a single enterprise or a single equipment system, and most of them have created theoretical models applicable to the research background.

Due to the rapid development of the new generation of information technology and intelligent manufacturing collaboration networks, these studies have not considered the network characteristics of MRO maintenance services, the uncertainty of large-scale random faults, and the fast response characteristics of services, nor have they solved the planned and real-time scheduling requirements of maintenance services in the MRO network environment, and they have not considered the collaboration characteristics of various participants in the network. Therefore, it is necessary to solve the theory and method of data acquisition and maintenance decision-making for the equipment operation network.

3.3.1 *Equipment Operation and Maintenance Network*

With the evolution of business models and service value chains, equipment users outsource the operation and maintenance of equipment to professional MRO (Maintenance Repair & Operation Technology Services Company has gradually formed an MRO service network that includes spare parts suppliers and their maintenance providers, equipment manufacturers and their maintenance providers, equipment users, domain experts (experts in the fields of whole machine and component operation, maintenance, fault diagnosis and repair from within and outside the company), and cloud platform providers.

Within the MRO service network, to solve the problems of remote, real-time, and timely operation and maintenance of equipment, equipment operating companies collect equipment operation data and upload it to the company data center (or company cloud). System software monitors and controls the equipment in real time online, and predictive maintenance of the equipment is carried out through data analysis, forming the Smart Predictive Maintenance (SPdM) network shown in Fig. 3.5 based on the (mobile) Internet (these networks include data acquisition

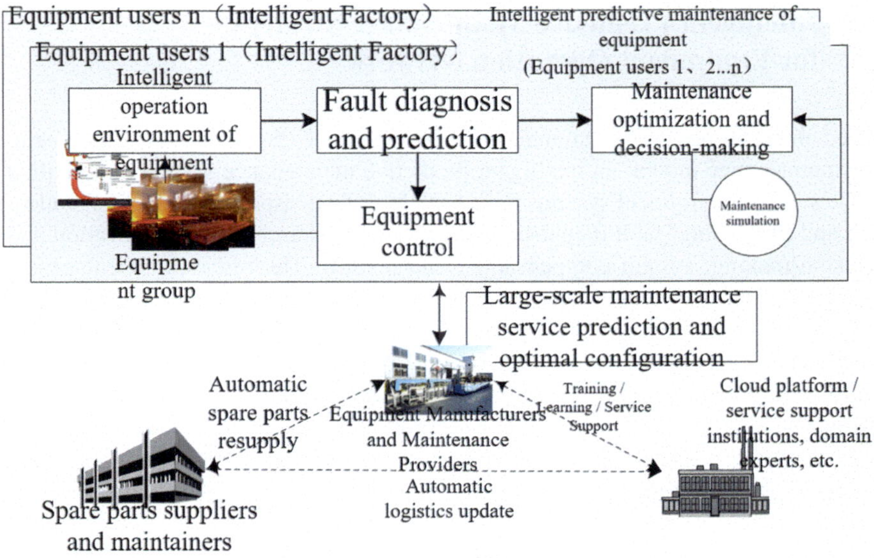

Fig. 3.5 Smart predictive maintenance (SPdM) network

networks, business support networks, social networks, data analysis and support networks) and the corresponding MRO big data support environment (including product status, equipment status, environmental status, business operation status, personnel status, social network data and customer feedback data). This reflects global (Internet connection), real-time (state detection supported by the industrial Internet) and timely (mobile terminal scheduling) characteristics, enabling knowledge of the entire lifecycle of the equipment to be efficiently and spontaneously generated and utilized. The SPdM network can also automatically connect to the cloud platform to search for suitable domain experts to handle problems based on the Internet and value chain networks, and experts can more effectively determine the optimal maintenance strategy and service through the integrated knowledge platform and mobile devices.

3.3.2 *Data Acquisition and Maintenance Decision Making for SPdM Network*

Germany's Industry 4.0, the United States' Industrial Internet and Cyber-Physical Systems, and China's Made in 2025 drive traditional manufacturing to use the Internet of Things and big data analysis to undergo knowledge-based intelligent transformation, providing customers with personalized products, services, and capabilities through the network. This breaks the traditional equipment maintenance field's reliance on high reliability requirements for systems, subsystems, and components.

Through the intelligent predictive maintenance theory driven by big data in the SPdM network of equipment, information related to maintenance, repair, and service is associated, realizing the accumulation and prediction of equipment life and the prediction and optimization of a wide range of maintenance services for the MRO network.

In the MRO service network composed of spare parts suppliers and maintenance providers, equipment manufacturers and maintenance providers, equipment users, domain experts, cloud platform providers, and other collaborative entities, intelligent predictive maintenance is a highly integrated, comprehensive system. It consists of the functions shown in Fig. 3.6, such as online health assessment of equipment, autonomous logistics support and offline fault diagnosis and prediction by equipment users, large-scale maintenance service prediction and optimization configuration by equipment operators, and interfaces of Cyber-Physical Systems (realizing multi-level interconnection of various functions). It also includes content oriented towards intelligent predictive maintenance of equipment. Intelligent predictive maintenance for the equipment operation network mainly focuses on high-quality acquisition and fusion methods of multi-source heterogeneous data for the SPdM network of equipment, big data-driven equipment fault diagnosis and prediction theory and methods, maintenance optimization and decision-making for intelligent factories, wide-range maintenance service prediction and optimization configuration for the MRO network, and process control based on the Cyber-Physical System CPS.

(1) High-quality acquisition and fusion methods of multi-source heterogeneous data for equipment operation network

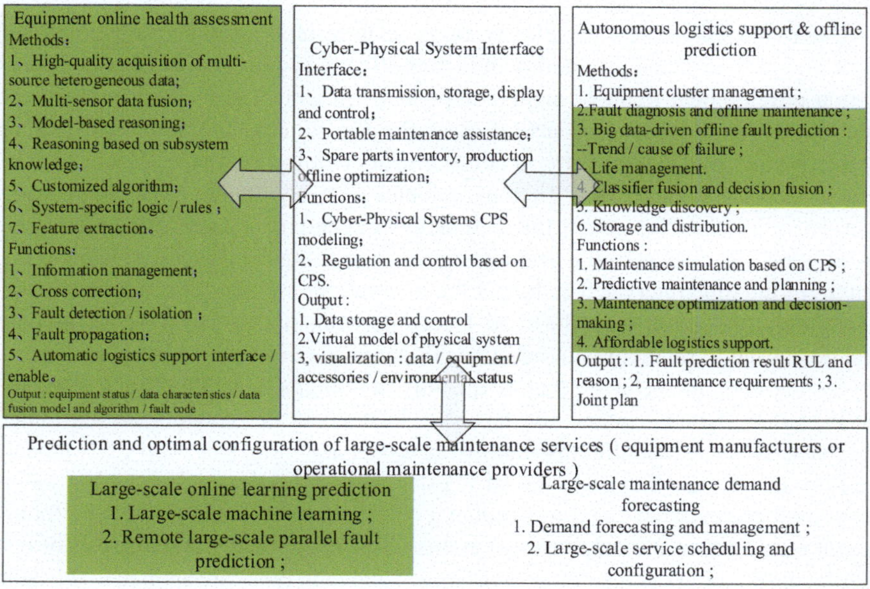

Fig. 3.6 Content of intelligent predictive maintenance for SPdM network

In the process of equipment operation and maintenance, the amount of data from the operation status of equipment and components, environmental information, operation business, maintenance social network, and customer's product and feedback is growing, gradually forming a big data application environment for equipment operation and maintenance. Generally speaking, equipment operation data and maintenance records are stored in the servers of equipment users, and manufacturers can't even access the data related to their own equipment, which is not conducive to continuous supervision and full lifecycle maintenance of production equipment. So, how to acquire these data information distributed in various users in high quality? How to search for valuable association information hidden in the data through algorithms from this big data environment, and then support health prediction, demand prediction, and service optimization decision modeling for equipment and components? Compared with traditional data acquisition and analysis methods, big data analysis pays more attention to the quantity of data, replacing traditional cause analysis with associated data analysis, and turning model-driven applications into evidence-driven applications. In this case, it is necessary to study the blockchain-based big data sharing and acquisition mechanism in the MRO network environment, as well as MRO big data analysis methods such as multi-source heterogeneous data processing analysis, association relationship analysis, big data clustering fusion, real-time stream processing, and parallel processing, to achieve fusion and classification of operation status data, operation business data, social maintenance network interaction data, and customer feedback information, and support the establishment of equipment fault prediction and decision theory.

(2) Big data-driven equipment fault diagnosis and prediction theory and methods

Due to the hierarchical complexity, fuzzy relationship, dynamic change randomness, and indicator data uncertainty of factors affecting equipment health, the degradation development trend of many components presents a non-stationary random process. Deep convolutional neural networks are suitable for automatic feature extraction and fault classification of non-stationary nonlinear signals. Markov prediction models are suitable for predicting problems with large random fluctuations and can also reflect the microscopic fluctuation rules of equipment degradation processes. In addition, the change in the health status of each component is also related to the length of the time interval. The longer the interval, the weaker the randomness and uncertainty, and the stronger the trend. Neural network models and Bayesian models can be chosen to predict the trend of health parameters of each component. Therefore, combining the behavior feature quantity of equipment health status, operation business data, dynamic interaction data of social maintenance network, and customer feedback data, etc., a multi-classifier fusion model that integrates deep CNN model, Markov model, and Bayesian model is established to diagnose and predict faults from multiple measures, and to conduct a comprehensive evaluation of the future health status fusion of components. On this basis, deep mining and analysis of big data such as equipment operating environment and fault types are applied to the fault prediction of cluster equipment.

(3) Maintenance optimization and decision-making for intelligent factories

To address the uncertainty maintenance requirements caused by faults in the network, this research focuses on the joint optimization model of single enterprise equipment production maintenance and spare parts inventory based on the fault prediction model, providing decision-making strategies for production, maintenance, and spare parts inventory within a single enterprise.

(4) Large-scale maintenance service prediction and optimization configuration for MRO networks

In response to the high uncertainty of maintenance requirements in the Internet environment, this research combines long-term prediction and real-time scheduling of services within the MRO network, intelligent agent-based maintenance service collaborative optimization scheduling, and real-time decision-making and scheduling of maintenance services within the Internet range. It establishes the basic theory of large-scale maintenance scheduling and optimization decision-making supported by the Internet for MRO networks, provides optimization conditions for equipment maintenance businesses and spare parts suppliers based on Internet collaboration, and equilibrium strategies for maximizing profits through negotiation among collaborative enterprises. Furthermore, it combines big data information such as maintenance task requirements, the geographical location of service personnel, and work task status to perform real-time optimization and decision-making for registered member groups in the MRO maintenance network.

(5) Process control based on Cyber-Physical Systems (CPS)

The information physical system in the manufacturing field has the following characteristics: the network and manufacturing equipment are highly integrated, locally physical, and globally virtual; all components of the manufacturing equipment are equipped with sensors and actuators, with online communication, remote control, and autonomous coordination functions among components; the sensors installed on the components form a closed-loop process based on event-driven control by perceiving and feeding back the equipment status and applying control decisions to the actuators; the system has self-learning, self-adaptation, and autonomous collaboration capabilities, highly autonomous, and satisfies real-time robustness. Based on these characteristics, this research studies the equipment operation control method based on the information physical system, proposes the equipment operation control framework based on the information physical system, realizes real-time adjustment and control of the operation process through equipment fault prediction and maintenance simulation; fully utilizes the rich computing resources of the offline virtual environment to implement big data analysis methods, realizes the collaborative joint optimization of maintenance tasks, spare parts inventory, and production plans based on intelligent agents, and achieves the maintenance decision with the smallest expected cost.

3.3.3 Intelligent Predictive Maintenance Technology System and Framework

Using big data, the Internet, and artificial intelligence and other new generation information technologies to build the intelligent predictive maintenance technology system and framework based on data mining as shown in Fig. 3.7, including data collection, processing, and analysis methods based on IoT, data-driven fault diagnosis and prediction methods, maintenance optimization and decision-making for intelligent factories, large-scale maintenance service prediction and optimization configuration for SPdM networks, and process control based on Cyber-Physical Systems (CPS).

The key research content of this book includes: ① In the part of data collection, processing, and analysis methods based on IoT, research on the perception resource management framework and model based on IoT at the equipment level, the WSN

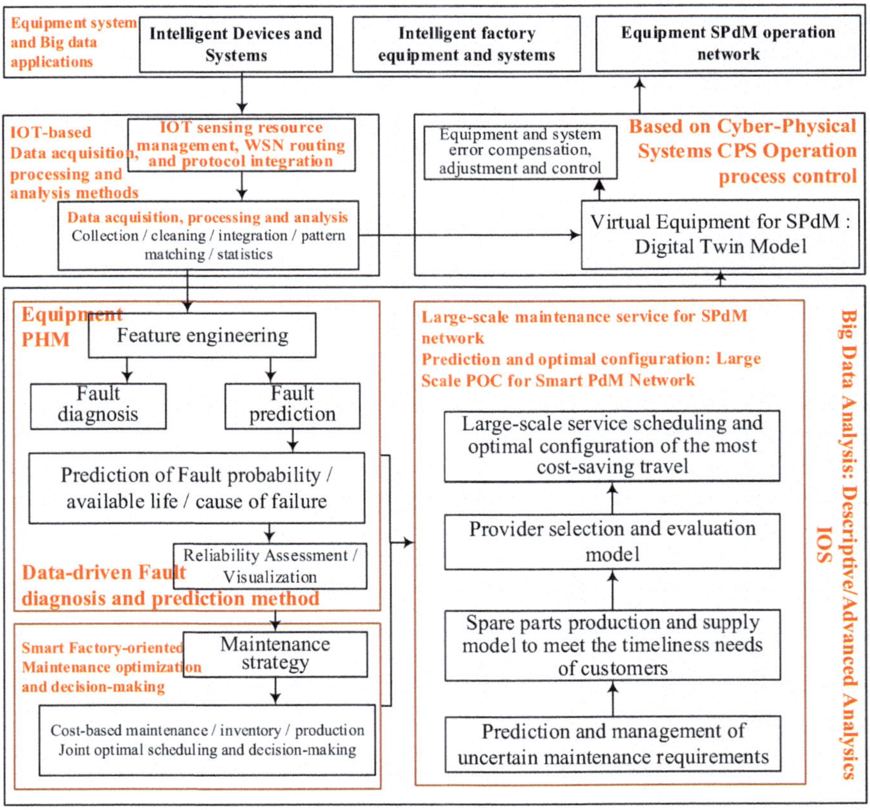

Fig. 3.7 Intelligent predictive maintenance technology system and framework based on data mining

routing protocol and model for complex manufacturing environments, and the corresponding software and hardware protocol integration methods, etc., to support the collection and preprocessing of equipment status monitoring data; ② In the part of data-driven fault diagnosis and prediction methods, research on feature extraction methods for various non-stationary non-linear signals, data-driven equipment fault diagnosis methods and models, data-driven fault prediction and health management (Prognostic and Health Management, PHM) theory and methods; ③ In the part of maintenance optimization scheduling and decision-making methods for intelligent factories, research on the joint optimization scheduling and decision-making model of equipment maintenance, spare parts inventory, and production process based on cost optimization for intelligent factories using equipment; ④ In the part of large-scale maintenance service prediction and optimization configuration methods for SPdM networks, research on uncertainty service demand management and prediction models in dynamic network environments, service spare parts prediction and management based on improved stochastic programming, service provider selection and evaluation, service resource configuration based on fuzzy stochastic programming and profit sharing model, etc., to build the theory and method of large-scale intelligent predictive maintenance service prediction and optimization configuration for equipment intelligent predictive maintenance operation network; ⑤ Research on the process compensation, system adjustment, and control methods based on Cyber-Physical Systems (CPS) for intelligent manufacturing equipment and systems.

References

1. Industrial 4.0 reference architecture [EB/OL]. http://www.innovation4.cn/library/r3738.
2. Industrial Internet reference architecture [EB/OL]. http://www.innovation4.cn/library/r1797.
3. Liu Min, Yan Junwei. Intelligent manufacturing concepts, systems, and modeling methods. Beijing: Tsinghua University Press, 2019.
4. Comparison of industrial Internet reference architectures in Germany, Japan, and China [EB/OL]. http://blog.sina.com.cn/s/blog_654887a50102x338.html.
5. LEE J, BAGHERI B, KAO H A. A cyber-physical systems architecture for industry 4.0-based manufacturing systems. Manufacturing Letters, 2015, 3: 18–23.
6. LEE J, ARDAKANI H D, YANG S, et al. Industrial big data analytics and cyber-physical systems for future maintenance & service innovation. 4th International Conference on Through-life Engineering Services (TESConf) (Procedia CIRP). Cranfield, England, 2015.
7. WANG K, DAI G, GUO Lanzhong. Intelligent predictive maintenance (IPdM) for elevator service through CPS, IOT & S and data mining. Proceedings of the 6th International Workshop of Advanced Manufacturing and Automation. Manchester, England, 2016.

Chapter 4
IoT-Based Perception Resource Management Framework and Model

The data collection and processing in the intelligent predictive maintenance framework based on the Internet of Things (IoT) uses intelligent sensors and other sensing technologies to transmit the obtained key component status and operating environment information to the equipment control center, realizing round-the-clock automatic monitoring of the equipment, effectively monitoring and predicting the operation status and faults of the equipment [1, 2], bringing great convenience to equipment maintenance, promoting the further development of equipment status monitoring and Condition based Maintenance, and improving the operational efficiency of enterprises. This chapter mainly discusses the IoT-based sensing resource management framework, the IoT resource security management model based on blockchain, and the data sharing model.

4.1 IoT-Based Perception Resource Model and Management Framework

With the development of the Internet of Things (IoT), comprehensive perception, various data transmission, and intelligent data processing have been fully developed. Comprehensive perception uses various bus technologies, ZigBee, RFID, cameras, intelligent sensors to perceive and measure device status parameters, various data transmissions are realized through UDP, TCP/IP, HTTP and other means to achieve information interaction and sharing, and intelligent data processing integrates and analyzes data and information from cross-industry departments to achieve intelligent device monitoring. However, the IoT system of the smart factory has many types of sensors and terminal devices (including temperature, pressure, speed, acceleration sensors, Android devices, RFID tags, etc.), diversified transmission methods (including Zigbee, Bluetooth, GPRS/3G/4G/5G, FF/Lonworks/Profibus/CAN/HART and other 40 + field buses, TCP/IP and other transmission protocols),

M. Liu et al., *Intelligent Predictive Maintenance*, Advanced and Intelligent Manufacturing in China, https://doi.org/10.1007/978-981-97-2677-6_4

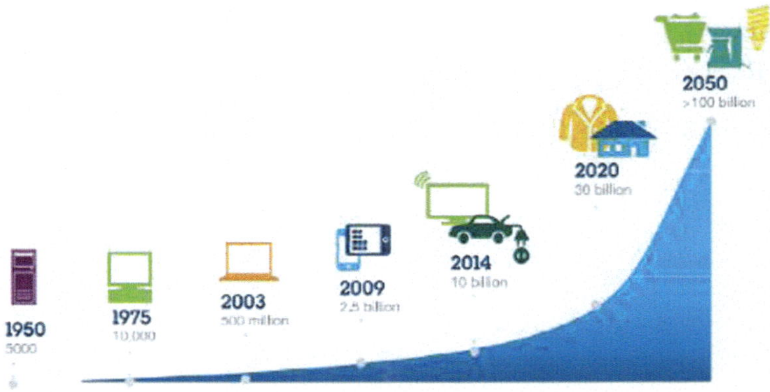

Fig. 4.1 Every inflection point in computing history triggers an explosion in the number of computing devices (IBM)

heterogeneous data and large data scale, and it is difficult for the data to be integrated with systems of different operating platforms and languages (Fig. 4.1).

To solve the above problems, it is necessary to break through the hierarchical structure of traditional control systems, focus on industrial needs, and systematically study the IoT resource management framework that supports barcodes, QR codes, RFID, wireless sensor networks, wireless local area networks, and various industrial buses [3], separating the functions of perception device management, data storage access, and data integration from the application layer, so that the upper-layer applications can focus more on their own services and functions, achieve low coupling between the upper-layer applications and the data collection perception devices used by the perception layer, reduce the development volume of the application layer and the difficulty of system development, thereby achieving data collection and processing.

In the intelligent predictive maintenance framework, data collection and processing based on IoT need to authenticate the interruption of massive data devices in the perception layer, need to handle massive data service interaction, distributed network deployment; at the same time, the network access layer needs to support multiple access methods, compatible with different terminal capabilities, able to support mobility, various deployment environments and multi-application platform interaction; the application layer needs to achieve multi-service integration, shield the differences of underlying perception terminals, support cross-application data sharing and reuse of basic business functions. Therefore, to meet the distributed heterogeneous network environment, drawing on some existing IoT middleware models, construct a perception resource management model and framework based on IoT, encapsulate business function modules into services, so that users can use such a framework to encapsulate and publish resource services, shield the heterogeneity of underlying perception layer data and provide a unified Web service interface for the upper application layer, in the hope of reducing the difficulty of system software development and reducing software coupling problems.

4.1.1 IoT Perception Resource Model

Model the IoT resource system, abstract the resource meta-model, analyze the resource hierarchy and its association relationship, and give resource categories such as entity resources, data resources, and method resources.

(1) Resource Meta-Model

In the equipment online monitoring system, any device and parts, sensor nodes, smart gateways, and data can be seen as a resource. The resource meta-model is a model that describes the main resource composition and constraints by defining the logical relationship and specific attributes of the key resources involved in the device monitoring process.

The essence of IoT perception resource management is the process of processing and transforming the resource model carrying business information in the process activities, changing its attributes according to certain operations, and passing the corresponding parameters and messages to other activities. To achieve the goal of business integration, it is first necessary to identify the resources in the existing dispersed heterogeneous systems and encapsulate them, register them on the resource integration platform and accept their unified management, thereby shielding the data differences between the systems, and then implementing resource interoperability in the form of Web services. Therefore, a unified IoT perception resource modeling method needs to be developed.

The meta-model (Meta Model) is used to define the construction (Construct) and rules (Rule) of the semantic model, usually referred to as the language defining the expression model. It describes how to build a model, the semantics of the model, or how models inherit and interoperate with each other. The abstraction level of the meta-model is higher than that of the model, and it has good universality and consistency. Establishing a resource meta-model is to analyze and solve the problems of resource model establishment and application in a multi-application environment from a higher level of abstraction.

Resource Meta Model (RMM) definition:

$$RMM =< RP, OS, RM, CS, RS, MI >$$

where, *RMM* is described as a set of properties (Property Set, PS), *RP* represents resource properties (Resource Property), *OS* represents the operation set (Operations Set), *RM* represents the rule model (Rule Model), *CS* represents the constraint set (Constraint Set), *RS* represents the rule set (Rule Set), *MI* represents management information (Management Information) and their mutual relationship R, its formal definition is as follows.

Definition 4.1 Property set PS:

$$PS = set < p_1, p_2, \ldots, p_n >$$

where, p_i = *set* < *PropName*, *DefaultValue*, *ValType*, *ConstrainExp*, *PropType* >; *PropName* represents the property name; *DefaultValue* represents the initial value of the property; *ValType* represents the type of property, such as integer, character, or time type, etc.; *ConstrainExp* represents the constraint on the property value, such as the range of numeric property values, the format of character property values, etc.; *PropType* is used to specify that the property is a static property or a dynamic property.

For example, the temperature property of a temperature sensor can be represented as:

$Sensor\_temperature =< Temperature,$

$78.00, Double, Range(0, 100), Dynamic >$

Definition 4.2 Resource attribute RP is defined as follows:

$$RP =< ResID, ResType, SProps, DProps >$$

where, $SProps = \{p_i|p_i \in PS$^$PropType\ (p_i) = static\}$, $DProps = \{p_i|p_i \in PS$^$PropType\ (p_i) = Dynamic\}$. *ResID* represents the unique identifier of the resource entity. *ResType* represents the category of the resource, which is a description of a type of resource with specific static attributes. The tree structure is a very orderly and easy-to-manage organization vertically, so the resource categories can be organized in a tree-like hierarchical manner, that is, the resource classification method of nested definition is adopted to form a resource classification tree. *SProps* represents the static attributes of the resource, the values of these attributes will not change after the resource object is created, such as the manufacturing date of the device. *DProps* represents the dynamic attributes of the resource, these attributes are used to describe the usage calendar, real-time status and lifecycle status of the resource object, such as the status of the sensor in the usage process (created, modified, deleted and queried, etc.). Dynamic attributes can describe the status of resources in business activities, so as to understand the current operating status of continuous casting equipment.

Definition 4.3 The formal definition of operation set OS is as follows:

$$OS = \{opi|opi =< oi, rpj >, rpj \in RP\}$$

where, *oi* is a function with resource object attributes and status as parameters, which can be mapped to specific attribute-related operations, such as creating, deleting, modifying, querying and other operations included in sensor resources; *rpj* represents the attributes related to this operation.

Definition 4.4 The formal definition of rule model *RM* is as follows:

$$RM = \{CS, RS\}$$

where, *CS* is a constraint set; *RS* is a rule set.

Definition 4.5 The formal definition of constraint set *CS* is as follows:

$$CS = \{ci|ci = f(si, sj), f \in Fc, si, sj \in Sn\}$$

where, *ci* represents a function mapping of resource state transition, used to reflect the reachability between resource states, such as the inability to access sensor data under other gateways from a sensor gateway.

Definition 4.6 The formal definition of rule set *RS* is as follows:

$$RS = \{ri|ri = f(si, sj, RS), f \in Fr, si, sj \in Sn\}$$

where, *ri* represents a function mapping from resource attributes to workflow status, used to reflect the resource state transition caused by resource attribute changes. It includes an initial state *si* and a reached state *sj*, as well as the set of resource attributes involved in the transition process. Organizing business rules around resources can arrange resource services by combining resource rules during process integration.

Definition 4.7 The formal definition of management information *MI* is as follows:

$$MI = \{SysInfo, PlatformInfo\}$$

where, *SysInfo* represents the information of the system where the resource is located; *PlatformInfo* represents the registration information of the resource in the integration platform, mainly used to determine the service characterization method of the resource object.

In summary, the XML Schema description of the resource meta-model is as follows, and the resource model can be described using an XML document that complies with this Schema, as shown in Fig. 4.2.

(2) Resource hierarchy analysis model

After establishing the meta-model of resources, it is necessary to carry out the overall resource level analysis, and regard the device, embedded gateway and perceived data as resources to facilitate the management and maintenance of the device. According to the resource access requirements and resource management requirements of the equipment online monitoring system, the resource level is divided into three levels shown in Fig. 4.3: DataPoint (sensor and intelligent terminal data), Device (terminal equipment, sensor or point inspection handheld terminal), Gateway (intelligent embedded gateway).

The Gateway (Intelligent Embedded Gateway) node is responsible for configuring the gateway's own information and managing the lower-level terminal device nodes. The gateway's own information includes gateway identification information, gateway status information, gateway description information, etc. Since the terminal devices connected to the gateway are different, the corresponding device nodes are configured

```xml
<?xml version="1.0" encoding="UTF-8"?>
<xs:schema xmlns:xs="http://www.w3.org/2001/XMLSchema" elementFormDefault="qualified" attributeFormDefault="
unqualified">
    <xs:element name="ResouceMetaModel" type="rmm">
        <xs:complexType name="state">
            <xs:sequence>
                <xs:element name="stateName" type="string"></xs:element>
            </xs:sequence>
        </xs:complexType>
        <xs:complexType name="propertySet">
        <xs:complexType name="resourceProperty">
        <xs:complexType name="propertySet">
        <xs:complexType name="activitySet">
        <xs:complexType name="ruleModle">
        <xs:complexType name="constraint">
        <xs:complexType name="rule">
            <xs:sequence>
                <xs:element name="fromState" type="state"/>
                <xs:element name="toState" type="state"/>
                <xs:element name="condition" type="ruleCondition"/>
            </xs:sequence>
        </xs:complexType>
        <xs:complexType name="resourceProperty">
        <xs:complexType name="ruleCondition">
        <xs:complexType name="conditionSet">
        <xs:complexType name="ruleSet">
        <xs:complexType name="managementInfo">
            <xs:complexType name="rmm">
            <xs:sequence>
                <xs:element name="RP" type="resourceProperty"/>
                <xs:element name="OS" type="operationSet" nillable="true"/>
                <xs:element name="RM" type="ruleModle" nillable="true"/>
                <xs:element name="MI" type="managementInfo"/>
                <xs:element name="RV" type="resourceView" nillable="true"/>
            </xs:sequence>
        </xs:complexType>
    </xs:element>
</xs:schema>
```

Fig. 4.2 Schema description of resource metamodel

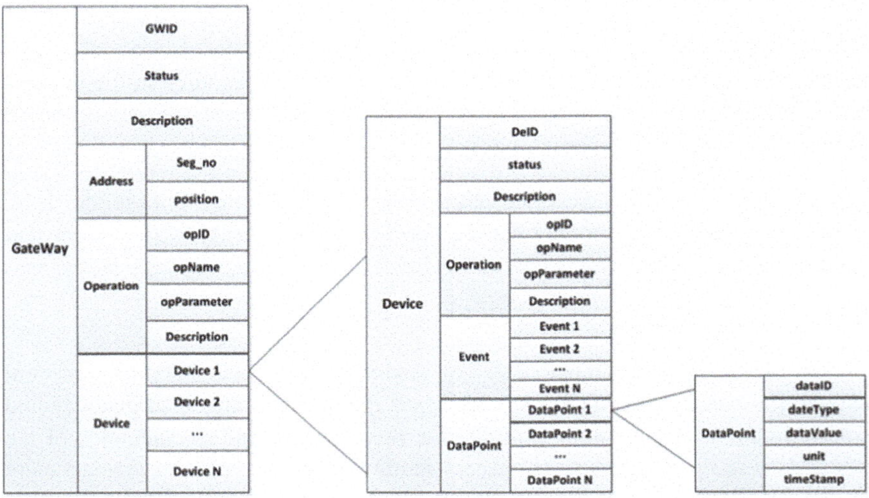

Fig. 4.3 Resource hierarchy analysis model

under each gateway node for device description. The gateway nodes in the system are distinguished by the IP address, port number and device ID of the gateway. The specific resource description information is shown in Table 4.1.

The Device (terminal device, sensor or spot check handheld terminal) node resource is responsible for configuring the terminal device information in the system

Table 4.1 Gateway resource description information

Parameter		Description	Necessity
DeID		Gateway device identifier	YES
Status		Gateway status	YES
Description		Gateway information description	No
Address	Seg_no	Fan segment numbering	YES
	Position	Position	No
Operation	opID	Operation ID	YES
	opName	Operation name	No
	opPrameter	Operation parameter	No
	Description	Operation description	No

and managing the terminal device capabilities. Terminal device information includes device identification information, device status information, device description information, etc. Since each terminal device has different data capabilities and control capabilities, each device node needs to configure corresponding capabilities and describe the capabilities of this node. Devices in the system are distinguished by system description information Description. The specific resource description information is shown in Table 4.2.

Table 4.2 Perception terminal resource description information

Parameter		Description	Necessity
GWID		Perception device identifier	Yes
Status		Perception device status	Yes
Description		Perception device information description	No
Operation	opID	Operation ID	Yes
	opName	Operation name	No
	opPrameter	Operation parameter	No
	Description	Operation description	No
Event	EventID	Event ID	Yes
	EventName	Event name	No
	EventPrameter	Event parameter	No
	Description	Event description	No
DataPoint	dataID	Data ID	Yes
	dataType	Data type	No
	dataValue	Data value	No
	unit	Data unit	No
	timeStamp	Collection time	No

DataPoint (sensor and smart terminal data): Due to the diverse data collection capabilities of continuous casting equipment terminals, including temperature, pressure, roller speed, frame displacement, and spot check information, it is necessary to abstract these capabilities for easy service access and expansion. Data in the system can be abstracted into a data resource, with main attributes such as data identifier, data type, data value, data unit, collection time, etc.

(3) Resource association relationship

By analyzing the relationship graph of resources in the online monitoring system of equipment, the Uniform Resource Identifier (URI) can be designed according to the mutual constraints between resources. In summary, the relationship between resources can be divided into inheritance, aggregation, combination, and association. Table 4.3 explains the mapping rules of resource object URIs.

Based on the resource meta-model and resource association relationship, the resource relationship of the equipment can be designed into a resource model system with inheritance relationship. Its structure is a tree structure, with the meta-model resource named as MetaModel, and the outer resource as XML identifier. The resource meta-model contains all the basic attributes of the resources, and other resources can inherit from the parent resource model, build its meta-model according to the structure of the parent resource, and then divide the resource types on the basis of the inheritance framework, and extend the original framework according to the resource types. The association relationship of resources is shown in Fig. 4.4.

① Entity resources: Entity resources represent the constituent equipment of continuous casting equipment and the sensing equipment installed on each equipment. The focus of entity resources is on the expansion of resource attributes.
② Data resources: The data resource meta-model extends the state range in resources, mainly used to characterize the temperature, pressure, and other state information of the continuous casting equipment during operation.
③ Method resources: In the REST-style architecture, it is stipulated to access resources with HTTP standard methods. The advantage is that it can standardize

Table 4.3 URI mapping rules

Resource model relationship	Mapping rules	URI example
Inheritance	":" represents inheritance relationship	/people:student
Aggregation	① "/" represents aggregation relationship ② If the local resource can exist independently, it needs to provide an independent URI for it	/device/{did}/data/{dataID} /device/did/data/{dataID}
Combination	"/" is used to represent the combination relationship	/mall/{id}/shop/{sid}
Association	"–" is used to represent the association relationship	/user-role/{id}

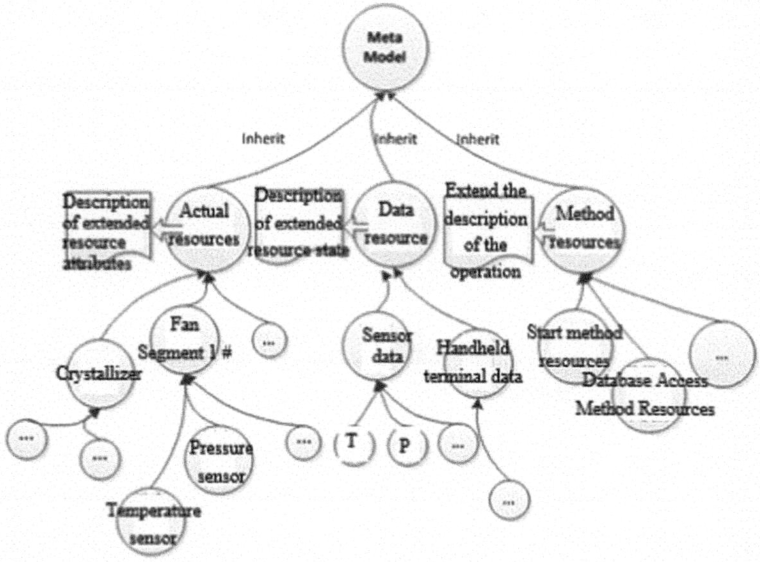

Fig. 4.4 Resource association relationship

the resource interface and increase universality. However, it is difficult to implement operations that require multiple steps, so method resources are needed to solve the problem of low support for complex operations by resources.

4.1.2 IoT-Based Perceptual Resource Management Framework

The layered architecture of the Internet of Things is shown in Fig. 4.5, divided into three layers: perception layer, network layer, and application layer.

The perception layer is composed of various sensing devices, such as RFID readers and electronic tags, temperature and humidity sensors, various gas concentration sensors, barcode recognizers, cameras, etc. It forms the basis of the Internet of Things, and its function is to identify networked items and collect various monitoring information.

The network layer uses various networking technologies to transmit the various information obtained by the perception layer. It includes local area networks, ZigBee wireless sensor networks, CAN bus networks, Internet technology, etc. At present, this layer is relatively mature, and what needs to be done in this layer is to make necessary improvements and optimizations to the network in combination with the application of the Internet of Things.

The application layer is a variety of intelligent applications developed based on the various data collected by the perception layer through the network layer. It faces

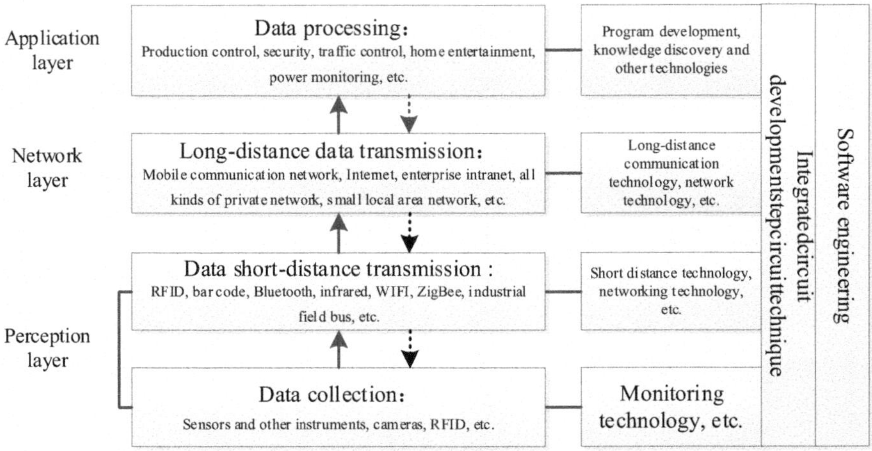

Fig. 4.5 Layered architecture of the internet of things

different users, combines with different industries, and is the fundamental purpose of the development of the Internet of Things. Its fundamental implementation is to integrate different industries, comprehensively develop and utilize information resources, propose high-quality and low-cost solutions, provide effective information security guarantees, and effectively develop and utilize commercial models.

The IoT-based perceptual resource management framework should have the following characteristics.

- Support for the reconstruction of web services in heterogeneous systems. After decomposing the existing business, it can be encapsulated into resource services and published to the framework for management.
- Support for resource model management. Various resource models can be defined in the framework based on the resource meta-model and managed by the platform. It can manage different resources in various heterogeneous systems with a unified view, achieving data-level information sharing between heterogeneous systems.
- Supports the processing of massive data. It has a complete caching and processing mechanism for large amounts of data, and a transaction concurrency processing mechanism for real-time databases, ensuring the timeliness and accuracy of data.
- Supports the registration, publication, and management of resource services. It supports synchronous and asynchronous interaction between heterogeneous systems, ensuring the security and reliability of message transmission.

Based on the above framework requirements, the IoT-based perceptual resource management framework shown in Fig. 4.6 was constructed, divided into the control access management layer, service resource management layer, and service resource bus layer.

(1) Service Resource Bus Layer

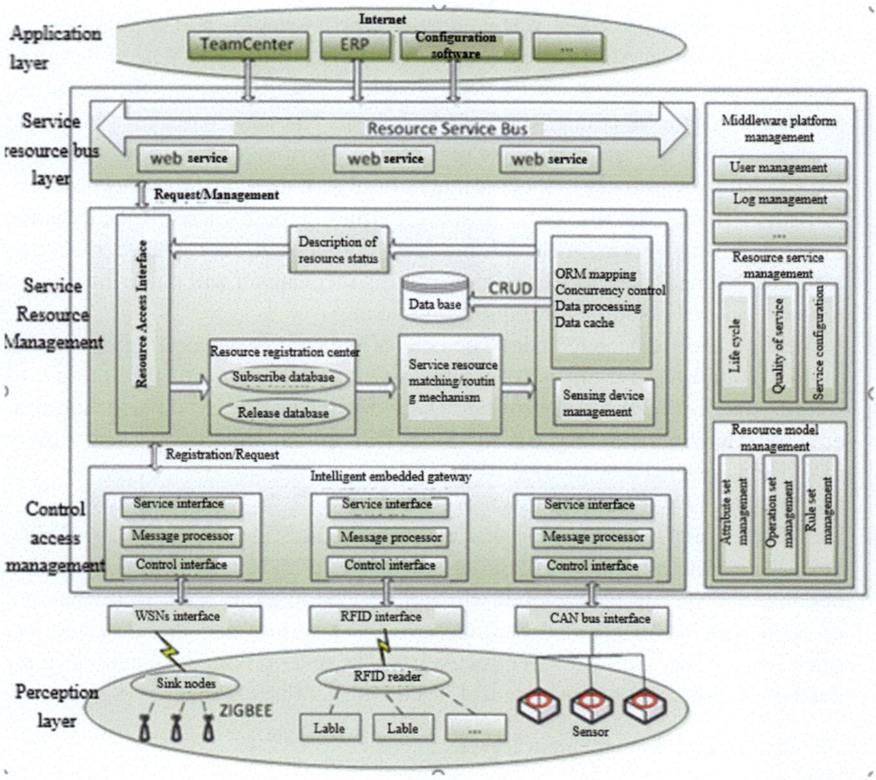

Fig. 4.6 IoT-based perceptual resource management framework

The main function of the service resource bus layer is to expose resource addresses and unified REST services defined GET, POST, DELETE, PUT, etc., to upper-layer applications as a C/S or B/S resource publishing platform. The abstracted perceptual data or operation data after XML is provided to the upper-layer applications in the form of Web services.

(2) Service Resource Management Layer

The main functions completed in this middleware platform include: service registration and matching, service routing, parsing and storing service information in different XML formats, database CRUD operations, returning the results of service resource operations, i.e., resource status description information.

The service resource management layer mainly contains four parts.

① Resource Access Interface The resource access interface needs to handle resource access requests or resource registration requests from the control access management layer and upper-layer applications. When resource access is allowed and exists, it needs to return the correct HTTP 200 status response code. When the accessed resource does not exist, it needs to return the HTTP 404 resource not found status response code. When encountering a resource registration request, it needs to judge whether the resource already exists. If it exists, it needs to return an error message. If the resource does not exist, it needs to create the resource.

② Service Resource Matching/Routing Mechanism The service resource matching mechanism will access the resource registration information in the resource registration center according to a specific matching algorithm, find the service resource provider that matches the service resource requester, perform real-time dynamic matching of the service provider and service requester, and when a matching resource is found, the service resource routing mechanism will guide the request to the requested resource for request processing.

③ Perception Device Management The core of the continuous casting equipment online monitoring system is to use the sensing technology in the Internet of Things, the ZigBee WSN transmission network, the wired CAN Bus technology, etc., for data collection and transmission. Due to the large number and variety of sensing devices, how to effectively organize and manage devices is a problem that needs to be solved in the system.

④ Data Processing The amount of data collected in the system is quite large. When the data collection frequency of the system increases and the number of collection nodes in the system increases, it will further lead to an increase in the amount of data, which will affect the efficiency of the system and the accuracy and timeliness of the data. How to effectively cache and process data, and optimize database access concurrency control, needs further discussion.

(3) Control Access Management Layer

The control access management layer mainly abstracts and manages the underlying sensing network, which can be divided into three modules: control interface layer, message processor, and service interface. The control sub-layer implements the initialization of all active networks and manages data access in the gateway. The message processor is responsible for standardizing the heterogeneous data of the message into XML. The service interface is responsible for calling the interface of the service resource management layer, mainly the management operation data of the device, the sensing data collected by the sensor, and the spot check information data collected by the spot check personnel on the spot.

4.1.3 IoT-Based Data Collection and Processing Framework

In order to achieve real-time monitoring of equipment and effective fault prediction, data processing is very important. The data of the online monitoring system has the following characteristics.

- Heterogeneity: Data is collected by various sensors at the device layer and transmitted through different transmission protocols at the physical layer. In order to obtain complete and accurate perception, it is necessary to comprehensively use these heterogeneous data to obtain useful information and provide information services.

- Large amount: There are various networks and a large number of sensors in the system, which will generate monitoring big data. How to improve existing technologies and methods, or propose new technologies and methods to effectively manage these massive data, is very important for further data integration, inference, and decision-making.
- Uncertainty: Measurement data will change with the sensing environment, and data may change during transmission. It is necessary to delete these erroneous data and extract accurate environmental parameters. Therefore, the data in the IoT system has obvious uncertainty characteristics, including the uncertainty of the data itself, the uncertainty of semantics, and the uncertainty of analysis, etc.

The IoT-based data collection and processing framework is shown in Fig. 4.7, which is divided into six layers.

(1) Device Layer

The device is composed of various sensors and Radio Frequency Identification (RFID) tags, which are used to measure the parameters of the device (such as temperature, pressure, etc.). The device layer is the bottom layer of the system's logical structure, implementing raw data collection and system commands.

(2) Physical Layer

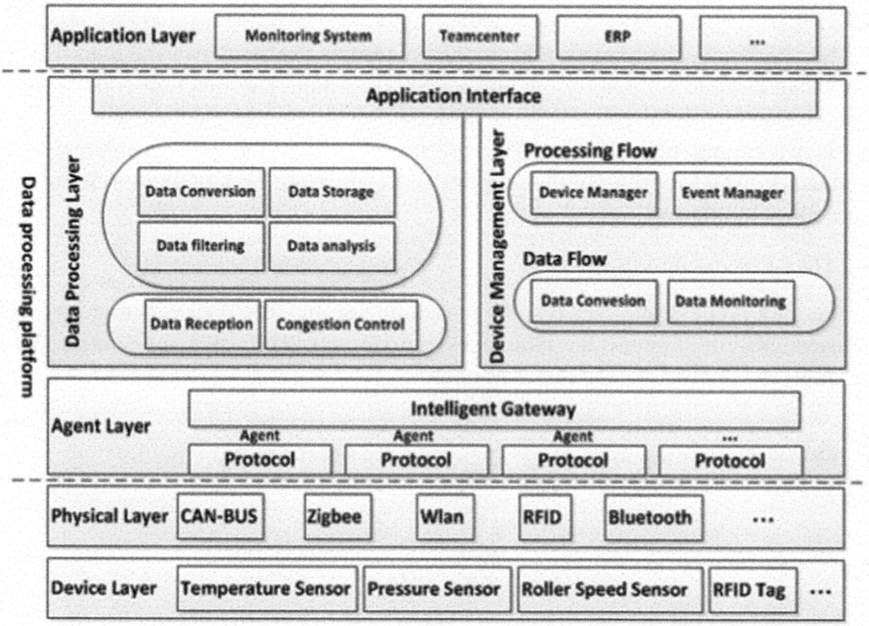

Fig. 4.7 IoT-based data collection and processing framework

Due to different application environments and different needs, the physical layer is responsible for communication between the device layer and the upper layer, including different physical transmission methods (such as ZigBee, RFID, etc.) and protocols.

(3) Agent Layer

The Agent layer can be seen as a protocol conversion layer between the physical layer and the upper layer. The physical communication protocol used by any type of device is converted into a physical protocol by the smart gateway, which has the ability to carry the Modbus/TCP protocol in the agent layer. Therefore, transforming a complex heterogeneous network into a relatively simple homogeneous network, the agent layer can be seen as a part of protocol conversion.

The data collected from the physical layer is composed of different protocols, which will increase the difficulty of operation, management, and maintenance of the online monitoring system. The smart gateway added to the system can convert data from different protocols into data based on the Modbus/TCP protocol. The Modbus/TCP protocol is a protocol based on TCP/IP. The application program in the application layer can communicate with the sensor nodes in the physical layer and the device layer through the Modbus messages sent by the smart gateway. The Modbus/TCP protocol has become a standard protocol used in industrial networks.

The rule of the Modbus/TCP protocol is that each sensor node and hardware device must have its own address, so the system can identify the message through the address contained in the Modbus message. As shown in Fig. 4.8, the Modbus/TCP data frame includes MBAP (header), Function Code (function code), and Data (data area).

MBAP (Modbus Application Protocol) represents the header of the data frame, which can be used to identify the application data unit. The function code is used to determine the performance that the device needs to use and indicates whether the data is in a normal response state. The data area is used to store request and response parameters. ADU represents the Application Data Unit, and PDU represents the simple Protocol Data Unit.

(4) Data Processing Layer

The data processing layer mainly processes the massive data generated by the online monitoring system. Its main tasks include network congestion control, data filtering, data analysis, and data conversion.

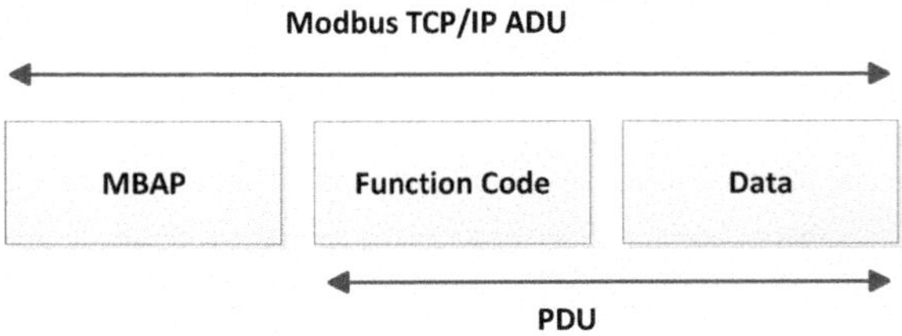

Fig. 4.8 Modbus/TCP data frame

To solve the system's massive data, effective data processing methods must be adopted. The data processing process is as follows.

① Congestion Control: The large amount of data in the system can easily cause congestion. If effective algorithms are not adopted, it will lead to a vicious cycle, making the congestion more severe. The Random Early Detection (RED) algorithm is often used to control congestion. The algorithm monitors the length of data packet queuing in the router. Before the cache is filled, once congestion is detected, data packets will be discarded at a certain probability, thereby improving network throughput. The RED algorithm consists of two parts, one is to monitor the average queue length, and the other is to judge whether the data packet should be discarded.

The average queue length is calculated by the Exponentially Weighted Moving Average (EWMA):

$$Q_{avg} = (1 - w_q)Q_{avg} + w_q q \tag{4.1}$$

In the formula, w_q is the weight; q is the queue length measured by sampling. Packet drop probability:

$$P = \begin{cases} 0, & Q_{avg} < Q_{\min} \\ \frac{Q_{avg} - Q_{\min}}{Q_{\max} - Q_{\min}} \times P_{\max}, & Q_{\min} \le Q_{avg} < Q_{\max} \\ 1, & Q_{\max} \le Q_{avg} \end{cases}$$

In the formula, $Q_{\min}$ is the minimum threshold; $Q_{\max}$ is the maximum threshold; $P_{\max}$ is the maximum drop probability.

The above formula indicates that when a packet arrives at the queue, it can be divided into the following three situations:

a. If the average queue length is less than the minimum threshold, the data packet enters the queue;
b. If the average queue length is within the range of the maximum and minimum thresholds, whether the data packet is discarded depends on the probability P;
c. If the average queue length exceeds the maximum threshold, the data packet will be directly discarded.

The RED algorithm can keep the average queue length at a lower level, and early discarding of data packets is very useful for absorbing short-term burst data streams. More importantly, the RED algorithm can avoid the global synchronization problem of the tail-drop algorithm.

② Data filtering. The data collected from the underlying hardware is very large, but only a part of the data is truly meaningful. If redundant data is not filtered out, it will bring burdens in three aspects: increasing the burden on network bandwidth, because the network needs to transmit a large amount of data; increasing the burden on the data processor, because the processor needs to process a large amount of data; increasing the burden on data storage, because the database needs to store a large amount of additional data.

In order to filter out redundant data, two types of filters are introduced in the data processing framework of the online monitoring system.

a. Duplicate data filter. The duplicate data filter filters out many duplicate data collected from sensors and underlying devices. The collected data can be described as a set of 3 attributes: NODE_DATA = {Node_ID, Datavalue, Timestamp}. Here, NODE_DATA is the data structure collected from the underlying devices of the online monitoring system, Node_ID is the unique identifier of the sensor or device in the system, Datavalue is the value measured by the sensor, and Timestamp is the acquisition time of the data. During the data filtering process, data is stored in a hash table, and Node_ID is used as the key of the hash table.

Define a time interval T_{interval}. When a new node data is obtained, check whether there is the same Node_ID in the hash table. If Node_ID does not exist, the node data will be inserted into the hash table and the node data will be output at the same time; if there is a same Node_ID, and the collection time interval of the two node data is less than the defined time interval value, it will be regarded as a duplicate node data, the duplicate data should be filtered out, and the timestamp of the node data will be updated at the same time; if there is a same Node_ID and the time interval of the two node data is greater than the defined time interval, the new node data is considered to be a valid new node data, the data should be output and the value and timestamp data of the data node should be updated in the hash table at the same time. The flowchart of the duplicate data filter is shown in Fig. 4.9.

d. Event filter. The event filter will select valuable data of interest to the user based on user settings. For example, in the application of IoT (Internet of Things), users need to know the number of sensor nodes in a specific time and area, how many nodes are left, and how many new nodes enter. The event filter is designed to achieve this goal.

The event filter is mainly used to manage sensor nodes, mainly dealing with three types of sensor nodes: new sensor nodes, leaving sensor nodes, and current active sensor nodes. Define a time interval parameter $T_{\text{persisttime}}$ as the basis for determining whether the sensor node leaves or stays active, define the queue Q_{addnode} as the recovery of new sensor nodes, define the queue $Q_{\text{deletenode}}$ as the recovery of leaving sensor nodes.

When a new node data is obtained, get the Node_ID and check whether there is the same Node_ID in the hash table. If there is, go to check whether the collection time interval of the two node data is less than the defined time interval; if the time interval is less than $T_{\text{persisttime}}$, then update the timestamp of the node data, which means to treat this sensor node as an active node; if the time interval is greater than $T_{\text{persisttime}}$, delete the node data from the hash table and add $Q_{\text{deletenode}}$ node data, which means that the sensor node does not respond within the predetermined time, define this type of sensor node as a dead node of a node; if not, add the new node

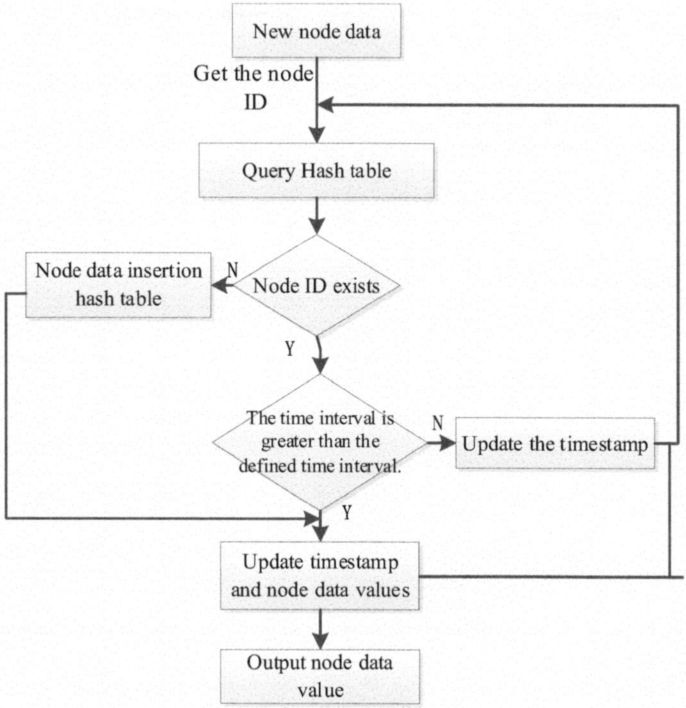

Fig. 4.9 Flowchart of the duplicate data filter

to Q_{addnode} hash table queue, which means that this sensor node has not appeared before, define this type of node as a new node.

③ Data analysis. Data analysis includes parsing the MBAP header and Modbus PDU. The data packet parsing algorithm flow is shown in Fig. 4.10.

a. MBAP header parsing. The MBAP header contains the sensor address and part of the Modbus protocol. If the MBAP header contains the correct address and Modbus protocol, it is considered to be consistent with the Modbus protocol.
b. PDU parsing. If the data is Modbus protocol data, then the PDU part of the data needs to be parsed. If the PDU conforms to the normal response format, then extract the data and response content. Otherwise, consider it as an abnormal Modbus protocol data, and need to deal with the exception problems that occur in the program or failure in the underlying sensor.

④ Data conversion. After data analysis, the required information can be obtained, but the data obtained from the data processing layer is difficult to integrate with systems built with different languages and different operating systems. Because existing application systems are distributed, heterogeneous, and closed, it is easy to lead to a problem, that is, the underlying information cannot be shared by

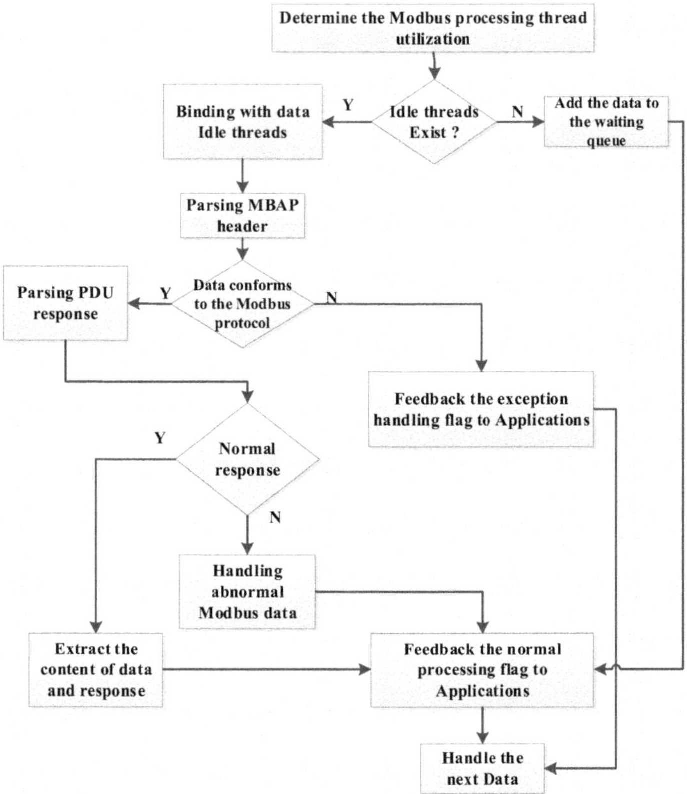

Fig. 4.10 Data packet parsing algorithm flow

the upper layer. The well-structured, platform and application independent, self-descriptive, simplicity and many other features of XML make it independent of any application system and platform.

As shown in Fig. 4.11, data A and data B represent the collected data. Each data may have its attributes, including system ID attributes, sensor location attributes, data value attributes, and acquisition time attributes, etc.

Schema is a language used to describe and specify the logical structure of XML documents, and is used to verify the correctness of the logical structure of XML documents. Use Schema to establish the mapping and constraints between data attributes and schema elements. The role of DOM and SAX is to convert data consistent with Schema rules into standard XML data, so that applications in the application layer can obtain the parsed data by querying the database or calling the interface.

(5) Device management layer

The device management layer is composed of data flow and process flow. The function of the data flow is to convert heterogeneous data obtained from the physical

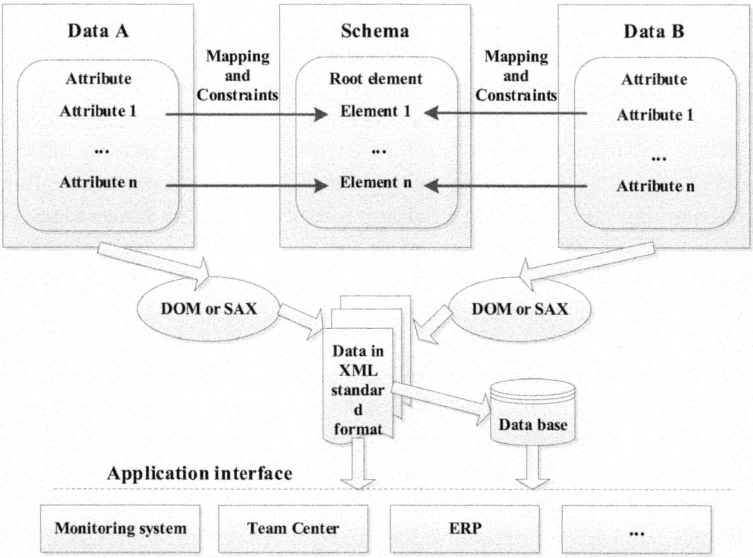

Fig. 4.11 Data conversion process

layer through different protocol technologies into unified format data. The function of the process flow is to respond to services from the upper layer and then convert them into commands that can be executed by different underlying technologies.

(6) Application layer

The application layer is composed of applications in different platforms and languages. Applications can obtain data and control underlying devices and sensors through the application programming interface.

4.2 Blockchain-Based IoT Resource Security Management Model

With the exponential growth of devices deployed in the Internet of Things in smart factories, traditional centralized proxy communication modes or server/user-end modes are no longer suitable for the construction of the IoT ecosystem. The increase in device connections has increased the risk for device manufacturers and smart factories, posing significant challenges to the privacy, security, and fault tolerance of the IoT. In 2014, IBM proposed the concept of device democracy, decentralization, and self-governance of the IoT in the "IBM IoT White Paper", building the IoT through blockchain [4], establishing secure connections between heterogeneous systems or open source systems, which is exactly the challenge brought by the networking and intelligence of Industry 4.0 and manufacturing transformation and upgrading.

Therefore, introducing blockchain technology in smart factories can establish low-cost direct communication bridges between devices (machine-to-machine dialogue, transactions, and even payments), and improve the security and privacy of the system through a decentralized consensus mechanism. In addition, blockchain technology superimposed on smart contracts can turn each smart device into an independent network node that can self-maintain and regulate. These nodes can execute functions such as exchanging information or verifying identity with other nodes based on pre-set or implanted rules, and help asset owners who do not know each other negotiate asset usage rights transactions and realize P2P payments, coordinate transactions between devices, and realize data sharing between devices and devices, and devices and people.

In order to solve the problems of device reliability and security and island-style connections in the implementation of IoT and the construction of the IoT ecosystem, using blockchain technology and smart contracts, a blockchain-based IoT resource security management model (Blockchain-based Resource Safe Management Model for Internet of Things, BRSMMIoT) was constructed.

The BRSMMIoT model integrates the physical space workshop equipment into the cloud environment, supports the development of decentralized, end-to-end manufacturing applications. The foundation of the BRSMMIoT platform is a blockchain network embedded with smart contracts. Smart contracts serve as protocols between service consumers and manufacturing resources, providing on-demand manufacturing services. In order to reduce the load and latency of the network, the platform is designed as a network architecture composed of on-chain networks and off-chain networks. All transactions are conducted in the on-chain network, such as digital signatures based on access control, programmable permissions, etc. The off-chain network handles problems that blockchain technology cannot solve, such as storage, complex data processing, etc.; and elaborates on several basic protocols of secure multi-party computation, and based on this, analyzes and studies the secret sharing mechanism and data access methods. In addition, for the interconnection problem between the on-chain network and the off-chain network, the blockchain ledger identification management method and the corresponding link protocol are introduced. Through the BRSMMIoT platform, smart devices in smart factories can carry out decentralized, trustless, end-to-end network interaction and transactions, which helps to open up the horizontal industrial chain of the Internet of Things and the data channel of vertical Internet of Things devices, establish and maintain the consensus of the industrial Internet of Things ecosystem, and promote the use of data in the entire ecosystem. Finally, the BRSMMIoT and smart contracts are verified with the application of smart diagnosis and equipment maintenance as an example.

4.2.1 Relationship Between Blockchain and IoT

As shown in Fig. 4.12, the security threats of the Internet of Things mainly come from the application layer, network layer, and perception layer. Table 4.4 lists the industry pain points in the development of the Internet of Things, among which, the biggest challenge is the current IoT server/client mode of the ecological architecture, where devices are connected and identified through the central server. This architecture mode cannot adapt to today's increasingly powerful IoT ecosystem.

Blockchain technology can improve the current state of Internet of Things (IoT) security. Applying the characteristics of blockchain technology to IoT security can promote the development of IoT and reduce the cost of IoT applications. Blockchain, with its features of peer-to-peer, public transparency, secure communication, tamper resistance, and multi-party consensus, will have a significant impact on IoT: ① The multi-center, weak centralization characteristics will reduce the high operation and

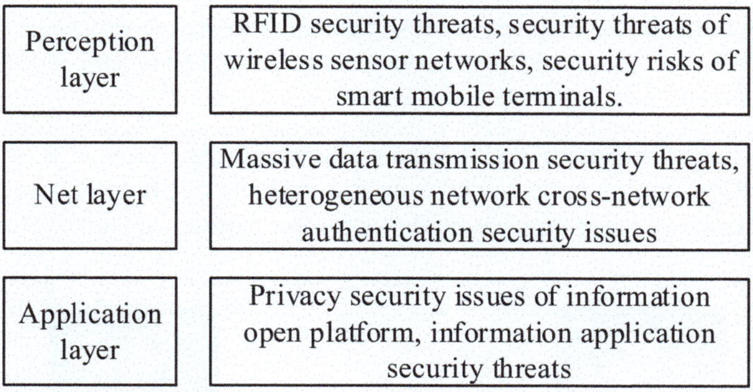

Fig. 4.12 Main security threats of the internet of things

Table 4.4 Industry pain points in the development of the internet of things

Industry pain points	Specific description
Device security	The Mirai botnet has infected more than 2 million IoT devices such as cameras, and the DDoS attack it launched caused the American domain name resolution service provider Dyn to paralyze, and many websites such as Twitter and Paypal were inaccessible at that time
Privacy protection	The main issue is that the centralized management architecture cannot prove its innocence, and incidents of personal privacy data leakage occur from time to time
Rigid architecture	IoT data flows are all pooled into a single central control system. With the continuous evolution of low-power wide-area technology (LPWA), IoT devices will grow geometrically in the future, and the cost of centralized services will be difficult to bear
Communication compatibility	The IoT platform lacks a unified language, which hinders communication between IoT devices and produces different standards and platforms
Multi-entity collaboration	At present, many IoTs are self-organized networks within operators and enterprises. When it involves cooperation across multiple operators and multiple peer entities, the cost of establishing credit is very high

maintenance costs of centralized architecture, and the characteristics of information encryption and secure communication will help protect privacy; ② Identity authority management and multi-party consensus can help identify illegal nodes, and timely prevent the access and malicious actions of malicious nodes; ③ Relying on the chain-like structure can help build verifiable and traceable electronic evidence storage; ④ The characteristics of distributed architecture and peer-to-peer can help break the shackles of multiple information islands existing in IoT, and promote the horizontal flow of information and multi-party cooperation.

From the development trends at home and abroad and the development and evolution path of blockchain technology, the development of blockchain technology and applications not only requires new generation information technologies such as cloud computing, big data, and IoT as infrastructure support, but also plays an important role in promoting the development of new generation information technology industries. Figure 4.13 illustrates the relationship between blockchain and new generation information technology.

(1) Application Challenges of Blockchain + IoT The application of Blockchain + IoT will encounter challenges in many aspects, including resource consumption, data expansion, performance bottlenecks, and partition tolerance. In terms of

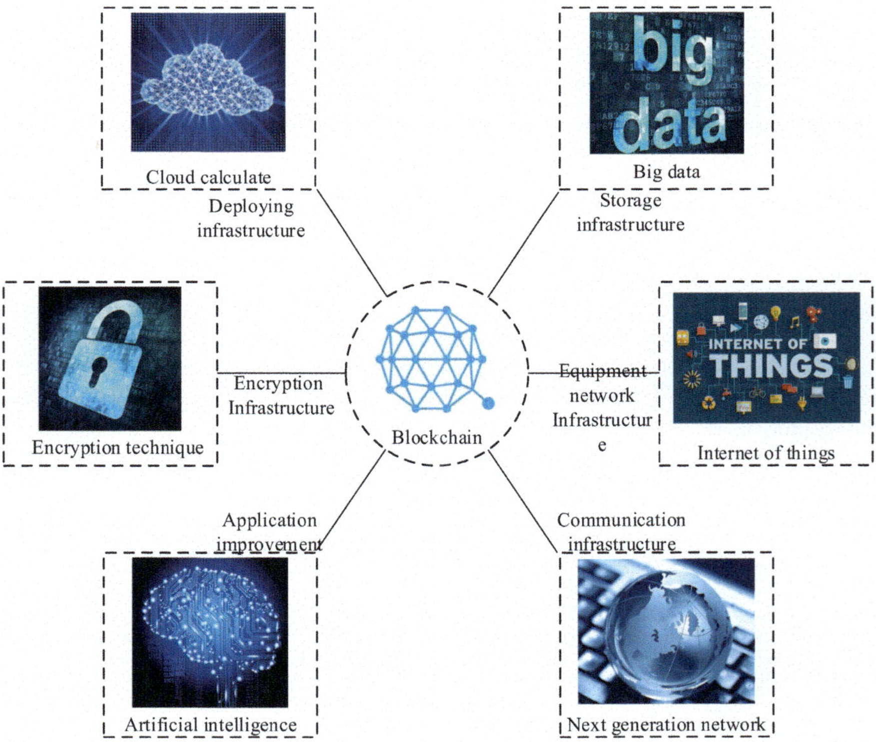

Fig. 4.13 Blockchain and new generation information technology

resource consumption, IoT devices generally have low computing power, weak networking capabilities, and short battery life. The Proof of Work (PoW) mechanism of Bitcoin consumes too many resources and is obviously not suitable for deployment in IoT nodes, but can be deployed in IoT gateways and other servers. Moreover, blockchain 2.0 technologies such as Ethereum are also PoW + PoS, and are gradually switching to PoS (Proof of Stake). The distributed architecture requires a consensus mechanism to ensure the final consistency of the data. However, compared with the centralized architecture, the consumption of resources is not negligible. In terms of data expansion, blockchain is a data storage technology that can only be appended and cannot be deleted. As the blockchain continues to grow, do IoT devices have enough storage space? For example, Bitcoin has been running up to now, requiring 100 GB of physical storage space. In terms of performance bottlenecks, the traditional Bitcoin transaction is 7 transactions per second, and with consensus confirmation, it takes about an hour to write into the blockchain. This kind of delay caused by feedback delay and alarm delay is not feasible on the delay-sensitive industrial internet. In terms of partition tolerance, the industrial IoT emphasizes that nodes are "always online", but it is common for ordinary IoT nodes to fail and frequently join and exit the network, which can easily cause network oscillations that consume a large amount of network bandwidth, and even the phenomenon of "network partitioning". (2) Improvement Ideas for Blockchain + IoT From the improvement aspect, it can be measured from both blockchain and IoT. ① From the perspective of blockchain a. For resource consumption, you can not use the mining-based, resource-consuming consensus mechanism, use the voting consensus mechanism [such as PBFT (Practical Byzantine Fault Tolerance) etc.], reduce the notification of resource consumption, and can effectively increase the transaction speed and reduce the transaction delay. Of course, there will be some loss in terms of node scalability, which requires a balance oriented towards business applications. b. For data expansion, you can use the Simple Payment Verification (SPV) method, compress transaction records through the Merkle tree. In terms of system architecture, support for heavy nodes and light nodes. Heavy nodes store all data of the blockchain, and light nodes only store the 256 hash value of the Merkle tree root node and only do verification work. c. For performance bottlenecks, many blockchain software platforms oriented towards IoT have made improvements. For example, IOTA proposes not to use a chain-like structure, adopts a Directed Acyclic Graph (DAG) data structure, on the one hand, it improves transaction performance, on the other hand, it also has anti-quantum attack characteristics. Lisk uses cross-chain technology such as main chain-side chain, manages partitioning and slicing, and has made many breakthroughs in performance. d. For partition tolerance, in response to possible network partitioning, you can choose to support on-chain and off-chain transactions, especially offline transactions, and support multiple CPS clusters in system design. ② From the perspective of IoT a. For resource consumption, with the development of eMTC, NB-IoT, LoRA and other Low Power Wide Area Network (LPWA) technologies, problems such as transmission quality, transmission distance, power consumption, and battery life will be gradually resolved. b. For data expansion, according to Moore's Law and Super Moore's Law, IoT storage capacity continues to rise. c. For performance bottlenecks, with the continuous maturity of new technologies, new processes, and new architectures such as MEMS sensors and SiP packaging processes, and the reduction of costs, small-volume, low-power sensor nodes are expected to be widely used.

4.2.2 Blockchain-Based Industrial IoT Platform

This section designs the architecture of the Blockchain-based Platform for Industrial Internet of Things (BPIIoT), as shown in Fig. 4.14. The foundation of the BPIIoT platform is a blockchain network embedded with smart contracts. Smart contracts serve as the protocol between service consumers and manufacturing resources, providing on-demand manufacturing services, supporting the development of decentralized, end-to-end manufacturing applications (DApps).

The deployment and implementation of blockchain technology requires the participation of multiple nodes. Under the conditions of the Internet of Things (IoT), the

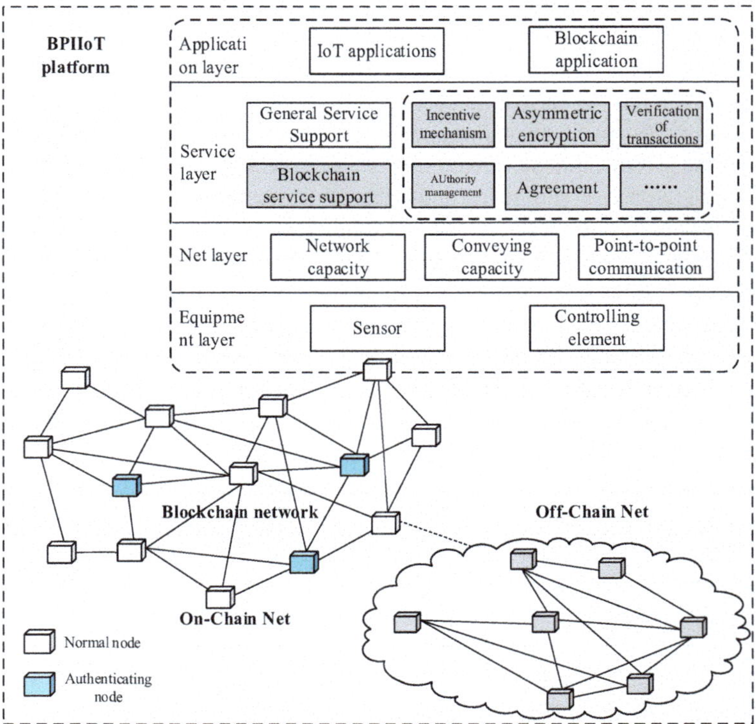

Fig. 4.14 BPIIoT architecture

computing power of each smart device is very limited. Compared with traditional blockchain mining nodes, its hash computing power is even less than one thousandth of the GPU system. In practical applications, the power consumption of IoT devices is also a tricky issue, and the existing blockchain technology cannot be directly applied to the industrial Internet of Things. Therefore, the BPIIoT platform is designed as a network architecture composed of on-chain and off-chain networks to reduce network load and latency. All transactions are conducted in the on-chain network, such as digital signatures based on access control, programmable permissions, etc.; correspondingly, the off-chain network handles problems that blockchain technology cannot solve, such as storage, complex data processing, etc. BPIIoT is a decentralized platform, compared with blockchain, the privacy and scalability of the BPIIoT platform are better. The off-chain network avoids the participation of third parties by introducing Secure Multi-Party Computation (SMPC), data queries and calculations are distributed, data is distributed on different nodes, these nodes participate in the calculation without leaking information. In addition, it is not required that each node in the off-chain network repeatedly perform data calculation and storage to meet higher computing requirements.

1. On-Chain Network

The on-chain network (blockchain network) is composed of ordinary nodes and verification nodes. Ordinary nodes (sensor and microcontroller nodes) do not participate in ledger recording, that is, smart devices (IoT units) do not participate in PoW calculations, but only encrypt and transmit data, and broadcast data transmission as transactions to the entire blockchain network. Verification nodes are specially deployed for PoW calculations, have strong computing power, are responsible for access control management and settlement payment of ordinary nodes, and these verification nodes can be built by multiple different IoT service providers using mainstream PC servers. Compared with the traditional IoT central server architecture, the investment in computing resources of a small number of verification nodes will be much smaller than the cost of the original service cluster. Since these verification nodes do not save user data, there is no possibility of user data leakage and use. The primary function of the on-chain network is to provide communication services, and data transmission is completed between nodes through the blockchain. There are two ways to store data on the blockchain, one is to add data to transactions, such as Bitcoin, and the other is to write data into contracts, such as Ethereum. Both methods store data by sending transactions to the blockchain, including transfer information and other arbitrary data. When the transaction is completed, the data contained in the transaction is open to the entire BC network. The on-chain network supports multiple blockchain services, such as incentive mechanisms, permission management, transaction verification, asymmetric encryption, etc. Transaction verification: Generally, the node that receives the transaction executes the transaction verification process. In Bitcoin, transaction verification depends on two scripts in the transaction output. The lock script stipulates the conditions for spending Bitcoin, and the corresponding unlock script must meet the conditions listed by the lock script. When verifying a transaction, execute the unlock script in each input to check whether the spending conditions are met. If the result of the script execution is "TRUE", the transaction is valid, which means that the unlock script successfully matches the spending conditions; otherwise, the transaction is invalid. Permission management: Blockchains can be divided into permissioned chains and non-permissioned chains. Permission management services are only available in permissioned chain networks. In addition to verifying permissions, other basic permissions such as joining the network and submitting transactions are also managed by permission management. Once you join the blockchain network, the joiner naturally has the read permission of the network, because all recorded information is public.

2. Off-chain network

In the off-chain network, data storage and calculation processing are mainly carried out. The blockchain is not a general-purpose database. The off-chain network deploys a decentralized off-chain database, that is, a Distributed Hash-Table (DHT), which can be accessed by the blockchain. The blockchain saves the index of the data instead of the data itself. The data is encrypted on the blockchain and stored in the DHT, and the access control protocol is written on the blockchain. The off-chain network provides an API interface. The nodes in the off-chain network form a distributed database. Each node is allocated a certain proportion of encrypted data according to the share, ensuring the confidentiality and fault tolerance of the calculation process. Nodes can also store a large amount of unencrypted public data and link it to the blockchain (Fig. 4.15). Distributed storage is implemented based on the Kademlia DHT protocol. This protocol considers the preference probability of nodes, which is conducive to share storage. In terms of calculation processing, the off-chain network can ensure the correctness of code execution and will not leak the original data to any node. In addition to the superiority in privacy, the off-chain computing network can broadcast complex calculations through the blockchain.

3. IoT unit

The IoT unit is the most important component in the BPIIoT platform. As a bridge between machine equipment and the cloud and the blockchain network, it

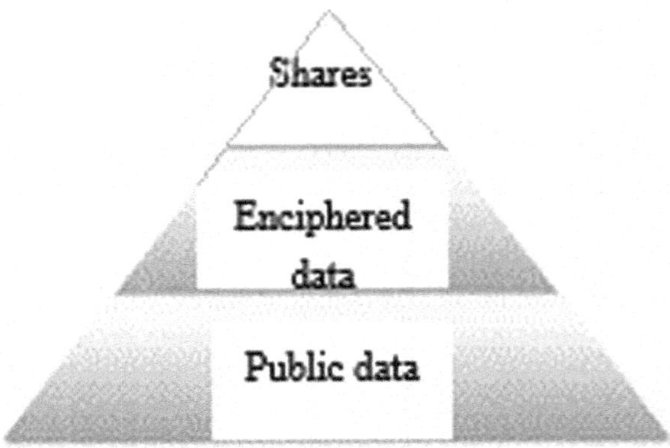

Fig. 4.15 Storage schematic diagram of Off-chain network

provides a plug-and-play solution. Through the IoT unit, machine equipment can upload operation data to the cloud, send transactions/transactions to related smart contracts, and receive transactions from other nodes in the blockchain network. Each node (IoT unit) in the blockchain network contains four layers, namely the device layer, network layer, service layer, and application layer. The device layer is mainly sensors and actuators deployed in the industrial field, connected to machine equipment through the interface board. The node communicates with the blockchain network through the network layer, sending/receiving transactions. The service layer is composed of general service support and blockchain service support. General service support includes device managers, I/O interfaces, controller services, and microcontroller functions, etc. Blockchain service support covers incentive mechanisms, asymmetric encryption, transaction verification, permission management, contract and other services.

4.2.3 Secret Sharing Mechanism and Data Storage Method

Secure Multi-party Computation (SMPC) is a solution to the problem that two or more participants can cooperate to perform a certain calculation task without revealing their private input information in a multi-user network that does not trust each other. SMPC is very important in cryptography and is the basis of cryptographic protocols such as electronic voting, threshold signatures, and electronic auctions. Yao [5] first proposed a secure two-party computation protocol in 1982, and Goldreich and others [6] extended it to multi-party computation problems, proposed a secure multi-party computation protocol based on the cryptographic security model, and proved the existence of n-Secure under passive attack conditions. The protocol, and the existence of (n-1)-Secure protocol under active attack conditions. For general multi-party computation (MPC) problems, its computation protocol is composed of basic MPC threshold circuits. There are mainly two types of solutions: based on Boolean circuits and MPC methods based on secret sharing. The latter is more common in production systems. [7].

The mathematical model of secure multiparty computation can be summarized as follows: in a distributed network, there are n untrusted participants $P_1, P_2, \ldots, P_n$, each participant P_i secretly inputs x_i, they need to jointly execute the function:

$$F : (x_1, x_2, \ldots, x_n) \rightarrow (y_1, y_2, \ldots, y_n) \tag{4.2}$$

In the formula, y_i is the corresponding output obtained by P_i. During the calculation of function F, it is required that any participant P_i cannot obtain any input information of other participants P_j $(j \neq i)$ except for y_i. Since in most cases $y_1 = y_2 = \ldots = y_n$, the function can be simply represented as F: $(x_1, x_2, \ldots, x_n) \rightarrow y$.

(1) Basic Protocol

The basic protocol of secure multiparty computation refers to those secure protocols that compute basic functions. These basic functions are often used as underlying tools frequently called by high-level protocols.

Protocol 4.1 Oblivious Transfer Protocol (OTP). The problem that Oblivious Transfer wants to solve is described as follows: Bob has n pieces of data b_1, b_2,..., b_n, Alice needs to select k ($k < n$) pieces of data from these n pieces of data, but for some reason, Bob only allows Alice to know the k pieces of data she chose and knows nothing about the other $n - k$ pieces of data, Alice does not want Bob to know which k pieces of data she got.

The Oblivious Transfer Protocol is an extremely important basic protocol for SMPC. Theoretically, the problem of secure multiparty computation in a general model can be solved through the Oblivious Transfer Protocol. The concept of Oblivious Transfer (OT) was first proposed by Rabin and others in 1981. Many variants of OT have been produced, such as 1-2-OT, 1-n-OT, etc.

① 1-2-OT Any kind of public key encryption mechanism can construct 1-2-OT. Suppose Alice wants to get b_i ($i = 0,1$), Alice randomly selects plaintext m_i, and notes the final bit d_i, then uses Bob's given public key to encrypt $E_{k_{Bob}}(m)$ to get c_i, then randomly selects another ciphertext $c_{1\text{-}i}$, and sends the ciphertext pair (c_0, c_1) to Bob; Bob decrypts the ciphertext pair (c_0, c_1) to get the plaintext pair (m_0, m_1), let m_0 and m_1 's last bit be d_0 and d_1 respectively, set $b_i = b_{1-I}\ d_i$, obviously, Bob cannot know what Alice's specific choice is. It should be noted that by requiring c_0 and c_1 to satisfy a certain specific relationship, Alice can be ensured to know nothing about m_{1-i}, thus preventing Alice's deception.

f) ② 1 -n-OT In the ElGamal encryption mechanism, the 1-n-OT system is set as follows: p is a public large prime number, g and h are generators of Z_p^*, $\log_p h$ is unknown to anyone. Suppose Bob has data m_1, …, $m_n \in G$ (p), Alice wants to get m_i ($i \in \{1, 2, \ldots, n\}$). First, Alice randomly selects $r < p$, calculates and sends $y = g^r\ h^i \bmod p$ to Bob; after receiving y, Bob calculates (α_1, β_1), …, (α_n, β_n) and sends them to Alice, where $\alpha_j = g^{k_j} \bmod p$, $\beta_j = m_j\ (y/h^j)^{k_j} \bmod p$ ($k_j \in Z_p^*$, $j = 1, \ldots, n$); Alice calculates β_i/α_i^r to get m_i.

$$\beta_j/\alpha_j^r = m_j h^{k_j(i-j)} = \begin{cases} m_i (j = i) \\ Z_p^* (j \neq i) \end{cases} \tag{4.3}$$

Protocol 4.2 Secure Multiplication Protocol (MULP). The traditional data form of the secure multiplication protocol is described as follows: Alice has a private input x, Bob has an input y, $x, y \in R$, both parties need to cooperate to calculate the product, so that Alice gets u, Bob gets v, satisfying $u + v = xy$, where u or v is a number randomly selected from the corresponding number field. The protocol ensures that the private inputs of both parties are not leaked to each other, the final product is shared in an additive manner, and no party can know the result alone. The essence of this protocol is to transform the multiplication sharing data into addition sharing data to support further calculations.

Protocol 4.3 Secure Two-Party Scalar Product Protocol (STSP). The dot product problem is described as follows: Alice has a private input vector $(x_1, x_2,\ldots, x_n)$, Bob has a private input vector $(y_1, y_2,\ldots, y_n)$, x, y satisfy the corresponding number field, $n \geq 2$. Both parties cooperate to calculate the dot product of the two vectors, that is, Alice gets u, Bob gets v, satisfying $u + v = \sum_{i \leq n} x_i y_i$, where u or v is a number randomly selected from the corresponding number field. It is required that the private inputs of both parties are not leaked to each other, and no party can know the result alone.

Protocol 4.4 Secure Sum Protocol. The secure sum protocol is especially widely used in the field of privacy-preserving data mining. The goal of the secure sum protocol is to calculate the sum of the inputs of all parties and requires no leakage of any input. There are two simple methods to complete the secure sum. One method is for one party to add a random number and then pass the input to all parties in turn. Each party adds their own input and passes it to the next participant. Finally, subtract the random number to get the result. Another method is for each participant to randomly split their input into n parts and send them to other participants, while calculating the sum of the data received from other participants, and finally summing them up in turn. The second method can resist the possibility of collusion.

Protocol 4.5 Permutation Protocol. The permutation protocol can be described as follows: The two parties involved in the calculation, Alice and Bob, Alice has a vector $X = (x_1,\ldots, x_n)$, Bob has a vector $Y = (y_1,\ldots, y_n)$ and a random permutation π [for any vector Z, π (Z) is also a vector, with elements the same as those in Z, but in a different order]. The protocol requires Alice to obtain the vector π $(X + Y)$ without knowing π and Y, and without revealing X.

Protocol 4.6 Secret Comparison Protocol (Secure Comparing). Secret data comparison is a basic operation of secure multi-party computation, where each of the two computing parties inputs a value, and they want to compare the sizes of these two numbers without revealing their own data to each other. When these two numbers are not equal, neither party can know any information about the other's data. This problem plays a key role in designing efficient secure multi-party computation protocols.

There are currently two types of secret comparison protocols: the first type of secret comparison protocol determines whether two data are equal, and if they are not equal, neither party can know any information about the other's data; the second type of secret comparison protocol can determine the size relationship of the two inputs.

(2) Secret Sharing Mechanism

Define a threshold cryptosystem (Threshold Cryptosystem), that is, $(t + 1, n)$-threshold, n is the number of all participants, $t + 1$ is the minimum number of participants needed to decrypt a secret that has been threshold encrypted. Secret sharing is an instance of a threshold cryptosystem. If the secret s is shared by n participants, then at least $t + 1$ participants can recover the shared secret s, and any subset of t participants cannot know the secret. The linear secret sharing mechanism

(Linear Secret-sharing Scheme, LSSS) splits a secret into multiple shares, which form a linear combination of the secret. For example, Shamir's Secret Sharing (SSS) [8] uses polynomial interpolation and is secure over the finite field F_p. For the secret s, it can be represented by a t-degree polynomial $q\ (x)$:

$$q(x) = a_0 + a_1 x + \ldots + a_t x^t \tag{4.4}$$

$$a_0 = s, a_i \sim U(0, p-1) \tag{4.5}$$

Then the shares can be represented as $\forall\ i \in \{1,\ldots, n\}$: $[s]_{p_i} = q\ (i)$.

Given t+1 shares, you can use Lagrange interpolation to reconstruct $q(x)$, let $s = q(0)$, then the secret s is recovered. Since SSS is linearly homogeneous, scalar operations of addition and multiplication can be performed directly on the shares. For example, the process of addition operation can be represented as follows:

$$cs = reconstruct\left(\{\ c[s]_{p_i}\}\ {}_{i \in n}^{t+1}\right) \tag{4.6}$$

$$s_1 + s_2 = reconstruct\left(\{\ [s_1]_{p_i} + [s_2]_{p_i}\}\ {}_{i \in n}^{t+1}\right) \tag{4.7}$$

The multiplication calculation of two secrets s_1 and s_2 is more complex. If a participant tries to calculate the product of two secrets, they will get a $2t$ degree polynomial, which requires a polynomial reduction step ($2t \rightarrow t$), adding a majority honesty constraint to the privacy and correctness of the result (i.e., $t < n/2$).

(3) Data Access

In the BRSMMIoT system, there are three different types of decentralized databases that can be accessed through a global dictionary.

Public Ledger: Use L to access and control the blockchain public ledger. For example, $L\ [k] \leftarrow 1$ means to update the keys of all nodes to k. Since the ledger is completely open and appendable, storing the entire history of data, data can be accessed through *L.get* (k, t).

DHT: Off-chain data is stored in the DHT, and its data access method is similar to the public ledger. By default, data is encrypted locally before being sent, and only the entity with the signature can request the data to be sent back. Otherwise, use *DHT.set* (k, v, p), where k is the key, v is the value of the data, p is the predicate, i.e., p: $X \rightarrow \{0,1\}$, only when p is satisfied will the data be accessed through the key k.

MPC: Syntactically, the use of MPC is similar to DHT. Specifically, when executing *MPC.set* (k, v, p) to share the value v, the shares are distributed to potential computing parties and stored locally. The predicate p is used to determine which participant can reference the data for computation without revealing v, i.e., $v_{\text{ref}} \leftarrow$ *MPC* [k]. By default, only the original owner can request the original data back through $v \leftarrow$ *MPC.declassify* (k), and any participant can reference the data for computation.

4.2.4 *Identity Management and Link Protocol*

The on-chain network and the off-chain network are interconnected, thanks to the important protocol that connects the blockchain to off-chain resources. This section will introduce how identities are formed and stored on the ledger, and how off-chain storage (DHT) and computation (MPC) requests are looped through the blockchain.

(1) Identity Management

The blockchain can be divided into fully decentralized permissionless chains and semi-centralized permissioned chains. However, by connecting to the blockchain through off-chain technology, a balance can be maintained between the pseudonymous system and the semi-centralized system while retaining the decentralized structure. First, define the Shared Identity as an extension of multiple identities and their semantics, represented by a $(2\,n + 1)$-tuple:

$$SharedIdentity_P = \left(addr_P, pk_{sig}^{(p_1)}, pk_{sig}^{(p_2)}, \ldots, k_{sig}^{(p_n)}\right) \tag{4.8}$$

In the formula, n is the number of participants. When $n = 1$, it is a special case of pseudonym.

The introduction of metadata represents the potential semantics of the identity, including public access control rules and related public or private data. For example, Alice doesn't want to tell Bob her exact height, but allows Bob to use her height information for joint query calculations. At this time, Alice can call a smart contract, use the MPC predicate (MPC Predicate), that is, $MPC\,[\text{'}alice\_height\text{'}] = alice\_height$ to share her height data, Bob refers to this data for calculation without knowing Alice's specific height. This MPC predicate sets up a sharing information owner Alice, and a sharing party Bob with restricted access rights. The MPC predicate (including the address of the shared identity list and the reference of the data) is saved on the blockchain, which defines the public metadata, that is, the non-sensitive information related to the identity, which can be used to verify access rights. Other additional metadata is private, that is, Alice, Bob and others with access rights can access it, and it is saved in the off-chain network through DHT.

Public transactions are verified by the blockchain. Since the ledger stores the shared identity and the predicate managing access control, the blockchain can access any off-chain resources. For private metadata involving privacy, the off-chain network can act as a trustless privacy protection verifier.

(2) Link Protocol

The access control protocol includes Protocol 4.7 and Protocol 4.8. Protocol 4.7 describes the process of building a shared identity, while Protocol 4.8 executes the contract to verify whether the judgment is satisfied.

Protocol 4.7: Building a Shared Identity.
Input: $P = \{\, p_i \,\}_{i=1}^{N}$ parties, $A = \{\, POLICY_{p_i} \,\}_{i=1}^{N}$
Output: Ledger L stores reference to the shared identity

$addr\ p = 0$
$AC\ L = \emptyset$
for $p_i \in P$ do.
$(p\ k_{sig}^{(p_i)}, s\ k_{sig}^{(p_i)}) \leftarrow \zeta_{sig}\ ()$
$addr\ p = addr\ p\square\ p\ k_{sig}^{(p_i)}$
$AC\ L\ [pk_{\ sig}] \leftarrow A\ [p_{\ i}]$
end for
$m \leftarrow (addr\ p, AC\ L\ )$
send signed $tx\ (m)$ to the network
procedure StoreIdentity $(addr\ p, AC\ L)$
$L\ [addr\ p] \leftarrow AC\ L$
end procedure
Protocol 4.8 Blockchain Permission Verification.
Input: $p\ k_{sig}^{(p_i)}$ the requesting party signature, $addr\ p$ the shared identity's address, q -a predicate verifying if $p_{\ i}$ has sufficient access rights.
Output: $s \in \{0,1\}$.
procedure $CheckPermission\ (p\ k_{sig}^{(p_i)}, addr\ p, q)$
$s \leftarrow 0$
if $L\ [addr\ p] \neq \emptyset$ then
$AC\ L = L\ [addr\ p]$
if $q\ (AC\ L, p\ k_{sig}^{(p_i)})$ then
$s \leftarrow 1$
end if
return s
end procedure

Protocol 4.9 introduces the process of directly storing and loading data through DHT. When storing data, the given q_{store} predicate is used to check write permission, and the data storage party provides a client predicate to check who can read the data.

Protocol 4.9—Storing or loading data.
Input: $p\ k_{sig}^{(p_i)}$, $addr\ p$, x (data), $q_{read}^{(x)}$—a predicate for verifying future read access.
Output: if successful, returns a_{x-} the pointer to the data (predicate), or Ø o.w
procedure $Store\ (p\ k_{sig}^{(p_i)}, addr\ p, x, q_{read}^{(x)})$
if $CheckPermission\ (p\ k_{sig}^{(p_i)}, addr\ p, q_{\ store}) = True$ then
$a_{\ x} = H\ (addr\ p \parallel x)$
$L\ [a_{\ x}] \leftarrow\ q_{read}^{(x)}$
$DHT\ [a_{\ x}] \leftarrow x$
return a_x
end if
return Ø
end procedure
Input: $p\ k_{sig}^{(p_i)}$, $addr\ p$, a_{x-} the address of the data(predicate)
Output: if successful, returns the data x, or Ø o.w
procedure $Load\ (p\ k_{sig}^{(p_i)}, addr\ p, a_x)$

$q_{read}^{(x)} \leftarrow L\,[a_x]$
if $CheckPermission\,(p\,k_{sig}^{(p_i)}, addr\,p, q_{read}^{(x)}) = True$ then
return $DHT\,[\,_x]$
end if
return Ø
end procedure

The computation and sharing protocol is equivalent to the storage and loading protocol of MPC. As shown in Protocol 4.10, the secure computation and sharing protocol stores and loads shares from the DHT, ensuring data security by computing through data references.

Protocol 4.10 Secure Computation and Secret Sharing Protocol.

Input: $p\,k_{sig}^{(p_i)}$, $addr\,p$, x (data), x_{ref} -reference for computation, $q_{compute}^{(x)}$—predicate verifying computation rights

Output: if successful, returns pointer to x_{ref} for future computation, or Ø o.w

procedure $Share\,(p\,k_{sig}^{(p_i)}, addr\,p, x, x_{ref}, q_{compute}^{(x)}, n, t)$
$[x]_p \leftarrow VSS\,(n, t)$
$peers \leftarrow$ sample n peers
For $peer \in peers$ do
send $[x]_p^{(peer)}$ to $peer$ on a secure channel
end for
return $Store\,(p\,k_{sig}^{(p_i)}, addr\,p, x_{ref}, q_{compute}^{(x)})$
end procedure

Input: $p\ k_{sig}^{(p_i)}$, $addr\ p$, $a_{x_{ref}}$—reference data address, f—unsecure code to be rewritten as a secure protocol

Output: if successful,returns $f\,(x)$ without revealing x, or Ø o.w

procedure $Compute\,(p\,k_{sig}^{(p_i)}, addr\,p, a_{x_{ref}}, f)$
$x_{ref} \leftarrow Load\,(p\,k_{sig}^{(p_i)}, addr\,p, a_{x_{ref}})$
if $a_{x_{ref}} \neq Ø$ then
$f_s \leftarrow$ generate secure computation protocol from f
return $f_s\,(x_{ref})$
end if
return Ø
end procedure

4.3 Blockchain-Based IoT Data Sharing Model

Blockchain applications in the financial and medical fields are relatively mature, and there have been some practices in decentralized information recording systems. To solve the problem of sharing manufacturing equipment data, the blockchain-based manufacturing equipment data sharing model uses blockchain technologies such as Ethereum smart contracts and proof of work mechanism to realize data circulation and sharing among multiple parties, real-time tracking and management of data flow, and effectively reduce the management cost of data sharing process through access control.

(1) Data Sharing Implementation Model

In traditional production environments, the types of data from manufacturing equipment are complex and stored in independent systems. These systems may belong to different service providers, and manufacturing companies do not have direct control over their data and cannot understand the real value of massive data. For example, equipment operation data and maintenance records are stored in the servers of equipment manufacturers or maintenance service providers. If the service provider is changed, it will cause historical data loss and discontinuity, which is not conducive to continuous supervision and full lifecycle management and maintenance of manufacturing equipment. E-chain can support manufacturing companies to achieve supply chain multi-party collaboration, intelligent production management and distributed production collaboration, promoting the sharing of R&D technology, production equipment and manufacturing services. The sharing of equipment not only includes the sharing of equipment capacity but also the sharing of equipment data. Equipment data sharing makes design R&D, production manufacturing and circulation audit more effective, which is beneficial for manufacturing companies to reduce operating and manufacturing costs and enhance competitive advantages.

Figure 4.16 shows the blockchain-based manufacturing equipment data sharing model. The manufacturing equipment in the factory registers and uploads data through Ethereum smart contracts after registration. The validation of data uses multi-point consensus and encryption technology to ensure the consistency and security of data. The validated data is organized into a standardized readable and writable data format stored in the off-chain database using big data and AI technology, and a summary index is generated and sent to the blockchain network. The content of the block records data ownership, browsing permission and other information. Different node access permissions are defined through the blockchain, and data sharing from equipment nodes to other nodes can be supported by sending transactions using smart contracts. The proof of work mechanism is introduced to provide incentive feedback in the form of tokens (Token). The transaction can include binary data and tokens, forming a peer-to-peer network with both private, central, open, and transparent features.

Node: Network nodes are divided into equipment nodes (data contributors) and service provider nodes (data beneficiaries). Each node's software components are the same, including administrators, backend API libraries, Ethereum clients (Eth Client) and databases. The Ethereum client provides account management, mining, transfer, deployment and execution of smart contracts and other functions. After registering through a smart contract, nodes participate in network activities with a unique digital identity, owning an external account and a contract account that stores code (sharing an address space). The address of the external account is determined by the public key, and the address of the contract account is determined when the contract is created. Each account has a token balance, which can be changed by sending a transaction with tokens to it.

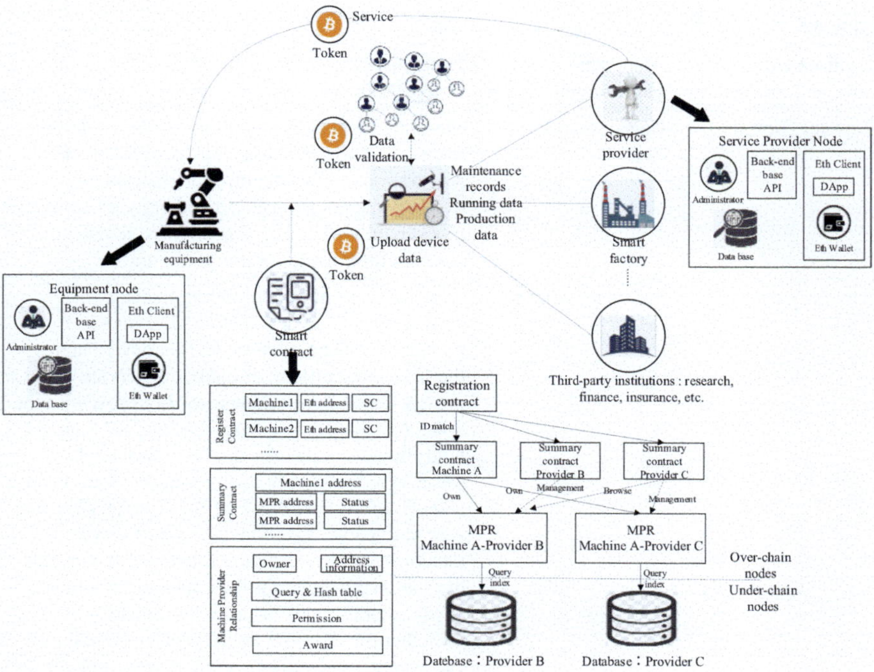

Fig. 4.16 Blockchain + manufacturing equipment data sharing model

Transaction: The records of equipment data and digital assets exist in the blockchain network. Transaction information includes the registration alias of the equipment (for privacy protection), data type, metadata tags of the original transaction data, complete metadata index, encrypted link of the transaction record, and timestamp of the transaction generation. All information is encrypted and digitally signed to ensure authenticity and accuracy.

(2) Types and Functions of Smart Contracts

The management of network nodes and API applications are implemented through smart contracts, which automatically track specific state changes, such as changes in access permissions, updates of data records, etc. Based on the generic business models and processes of different business domains, a generic smart contract template is formed. This book defines the Register Contract (RC), Summary Contract (SC), and Machine Provider Relationship (MPR). Table 4.5 describes these three types of smart contract templates and their functions. Each smart contract is managed as an on-chain asset throughout its lifecycle, with complete and controllable process management for the submission, deployment, use, and cancellation of smart contracts.

Using the MPR contract to build the relationship between the device and the service provider, associating the device's data records with permissions, data pointers, etc., and entering them into an external database. Service providers can add new data records for the device, and the device can authorize the sharing of data records

Table 4.5 Smart contract templates and functions

Contract template	Name and abbreviation	Function description
Register contract	Register contract (RC)	Maps the identity string of the participant to the corresponding Ethereum address, supports the registration of new identities or changes to existing mappings; makes the identity string and blockchain form a mapping address, finds and matches the corresponding SC contract
Summary contract	Summary contract (SC)	Contains a list of MPR contract indices, representing all nodes' historical business participation with other nodes; for device nodes, the SC contract holds all indices of service providers that have business relationships with the node; for service provider nodes, the SC contract holds indices of devices that have provided services or third-party institutions that have been authorized to share data; the SC contract supports user notification functions; the status variable of the device-service provider relationship indicates the status of the relationship, such as "whether it is newly established", "waiting for update", "whether the device is authorized"; device nodes can accept, reject, delete relationships, and decide which data records to authorize for sharing
Machine provider relationship contract	Machine provider relationship (MPR)	Defines the combination of data pointers and their associated access permissions to identify the data records saved by the service provider; each pointer consists of a query string that can return a subset of data results, with the query string tagged with the hash of that data subset to ensure the data is not tampered with; the hash table can match the query address with the attached query string list, thus sharing data records with other nodes

between various service providers. Participants automatically receive notifications and perform verification before operation changes, ensuring that all participants are informed and participate in the maintenance of data records. Use public key encryption algorithms to manage the digital identity of network nodes, and match the digital identity of nodes with Ethereum addresses. After the blockchain network confirms the command, the databases of the off-chain devices and service providers are synchronously updated.

(3) Reward Mechanism

Introducing the mining reward mechanism of the blockchain, incentivizing data beneficiaries such as service providers or third-party institutions to contribute computing power as miners, using the proof-of-work mechanism to establish consensus, and jointly building a trustworthy blockchain data sharing network.

Service providers or third-party institutions participating in the management of the blockchain network can receive reward feedback, that is, the device grants them data access permissions. The MPR contract has a built-in mining reward function. Service providers attach reward queries to all transactions that update the MPR contract. If the block containing the record update transaction is mined, the mining reward function assigns the ownership of the reward query to the miner, who then sends a request to the corresponding service provider database administrator to receive the reward. By issuing Token tokens, the circulation of data and value in the network is realized. With the device's authorization, the authorized party has read and write permissions to the device data. For example, device nodes can earn Token rewards by uploading data; service provider nodes and other nodes can also earn Token rewards by verifying the uploaded device data; the Tokens earned by device nodes can be used to purchase maintenance services, etc.

Service providers include equipment manufacturers and maintenance service providers. Maintenance service providers can provide more comprehensive equipment health management plans, formulate more efficient and reasonable maintenance scheduling plans, and establish a complete evaluation and assessment standard system; for equipment manufacturers, analyzing equipment operation data and production data can help improve processes, accelerate the R&D cycle, bring about efficiency improvements throughout the lifecycle, and bring about business model changes from selling machines to selling services, personalized customization, and industrial chain finance.

Smart factories can master all the data of the factory's equipment through the API docking with the enterprise's internal information system and the equipment data sharing network, achieving online operation and monitoring; machine equipment networking can provide real-time data feedback and use the computing processing capabilities of the cloud to optimize equipment operation, adjust production scheduling in a timely manner, bringing efficiency improvements and cost reductions.

Blockchain and multi-party computing support the sharing of factory production data without leaking secrets, providing statistical research data for universities, research institutions, and other third-party institutions, helping to coordinate and plan production resources, plan capacity, and provide production plans and suggestions.

By creating a manufacturing equipment data sharing network, privacy information is retained in the account of the device node, and equipment operation data, maintenance records, and production data are all stored in the off-chain database. After anonymization, the data index is extracted and transmitted to the blockchain network, and other nodes use smart contracts to obtain relevant information and perform big data analysis and mining through artificial intelligence technology.

4.4 Case Study

Taking the application of intelligent diagnosis and equipment maintenance as an example, the BRSMMIoT and smart contracts are verified. Different sensors are deployed and installed on the equipment to monitor the temperature and vibration levels at different locations. A smart contract for equipment maintenance and spare parts is constructed, which acts as a protocol between the device and the service provider/supplier, scheduling service requests for the device or handling part replacement instructions. The pseudocode for the Machine Service and Part Replacement smart contracts is as follows (Table 4.6).

Table 4.6 Smart contract machine service and part replacement pseudocode

```
from ethjsonrpc import EthJsonRpc
import serial
ser = serial.Serial('/dev/ttyACM0', 38,400)
#Wait for some time to initialize serial port
time.sleep(10)
#Connect to Blockchain network
c = EthJsonRpc('192.168.1.20', 8101)
temCount = 0
vibrationCount = 0
tempThreshold = 200
vibrationThreshold = 100
tempCountThreshold = 1000
vibrationCountThreshold = 1000
while True:
line = serialTTL.readline()
data = line.split(',')
vibration = data[0]
temperature = data [1]
if(vibration > vibrationThreshold); vibrationCount = vibrationCount + 1
if(temperature > tempThreshold);
tempCount = tempCount + 1
if(vibrationCount > vibrationCountThreshold);
#Send a transaction to requestService function of MachineService contract with Machine-ID
c.call_with_transaction(c.eth.coinbase(),
'0xc421d5e214ddb07a41d28cf89ee37495aa5edba7',
'requestService(uint256)', [987])
vibrationCount = 0
if(tempCount > tempCountThreshold);
#Send a transaction to orderPart function of PartReplacement contract with Part-ID & Qty
c.call_with_transaction(c.eth_coinbase(),
'0 × 20935bc944dba7b4d230b236edb36f8f31be3d9e'
'orderPart(uint256, uint256)', [123, 1],
value = 50,000,000,000,000,000)
tempCount = 0
time.Sleep(1)
```

Controller services typically monitor vibration levels at different locations on the equipment. The controller service defines different rules to determine whether to execute service requests or part replacement instructions. For example, if the vibration level of the equipment exceeds a predetermined threshold a certain number of times, the controller service will send out a device service request, that is, send a transaction through the requestService function of the smart contract Machine Service (between the device and the service provider). Similarly, if the temperature of a part on the device exceeds a predetermined threshold a certain number of times, it triggers a spare part replacement instruction, and the device sends a transaction through the orderPart function of the smart contract Part Replacement (between the device and the supplier), and pays for the spare parts with cryptocurrency.

Deploy smart contracts, namely Machine Service and Part Replacement, through Ethereum (Ethereum) Go Client (geth). After deploying the smart contract on the blockchain network, it is assigned an address. In this case, a private Ethereum blockchain network is established, and any user in the network who knows the contract address and interface can send transactions through the contract. When the transaction is sent to the blockchain network, it is written into a block with other unfinished transactions. The miners in the network verify the validity of the transaction information to form a consensus in the block and link it to the main chain. Afterward, the new block is broadcast to the entire network. When the transaction is mined, the contract is executed.

References

1. LEE J, ARDAKANI H D, YANG S, et al. Industrial big data analytics and cyber-physical systems for future maintenance & service innovation. Procedia CIRP, 2015 (38): 3-7.
2. WANG K, DAI G, GUO L. Intelligent predictive maintenance (IPdM) for elevator service through CPS, IOT & S and data mining. International Workshop of Advanced Manufacturing and Automation (IWAMA 2016), 2016.
3. Li Nan, Yan Junwei, Liu Min. Industrial Internet of Things framework for steel continuous casting equipment maintenance and repair. Computer Integrated Manufacturing Systems, 2011, 17 (2): 413-418.
4. BRODY P, PURESWARAN V. IBM Internet of Things White Paper: Device Democracy, Decentralized, Autonomous Internet of Things. Translated by Shao Ping. http://www.8btc.com/device-democracy-saving-the-future-of-the-internet-of-things, 2014.
5. YAO A C. Protocols for secure computations. 23rd Annual Symposium on Foundations of Computer Science (sfcs 1982) (FOCS), 1982, 160–164.
6. GOLDREICH O, MICALI S, WIGDERSON A. How to play ANY mental game//Nineteenth ACM Symposium on Theory of Computing. ACM, 1987: 218–229.
7. BEN-DAVID A, NISAN N, PINKAS B. FairplayMP: a system for secure multi-party computation//ACM Conference on Computer and Communications Security, CCS 2008, Alexandria, Virginia, Usa, October. DBLP, 2008: 257–266.
8. SHAMIR A. How to Share a Secret. Communications of the Acm, 2011, 22 (22): 612-613.

Chapter 5
Wireless Routing Model and Algorithm for Complex Manufacturing Environment

The wireless communication network technology of IoT can provide high-bandwidth wireless data links and flexible network topology structures for the status monitoring data collection of on-site intelligent devices, mobile robots, and automated equipment in the factory workshop environment, which has advantages that wired technology cannot or is difficult to replace. However, the complex layout of the factory workshop, the existence of a large number of fixed and non-fixed metal isolation materials and materials and various obstacles, seriously affect the communication between low-power wireless sensors, have higher requirements for the Quality of Service (QoS) of the routing network, and need to consider QoS in the network routing protocol. This chapter introduces the QoS model of wireless sensor networks in view of the high requirements for network energy consumption, convergence speed, and global optimal solution in complex manufacturing application environments, combines the advantages of Quantum-inspired Evolutionary Algorithm (QEA) and Ant Colony Optimization (ACO) in network routing optimization, constructs a WSN routing algorithm based on Quantum-inspired Ant-Based Routing (QABR), and carries out simulation verification.

5.1 Network Routing Protocol

In recent years, various routing protocols based on different application goals have been proposed, and different applications have been classified and compared [1–4]. Wireless sensor network routing technology can be divided into four categories according to applications: data-centric routing protocols, hierarchical routing protocols, geography-based routing protocols, and quality of service-based routing protocols.

Data-centric routing protocols were first proposed, and this type of routing protocol reduces communication traffic through data fusion. In this type of routing

M. Liu et al., *Intelligent Predictive Maintenance*, Advanced and Intelligent Manufacturing in China, https://doi.org/10.1007/978-981-97-2677-6_5

protocol, the most classic protocols are Flooding and Gossiping [5], which are not only the earliest proposed, but also relatively simple and traditional. These two protocols do not require any algorithms and do not need to maintain routing information. Although simple, their scalability is relatively poor. The SPIN (Sensor Protocols for Information via Negotiation) protocol [6] was proposed by W. Heinzelman and others. This protocol communicates through three types of messages, thereby avoiding the problem of resource utilization blindness, but the reliability of SPIN is poor. The Directed Diffusion protocol [7]

Proposed by Intanagonwiwat and others, it is a query-driven, data-based protocol. Its advantage lies in using a query-driven mechanism to establish routing on demand, adopting multiple paths, strong robustness, and avoiding the storage of overall network information. However, it is not suitable for networks with multiple aggregation nodes, because the establishment of gradients occupies a large amount of overhead. Rumor Routing protocol [8] was proposed by Braginsky and others. It improved the Directed Diffusion protocol by introducing a unicast random forwarding mechanism for query messages, overcoming the defect of the Directed Diffusion protocol not being suitable for networks with multiple aggregation nodes. However, the Rumor Routing protocol consumes a lot of resources in busy situations; at the same time, due to its mechanism of randomly generating paths, it has poor ability to find the optimal solution, and in the worst case, routing loops may occur.

With the research of scholars at home and abroad in the field of wireless sensor networks, the hierarchical structure of wireless sensor networks has been proposed. The mechanism of this type of routing protocol is to enable nodes to participate in multi-hop communication within specific node clusters, and then the cluster heads perform data fusion, reducing the number of messages sent to the aggregation nodes, thereby improving scalability and saving energy. The LEACH (Low Energy Adaptive Clustering Hierarchy) protocol [9] was proposed by Heinzelman and others. It is the earliest proposed hierarchical routing protocol for data aggregation. This protocol avoids the problem of consuming too much energy by electing cluster heads randomly, thereby extending the network life cycle. However, its scalability is poor and it is not suitable for large-scale networks, and it has high requirements for nodes, such as strong power communication capabilities. The PEGASIS (Power-efficient Gathering in Sensor Information System) protocol [10] was proposed by Lindsey. This protocol improved the LEACH protocol. The PEGASIS protocol elects cluster heads without communication, thereby avoiding the communication overhead caused by frequent election of cluster heads. Although this protocol greatly extends the network life cycle, the single cluster head mechanism makes the cluster head a key point. If the cluster head fails, it will lead to routing failure. The TEEN (Threshold Sensitive Energy Efficient Sensor Network Protocol) protocol [11] was proposed by Manjeshwar. It monitors hotspot areas and sudden events through two thresholds, thereby reducing network communication traffic.

Due to the location relevance of wireless sensor networks, geographically based routing protocols have been proposed. This type of routing protocol uses the location information of nodes to forward data or query commands to the required areas instead of the entire network, reducing the data transmission area and thereby reducing energy consumption. At the same time, the mobility factor of nodes was considered during design. The GEAR (Geographic and Energy Aware Routing) protocol [12] was proposed by Yu and others. This protocol gathers divergent information in the network to appropriate location areas, thereby reducing the number of relay nodes, reducing the overhead of data transmission and path finding, and thereby extending the network life cycle.

With the development of the network, some new problems have also followed. The most critical issue is to ensure the quality of service (QoS) [13] of the network while meeting application requirements. For application requirements such as delay, bandwidth, and delay jitter, existing network services obviously cannot meet them. Therefore, quality of service-based routing protocols have been proposed. These protocols meet the requirements of communication service quality while providing data routing functions. The SAR (Sequential Assignment Routing) protocol [14] is the earliest proposed quality of service-based routing protocol, but because there is too much redundant routing information in the SAR protocol and it needs to maintain this information, it causes a large overhead. The SPEED protocol [15] was proposed by T. He and others. It guarantees end-to-end transmission rate, load balancing, and network congestion control based on real-time.

Now the routing protocol of wireless sensor networks is beginning to develop in the direction of intelligence. The recently proposed routing solution is based on swarm intelligence (Swarm Intelligence, SI) [16]. The original definition of swarm intelligence is: a group composed of many simple individuals achieves a certain function and completes a certain task through simple cooperation with each other. Intelligent swarms are applied to autonomous distributed systems, emphasizing bottom-up design, and they perform well in adaptability and robustness. Saleem summarized the basic rules of SI routing based on some existing SI algorithms, and proposed a new classification method for wireless sensor network routing protocols. He used this method to classify the algorithms he had studied [17]. Zhang developed three different ant colony algorithm-based wireless sensor network routing protocols [18], namely the SC (Sensor-drive and Cost-aware ant routing) ant colony routing algorithm, the FF (Flooded Forward) ant colony routing algorithm, and the FP (Flooded Piggybacked) ant colony routing algorithm. These algorithms have great advantages when initially established. However, these routing algorithms only focus on the establishment of initial pheromones, which makes these algorithms have some defects. The SC and FF ant colony routing algorithms are not very effective in solving latency. The FP ant colony routing algorithm consumes too much energy while achieving a high data delivery rate. Camilo proposed an energy-based ant colony routing algorithm (Energy-Efficient Ant-based Routing Algorithm, EEABR) [16]. The EEABR protocol can establish a routing tree, and this routing tree has the optimal energy branch. However, when ants choose the next hop node, they only consider the energy of the neighbor node, not the energy balance of the entire network. Load balancing is very important for wireless sensor networks, so the performance of the EEABR protocol is not very good.

In summary, the research on wireless sensor network routing protocols has gradually become a hotspot. The research on routing protocols has evolved from simple to complex, from data-centric to high-quality requirements, and is developing in the direction of intelligence. How to establish an efficient and adaptive sensor routing protocol has become a hotspot in current research.

5.2 QoS-Based Wireless Sensor Network Routing Model

5.2.1 QoS Measurement Parameters

The QoS measurement value of routing is calculated by quantifying various QoS indicators. These measurement values can be one or more parameters, such as delay, bandwidth, delay jitter, etc.

(1) Delay

Delay, also known as latency, refers to the transmission delay between two reference points. In wired networks, delay is mainly caused by congestion, while in wireless networks, it is not only due to congestion, but also due to some events that may cause data transmission delay, such as queue delay, propagation delay, node traffic competition, and queue entry and exit delay. Because it is impossible to get a guarantee from every node in the route, the actual end-to-end delay is difficult to calculate. A feasible solution is to prevent the establishment of routes that cannot meet the QoS delay requirements. QoS delay-supporting routes can be implemented by adding initial delay and maximum delay fields to the forward data packets. When searching for routes, each node needs to notify its neighboring nodes of the delay field situation. Through each node, due to the node's delay, the delay field needs to be reduced. If the delay field value becomes zero before reaching the destination, an error message needs to be sent to the source node. When it reaches the destination node, it should send information to update the routing table to the source node, indicating that this route is feasible. (2) Bandwidth Bandwidth is one of the important metrics of QoS routing protocols, specifically referring to the available bandwidth of the path from the source node to the destination node. However, the bandwidth will also change when the node moves. To simplify the routing algorithm that includes bandwidth metrics, the routing algorithm should be able to confirm the bandwidth of each path when the routing table and topology change. Many routing protocols are striving to find the route with the maximum bandwidth, but when there are several paths to choose from, the path with the smallest delay should be chosen at this time. This widest route problem can be solved by some shortest path algorithms. Bandwidth support needs to be represented in the routing table as the minimum available bandwidth and requested bandwidth guarantee. The minimum available bandwidth refers to the requested bandwidth of the specified route. If a node finds that its link capacity cannot meet the request, it will ignore this request. If the QoS request is feasible, then this request will be continuously broadcast to the destination and update the routing table of each node. When the minimum available bandwidth of a node is no longer satisfied, this route is also considered an invalid route. (3) Delay Jitter Delay jitter, also known as variable delay, refers to the time difference between data packets in a group of data streams sent on the same route. Applications that are usually affected by delay will also be affected by jitter. In some real-time systems that transmit audio and video, if jitter cannot be resolved, the transmission performance will not be ideal. (4) Packet Delivery Rate and Packet Loss Rate The packet delivery rate refers to the ratio of the data packets received by the target node to the data packets sent by the application layer of the source node, that is, the statistical measure of correctly transmitted data packets, mainly reflecting the reliability of the network, network congestion/communication conditions and other characteristics. The packet loss rate refers to the highest ratio of discarded data packets when transmitting data packets in the network. Packet loss is generally caused by network congestion. (5) Energy Efficiency In most cases, the nodes in the sensor network are powered by batteries. The limited battery capacity limits the survival time of the nodes. If the nodes in the network cannot work due to energy exhaustion, it will bring about changes in the network topology and the re-establishment of routes, and may even cause the network to be disconnected and interrupt communication. How to improve energy efficiency is an important issue in the design of wireless sensor networks. First of all, in terms of functions, since wireless sensor networks are mostly designed for a specific purpose, removing unnecessary functions can save energy and extend the survival time of nodes. Therefore, wireless sensor networks need to consider two design principles: extending the working time of the network, reducing unnecessary functions, and highlighting specificity. Secondly, special protocols and technologies can be designed to improve the energy efficiency of sensor networks, which involve all levels of the network, such as the physical layer can adopt ultra-wideband wireless communication technology, the MAC layer can adopt access protocols suitable for nodes to switch between sleep and work states, and the network layer can use node energy as a routing metric, etc. In addition, cross-layer design can also be used to improve the energy efficiency of the network. There are many QoS indicators, but how a QoS-supporting route supports various QoS to maximize the satisfaction of various QoS is an important issue in the network. Different network applications will have different QoS expressions, and based on this, different strengths of requirements will be put forward for various QoS, thereby reflecting the certain needs of users and applications for the network. Overall, the QoS in wireless sensor networks involves parameters such as delay and bandwidth, and should also include energy effectiveness, robustness of network self-organization, etc. It is not enough to study a single QoS indicator, and it may even affect some other QoS parameters. It is necessary to study several QoS parameters comprehensively, balance the role of various QoS parameters in wireless sensor networks and their impact on algorithms, and need to integrate various QoS indicators. The purpose of QoS-based routing is to find the optimal path in the network, requiring departure from the source node, arrival at the destination node, and satisfaction of all QoS constraints to achieve a specific service level. Here, the specific service level does not refer to a single indicator but a balance of various QoS indicators.

5.2.2 Network Routing Protocols and Models with QoS

Typically, a QoS routing protocol needs to complete the following functions. ① Determine the path. Determine the optimal path that can meet the QoS requirements among many paths. ② Resource reliability. Even without the ability to reserve network resources, maximize the use of network resources. Usually, some resource reservation mechanisms can be added in the upper layer to control QoS, such as the Resource Reservation Protocol (RSVP). ③ Path maintenance. Prevent the performance of the QoS-compliant route from suddenly dropping. QoS routing protocols will search for QoS-compliant paths with sufficient resources. The differences between a routing protocol with QoS and a routing protocol without QoS are as follows. ① A routing protocol with QoS can handle different types of services, while a routing protocol without QoS only supports one type of service at the same time. ② A routing protocol with QoS cannot change routes immediately, while a routing protocol without QoS can change immediately as soon as a better route is found. ③ When a calculated optimal route cannot be used, a routing protocol with QoS can find another alternative route, while a routing protocol without QoS cannot. A QoS-guaranteed routing protocol refers to considering QoS parameters such as delay, packet loss rate, and bandwidth when establishing a route, and choosing the most suitable route for QoS application requirements from many feasible routes. To facilitate the discussion of routing and QoS issues, first establish a model of a wireless sensor network. A wireless sensor network can be defined as an undirected graph G (V, E), where V It is a collection of sensor nodes with wireless connections, with n + 1 nodes in the network, $V = \{v_0, v_1, v_2, v_3, \ldots, v_\text{n}\}$. Each node has a wireless communication radius of r_i, a communication area of A_{vi}, and an edge $e = (v_i, v_j) \in E$ represents a bidirectional wireless connection between a pair of nodes (v_i, v_j). A path P (v_1, v_n) in G is an ordered combination sequence of edges, $P(v_1, v_n) = ((v_1, v_2), (v_2, v_3), \ldots, (v_{i-1}, v_i), \ldots (v_{n-1}, v_n))$. Where $v_i \in V, 2 \le n \le |V|$, so $P(v_1, v_n)$ is a multi-hop path. If $d(v_1, v_n)$ is defined as the number of edges in $P(v_1, v_n)$, it indicates the hop distance from the source node v_1 to the destination node v_n, also known as the hop distance. Each node in the path can be considered as an independent router, the first node of the path is the source node, and the last node is the destination node, which can be denoted as v_s and v_d respectively. Each node has its neighboring nodes, all edges $e = (v_i, v_j) \in E$ represent that nodes v_i and v_j are neighboring nodes. For a node v_i, the node set $N_{vi} = \{v_j | e = v_i, v_j \in E, i \ne j\}$ is the set of neighboring nodes of node v_i. It should be noted that: N_{vi} is discovered by the neighbor discovery mechanism, also known as HELLO information exchange. When the HELLO information is exchanged, the node includes the relevant QoS information of the node in the HELLO information. For any path $P(v_s, v_d)$, a comprehensive QoS indicator can be defined. Here we consider delay, bandwidth, and packet loss rate. The QoS of path $P(v_s, v_d)$ can be reflected on each node and each link. For any network node $v \in V$, respectively for the delay function *delay* (v), bandwidth function *bandwidth* (v), packet loss rate function *packet loss* (v), delay jitter function *delayjitter* (v) and cost function *cost* (

v). Correspondingly, for any link $e = (v_i, v_j) \in E$ in the network, there are also corresponding QoS indicators, respectively for the delay function *delay* (e), bandwidth function *bandwidth* (e), packet loss rate function *packet loss* (e), delay jitter function *delayjitter* (e) and cost function *cost* (e).

By defining the QoS indicators of nodes and links, the QoS indicators of the path can be calculated. For a given source node $v_s \in V$, destination node $v_d \in V$, a path $P(v_s, v_d)$ corresponding QoS indicators have the following relationship:

$$delay(P(v_s,v_d)) = \sum_{v\in P(v_s,v_d)} delay(v) + \sum_{e\in P(v_s,v_d)} delay(e) \tag{5.1}$$

$$bandwidth(P(v_s, v_d)) = \min_{e\in P(v_s v_d)}\{bandwidth(e)\} \tag{5.2}$$

$$packet\ \ loss(P(v_s, v_d)) = 1 - \prod_{e\in P(v_s,v_d)} (1 - packet\ \ loss(e)) \tag{5.3}$$

$$delayjitter(P(v_s, v_d)) = \min_{e\in P(v_s,v_d)}\{delayjitter(e)\} \tag{5.4}$$

$$cost(P(v_s,v_d)) = \sum_{v\in P(v_s,v_d)} cost(v) + \sum_{e\in P(v_s,v_d)} Cost(e) \tag{5.5}$$

If the path $P(v_s, v_d)$ is a path that satisfies all QoS indicators, it must meet the following conditions:

$$delay(P(v_s,v_d)) = \sum_{v\in P(v_s,v_d)} delay(v) + \sum_{e\in P(v_s,v_d)} delay(e) < D \tag{5.6}$$

$$bandwidth(P(v_s, v_d)) = \min_{e\in P(v_s,v_d)}\{bandwidth(e)\} > B \tag{5.7}$$

$$packet\ \ loss(P(v_s,v_d)) = 1 - \prod_{e\in P(v_s,v_d)} (1 - packet\ \ loss(e)) < PL \tag{5.8}$$

$$delayjitter(P(v_s, v_d)) = \min_{e\in P(v_s,v_d)}\{delayjitter(e)\} < DJ \tag{5.9}$$

where, D, B, PL and DJ are delay constraints, bandwidth constraints, packet loss constraints, and delay jitter constraints, respectively.

After defining the QoS functions of the wireless sensor network routing model, a comprehensive QoS index applied to each path is established. The premise of this index calculation is that each QoS index meets its QoS constraint requirements. If one does not meet, the contribution of the index to the comprehensive QoS will be greatly reduced, and it will have a negative and punitive impact on the comprehensive QoS.

The problem solved by the QPSO algorithm is: to find a path that satisfies the constraints of bandwidth, delay, delay jitter, and packet loss rate, and makes the path cost as low as possible. When designing the comprehensive QoS function, delay, delay jitter, packet loss rate, cost, etc. are assigned their own weight coefficients, respectively representing the weight of each constraint variable in the function.

For example, for the delay index, if *delay* $(P(v_s, v_d) < D$, that is, the delay of the path satisfies the constraint condition, then

$$f_{delay} = 1 - \frac{(1-k)delay(P(v_s,v_d))}{D} \tag{5.10}$$

In the formula, k is the weight coefficient, close to 1, such as 0.9, then the value of f_{delay} will be between 0.9 and 1. If $delay(P(v_s, v_d)) > D$, it indicates that the delay index of the path cannot meet the application's delay constraint requirements, then

$$f_{delay} = (1-k) \div \frac{delay(P(v_s,v_d))}{D} \tag{5.11}$$

In this way, the value of f_{delay} will be very small, between 0 and 0.1. The same treatment is applied to the indicators bandwidth, delay jitter, and packet loss rate. The final comprehensive QoS function is the comprehensive product of the four QoS indicator functions.

$$QoS = f_{delay} f_{bandwidth} f_{delayjitter} f_{packetloss} \tag{5.12}$$

As can be seen from the above formula, the comprehensive QoS is between 0 and 1. Under the premise of satisfying QoS, the path with small bandwidth, delay, delay jitter, and packet loss rate, its value is closer to 1.

5.3 Routing Optimization Algorithm Based on Quantum Ant Colony Algorithm

5.3.1 Optimization Mechanism of Ant Colony Algorithm and Quantum Evolution Algorithm

(1) Ant Colony Algorithm

The ant colony algorithm is an important branch of swarm intelligence algorithms. This concept first appeared in the doctoral thesis of Italian scholar Dorigo [19], used to solve the famous Traveling Salesman Problem (TSP). As the ant colony algorithm is continuously improved, it has now been successfully applied to solve many combinatorial optimization problems. Ants release pheromones while looking for food. Subsequent ants can detect these pheromones during the process of finding a path, and can perceive the concentration of pheromones. These ants will choose the path with a higher concentration of pheromones as their own path. Gradually, some of the pheromones will be naturally evaporated. When the concentration of pheromones on this path is small, the probability of subsequent ants choosing this path will decrease. Gradually, the concentration of pheromones gathered on this path will become less and less. On the contrary, if the concentration of pheromones gathered on a certain path is large, more ants will choose this path, causing the concentration of pheromones gathered on this path to become larger and larger, which is the process of pheromone positive feedback. With the help of pheromone positive feedback, ants can walk from the ant nest to the food source in the shortest path.

In addition, bionics scholars have found that not only can ants search for the shortest path from the ant nest to the food, but also if obstacles are set on this path, making it impossible for ants to pass, ants can still adapt to this change at the fastest speed and re-establish a shortest path. Figure 5.1 is the flowchart of the ant colony optimization algorithm.

Since successfully solving the famous TSP problem, the ant colony algorithm has gradually been applied to other fields, including sensor network routing protocols. The basic steps of the ant colony algorithm are as follows.

① Every once in a while, each sensor node will generate a forward ant k, whose task is to find the path to the destination node. Ant k will save the information of each visited node in memory M_k.

② For each node r, the forward ant will choose the next hop node according to the probability formula in ACO, as shown in Eq. (5.13).

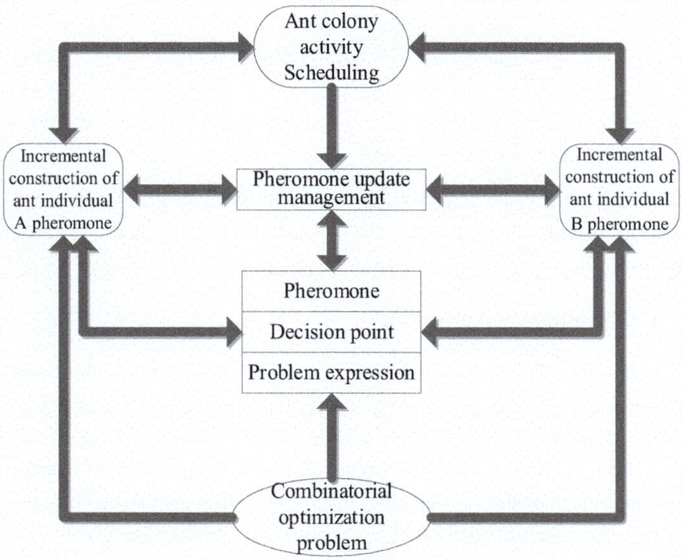

Fig. 5.1 Ant colony optimization algorithm flowchart

$$P_k(r,s) = \begin{cases} \frac{[T(r,s)]^\alpha [E(s)]^\beta}{\Sigma [T(r,s)]^\alpha [E(s)]^\beta}, & s \not\subset M_k \\ 0, & others \end{cases} \tag{5.13}$$

In the formula, P_k (r, s) is the probability of ant k moving from node r to node s; T is the routing table existing in each node, which stores the pheromone concentration on the (r, s) link; E is the energy function of $1/(C - e_s)$ (where C is the initial energy of node s, and e_s is the actual energy of node s); α and β are the weight parameters of pheromone and energy respectively. The meaning of pheromone is that the more pheromone on the (r, s) path, the greater the probability of this path being chosen; the meaning of energy is that the more energy a node has, the greater the probability of being chosen. α and β can balance these two factors well.

③ When the current ant reaches the destination node, it will transform into a backward ant. The task of the backward ant is to update the pheromone on the path it has visited.

④ Before the backward ant departs, the destination node will calculate the amount of pheromone to be released. The calculation formula is shown in Eq. (5.14).

$$\Delta T_k = 1/(N - Fd_k) \tag{5.14}$$

In the equation, N is the number of nodes; Fd_k is the distance of ant k's journey, which is the number of nodes stored in its memory.

⑤ When node r receives the backward ant sent by its neighbor node, it immediately updates its routing table according to Eq. (5.15).

$$T_k(r,s) = (1-\rho)T_k(r,s) + \Delta T_k \tag{5.15}$$

In the equation, ρ is a coefficient, $1 - \rho$ represents the evaporation amount of pheromone $T_k(r, s)$ after the last update.

⑥ When the backward node reaches its source node, each node will know which is the best neighbor node, that is, it will send data packets to the destination node through the path given by the optimal solution.

(2) Quantum Evolutionary Algorithm

Kuk-Hyuan Han and others proposed the Quantum Evolutionary Algorithm (QEA) [20], which is an evolutionary algorithm based on quantum computing. It absorbs the coherence, superposition state, and entanglement mechanisms in quantum computing [21], thus making the quantum algorithm break through the limit of traditional algorithms and show better performance [22]. Currently, this algorithm has been widely used in various optimization problems, especially in solving NP (Non-deterministic Polynomial) hard problems in classical computing [23].

① The basic unit of quantum computing—The storage unit of a classical computer is a bit, which has only two states, 0 or 1. The most basic storage unit of a quantum computer is a Q-bit, which is any quantum system with a two-dimensional Hilbert space. Its state space has two bases, denoted as $|0\rangle$ and $|1\rangle$, where "|" is a representation of a quantum state. Unlike the bits in a classical computer, the state of a Q-bit can be any normalized state: | where "|" is a representation of a quantum state. Unlike the bits in a classical computer, the state of a Q-bit can be any normalized state: $\alpha|0\rangle + \beta|1\rangle$, where α and β are any complex numbers that satisfy the normalization condition $|\alpha|^2 + |\beta|^2 = 1$. Therefore, a Q-bit contains more information than a bit in a classical computer. The states of each Q-bit can be independent or entangled. To illustrate the entanglement property between Q-bits, let's take two Q-bits as an example (distinguished by subscripts A and B). If the total state of the two Q-bits $|\phi_{\mathrm{AB}}\rangle$ can be written as the direct product of the state of A-bit $|\phi_{\mathrm{A}}\rangle$ and the state of B-bit $|\phi_{\mathrm{B}}\rangle$, that is, $|\phi_{\mathrm{AB}}\rangle = |\phi_{\mathrm{A}}\rangle \otimes |\phi_{\mathrm{B}}\rangle$, then the state $|\phi_{\mathrm{AB}}\rangle$ is called a separable state, at which time the states of A-bit and B-bit are independent. On the contrary, if $|\phi_{\mathrm{AB}}\rangle$ cannot be written as the direct product of the states of A and B, A and B are said to be in an entangled state $|\phi_{\mathrm{AB}}\rangle$, such as the so-called EPR state, as shown in Eq. (5.16).

$$|\phi_{\mathrm{AB}}\rangle = \frac{1}{2}(|0\rangle_A \otimes |1\rangle_B - |0\rangle_A \otimes |1\rangle_B) \tag{5.16}$$

② Parallelism in Quantum Computing Quantum computing is relative to classical algorithms, and its most essential feature is the full utilization of the superposition and coherence of quantum states, as well as the entanglement between Q-bits. It is a product of quantum mechanics directly entering algorithm theory. The main difference between quantum algorithms and classical algorithms is that they have quantum parallelism.

The parallelism of quantum computing means that it can not only operate $f(0)$ and $f(1)$, but also operate on any superposition state of $|0\rangle$ and $|1\rangle$, and then extract the overall information of function f, that is, it depends on $f(0)$ and depends on $f(1)$ information. When the function f is defined on N bits, that is, the domain of f is $\{0, 1, \ldots, 2^N - 1\}$, but what is desired is only a holistic property of f, such as whether f is a periodic function, what is the period, etc. When you want to obtain the overall properties of f, the classical algorithm needs to run 2^N times, calculate $f(x)$ ($x = 0, 1, \ldots, 2^N - 1$) separately, and then make judgments based on these function values. Using quantum algorithms, making full use of their parallelism, can quickly give results.

5.3.2 *Quantum Ant Colony Algorithm*

The recently proposed quantum ant colony algorithm is a probabilistic optimization method based on the principles of quantum computing. It is based on the ant colony algorithm, some concepts and theories of quantum computing, and uses quantum bit encoding to represent pheromones and quantum gate rotation to complete pheromone updates. Its advantages are the ability to choose the currently optimal path and explore new paths, the size of the population can be adjusted without affecting the performance of the algorithm, and the convergence speed is relatively fast [24].

(1) Quantum encoding represents pheromones

The quantum evolutionary algorithm uses the representation method of Q-bits, which is a probabilistic representation method. In the quantum evolutionary algorithm, the Q-bit is the smallest computing unit, which is composed of a pair of numbers (α, β), where $|\alpha|^2 + |\beta|^2 = 1$, $|\alpha|^2$ represents the probability of the Q-bit being in the "0" state, and $|\beta|^2$ represents the probability of the Q-bit being in the "1" state. A Q-bit can be in the "0" state, the "1" state, or a superposition state between the two. Therefore, a Q-bit of length m can have 2^m different states. $\xi(\xi \subset (-\pi, \pi])$ represents the phase of the Q-bit, so the phase of the ith Q-bit is $\xi_i = \arctan(\beta_i/\alpha_i)$. Figure 5.2 represents the position of ξ_i in the coordinates. This method can greatly reduce the time and space used to find the optimal solution.

In a Q-bit of length m, the representation of the jth individual is formula (5.17):

$$P_j = \begin{bmatrix} \alpha_1 \ \alpha_2 \ \ldots \ \alpha_m \\ \beta_1 \ \beta_2 \ \ldots \ \beta_m \end{bmatrix} \tag{5.17}$$

In the formula, $|\alpha_i|^2 + |\beta_i|^2 = 1, i = 1, 2, \ldots, m$. In quantum algorithms, Q-bits are used to represent pheromones, assuming the population size is n, its pheromones are represented by quantum bits as $p = (p_1, p_2, \ldots, p_n)$.

(2) Rotation gate updates pheromones

Q-bit individuals are updated by the rotation gate, which is an update operation in quantum evolution theory. It satisfies the normalization condition $|\alpha|^2 + |\beta|^2 = 1$.

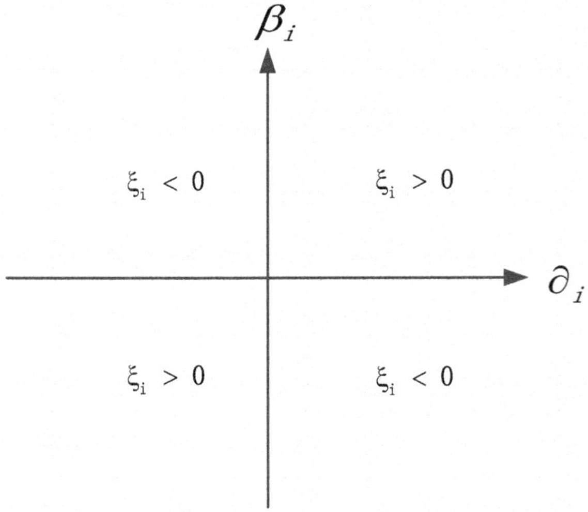

Fig. 5.2 Position of ξ_i

Formula (5.18) is the basic rotation gate in quantum evolution theory:

$$\begin{bmatrix} \alpha_i' \\ \beta_i' \end{bmatrix} = \begin{bmatrix} \cos(\theta_i) & -\sin(\theta_i) \\ \sin(\theta_i) & \cos(\theta_i) \end{bmatrix} \begin{bmatrix} \alpha_i \\ \beta_i \end{bmatrix} \tag{5.18}$$

In the formula, $i = 1, 2, \ldots,$ m; $[\alpha_i, \beta_i]^T$ is the i is the probability amplitude of the i-th quantum bit; θ_i is the rotation angle of the i-th quantum bit, the size and direction of which are obtained by looking up a pre-determined adjustment strategy. Here, the size and direction of θ_i are automatically adjusted by Eq. (5.19).

$$\theta_i = \Delta\theta f(\alpha_i, \beta_i) \tag{5.19}$$

In the equation, $\Delta = 5\exp(-t/t_{\max})$, is a variable related to the number of iterations; $\Delta\theta$ controls the size of the rotation angle θ_i; t is the number of iterations; $t_{\max}$ is a constant defined according to the specific problem, representing the maximum number of iterations; $f(\alpha_i, \beta_i)$ is a function related to α_i and β_i, it can adjust the rotation direction of θ_i, so that the algorithm can find the direction of a better solution, as shown in Eq. (5.20).

$$f(\alpha_i, \beta_i) = (d_{i\text{best}}/d_{i\text{now}})(\xi_{i\text{best}} - \xi_{i\text{now}}) \tag{5.20}$$

In the equation, $d_{i\text{now}} = \beta_{i\text{now}}/\alpha_{i\text{now}}$, $d_{i\text{now}}$ is the current solution of the i-th quantum bit, $\alpha_{i\text{now}}$ and $\beta_{i\text{now}}$ are the probability amplitudes of the current solution of the i-th quantum bit; $d_{i\text{best}} = \beta_{i\text{best}}/\alpha_{i\text{best}}$, $d_{i\text{best}}$ is the best solution found for the i-th quantum bit, $\alpha_{i\text{best}}$ and $\beta_{i\text{best}}$ are the probability amplitudes of the best solution found for the i-th quantum bit.

$$\xi_{i\text{best}} = \arctan(\beta_{i\text{best}}/\alpha_{i\text{best}}) \tag{5.21}$$

$$\xi_{i\text{now}} = \arctan(\beta_{i\text{now}}/\alpha_{i\text{now}}) \tag{5.22}$$

In the equation, $\xi_{i\text{best}}$ is the phase of the best solution of the i-th quantum bit; $\xi_{i\text{now}}$ is the phase of the current solution of the i-th quantum bit. When $f(\alpha_i, \beta_i) < 0$, θ_i rotates in the clockwise direction. When $f(\alpha_i, \beta_i) > 0$, θ_i rotates counterclockwise. Equation (5.23) describes the update operation of the quantum rotation gate.

$$P_j^{t+1} = G(t)P_j^t \tag{5.23}$$

In the equation, t is the number of iterations, $G(t)$ is the quantum rotation gate of the tth iteration, P_j^t is the individual probability amplitude of the tth iteration, P_j^{t+1} is the individual probability amplitude of the $t + 1$th iteration. The expressions of P_j^t and P_j^{t+1} are as shown in Eq. (5.17).

(3) Metrics of the routing algorithm

Metrics play a very important role in routing protocols and algorithms. The algorithm usually determines the performance of the optimal path. Therefore, it is necessary to consider carefully when choosing a routing algorithm. In general, the following metrics need to be considered comprehensively [25].

① Simplicity: The algorithm is concise, uses the least overhead, and provides the most effective functionality.
② Optimization: The routing algorithm should have the ability to calculate the global optimal path.
③ Fast convergence: Convergence refers to all routers unanimously considering a certain path as the best path. When some nodes join the network or some nodes fail, the router sends out update information. The routing update information spreads throughout the network, thereby recalculating the best path, and finally reaching the best path unanimously recognized by all routers.
④ Flexibility: The routing algorithm can quickly and accurately adapt to various network environments. For example, when a certain network segment experiences a sudden situation, the routing algorithm should be able to discover the sudden situation at the fastest speed and choose another optimal path for all routes using that network segment.
⑤ Robustness: The routing algorithm can run correctly when it is in a sudden or unpredictable situation, such as high load, hardware failure, or operation error. Since routers are distributed at network connection points, if a router fails, it will have serious consequences. The best router algorithms often need to run continuously for a long time and can work normally under various network environments.

A single metric cannot fully describe network performance, so multiple types of metrics are used by routing algorithms to determine the optimal path. Multiple metrics are used by complex routing algorithms. Through standardized weighted operations, they are integrated into a single composite metric, which is finally used as the standard for searching paths and filled into the routing table [26]. The commonly used metrics include: reliability, path length, bandwidth, delay, communication cost, and load, etc.

(4) Multi-objective routing algorithm design strategy

Many problems in the real world are usually composed of multiple conflicting objectives. When optimizing one of the objectives, it may cause a decline in the performance of other objectives. The multi-objective optimization problem [27] is quite different from traditional single-objective optimization. It is not to find a single optimal solution, but to search for a set of non-inferior solutions. This set of non-inferior solutions can provide support for the final decision. For multi-objective problems, it is common to find that a solution obtained is better for one objective but may be worse for other objectives. Therefore, there exists a set of compromise solutions, called the Pareto optimal solution set (Pareto-optimal Set) or non-dominated solution set (Non-dominated Set). The main goal of all these multi-objective optimization algorithms is to make the non-inferior solution set obtained by the algorithm as close as possible to the Pareto front, and the distribution of non-inferior solutions is as wide and uniform as possible [28].

Currently, most routing protocols mainly consider energy consumption factors, with a single research objective, lacking a multi-objective examination and evaluation of the overall network performance. In fact, when evaluating wireless sensor network routing algorithms, more than one objective needs to be considered. In addition to energy saving, the network's delay, transmission efficiency, robustness, and other performance indicators should also be considered.

The basic ant colony algorithm was first used to solve the TSP problem. The TSP problem has a single objective characteristic, which is to find the shortest path [29, 30]. However, wireless sensor networks are not that simple. They need to optimize multiple objectives, including network delay, energy consumption, reliability, and transmission efficiency. Therefore, this algorithm uses a multi-objective evaluation function to comprehensively evaluate the routing tree established during the ant's search for the optimal solution. The process of the ant constantly moving is the formation process of the multi-objective routing tree. Any ant starts from the source node and reaches the gradually forming routing tree, completing a trip. After a routing tree is formed, data will be transmitted from the source node along this tree to the Sink (convergence) node. Equation (5.24) gives the evaluation function of the routing tree, which includes energy consumption factors, delay factors, and routing tree load balancing factors.

$$f(t) = \frac{C_0}{[Q_1(t)]^{C_1}[Q_2(t)]^{C_2}[Q_3(t)]^{C_3}[Q_4(t)]^{C_4}} \tag{5.24}$$

$Q_1(t)$ is the energy consumption factor in the evaluation function, as shown in Eq. (5.25). C_1 is the weight of the energy consumption factor.

$$Q_1(t) = \sum_{rs} Kd_{rs}^{n}(r,s) \in Tree(t) \tag{5.25}$$

In Eq. (5.18), d_{rs} for the length of the path (r, s); $2 < n < 4$, usually n is close to 4.

$Q_2(t)$ is the delay factor in the evaluation function, represented by the longest path from the source node to the Sink node, as shown in Eq. (5.26). C_2 is the weight of the delay factor.

$$Q_2(t) = \max F_{dk}(t) k = k_1, k_2, \ldots k_n \tag{5.26}$$

In Eq. (5.26), F_{dk} is the length of one trip of ant k. In the wireless sensor network of multi-level synchronous data fusion, the delay increases with the depth of the tree, so Eq. (5.26) can reflect the delay performance of the network well.

$Q_3(t)$ and $Q_4(t)$ combine to form the load balancing factor of the evaluation function's routing tree, represented by the mean and standard deviation of the load, as shown in Eqs. (5.27) and (5.28). C_3 and C_4 are the weights of the load balancing factor.

$$Q_3(t) = E_r(t)r = r_1, r_2, \ldots, r_n \quad (5.27)$$

$$Q_4(t) = \sigma_r(t)r = r_1, r_2, \ldots, r_n \quad (5.28)$$

For any node r, search all its child nodes until all its leaf nodes are reached, the number of all leaf nodes (i.e., the number of source nodes) is the load of r. On the routing tree, the load of the Sink node is the largest, which is the total number of source nodes; the load of the source node is the smallest, which is 0. Here, the mean and standard deviation of node load are used to measure the load balancing degree of the routing tree.

5.3.3 *Quantum Ant Colony Multi-objective Routing Algorithm*

(1) Routing establishment preparation phase

① The aggregation node broadcasts the initialization signal, so that all nodes get the information of their neighbor nodes, and add the neighbor nodes to their own routing table, and set the pheromone of the neighbor node link to 1. Each node generates a forward ant packet of the current node, which includes the source node number, the passed node number, and the aggregation node number. That is, initialize the population $P(t) = (P_1^t, P_2^t, \ldots, P_k^t)$, where $P_j^t(j = 1, 2, \ldots, k)$ is the jth individual in the tth iteration, as shown in Eq. (5.29).

$$P_j^t = \begin{bmatrix} \alpha_1^t \; \alpha_2^t \ldots \alpha_m^t \\ \beta_1^t \; \beta_2^t \ldots \beta_m^t \end{bmatrix} \quad (5.29)$$

where, m is the number of qubits. Initialize all α_i and β_i to $1/\sqrt{2}$. Set the maximum number of iterations to $t_{\max}$. Initialize the current iteration number t to 0.

② Perform binarization, according to the values of the probability amplitude in $P(t)$, construct $R(t) = (r_1^t, r_2^t, \ldots, r_k^t)$, where $r_j^t(j = 1, 2, \ldots, k)$ is a binary string of length m. In $P_j^t(j = 1, 2, \ldots, k)$, each element is composed of $\left|\alpha_j^t\right|^2$ and $\left|\beta_j^t\right|^2$ Decision. Set a random number w ($w \subset [0, 1]$). If $\left|\alpha_j^t\right|^2 > w$, let $r_i^t = 0$. Otherwise, let $r_i^t = 1$ ($i = 1, 2, \ldots, m$). (2) Route establishment phase On each source node, place k ants. Each ant repeatedly applies the state transition rule (pseudo-random probability transition rule), establishes a path according to formula (5.30), and releases pheromones using the method shown in formula (5.14) every time it takes a step, until all ants have completed the construction of the solution path.

$$b = \begin{cases} \text{argmax}_{s \in U}\{\ [\tau(a,s)][\eta(a,s)]^{\gamma}\ \}\ ,\ q \le q_0 \\ b', \qquad\qquad\qquad\qquad \text{other} \end{cases} \tag{5.30}$$

Formula (5.30) determines the transition of ant k from node r to node s. $T(a, b)$ is determined by α and β, it is the amount of pheromone on the (a, b) path. E is the energy function of $1/(C - e_b)$ (C is the initial energy of node b, e_b is the actual energy of node b), α and β are the weight parameters of pheromone and energy respectively. The meaning of pheromone is that the more pheromone on the (a, b) path, the greater the probability of this path being selected; the meaning of energy is that the more energy a node has, the greater the probability of being selected. α and β can balance these two factors well. U is a set that contains all nodes that can be used as the next hop. b' It is the transition probability of an ant from node a to node b, represented by Eq. (5.26). q is a uniformly distributed random number between 0 and 1, which can determine the probability of the next hop. The size of the parameter q_0 determines which is more important for the ant to use prior knowledge or to explore new paths: whenever an ant located at node a chooses the next hop to reach node b, it randomly selects a number q, if $q \le q_0$, the next hop node selects the best link according to Eq. (5.23). Otherwise, a link is selected according to the probability shown in Eq. (5.31). By adjusting the size of the parameter q_0, the algorithm's exploration of new paths can be adjusted, thus determining whether the algorithm should focus on searching in the area near the best path so far, or should explore unexplored areas. That is to say, the probability of the ant choosing the current best path is q_0. This optimal movement method is determined by the accumulation of pheromones and heuristic information values. At the same time, the ant explores each path with a probability of $1 - q_0$ in a biased manner.

$$P_k(a, b) = \begin{cases} \dfrac{[T(a,b)]^{\alpha}[E(b)]^{\beta}}{\sum\limits_{u \not\subset M_k} [T(a,b)]^{\alpha}[E(b)]^{\beta}},\ b \not\subset M_k \\ 0, \qquad\qquad\qquad \text{others} \end{cases}, \tag{5.31}$$

(3) Route optimization phase

When the aggregation node receives the forward ant packet, the aggregation node will calculate the optimal solution. According to the routing tree evaluation strategy, calculate the evaluation function, that is, the optimal solution of Eq. (5.24), the calculation method is given by Eq. (5.32).

$$f_{\text{best}} = \max f(n)(n = 0, 1, \ldots, t) \tag{5.32}$$

If the current number of iterations is less than $t_{\max}$, the Sink node generates the backward ant packet for each node based on its forward ant packet. When a node receives its own backward ant packet, it updates the pheromone on the current optimal solution path.

5.3.4 Performance Analysis

The symbol definition for the proof of convergence performance is as follows: T_k(r, s) is the pheromone value released by the kth ant on the path (r, s), m is the total number of ants, P_k(r, s) is the probability of ant k moving from node r to visit node s, P_k^* is the optimal value of the probability of ant k choosing the next visit node, and P' is the feasible probability of the next visit node in path selection.

First, we prove that in the process of ants finding paths, the pheromone on each link cannot exceed a maximum value T_{max}, as follows.

For edge (r, s), let $T_0(r, s) = T_0$,

when k = 1, that is, the first ant, we have $T_1(r, s) = (1-\rho)T_0(r, s) + \rho\Delta T(r, s)$,

It can be seen that,

$$\begin{aligned} T_2(r, s) &= (1-\rho)T_1(r, s) + \rho\Delta T(r, s) \\ &= (1-\rho)\left[(1-\rho)T_0 + \rho\Delta T(r, s)\right] + \rho\Delta T(r, s) \\ &= (1-\rho)^2 T_0 + \rho(1-\rho)\Delta T(r, s) + \rho\Delta T(r, s) \end{aligned}$$

When it cycles to the kth ant, we have

$$\begin{aligned} T_k(r,s) &= (1-\rho)^k T_0 + \rho(1-\rho)^{k-1}\Delta T(r,s) + \cdots + \rho(1-\rho)\Delta T(r,s) + \rho\Delta T(r,s) \\ &= \sum_{n=1}^{k} \rho(1-\rho)^{n-1}\Delta T(r,s) + (1-\rho)^k T_0 \end{aligned}$$

When k tends to infinity, we have

$$\begin{aligned} \lim_{k\to\infty} T_k(r,s) &= \lim_{k\to\infty}\left[\sum_{n=1}^{k} \rho(1-\rho)^{n-1}\Delta T(r,s) + (1-\rho)^k T_0\right] \\ &= \Delta T(r,s) \le T_{\max} \end{aligned}$$

Secondly, prove that there exists any ε, when $k \to \infty$, there is $P_k^* \ge 1-\varepsilon$, and converges to $\lim_{k\to\infty} P_k^* = 1$, the proof is as follows.

Suppose the global minimum of the pheromone concentration $T_k(r, s)$ is $T_{\min}$, the lower bound of the feasible solution probability is $P_{\min}$, there is $T_{\min} \le T_k(r, s) \le T_{\max}$, according to formula (5.19), the lower bound of $P_{\min}$ is given as

$$P_{\min} \ge \frac{T_{\min}^{\alpha}}{(m-1)T_{\max}^{\alpha} + T_{\min}^{\alpha}} \ge \frac{T_{\min}^{\alpha}}{mT_{\max}^{\alpha}} = P'_{\min}$$

The lower bound of P_k^* is $P_k^* = 1-(1-P')^k$.

Therefore, it can be known that $P_k^* \ge 1 - \left\{1 - \left(\frac{T_{\min}^{\alpha}}{mT_{\max}^{\alpha}}\right)^m\right\}^k$.

Therefore, $P_k^* \ge P'_k \ge 1-\varepsilon$, when $k \to \infty$, $\lim_{k\to\infty} P_k^* = 1$.

From the above two proofs, it can be seen that the QABR algorithm has convergence.

5.4 Experiments and Result Analysis

MATLAB simulation platform is used for the simulation and analysis of the experiment.

In order to verify the performance of the QABR algorithm, this algorithm was compared with the traditional ant colony algorithm. In order to study the impact of parameters such as C_1, C_2, C_3 on the performance of the algorithm, the scenarios of different values of C_1, C_2, C_3 were first simulated. To reduce randomness, each experiment was conducted 50 times, and the final experimental results were averaged, as shown in Table 5.1.

(1) Experiment One

In this experiment, the number of network nodes is fixed, with the aim of verifying the convergence of the QABR algorithm. Under the same parameters, the operation of the QABR algorithm and the traditional ant colony algorithm are observed, the network topology is observed, and the optimal paths obtained by the two algorithms are displayed. At the same time, a curve graph of the convergence of the two algorithms with the increase of iterations is drawn. Within a range of 1000 m × 1000 m, 50 nodes are randomly distributed. The maximum number of iterations is 400, and the number of ants is 40. The simulation parameter list is shown in Table 5.2.

The network topology can be randomly generated by MATLAB. Figure 5.3 is the WSN network topology structure randomly generated according to the k-means clustering algorithm. The lines represent communication between nodes, which can

Table 5.1 Simulation parameter list

C_1	0.5
C_2	0.1
C_3	0.1
q_0	0.8

Table 5.2 Parameter list for experiment one

Number of network nodes	50
Network range	1000 m × 1000 m
C_1	0.5
C_2	0.1
C_3	0.1
q_0	0.8
$t_{\max}$	400

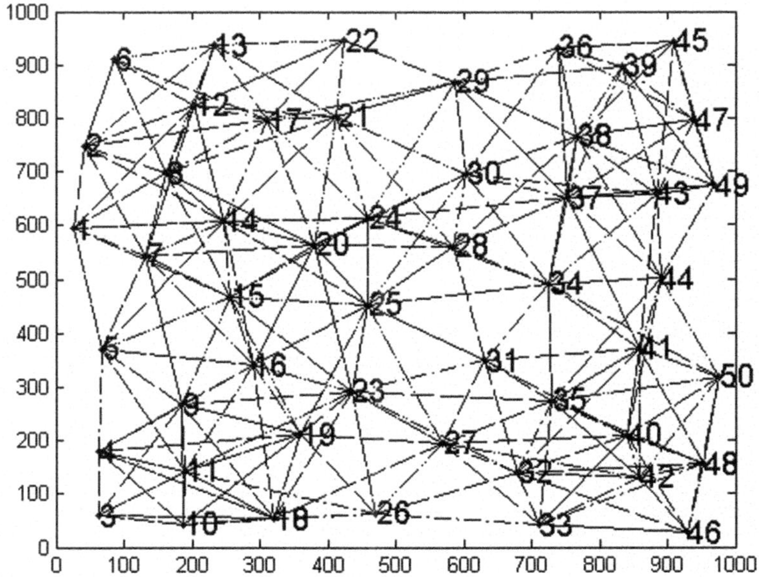

Fig. 5.3 Network topology (electronic version)

effectively simulate the changing monitoring environment and provide support for simulation experiments. Figure 5.4 marks the optimal paths obtained by QABR and the traditional ant colony algorithm in the network topology of Fig. 5.3. Node 1 is the starting point, node 50 is the end point, the red line is the QARB path, and the blue line is the traditional ant colony algorithm path.

Figure 5.5 is the convergence curve of the optimal solutions of the QABR algorithm and the traditional ant colony algorithm with the number of iterations. It can be seen that the optimal values of the two algorithms in each iteration process. From the figure, it can be seen that QABR converges at the 194th iteration, while the traditional ant colony algorithm only converges at the 348th iteration. It can be proved that the convergence speed of QABR is greater than that of the traditional ant colony algorithm. In addition, it can be seen from the figure that the traditional ant colony algorithm has a premature phenomenon, that is, it has converged to a relative non-optimal solution before finding the optimal solution. Although QABR is built on the traditional ant colony algorithm, it overcomes this defect very well. In summary, the QABR algorithm has a fast convergence speed and a high optimal value.

(2) Experiment Two

In this experiment, the optimal solutions of QABR and the traditional ant colony algorithm are compared according to the change in the number of nodes. Under the same parameters, the operation of the QABR algorithm and the traditional ant colony algorithm are observed, the network topology is observed, and the optimal paths obtained by the two algorithms are displayed. At the same time, a curve graph of the

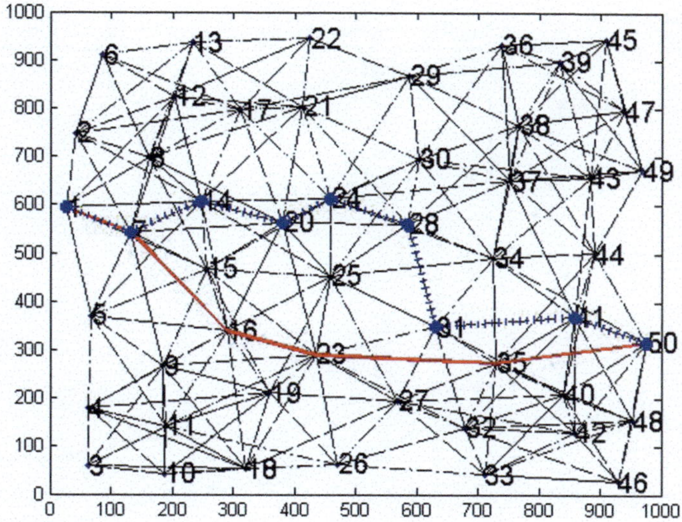

Fig. 5.4 Optimal paths obtained by the two algorithms (electronic version)

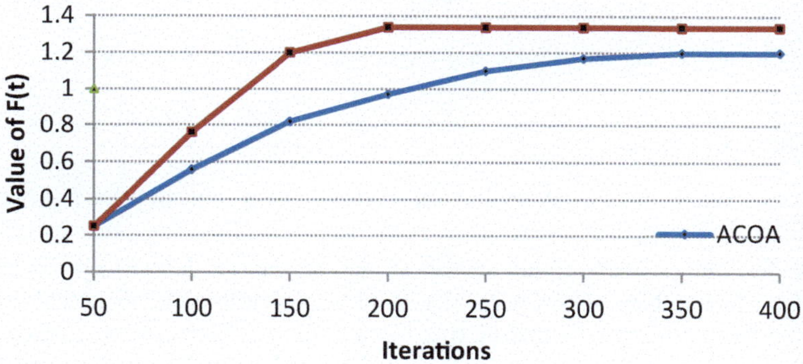

Fig. 5.5 Convergence curve of optimal solutions with iterations

changes of the two algorithms with the increase of network nodes is drawn. Within a range of 1000 m × 1000 m, 10, 20, 30, …, 100 nodes are randomly distributed. The maximum number of iterations is 400, and the number of ants is 40. The simulation parameter list is shown in Table 5.3.

Figure 5.6 shows the convergence curves of the optimal solutions of the QABR algorithm and the traditional ant colony algorithm with the number of network nodes. It can be seen that the values of the optimal solutions of the two algorithms in different network node environments. As the number of network nodes increases, the optimal solution decreases accordingly. Since the QABR algorithm takes into account both energy factors and network load factors, its decrease is less than that of the traditional ant colony algorithm.

Table 5.3 Parameter list for experiment two

Number of network nodes	10, 20, …, 100
Network range	1000 m × 1000 m
C_1	0.5
C_2	0.1
C_3	0.1
q_0	0.8
$t_{\max}$	400

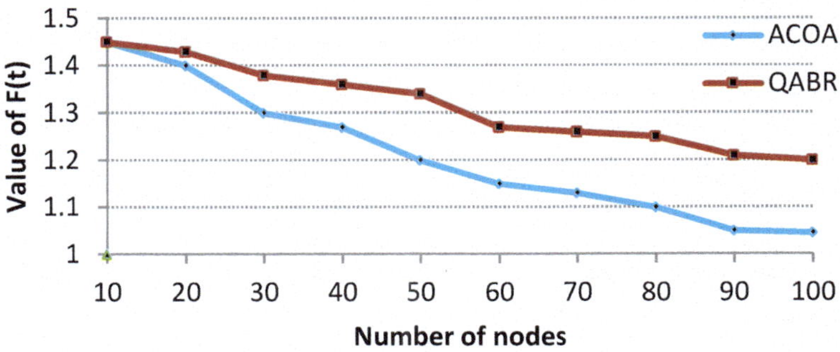

Fig. 5.6 The curve of the optimal solution changing with the number of network nodes

(3) Experiment Three

Considering that the routing algorithm affects node energy consumption and network life, which is related to the feasibility of the network. Therefore, this experiment analyzes the energy consumption and life of wireless sensor networks under certain conditions. In wireless sensor networks, routing algorithms are all aimed at optimizing energy saving and extending life. Usually for a WSN network, it must work normally for at least half a year to two years to be valuable. Therefore, the experiment analyzes node energy consumption, and the experimental parameters are shown in Table 5.4.

The energy consumption of nodes and even the life of the network are related to the choice of path and its speed. The comprehensive QoS proposed here can well reflect the relationship between them. Figure 5.7 shows the process from the death

Table 5.4 Parameters list of experiment three

Number of network nodes	50
Network range	5000 m × 5000 m
Communication range	5000 m
Maximum data rate	512 Kbps
Packet size	64 bytes
Link bandwidth	10 Kbps

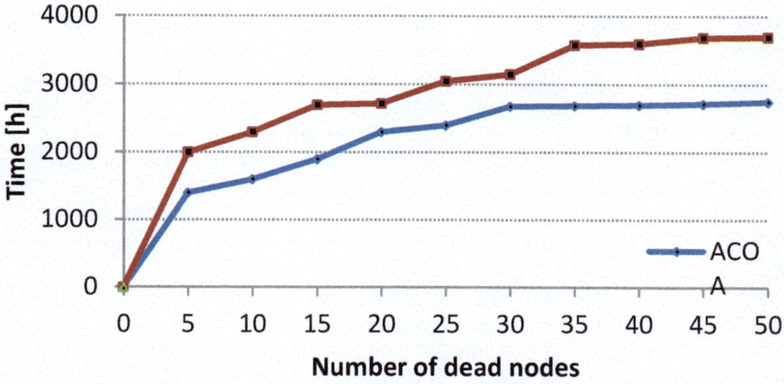

Fig. 5.7 The curve of the number of node deaths changing with time

of the first node to the shutdown of the entire network, with each node's energy consumption as the object. The QABR algorithm has the fastest convergence speed and the most optimal path while avoiding local convergence, thereby effectively reducing the energy consumption of the network as a whole. The time when the first node dies and the life of the entire network are significantly longer than the traditional ant colony algorithm, showing the superiority and feasibility of the QABR algorithm.

References

1. TANG L, LIU M, WANG KC, et al. Study of path loss and data transmission error of IEEE 802.15.4 compliant wireless sensors in small-scale manufacturing environments. International Journal of Advanced Manufacturing Technology, 2012, 63(5–8): 659–669.
2. LIU M, XU S, SUN S. An agent-assisted QoS-based routing algorithm for wireless sensor networks. Journal of Network and Computer Applications, 2012, 35(1): 29–36.
3. JIANG H, WANG M, LIU M, et al. A quantum-inspired ant-based routing algorithm for WSNs. IEEE 16th International Conference on Computer Supported Cooperative Work in Design, 2012, 609–615.
4. LIU Min, XU Shi-jun, SUN Si-yi, et al. A wireless sensor network routing method based on QoS-PSO. Journal of Tongji University, 2010, 39(12): 1846–1850.
5. HEDETNIEMI S, LIESTXNAN A. A survey of gossiping and broadcasting in communication networks. Networks, 1988, 18(4): 319–349.
6. HEINZELMAN W R, KULIK J, BALAKRISHNAN H. Adaptive protocols for information dissemination in wireless sensor networks. Proceedings of 5th ACM/IEEE Mobicom Conference (MobiCom '99), Seattle, WA, 1999: 174–185.
7. INTANAGONWIWAT C, GOVINDAN R, ESTRIN D. Directed diffusion: a scalable and robust communication paradigm for sensor networks. Proceedings ACM Mobi-Com'00, Boston, MA, 2000: 56–67.
8. BRAGINSKY, D, ESTRIN, D. Rumor routing algorithm in sensor networks. Proceedings ACM WSNA, in Conjunction with ACM MobiCom'02, Atlanta, GA, 2002: 22–31.
9. HEINZELMAN W, CHANDRAKASAN A, BALAKRISHNAN H. Energy efficient communication Protocol for wireless microsensor networks. Proceedings of the 33rd Hawaii International Conference on System Sciences. Maui: IEEE Computer Society, 2000: 3005–3014.

10. LINDSEY S, RAGHAVENDRA C S. PEGASIS: power efficient gathering in sensor information systems. Proceedings of IEEE Aerospace Conference, 2002: 1–6.
11. MANJESHWAR A, AGAWAL D P. TEEN: a routing protocol for enhanced efficiency in wireless sensor networks. Proceedings of the 15th Parallel and Distributed Processing Symposium. San Francisco: IEEE Computer Society, 2001: 2009–2015.
12. YU Y, GOVINDAN R, ESTRIN D. Geographical and energy aware routing: a recursive data dissemination protocol for wireless sensor networks. Marine Pollution Bulletin, 2011, 20(1): 48.
13. CRAWLEY E, NAIR R, RAJAGOPALAN B. A framework for QoS-based routing in the internet. RFC2386[EB/OL]. http://www.ietf.org/rfc/rfc.2386.txt, 1998.
14. SOHRABI K, GAO J, AILAWADHI V, et al. Protocols for self-organization of a wireless sensor network. IEEE Personal Communications, 2000, 7(5): 16–27.
15. HE T, STANKOVIC J, LU C, et al. SPEED: a stateless protocol for real-time communication in sensor networks. Proceedings of the 23rd International Conference on Distributed Computing Systems, 2003.
16. CAMILO T, CARRETO C, SILVA J. et al. An energy-efficient ant-based routing algorithm for wireless sensor networks. the 5th International Workshop on Ant Colony Optimization and Swarm Intelligence, 2006.
17. SALEEM M, CARO G, FAROOQ M. Swarm intelligence based routing protocol for wireless sensor networks: survey and future directions. Information Sciences, 2010.
18. ZHANG Y, KUHN L, FROMHERZ M. Improvements on ant routing for sensor networks. Ants 2004, Int Workshop on Ant Colony Optimization and Swarm Intelligence, 2004.
19. COLORNI A, DORIGO M, MANIEZZO V, Distributed optimization by ant colonies. Proceedings of the First European Conference on Artificial Life, Paris, France, Elsevier Publishing, 1991, 134–142.
20. HAN K H, KIM J H. Quantum-inspired evolutionary algorithms with a new termination criterion, H Gate and Two-Phase Scheme. IEEE Transactions on Evolutionary Computation, 2004, 8(2): 156–169.
21. HAN K H, KIM J H. Genetic quantum algorithm and its application to combinatorial optimization problem. Proceedings of the 2000 IEEE Congress on Evolutionary Computation, 2000, 1354–1360.
22. Wu Shengjun, Zhou Jindong, Zhang Yongde. Introduction to Quantum Algorithms. University Physics, 1999, 18(12): 1–5.
23. Tang Yilong, Pan Wei, Li Nianqiang. Research on Wireless Sensor Network Routing Based on Quantum Genetic Algorithm. Sensors and Microsystems, 2011, 30(12): 68–70.
24. Li Panchi, Li Shiyong. Quantum Ant Colony Algorithm for Solving Continuous Space Optimization Problems. Control Theory and Applications, 2008, 25(2): 237–241.
25. Xie Jinxing, Xing Wenxun. Network Optimization. Beijing: Tsinghua University Press, 2009.
26. NIKI T, BHASKAR K. Sensor network algorithms and applications. Philosophical Transactions of the Royal Society: Mathematical, Physical and Engineering Sciences, 2012, 370(1958): 5–10.
27. COELLO C. Evolutionary multi-objective optimization: a historical view of the field. IEEE Computational Intelligence Magazine, 2006, 1(1): 28–36.
28. Jiao Licheng. Multi-objective Optimization Immune Algorithm, Theory and Application. Beijing: Science Press, 2010.
29. Li Yueguang, Zhao Junsheng, Zhang Yuanping. Improved Quantum Ant Colony Algorithm for Solving TSP. Computer Engineering and Design, 2009, 30(16): 3843–3874.
30. Yan Chen, Wang Zhijie. Current Status and Prospects of Combinatorial Optimization Problems Represented by TSP. Computer Simulation, 2007, 24(6): 171–174.

Chapter 6
Protocol Integration and Design Case of Data Collection

In order to better provide equipment production and maintenance services for enterprises, it is urgent for enterprises to use high-tech and advanced applicable technologies to technically transform and upgrade equipment. This chapter introduces the on-site status parameter collection and monitoring of key parts of the equipment, and optimizes the maintenance and overhaul process of steel continuous casting equipment, in combination with the use, maintenance and management of continuous casting equipment in steel enterprises [1–5].

6.1 Data Collection Framework

Selecting one fan-shaped section of the 4CC horizontal section of the steel continuous casting line as the pilot object, through the monitoring of the operating status of the four objects of cylinder pressure, bearing temperature, flat roll speed, and frame displacement, it provides early warning for bearing failures and provides technical support for the transition from post-maintenance to preventive maintenance for bearing maintenance work; evaluates the impact of different factors on the bearing, provides a basis for improving the operating status of the bearing and frame, and improving the quality of slab production, as follows. Cylinder pressure: Monitor the overall force situation of the fan-shaped section, and indirectly obtain the average force situation of each flat roll. Bearing temperature: Monitor whether the bearing rotation and lubrication condition is good, and then judge the rotation condition of the flat roll. Flat roll speed: Monitor whether the flat roll rotation is normal, and judge whether there is a "jamming" phenomenon. Frame displacement: Monitor the position offset of the upper and lower frames of the fan-shaped section under working conditions (including the upper and lower spacing and horizontal offset).

M. Liu et al., *Intelligent Predictive Maintenance*, Advanced and Intelligent Manufacturing in China, https://doi.org/10.1007/978-981-97-2677-6_6

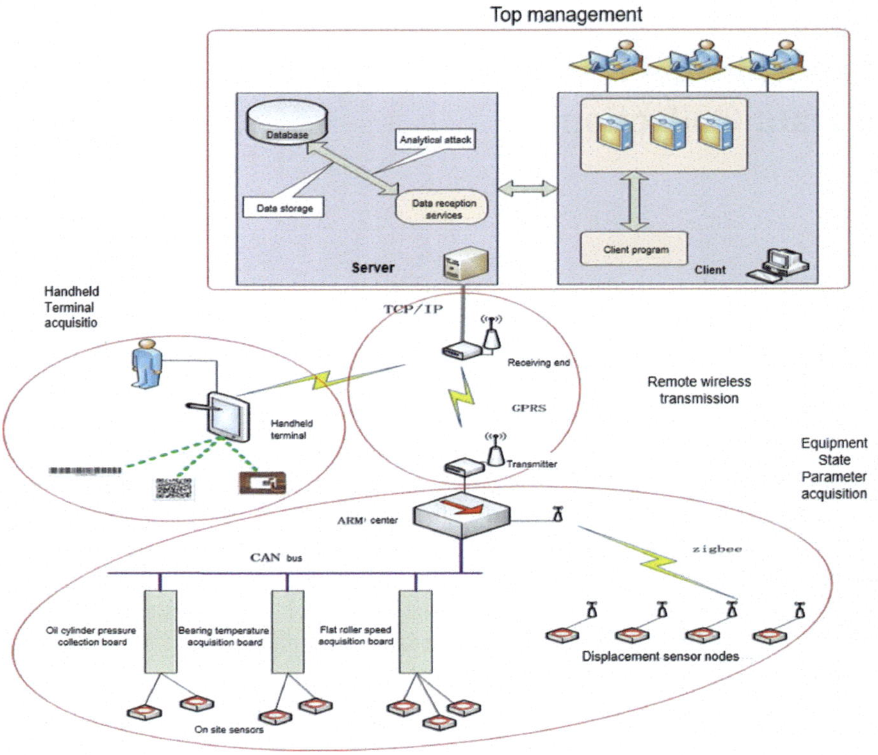

Fig. 6.1 Overall framework of equipment status monitoring

Through the measurement of cylinder pressure, bearing temperature, flat roll speed and frame displacement, the bearing failure is reflected and evaluated from different angles. The overall framework of status monitoring is shown in Fig. 6.1.

The equipment status information collected by the collector and the spare parts information and spot check information collected by the handheld terminal are transmitted to the remote server through remote wireless transmission. The upper-level management client interacts with the server through the company's existing Ethernet to realize the management and maintenance of the underlying equipment status, spare parts inventory information and offline spot check information. The equipment status parameters are converted into digital signals in the collection circuit according to the access signals of different sensors, and the information of each sensor is gathered to the ARM center through the on-site CAN bus or ZigBee wireless network, and then sent to the remote server. The handheld terminal part recognizes the corresponding information on the site through one-dimensional or two-dimensional codes and RFID, inputs the information into the handheld terminal, saves it locally on the handheld terminal on the one hand, and sends it to the remote server on the other hand. The equipment status parameters and the information collected and entered through the handheld terminal are transmitted to the remote server through GPRS remote wireless transmission. The upper-level management server receives data from the ARM center and the handheld terminal through the data receiving service, and stores the data in the database after parsing. The client is based on the existing Teamcenter/MRO platform, and realizes the maintenance and management of equipment online status monitoring, spare parts inventory management and spot check information through secondary development.

The network topology of the status monitoring system is shown in Fig. 6.2. The entire network adopts a fusion of ZigBee, CAN bus, GPRS and Ethernet. A collection device with wireless transmission function is installed on the fan-shaped section; the handheld terminal is used in the warehouse and maintenance area; the server is deployed in the mechanical factory office; the client is deployed in the on-site spot check office.

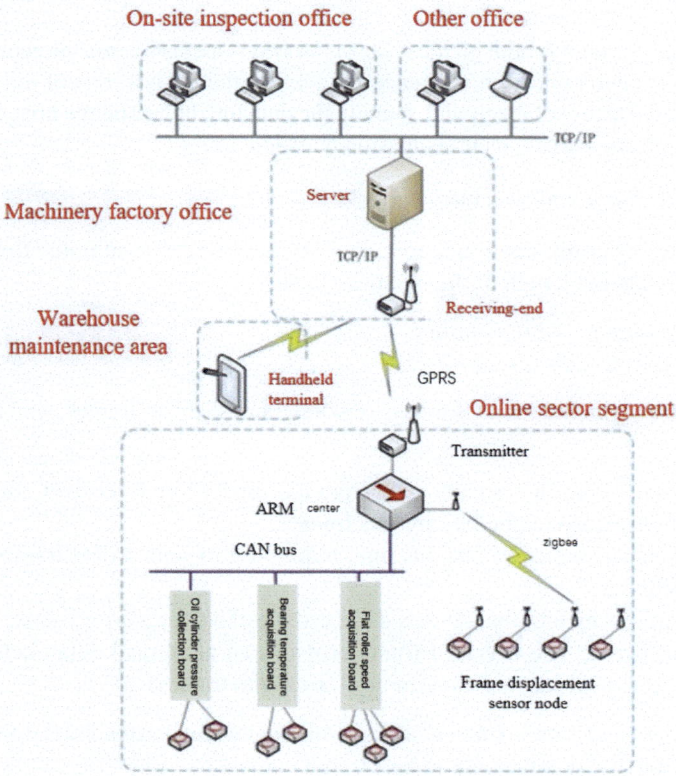

Fig. 6.2 Network topology of the status monitoring system

6.2 Monitoring Network Protocol

The status monitoring system involves: ① The characteristics, message content and structure of CAN bus technology; ② The main features and overall architecture of ZigBee technology; ③ The main features of GPRS technology, GPRS principles and structure, and GPRS networking methods.

6.2.1 CAN Bus Protocol

In the field of industrial control, CAN is one of the most important bus protocols and has formed an international standard. It is recognized as one of the most promising field buses.

The CAN protocol is built on the basis of the OSI (Open Systems Interconnection) 7-layer open interconnection reference model, but the CAN protocol only defines the bottom two layers of the model, namely the data link layer and the physical layer, only ensuring error-free data transmission between nodes.

(1) CAN message and structure

CAN message transmission frame format: There are two different frame formats, the difference being the length of the identifier field.

- Standard frame: 11-bit identifier.
- Extended frame: 29-bit identifier.

Frame types include data frames, remote frames, error frames, and overload frames:

- Data frame: Data frames carry data from the sender to the receiver, most of the frames transmitted on the bus are this frame;
- Remote frame: Issued by the bus unit, requesting to send a data frame with the same identifier;
- Error frame: Any unit that detects a bus error will issue an error frame;
- Overload frame: The overload frame provides an additional delay between the preceding and subsequent data frames (or remote frames).

Data frames and remote frames can use both standard and extended frame formats, separated from the previous frame by a frame interval.

Data frame structure: The data frame consists of seven different bit fields: frame start, arbitration area, control area, data area, CRC area, acknowledgment area, and frame end. The data frame structure is shown in Fig. 6.3.

The frame start (SOF) consists of only one dominant bit. All stations must synchronize with the first transmitted frame start edge.

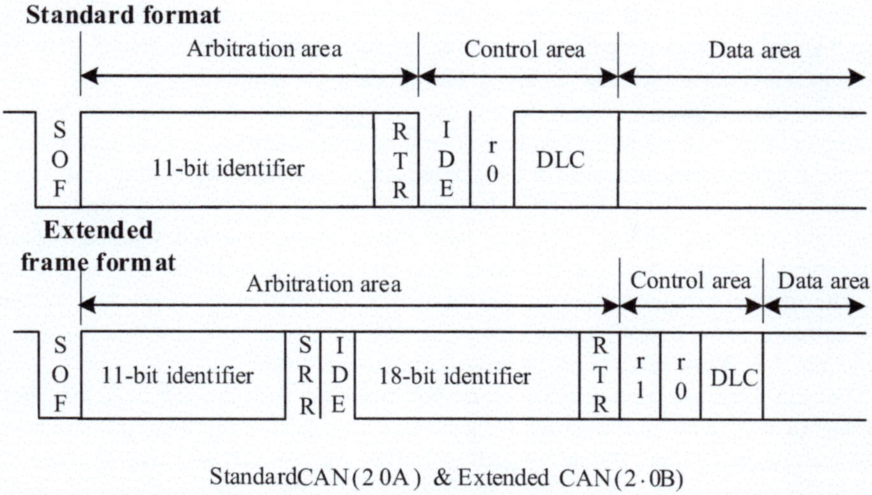

Fig. 6.3 CAN protocol standard frame and extended frame format

The arbitration area (standard format) consists of an 11-bit identifier ID28–ID18 and a remote transmission request bit RTR, where the high seven bits of the ID cannot all be 1 (recessive).

The arbitration area (extended format) consists of a 29-bit identifier ID28–ID0, SRR bit, identifier extension bit IDE bit, and RTR bit.

SRR is a recessive bit, used to replace the RTR bit of the standard format.

IDE = 1 (recessive) represents the extended format, which is located in the arbitration area in the extended format and in the control area in the standard format.

The control area consists of the 6 bits shown in Fig. 6.4, the length code DLC3–DLC0 is the number of bytes in the data area 0–8, other values are not allowed to be used. The reserved bits r1 and r0 must be 0, IDE (standard format) = 0. The data area has 0–8 bytes, 8 bits/byte, with the most significant bit (MSB) sent first. The CRC field consists of a 15-bit CRC sequence and a 1-bit CRC delimiter.

(2) Bit arbitration technology

As long as the bus is idle, any unit can start sending messages.

Fig. 6.4 Structure of control field in CAN protocol frame format

To process data in real time, the data must be transmitted quickly, which requires the physical transmission path of the data to have a high speed. When several stations need to send data at the same time, fast bus allocation is required.

If two or more units start transmitting messages at the same time, there will be bus access conflicts. This conflict can be resolved by using bit arbitration of identifiers.

The CAN bus transmits data in units of messages, and the priority of the message is combined in the 11-bit identifier. The identifier with the lowest binary number has the highest priority. This priority, once established during system design, cannot be changed. Conflicts in bus reading can be resolved by bit arbitration.

The arbitration mechanism ensures that neither information nor time is lost. When a data frame and a remote frame with the same identifier are initialized at the same time, the data frame takes precedence over the remote frame.

The CAN bus uses non-return-to-zero (NRZ) coding, and all nodes are connected to the bus in a "wire and" manner. If there is a node transmitting logic 0 to the bus, the bus presents a logic 0 state, regardless of how many nodes are sending logic 1. All nodes of the CAN network may attempt to send at the same time, but its simple arbitration rules ensure that only one node controls the bus and sends information. The low effective output state (0) plays a decisive role.

During arbitration, each sender compares the level of the sent bit with the monitored bus level. If the levels are the same, the unit can continue to send. If a "recessive" level (logic 1) is sent and a "dominant" level (logic 0) is detected, the unit loses arbitration and must exit the sending state.

(3) Error handling

The CAN controller has built-in TX and RX error counters, which increase or decrease each time a message is received. If the received message is correct each time, the counter decreases by 1. If a network-wide error occurs in the message, the counter increases by 1. If a local error occurs in the message, the counter increases by 8.

By querying the error counter value, you can know the quality of the communication network. This counter method ensures that a single fault node will not block the entire CAN network.

If a node has a local error, its count value will quickly reach 96, 127, or 255. When the counter reaches 96, it will interrupt the node microcontroller, indicating that the current communication quality is poor. When the count value reaches 127, the node assumes that it is in a "passive error state", that is, it continues to receive information and stops requesting the other party to resend information. When the count reaches 255, the node disconnects from the bus, stops working, and can only resume working after a hardware reset.

The types of errors are as follows.

① Active error state: The active error state is a state that can normally participate in bus communication. When a unit in an active error state detects an error, it outputs an active error flag.

② Passive error state: The passive error state is a state that is prone to errors. Although a unit in a passive error state can participate in bus communication, it cannot actively send error notifications during reception so as not to hinder other units' communication. Even if a unit in a passive error state detects an error, and other units in an active error state do not find an error, the entire bus is considered to be error-free. When a unit in a passive error state detects an error, it outputs a passive error flag.

6.2.2 ZigBee Protocol

ZigBee is a cheap, low-power, short-range wireless networking communication technology, specifically targeting certain control application needs in smart homes, smart buildings, and industrial automation. It is designed to achieve wireless networking communication capabilities at a rate of tens of kilobits per second and a distance of a few to tens of meters. Under such key indicator conditions, other technical requirements are determined—micro power consumption, low complexity, and thus low price, giving birth to a new wireless communication technology.

As a short-range, low-rate wireless communication technology, ZigBee has a unified technical standard. Its system architecture often uses layers to quantify its various simplified standards. Each layer in the protocol is responsible for completing its own tasks and providing corresponding services for the previous layer. The interface between layers provides corresponding services through defined logical links. Figure 6.5 shows the overall architecture of the ZigBee protocol. Its standard layered structure is based on the OSI model and is defined based on actual application requirements. The ZigBee protocol can generally be divided into the physical layer (PHY), medium access control layer (MAC), network layer (NWK), and application layer (APL). The IEEE 802.15.4—2003 standard defines the physical layer and medium access control layer, while the ZigBee Alliance defines the network layer and application layer architecture based on this protocol.

(1) Physical Layer (PHY)

The physical layer defines the interface between the physical wireless channel and the MAC sublayer, providing physical layer data services and physical layer management services. The physical layer data service receives and transmits data from the wireless physical channel, and the physical management service maintains a database composed of physical layer-related data. Physical layer functions: ① Activation of ZigBee; ② Energy detection of the current channel; ③ Receiving link service quality information; ④ ZigBee channel access method; ⑤ Channel frequency selection; ⑥ Data transmission and reception.

(2) Medium Access Control Layer (MAC)

The MAC layer is responsible for handling all physical wireless channel access and generating network signals, synchronization signals; supporting PAN connection and

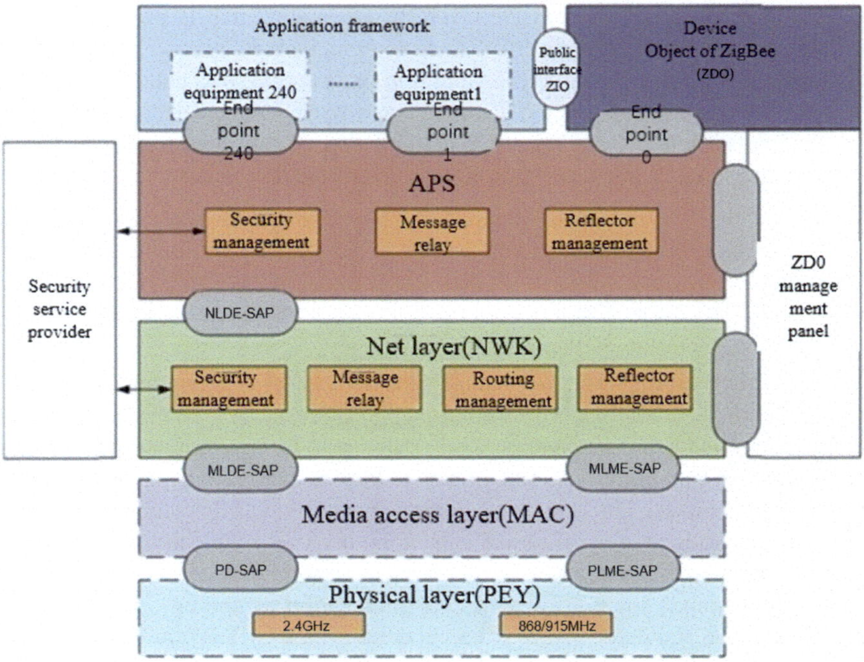

Fig. 6.5 ZigBee protocol overall architecture

separation, providing a reliable link between two peer MAC entities. MAC layer data service: Ensure that the MAC protocol data unit is correctly received and transmitted in the physical layer data service. The MAC layer management service maintains a database storing information related to the MAC sublayer protocol status. MAC layer functions: ① Network coordinator generates beacon; ② Synchronizes with beacon; ③ Supports the establishment and disconnection of PAN (Personal Area Network) links; ④ Provides support for device security; ⑤ Channel access method adopts conflict-free Carrier Sense Multiple Access with Collision Avoidance (CSMA-CA) mechanism; ⑥ Handles and maintains Guaranteed Time Slot (GTS) mechanism; ⑦ Provides a reliable communication link between two peer MAC entities.

(3) Network Layer (NWK)

In the ZigBee network, according to the communication capabilities of the nodes, they can be divided into Full Function Devices (FFD) and Reduced Function Devices (RFD). FFD devices can communicate with each other and with RFD devices. RFD devices cannot communicate directly with each other, they can only communicate with FFD devices, or forward data through an FFD device. This FFD device associated with the RFD is called the router of the RFD. In the network, nodes are divided into three types: coordinators, routers, and end nodes. One of the FFD devices is called the coordinator node, and each ZigBee network must have a coordinator node. Its main role is to participate in applications directly, and also to complete tasks such

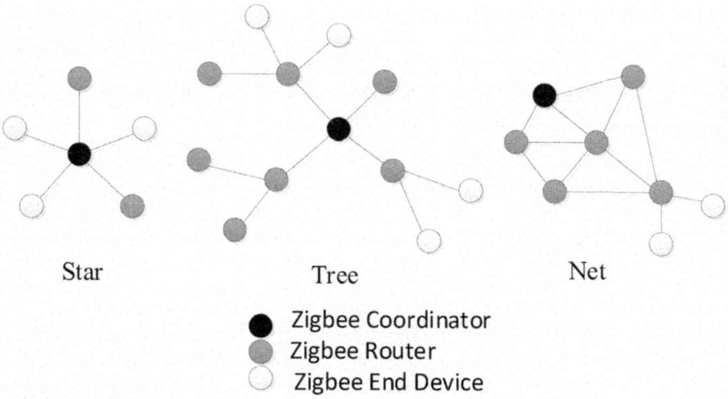

Fig. 6.6 ZigBee network topology diagram

as member identity management, link status information management, and packet forwarding. The ZigBee router is a routing node, providing routing information, and the ZigBee end node does not have routing functions. The network layer of the ZigBee protocol stack mainly implements functions such as node joining or leaving the network, receiving or discarding other nodes, routing lookup and data transmission, and supports various topologies such as star, tree, and mesh as shown in Fig. 6.6.

The functions of the network layer include network discovery, network formation, allowing device connections, router initialization, device connection with the network, direct device connection with the network, disconnecting network connections, resetting devices, receiver synchronization, and information library maintenance.

(4) Application Layer (APL)

The ZigBee application layer framework includes the application support layer (APS), ZigBee device object (ZDO), and manufacturer-defined application objects. The function of the application support layer includes maintaining the binding table and transmitting messages between bound devices. The function of the ZigBee device object includes defining the role of the device in the network (such as ZigBee coordinator and end device), initiating and responding to binding requests, and establishing security mechanisms between network devices. The ZigBee device object is also responsible for discovering devices in the network and deciding what application services to provide to them. In addition to providing some necessary functions and providing appropriate service interfaces for the network layer, an important function of the ZigBee application layer is that the applicant can define their own application objects at this layer.

(5) Application Framework (AF)

The application program running on the ZigBee protocol stack is actually the manufacturer-defined application object, which runs on endpoints 1–240 in compliance with the specifications. In ZigBee applications, two standard service types are provided: Key-Value Pair (KVP) or Message (MSG).

6.2.3 GPRS Protocol

GPRS stands for General Packet Radio Service. GPRS provides mobile users with faster data transmission speeds, suitable for bursty Internet and intranet communications. Unlike previous continuous transmission on the channel, GPRS is packet-based, and the fees users bear are calculated based on the units of data transmitted, not the use of the entire channel. GPRS transmission rates can be increased to 56Kbps, or even 114 Kbps, and can handle voice and data simultaneously, suitable for cutting-edge, frequent, large data transmission.

The GPRS system introduces new components to the GSM system (Global Mobile Communications System), including the GPRS Packet Control Unit (PCU), Serving GPRS Support Node (SGSN), and Gateway GPRS Support The wireless data transmission system is composed of nodes, using packet switching technology. It is compatible with GSM and can transmit high-speed data and signaling more effectively on the network, allowing users to send and receive data in an end-to-end packet mode. GPRS uses the same segments, bandwidth, burst structure, wireless modulation standards, frequency hopping rules, and the same TDMA frame structure as GSM. Therefore, when building a GPRS system on a GSM system, most of the components in the GSM system do not need hardware modifications, only software upgrades are required. The structure of the GPRS network is shown in Fig. 6.7. In Fig. 6.7, BSS is the base station subsystem, MSC is the mobile switching center, GMSC is the gateway mobile switching center, and PSTN is the public switched telephone network.

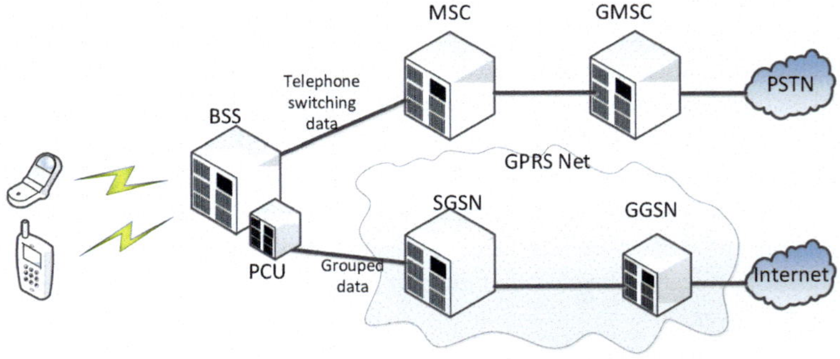

Fig. 6.7 GPRS network structure

Mobile phones or other mobile devices connect to the GPRS cellular phone via serial or wireless methods. The GPRS cellular phone communicates with the GSM base station, then sends it from the base station to the SGSN. The SGSN communicates with the GGSN; the GGSN processes the packet data accordingly, then sends it to the destination network, such as the Internet or X.25 network. IP packets from the Internet that are marked with a mobile station address are received by the GGSN, forwarded to the SGSN, and then sent to the mobile station.

6.3 Hardware Design

6.3.1 Sensor Selection

A sensor is a device or apparatus that can convert a certain physical quantity (mainly non-electrical physical quantities) into another parameter (usually an electrical parameter) that is easy to apply and process with a certain degree of accuracy according to a certain rule. For the ship computer monitoring system, the sensor generally converts non-standard physical quantities such as pressure, speed, flow, temperature, liquid level, displacement, and strong voltage (current) into standard current or voltage signals that can be processed and displayed by the computer through sensor collection and transmission. As the signal source of the monitoring system, the performance of the sensor directly affects the accuracy and reliability indicators of the system. The requirements for continuous casting equipment sensors are as follows.

① Accuracy: The output signal of the sensor must accurately reflect the changes in its input quantity. This requires that the output and input of the sensor must be a strict single-value function relationship, preferably a linear relationship.
② Stability: The single-value function relationship between the input and output of the sensor should not change with time and temperature, and the influence of external interference should be minimized, that is, the so-called no-drift property.
③ Sensitivity: A small change in the measured quantity can cause a significant change in the output signal of the sensor.
④ Corrosion resistance: Due to the special working environment of the continuous casting equipment fan-shaped section, it is generally required that the sensors used in the fan-shaped section have strong corrosion resistance, that is, the sensor will not degrade or even fail due to long-term work in water, oil, corrosive air and other environments.
⑤ High temperature resistance: Similarly, the fan-shaped section is an important equipment used for steel casting, and its working temperature is 0–300 °C.

In addition, some other characteristics, such as power consumption, volume, price, etc., will also affect the selection and use of sensors. The implementation object of the equipment online monitoring system is a fan-shaped section of the D section of

Table 6.1 Collection objects and parameters

Collection object	Collection parameter	Range	Sensor installation method	Remarks
Bearing	Temperature	– 20 to 120 °C	Install the sensor on the surface of the bearing seat or drill a hole in the bearing seat	Armored K type
Cylinder	Temperature	0–300 °C	The patch temperature sensor is attached to the surface of the cylinder	Armored K type
	Pressure	0–25 MPa	Use the pressure measuring joint reserved on the hydraulic valve	
Flat roll	Speed	Limit 2 m/min	Use the mounting hole already on the bearing end cover	Magnetic proximity switch
Upper and lower frame	Roll gap	200–400 mm	Install the connecting hole on the edge of the four corners of the upper and lower frames	Magnetostrictive displacement sensor

the 4CC continuous casting production line of a certain steel plant. The collection objects and parameters are shown in Table 6.1. The sensors include pressure sensors, speed sensors, temperature sensors, displacement sensors, etc.

6.3.2 Data Acquisition Network Layer Design

The network layer design includes two parts, namely the data collection of the underlying sensor network and the transmission to the server by the remote data transmission network. The underlying sensor network includes the CAN bus wired sensor network and the ZigBee wireless sensor network, as shown in Fig. 6.8.

(1) ARM Control Board

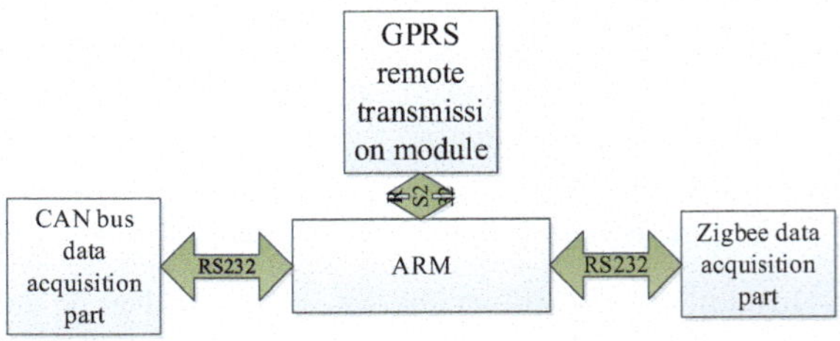

Fig. 6.8 Monitoring network layer diagram

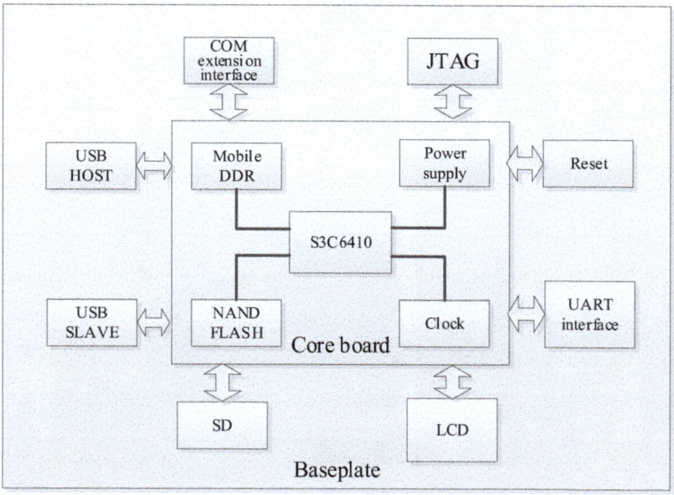

Fig. 6.9 ARM control board structure diagram

The ARM control board uses a control board with the ARM11 series S3C6410 microprocessor as its core. This board is mainly divided into two modules: the S3C6410 core board and the S3C6410 base board, as shown in Fig. 6.9. The S3C6410 core board integrates some important units, such as the S3C6410 microprocessor, large-capacity Mobile DDR, and NAND Flash external storage, etc. The S3C6410 base board expands modules such as LCD screen, USB, UART, JTAG, reset, etc.

The hardware of the development platform is shown in Fig. 6.10, using the Samsung S3C6410 processor. The S3C6410 is an RSIC processor launched by Samsung, based on the ARM1 core (ARM1176JZF-S). It is a low-power, high-cost-performance, high-performance processor widely used in mobile phones and general processing. It provides optimized H/W performance for 2.5G and 3G communication services, adopting a 64/32-bit internal bus architecture. This 64/32-bit internal bus structure consists of AXI, AHB, and APB buses, and also has an optimized interface connection to external memory. The memory system has dual external memory ports, DRAM, and FLASH/ROM/DRAM ports.

The ARM control board is connected to the CAN bus controller, GPRS transmission module, and ZigBee's Sink node using the RS232 serial port. RS232 is represented as a COM port on the ARM development platform.

(2) CAN Bus Control

The CAN control board uses the MCP2510 as the CAN controller, which is connected to the ARM control board through the RS232 serial port. The device operation data is transmitted to the ARM control board through the CAN controller. The hardware of this CAN control board is shown in Fig. 6.11.

The CAN bus controller uses the MCP2510 shown in Fig. 6.12. This device supports CAN1.2, CAN2.0A, active and passive CAN2.0B protocol versions, and

Fig. 6.10 ARM development platform hardware diagram

Fig. 6.11 CAN controller hardware diagram

can send and receive standard and extended messages. It also has acceptance filtering and message management functions. The device includes three transmission buffers and two reception buffers, reducing the management burden of the MCU. Figure 6.12 briefly shows the structure diagram of the MCP2510. The device mainly consists of a CAN protocol engine, control logic for configuring the device and its operation, transmission and reception buffers and acceptance mask registers and acceptance filters, and SPI interface logic.

The communication between each collection conversion module and the data processing center uses the CAN bus, and the network structure diagram is shown in Fig. 6.13. The bottom layer of the system is various data collection nodes. The

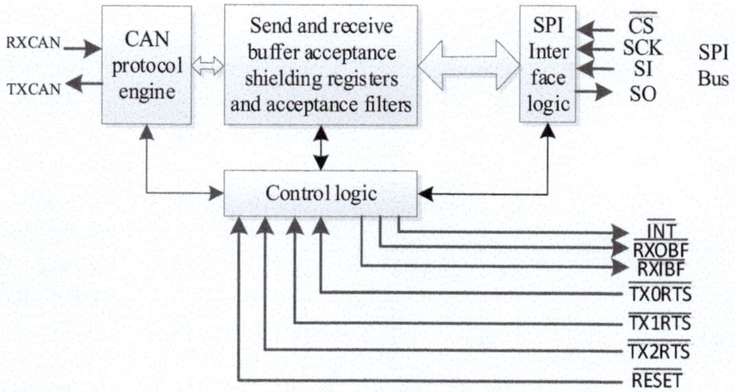

Fig. 6.12 MCP2510 structure diagram

Fig. 6.13 CAN bus network diagram

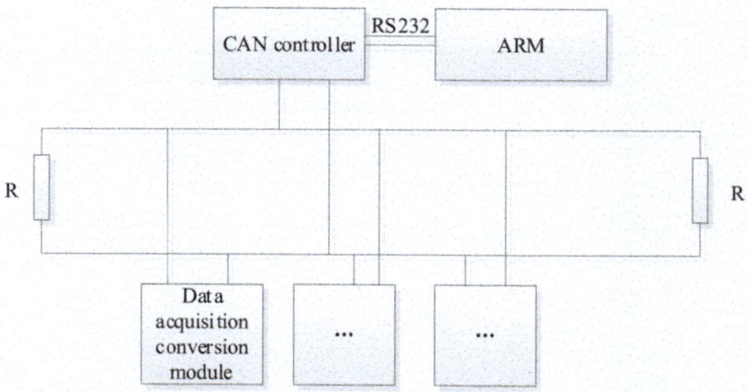

bottom layer data collection nodes are used to detect various sensor data, and at the same time, they communicate remotely with the upper layer host node through the CAN bus.

(3) CAN Bus Data Collection Conversion Module

The data collection module comes with a microprocessor, which converts analog input or pulse signals into digital signal output. It has the ability to collect multi-channel sensor data, the ability to self-identify in the system, and the ability to communicate with the field bus. The selection of the collection conversion module for the sensor is shown in Table 6.2. The collection module has two types: 4–20 mA analog input and pulse frequency measurement.

① 8-channel Analog Input Module

a. Input signal: 8-channel differential input, 4–20 mA current.
b. Input range: 0–5 V, 0–10 V, $\pm$ 5 V.
c. A/D resolution: 12-bit (as shown in Fig. 6.14 for K8512L)/16-bit (K8512H).
d. Conversion rate: 100 times/s.
e. Response time: The upper computer's 8-channel patrol cycle $\geq$ 100 ms.

Table 6.2 Collection conversion module selection table

Sensor	Output line	Output signal	Working voltage	Collection conversion board
Bearing temperature	2	4–20 mA	24V (12–30VDC)	8-channel Analog Input Module
Cylinder pressure	2	4–20 mA	24V (9–30VDC)	8-channel Analog Input Module
Cylinder temperature	2	4–20 mA	24V (10–30VDC)	8-channel Analog Input Module
Flat roll speed	2	Pulse	24V (10–33VDC)	7-channel Pulse frequency Measurement Module

- Module terminal diagram

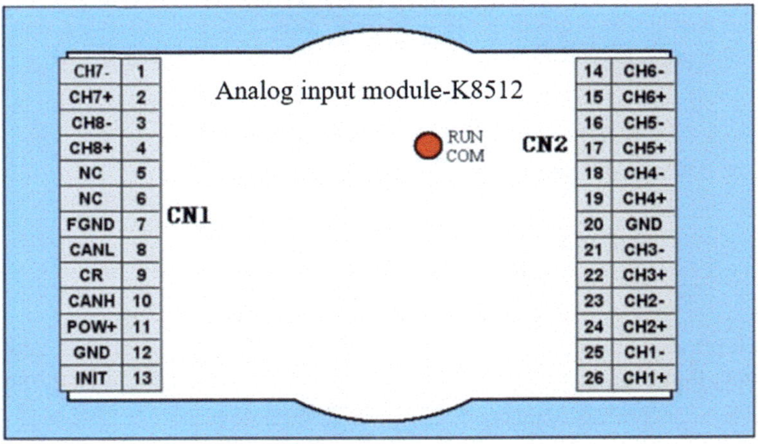

Fig. 6.14 Collection module

f. Data format: Hexadecimal.
g. Conversion accuracy: 0.1%FSR for 12-bit, 0.02% for 16-bit.

② 7-channel pulse frequency measurement module (Fig. 6.15)

a. Number of input channels: 7 channels.
b. Working mode: Frequency measurement.
c. Counter word length: 16-bit (2 bytes).
d. Input signal voltage range: 5–24 V (to be specified).

The data communication mode adopts the command/response mode shown in Fig. 6.16. In the general communication process, the command/response mode is often used. In the communication network based on the command/response mode,

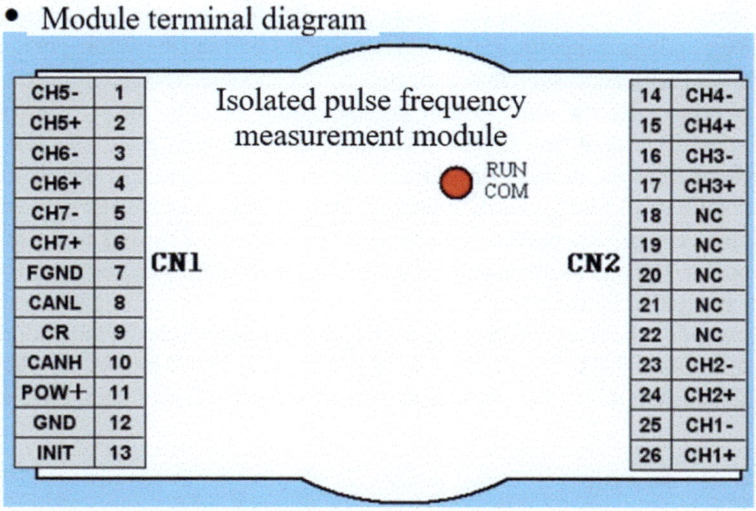

Fig. 6.15 Frequency measurement module

the master device sends a command frame, and the controlled device sends a response frame to the master device after receiving the command frame, thus realizing data exchange. In the CAN protocol, a frame of data can transmit up to 8 bytes, but the data needs to be stored in 16 bytes, so two frames are needed for transmission.

The technical specification CAN2.0A stipulates that the standard data frame has 11 identifier bits, and users can define their meanings and include the required information. In the collection system, each node has a unique address, and the address

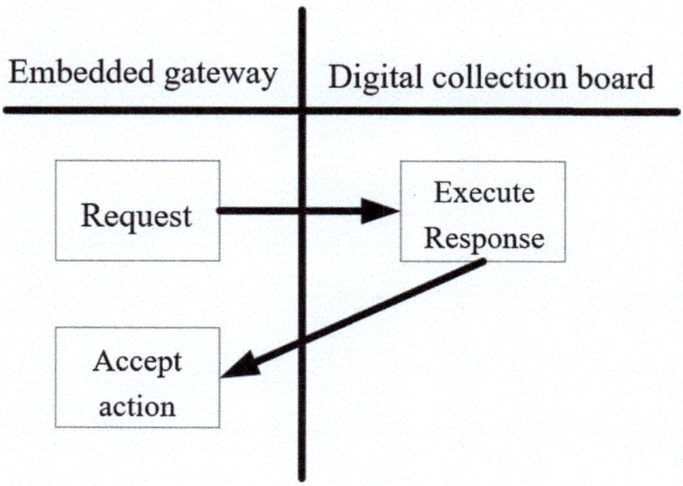

Fig. 6.16 Schematic diagram of command/response mode

code corresponds to the module one by one. The data transmission on the bus is also based on the address. Since the scale of this system is small and the number of nodes is less than 32, a 5-bit address code is allocated to each module. The same address code cannot be repeated in the same system, and the system is set by the program during initialization. The identifiers ID9–ID5 are defined as the source address, and ID4–ID0 are defined as the destination address. In this protocol, the destination address of the module is all filled with 0, indicating that the data is broadcast data, and all nodes can receive it. The destination address is determined according to the address of the module to be communicated. The data frame format of this communication protocol is detailed in Table 6.3.

Theoretically, the range of both the source address and the destination address is 0–31, but because the CAN protocol stipulates that the first 7 bits of the identifier cannot all be explicit bits, the source address cannot be 31. Therefore, there are only

Table 6.3 Data frame format

Data frame 1		Data frame 2	
ID10	Master 0 slave 1	ID10	Master 0 slave 1
ID9	Source address	ID9	Source address
ID8		ID8	
ID7		ID7	
ID6		ID6	
ID5		ID5	
ID4	Destination address	ID4	Destination address
ID3		ID3	
ID2		ID2	
ID1		ID1	
ID0		ID0	
RTR	Send request bit	RTR	Send request bit
DLC.3	Data length code	DLC.3	Data length code
DLC.2		DLC.2	
DLC.1		DLC.1	
DLC.0		DLC.0	
Data byte 1	Start bit	Data byte 1	High bit of year
Data byte 2	Segment address	Data byte 2	Low bit of year
Data byte 3		Data byte 3	Month
Data byte 4	Sensor address	Data byte 4	Day
Data byte 5		Data byte 5	Hour
Data byte 6	Data integer part	Data byte 6	Minute
Data Byte 7		Data byte 7	Second
Data Byte 8	Data decimal part	Data byte 8	End Bit

31 actual nodes (0–30), so each system contains no more than 31 modules. Therefore, the range of the source address and the destination address is reduced to 0–30. At the same time, the ARM central board also needs to occupy an address, so there are no more than 30 collection modules in the system. When designing, the node addresses are allocated from small to large according to the priority of the nodes. The 10th bit of the ID is defined as the main module identification code. This bit of the main module is an implicit bit, and the sub-module is an explicit bit, to ensure that the main module communication is prioritized. The address code of the module determines the priority of sending data. The data sent by the main module to the bus are of two types: one is broadcast data with all destination addresses filled in as 0; the other is non-broadcast data with a specific destination address.

The sub-module sends data to the bus in a broadcast manner, and recycles the data it sends. If it detects that the sent data does not match the received data, it immediately resends the previous frame of data. The order of sending information from the sub-module is determined by the instructions issued by the main module, to avoid the situation where the module with a lower priority cannot get the bus communication right when the bus communication is busy. The order of sending instructions is according to the address order of each sub-module, that is, the sub-module with a lower address gets the instruction first, and is able to send its own address, sensor sampling value, sampling time, etc. If a conflict occurs, the CAN controller automatically adjusts the sending order according to the priority of the module. There is a complete priority arbitration algorithm in the underlying protocol of CAN, so the application layer protocol does not need to consider this kind of problem.

The main module is the module with ID10 = 0, so it has the highest priority. After power-on, the main module first broadcasts its address code to the bus, and then sends instructions to the slave in order, waiting for the slave's answer. If the host does not receive any communication information after 1s, it is considered that the module is faulty and an alarm is issued.

The basic format of the protocol is as follows.

The format of the send request command (ARM central control board → data collection module) is shown in Table 6.4.

The format of the sampling frequency setting command (ARM central control board → data collection module) is shown in Table 6.5.

Sensor Data (Collection Board → ARM Central Control Board): Each data collected by the collection board is stored in 16 bytes. The CAN bus can only transmit 8 bytes of data in one frame, so this protocol uses a frame segmentation transmission

Table 6.4 Send request command format

Serial number	Number of bytes	meaning	Value	Explanation
1	1	Start byte	AAH	Hexadecimal code
2	1	Send request command code		Hexadecimal code

Table 6.5 Sampling frequency setting command format

Serial number	Number of bytes	meaning	Value	Explanation
1	1	Start byte	FFH	Hexadecimal code
2	1	Set sampling frequency command code		Hexadecimal code
3	1	Sampling frequency		

method, where the first byte of the first frame data area is set as the start bit, and the last byte of the second frame data area is set as the end bit. The data storage format is shown in Table 6.6.

Sensor address (Table 6.7): The first byte represents the board number, and the second byte represents the sequence number of a certain sensor on that board.

(4) ZigBee wireless module

There are many types of RF chips based on ZigBee, and many companies have developed ZigBee RF chips working in the 2.4GHz band, such as TI's CC2420, CC2430,

Table 6.6 Data storage format

<table>
<tr><th>Frame Segmentation</th><th>Serial number</th><th>Number of bytes</th><th colspan="2">Item name</th><th colspan="2">Value</th></tr>
<tr><td rowspan="7">Data frame 1</td><td>1</td><td>1</td><td colspan="2">Start bit</td><td colspan="2">7EH</td></tr>
<tr><td>2</td><td>2</td><td colspan="2">Segment address</td><td>See segment address table</td><td rowspan="6">BCD code</td></tr>
<tr><td rowspan="2">3</td><td rowspan="2">2</td><td rowspan="2">Sensor address</td><td>Acquisition board number (8bit)</td><td rowspan="2">See sensor address table</td></tr>
<tr><td>Sensor number (8bit)</td></tr>
<tr><td rowspan="2">4</td><td rowspan="2">3</td><td rowspan="2">Data</td><td>Integer part (16bit)</td><td></td></tr>
<tr><td>Decimal part (8bit)</td><td></td></tr>
<tr></tr>
<tr><td rowspan="8">Data frame 2</td><td>5</td><td>1</td><td colspan="2">High bit of the year</td><td colspan="2">Such as 20 is 14H</td></tr>
<tr><td>6</td><td>1</td><td colspan="2">Low bit of the year</td><td colspan="2">Such as 04 is 04H</td></tr>
<tr><td>7</td><td>1</td><td colspan="2">Month</td><td colspan="2">01H–0CH</td></tr>
<tr><td>8</td><td>1</td><td colspan="2">Day</td><td colspan="2">01H–1FH</td></tr>
<tr><td>9</td><td>1</td><td colspan="2">Hour</td><td colspan="2">00H–17H</td></tr>
<tr><td>10</td><td>1</td><td colspan="2">Minute</td><td colspan="2">00H–3BH</td></tr>
<tr><td>11</td><td>1</td><td colspan="2">Second</td><td colspan="2">00H–3BH</td></tr>
<tr><td>12</td><td>1</td><td colspan="2">End bit</td><td colspan="2">10H</td></tr>
</table>

Table 6.7 Sensor address

Data type	Board number	Sensor number
Bearing temperature	01H	1–16
Bearing load	02H	1–8
Cylinder temperature	03H	1–8
Cylinder pressure	04H	1–8
Flat roll speed	05H	1–8
Upper and lower frame roll gap	06H	1–8

CC2431 chip series, FreeScale's MC13192, MC13213, MC1322X chip series, and Ember, Jennic and Microchip have also launched their own ZigBee solutions and chip series. Choosing the ideal RF chip can reduce development difficulty, shorten development cycle, and reduce development costs. The system-on-chip (SOC) solution will be the trend of future development. SOC can greatly reduce the volume of ZigBee nodes, save the overall system cost, occupy a smaller PCB space, easy to assemble and debug, provide system reliability, reduce external stray noise, and the ZigBee-based SOC effectively integrates the RF transceiver and microprocessor. And SOC has low current working consumption and ultra-low power sleep mode, which can quickly switch from sleep mode to working mode. Based on the above analysis, the Sink node (coordinator), router, and sensor end node will all use the CC2430 system-on-chip launched by Chipcon. This system-on-chip integrates high-performance modules such as the CC2420 RF transceiver, enhanced industrial standard 8051MCU, 32/64/128KB flash memory, 8KB SRAM, and built-in ZigBee protocol stack. Plus ultra-low power consumption, it can form a ZigBee node at a very low cost. The Sink node communicates with the ARM control board through the RS232 serial port, so an expansion board is needed to provide a physical serial port and power interface.

(5) GPRS Module

The remote data transmission uses the Huawei GTM900-C module (Fig. 6.17) as the transmission terminal, which can quickly realize data transmission. The information from the GTM900 module's ARM side is transmitted to the MRO system server side. The GTM900 is a GSM/GPRS module product, supporting 800MHz, 900MHz, and 1800MHz bands. As a GSM access device, it provides external antenna interface, power interface, serial port, audio interface, SIM card interface, and control interface. Users can realize calling, short messages, supplementary services, GPRS, etc., by sending AT commands through the serial port. The main parameter features are shown in Table 6.8.

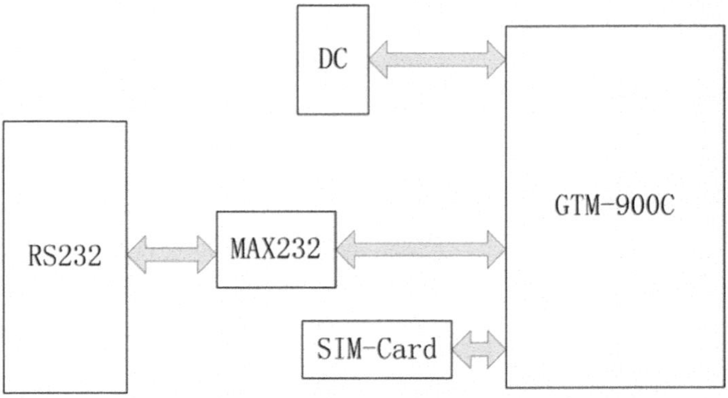

Fig. 6.17 GTM900-C structural diagram

Table 6.8 Technical parameters of GTM900-C

Parameter	Description
Operating band	EGSM900/GSM1800 dual band
Maximum transmit power	EGSM900 CLASS (2W) GSM1800 CLASS (1W)
Receiving sensitivity	<− 160 DBM
Operating temperature	− 30 to + 75 °C
Power supply voltage	3.4–3.7 V
Protocol	Supports GSM/GPRS Phase2/2+ Supports Huawei GT800 protocol
AT commands	GSM standard AT commands V.25AT commands Huawei extended AT commands
GPRS data services	Short message mode supports TEXT and PDU GPRS CLASS 10 Encoding method CS1, CS2, CS3, CS4 Maximum rate up to 85.6 Kbps Supports PBCCH Embedded TCP/IP protocol; supports multiple connections, provides large capacity buffer

6.4 Software Design

The system software part can be divided into two major parts, one is the program running on the node, and the other is the server program running on the PC. The program on the node is divided into the program running on the ZigBee node and the program running on the ARM central control board. The program running on the PC belongs to the server program. Since the server program is provided by the system, there is no need to develop it.

6.4.1 ZigBee Node Program Design

The software development of the ZigBee front-end node and coordinator uses the IAR Embedded Workbench, which is an integrated development environment developed by IAR. The program developed through IAR can be downloaded to the CC2430's ZigBee module after successful compilation through the CC2430 simulator. Z-Stack2006 is one of the ZigBee protocol specifications. This protocol stack includes many layers of components, including the media access control layer, physical layer, and ZigBee network layer in the IEEE 802.15.4-2003 standard. Each component provides an application with its own set of services and features. Since the MAC layer, PHY layer, and NWK layer of the Z-stack2006 protocol stack have provided open-source code, only the APL application layer needs to be developed.

The Sink node program includes: after receiving the data information sent by other nodes in the network, immediately send the relevant information of this data to the serial port; after receiving the command information sent by the serial port, parse this command and execute it. The program flowchart of the Sink node is shown in Fig. 6.18.

The sensor terminal node program includes: periodically trigger events to read the sensor values, and immediately send the read values to the Sink node; receive commands from the Sink node, either to terminate the sensor reading event or adjust the reading time. The program flowchart of the sensor terminal node is shown in Fig. 6.19.

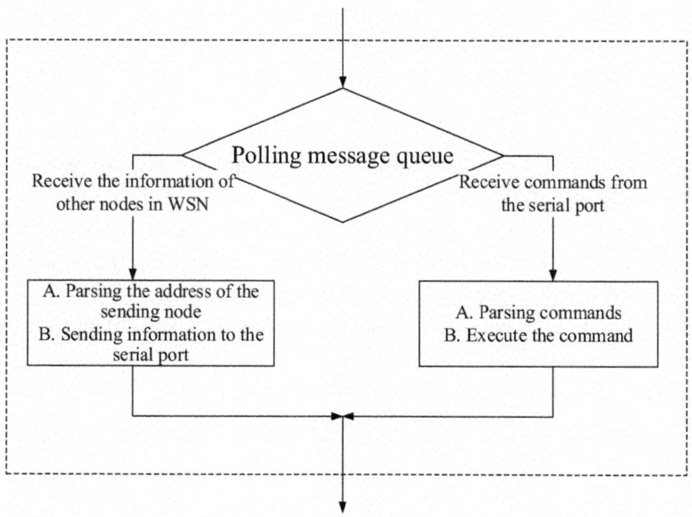

Fig. 6.18 Sink node program flowchart

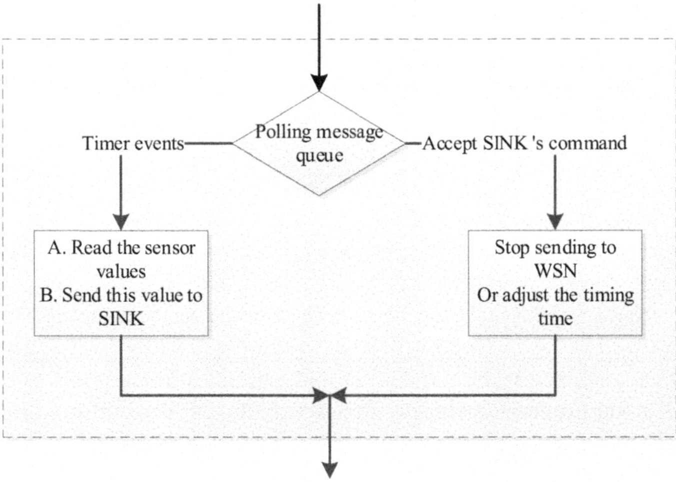

Fig. 6.19 Sensor terminal node program flowchart

6.4.2 *ARM Control Board Program Design*

Visual Studio 2005 was chosen to develop the program for the ARM central control end, which includes the development of programs including the server program and the PC client program, mainly checking and encoding the data collected from the CAN bus and ZigBee, and then forwarding the data through the GPRS module. The entire program is based on the embedded operating system wince6.0, developed in C language, and the development environment is VS2005. The ARM controller program flowchart is shown in Fig. 6.20.

① System hardware and software initialization enters the waiting state.
② GPRS module initialization.
③ Start two threads, one to receive data from the CAN bus and the other to receive data from ZigBee. Determine whether each data is correct. If the data is abnormal, discard the data and re-receive the next group of data.
④ Determine whether the public data area is full. If it is full, wait. Otherwise, put the data into the public data area.
⑤ The GPRS module sends data and clears the data area.

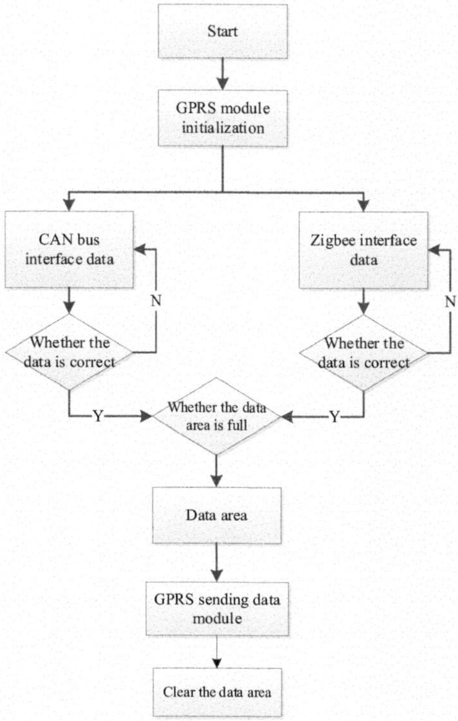

Fig. 6.20 ARM controller program flowchart

References

1. ZHANG Feng, LIU Min, ZHOU Zhuo, et al. An IoT based online monitoring system for continuous steel casting. IEEE Internet of Things Journal, 2016, 3(6):1355–1363.
2. Yin Jiubo, Yan Junwei, Liu Min. Intelligent spot-checking system for maintenance and overhaul of steel continuous casting equipment. Computer Integrated Manufacturing Systems, 2010, 16(12):2715–2719.
3. Zhang Yong, Yan Junwei, Liu Min. Knowledge model for maintenance and overhaul of continuous casting equipment. Computer Integrated Manufacturing Systems, 2010, 16(12):2718–2714.
4. Zhao Zhen, Yan Junwei, Liu Min, et al. Enterprise semantic integration and business collaboration platform for MRO of steel continuous casting equipment. Computer Integrated Manufacturing Systems, 2010, 16(10):2044–2056.
5. Li Nan, Yan Junwei, Liu Min. Industrial Internet of Things framework for maintenance of steel continuous casting equipment. Computer Integrated Manufacturing Systems, 2011, 17(2):413–418.

Chapter 7
Data-Driven Fault Diagnosis Methods

By mining and analyzing equipment maintenance, repair, and operation data, potential product failures can be diagnosed in a timely manner, maintenance needs can be generated, and intelligent predictive maintenance as shown in Fig. 1.2 can be proactively executed to achieve intelligent operation services. This chapter mainly introduces the feature extraction methods of non-stationary non-linear time series signals, and data-driven fault diagnosis methods based on convolutional neural networks, ensemble learning, and transfer learning.

7.1 Data-Driven Fault Diagnosis Method

Traditional fault diagnosis methods are divided into methods based on analytical models, signal processing, and knowledge. If a relatively accurate mathematical model of the monitoring object can be established, such as state estimation, parameter estimation, consistency testing, etc., the method based on the analytical model is preferred. The disadvantage is that due to the quality and capacity of the sample, the limitations of the model itself, the existence of noise, and the complexity of the system, the accuracy is poor. When the input and output signals of the controlled object can be obtained, but it is difficult to establish an analytical mathematical model, methods based on signal processing can be used, such as amplitude analysis, frequency domain analysis, wavelet analysis, adaptive time–frequency analysis, etc. This method is often combined with other methods for pre-processing of monitoring data. When it is difficult to establish a mathematical model of the controlled object, methods based on knowledge can be used, including expert system fault diagnosis, neural network fault diagnosis, fuzzy fault diagnosis, genetic algorithm fault diagnosis, rough set fault diagnosis, artificial immune algorithm fault diagnosis, fault tree diagnosis, support vector machine fault diagnosis, etc.

M. Liu et al., *Intelligent Predictive Maintenance*, Advanced and Intelligent Manufacturing in China, https://doi.org/10.1007/978-981-97-2677-6_7

With the development of mobile internet, Internet of Things, and cyber-physical systems, data-driven fault diagnosis methods are receiving more and more attention from the industrial and academic communities. Data-driven fault diagnosis methods usually include feature extraction, feature selection, and fault classification steps. Amplitude, time domain, frequency domain, and time–frequency domain signal processing methods are widely used in the feature extraction process of fault diagnosis; Principal Component Analysis (PCA), Artificial Neural Networks (ANN), Linear Discriminant Analysis (LDA), Kernel Discriminant Analysis (KDA) and other feature selection algorithms have reduced the feature dimension of faults, improved classification efficiency, and retained important features; Support Vector Machines (SVM), Back Propagation Neural Networks (BPNN), k-nearest neighbors and other methods based on traditional statistical learning and machine learning are used in the fault diagnosis process, and progress has been made. However, these methods have the following shortcomings:

① For non-stationary signal data, traditional feature extraction methods cannot effectively obtain sufficient fault information from time domain, frequency domain, and other state monitoring data. For vibration, impact pulse, and acoustic emission signals that are stationary, non-stationary, non-linear, multi-modal, and strongly correlated from product state monitoring, appropriate time–frequency analysis methods need to be used to extract fault features;
② The artificial feature selection algorithm for different diagnostic problems heavily depends on prior knowledge and specialized knowledge, and the feature selection is not comprehensive;
③ The shallow structure of traditional machine learning methods is difficult to accurately approximate the non-linear mapping relationship in complex systems;
④ The performance of traditional classifiers fluctuates with the quality of the training dataset, and their generalization ability is poor.

With the rapid development of big data and the new generation of artificial intelligence, the research of fault diagnosis methods faces the problem of automatic learning and processing of big data. Deep learning can automatically learn signal features from raw data and deep non-linear networks in a supervised, semi-supervised, or unsupervised manner. Some deep learning models, such as Stacked Autoencoder (SAE), Deep Belief Network (DBN), etc., have been applied to the fault diagnosis process.

Convolutional Neural Networks (CNN) have strong data processing capabilities in dealing with image and audio data, and have been widely applied to natural language processing, image recognition, audio recognition, and other fields, and have also achieved good results in some fault diagnosis tasks. The CNN model is not suitable for directly analyzing raw time-domain signals, and it is necessary to use wavelet transform, EMD, and other time–frequency analysis methods to convert raw vibration, acoustic emission, time series, and other signals into image data. Although time–frequency analysis fully retains fault features, the algorithm is complex and time-consuming. Some low-complexity methods are often used to improve the conversion efficiency and achieve online diagnosis, such as Continuous Wavelet Transform (CWT), etc.

In most research and applications of the CNN model, usually only the neurons of the last convolutional layer are used as the input vector of the classifier. Although the last layer contains more global and robust high-level features, there are still some precise detail information (intermediate features) in the previous layers that can be fully utilized. A CNN model with multi-layer multi-scale features can be constructed to improve classification accuracy. Sun et al. [1] combined the top-level features with some hidden neurons to improve the generalization ability of CNN. Ding et al. [2] combined the last convolutional layer with the pooling layer as the fully connected layer of CNN to obtain more robust multi-scale features and precise details. Lee et al. [3] proposed a CNN model with multi-layer multi-scale feature aggregation and proved its higher classification efficiency. It can be seen that the classification accuracy of the CNN model with multi-layer multi-scale feature aggregation is higher.

Traditional product fault diagnosis methods mostly target single components (such as bearings or axles), using single-modal data (such as vibration signals or temperature features) to build individual weak classifiers (such as axle crack diagnosis based on multi-layer neural networks, bearing fault diagnosis based on support vector machines, etc.) for offline fault diagnosis modeling. They do not consider the non-stationary non-linear (random signal), multi-modal (image, sound, vibration, impact, temperature and other different modal signals), and strong correlation (signals are interrelated and affect each other) of the real-time data features of the overall product MRO data, and cannot comprehensively and real-time discover potential faults of various components.

Integrated learning improves classification performance by combining multiple weak classifiers. Online integrated learning updates the parameters of some weak classifiers or replaces some weak classifiers during the classification process, and then dynamically integrates the diagnosis results of multiple weak classifiers for system-wide online prediction. In this case, even if a weak classifier makes a wrong prediction, other weak classifiers can dynamically correct the error, achieving dynamic integration of weak classifiers to adapt to target changes and background interference, such as Bootstrap aggregating (bagging), Boosting, Stacking, Bayes optimal classifier, Bayesian parameter averaging, etc. Moreover, the integration of multiple weak classifiers can also solve the learning and signal correlation problems of multi-modal data, thereby improving the classification accuracy of fault diagnosis in the MRO data environment.

Recent research has proven the superiority of the multi-random forest RF integrated classification method. Wang et al. [4] proposed an evaluation model based on RF, established a complex non-linear relationship between evaluation indicators and risk levels, and provided accurate evaluations. Therefore, in order to fully utilize the multi-layer multi-scale features of CNN and improve the quality, robustness, and accuracy of classification, an integrated learning method based on multi-layer multi-scale feature CNN and RF can be constructed for vibration, impact pulse, and acoustic emission signals from product status monitoring, or other time series signals, for online diagnosis.

7.2 Feature Extraction for Non-stationary Non-linear State Data

7.2.1 *Product Fault Modes and Signal Monitoring Methods*

Taking urban rail vehicle products as an example, Table 7.1 gives the fault modes, signal detection methods, and signal types of various components.

Currently, monitoring materials (including metals, ceramics, composites, rocks, concrete) and the acoustic emission during the manufacturing process is very important. There are many possible reasons for changes in the internal structure of product or component materials, such as the start of crack growth, dislocation movement, phase change, fiber fracture and decomposition, etc., which cause acoustic emission and most of which will cause harm. Acoustic emission is widely used in the following industrial inspection processes: detecting component defects and fatigue, monitoring welding and corrosion processes, understanding phase changes of metals and alloys, and metal processing processes (such as quenching, forging, extrusion, etc.). The difference between acoustic emission technology and other non-destructive testing technologies is that acoustic emission detects the situation inside the material, while general non-destructive testing focuses on the structure inside the material.

In practical applications, the standard wavelet transform (Wavelet Transform, WT) has certain limitations in its applicability to friction acoustic emission signals. The main reason is the friction coefficient (Coefficient of The randomness of the change in friction and the running time before failure results in high randomness in friction behavior. The extraction of acoustic features in parallel with early friction is completely different. Acoustic Emission (AE) signals have a random content and fluctuating energy, and developing a monitoring method that operates robustly under different friction behaviors requires not only studying the AE signal itself, but also studying the changes in AE content. These changes can be monitored using several wavelet-derived features, such as energy, entropy, and statistical information. Therefore, wavelet transform can be used for time–frequency analysis of the signal, and wavelet transform can be used to track changes in the signal. The corresponding time–frequency content is divided into independent non-overlapping frames (F), and the changes between the two signals are defined as the changes between the corresponding frames (F), calculated by the mean of a set of wavelet-derived features. These features are extracted from all frames, including energy, entropy, and statistical information of WP (Wavelet packet) in the frame. The change between two frames F is defined as the absolute difference of energy and statistical features, and entropy is divergence. In summary, using the changes in acoustic emission signals to track the indicators of metal surface friction behavior, contrary to COF behavior,

Table 7.1 Fault modes, signal detection methods, and signal types of various components of urban rail vehicles

Running gear components	Typical fault modes	Common offline detection methods	Online detection methods	Types of signals detected
Axle	Longitudinal cracks, transverse cracks	Electromagnetic flaw detection, ultrasonic testing	Acoustic emission	Non-stationary waveform signal
Rolling bearing	Material deterioration, part size reduction and thinning, fatigue damage, surface damage, and vibrations caused by imbalance, misalignment, self-excitation, etc.	Infrared, stress, etc.	Vibration, acoustic emission, impact pulse, temperature	Non-stationary non-linear waveform signal, digital signal
Wheel	Tread cracks, tread peeling, tread abrasions, tread groove wear, abnormal wear of the wheel rim	Ultrasonic testing and other trackside detection technologies	Vibration detection, infrared detection, impact pulse detection	Non-stationary non-linear waveform signal, digital signal
Axle box device	Tooth shape error, tooth surface wear, broken teeth, box resonance, shaft bending, shaft imbalance, axial crawling, bearing fatigue spalling and pitting, etc.	Ultrasonic testing and other trackside detection technologies	Vibration detection, infrared detection, impact pulse detection	Non-stationary non-linear waveform signal
Framework	Bolt loosening, falling off, breaking, cracks, gaps, foreign objects, cracks, etc.	Manual spot check	Trackside high-definition image recognition, vibration detection	Image signal, non-stationary waveform signal

AE tracking in the time–frequency domain shows similar behavior in the early wear of different samples.

7.2.2 *Feature Extraction Method of State Monitoring Signal*

Feature extraction is to transform the obtained signal through amplitude domain, time domain, frequency domain, and time–frequency analysis, and extract the most intuitive and sensitive fault feature information from different perspectives. In state monitoring, three types of data can be obtained: numerical data, waveform data, and multidimensional data. The most common waveform data are vibration signals and acoustic emission signals, other waveform data include impact pulse signals, ultrasonic signals, motor currents, partial discharge, etc. Time series signals are also waveform data, and image data is multidimensional data. The data processing of waveform data and multidimensional data is also called signal processing. Various signal processing techniques have been developed to analyze and interpret waveform data and multidimensional data, perform feature extraction, and facilitate fault diagnosis and prediction. However, many original waveform signals have non-stationary, non-linear, and multi-scale features. These features not only include trend information affected by stress conditions and time, but also include periodic information caused by control signals and variability information caused by sudden events and faulty components. It is difficult to accurately capture the performance features hidden in the multi-scale features. For example, impact pulse signals have non-stationary non-linear features, and acoustic emission and vibration signals are non-stationary multi-scale random signals. Therefore, extracting fault features containing strong background noise from these signals is a very tricky problem. Existing waveform signal processing techniques mainly fall into three categories: time domain analysis, frequency domain analysis, and time–frequency analysis. Time domain analysis is directly based on the time waveform itself. Traditional time domain analysis methods represent the features of time waveform signals as descriptive statistics such as mean, peak, peak-to-peak interval, standard deviation, peak factor, as well as higher-order statistics such as root mean square, skewness, kurtosis, etc. These features are usually referred to as time domain features. Time Synchronous Average (TSA) is a popular time domain analysis method that uses the overall average of the original signal in multiple processes to remove or reduce noise and influences from other sources and enhance the target signal. Advanced time domain analysis methods fit waveform data into parameter time series models to extract features. Common models include Autoregressive (AR) models and Autoregressive Moving Average (ARMA) models. There are also some other time domain analysis methods, such as pseudo-phase portraits, singular spectrum analysis, and correlation dimension, etc. Frequency domain analysis is based on frequency domain transformed signals, which can easily identify and separate certain frequency components of interest. Fast Fourier Transform (FFT) method for spectral analysis is currently the most widely used conventional frequency domain analysis method. FFT is a fast algorithm for discrete Fourier transform, which can transform a signal into the frequency domain. The core functions of Fourier transform are sine and cosine functions, which represent the speed of information changes contained in the original function with different periods. In Fourier transform, the reciprocal of the period (i.e., frequency) is usually used to describe this feature of the original function, and frequency is a concept unique to Fourier transform. After Fourier transform, the original function is transformed from function space to frequency space, realizing the time–frequency transformation of the function, thus gaining a clearer understanding of many features of the original function. The main idea of spectral analysis is to either observe the entire spectrum or carefully observe certain frequency components of interest, so as to extract features from the signal. The most commonly used tool in spectral analysis is the power spectrum. Fourier transform includes spectral analysis, time series analysis, transfer function analysis, envelope analysis, refined spectral analysis, and inverse spectral analysis, etc. It has excellent performance in analyzing linear, stationary signals. For non-stationary non-linear strong signals, its spectral resolution is limited, the time–frequency details are not well reflected, the random fluctuations are relatively obvious, and there may be some spectral distortion phenomena; Autoregressive Moving Average model uses time series for linear predictive modeling, which is one of the most common methods of spectral analysis, compensates for the shortcomings of Fourier transform, makes the spectrum smoother and the resolution improved, and does not require a high number of data points in data processing. However, some algorithms of modern spectral analysis extract feature signals through parameter models, which are complex in modeling, sensitive to signal-to-noise ratio, and difficult in order selection. One limitation of frequency domain analysis is that it cannot handle non-stationary waveform signals, which are very common when mechanical faults occur. Therefore, time–frequency analysis techniques are needed to handle non-stationary waveform signals from both time and frequency domains. Traditional time–frequency analysis methods use time–frequency distribution functions to reveal fault modes and perform diagnosis. Time–frequency distribution functions represent the energy or power of waveform signals with a two-dimensional function of time and frequency. Cohen class distribution is also based on Fourier transform, and the local autocorrelation function is processed accordingly. Different window functions can obtain time–frequency information with different properties, and can reflect the energy density of the signal on the time–frequency plane. Wigner distribution is one of the basic distributions of this method, which can well describe the energy distribution of the signal and obtain good time–frequency resolution effects, but the disadvantage is that there are cross terms in its distribution plane, and it is difficult to filter multi-component signals and noise intertwined together. Short-time Fourier transform (STFT) divides the signal into many smaller time intervals, and assumes that the signal is stationary within a short time interval. Fourier transform is used to analyze each time interval to determine the frequencies present in that interval. STFT is also an improved method of Fourier transform [5]. In order to simultaneously represent the characteristics of a function in the time domain and frequency domain, Gabor introduced the concept of windows in 1946, improved the Fourier transform, and developed the short-time Fourier transform, also known as the Gabor transform. Simply put, it is to multiply a long-time signal by a window function that is not zero for a period of time and then perform a Fourier transform, then move this window function along the time or space axis continuously, and finally arrange the series of spectra obtained to form a two-dimensional image. One axis of the image is time or space, and the other axis is frequency, so that all information in the time–space domain and frequency domain of the signal can be obtained simultaneously. Although the short-time Fourier transform has significantly improved the analysis of the time–space distribution of signal frequencies compared to the Fourier transform, the accuracy of the information obtained is very limited. In the short-time Fourier transform, the window function is predetermined according to the characteristics of the object being analyzed. Once the length and shape of the window function (fixed window function) are determined, the frequency resolution of the analysis is also fixed. That is to say, this method cannot accurately analyze the low-frequency and high-frequency components in the function at the same time, and is only suitable for the analysis of slowly changing signals or stationary signals. For non-stationary signals or time-varying signals, they can only be processed by segmenting and truncating.

Another transformation method for time–frequency analysis is the wavelet transform. The wavelet transform originated from Haar's work in the early twentieth century. Its idea is based on a window function that can automatically adjust the window shape (both the time window and the frequency window can be adjusted). It has a higher frequency resolution and a lower time resolution in the low frequency part, and a higher time resolution and a lower frequency resolution in the high frequency part. In the 1980s, the theory of wavelet transform (the discovery of continuous wavelet transform in 1975) and formulas (the establishment of the CWT algorithm in 1982) were gradually established, which greatly promoted the development of this technology. By the 1990s, the wavelet transform method had become very mature and widely used. Unlike time–frequency distribution, time–frequency distribution is a time–frequency representation of signals, and wavelet transform is a time-scale representation of signals.

In the continuous wavelet transform, the concept of the period of the trigonometric function is replaced by the scale of various wavelet functions, and the concept of displacement in the window transform is introduced at the same time, transforming the original function or signal from the function space to the scale and displacement space, so that the original function can be better understood in both the frequency and time–space domains [6]. The continuous wavelet transform not only retains the scale reflecting the rate of change of the function in the Fourier transform, but also reflects the time–space position where various scale components occur, which has great advantages. However, the continuous wavelet transform is an infinite series of two dimensions, and the amount of calculation is large. At the same time, the wavelet transform often only has numerical solutions, unlike in the Fourier transform, many commonly used functions have very simple analytical expressions for the Fourier transform, so there are still great limitations in specific applications. The Discrete Wavelet Transform (DWT) reduces the amount of computation in the transform by removing unnecessary repeated coefficients, and the method of computation is very regular, so the discrete wavelet transform is often used in practical applications. The wavelet transform is a technological innovation following the Fourier transform. It can not only locate short-time high-frequency components, but also analyze the low-frequency part, realize multi-resolution analysis, and has multi-scale characteristics and mathematical microscopic characteristics. It is widely used in image signal compression, noise removal, image boundary detection, and vibration signal analysis. At present, commonly used wavelets include Haar and Daubechies series, Biorthogonal series, Coiflets series, Symlets series, and Morlet, MexicanHat, Meyer and other wavelet functions. Wavelet basis functions that usually have orthogonality can provide better transformation performance. The wavelet transform also has some defects: ① The applicable range of each wavelet basis function is different. If the selected wavelet basis function is not suitable for the whole process, the transformation is constrained by the product of the time window and the frequency window being a constant, and it cannot achieve high accuracy in time and frequency at the same time, which will affect the analysis effect; ② The wavelet transform is restricted by the Heisenberg uncertainty principle and cannot accurately represent the change of frequency with time. The finite time width may cause the energy of the processed signal to leak and produce false harmonics; ③ The signal is mechanically distributed to the predetermined time–frequency structure, not determined by the characteristics of the signal to be processed, making the wavelet analysis lack adaptability.

Adaptive time–frequency analysis is developed for the processing of non-stationary non-linear complex signals, solving the non-adaptive problem of wavelet transform, and the most representative algorithm is the Hilbert-Huang transform (HHT) [7]. This algorithm uses Empirical Mode Decomposition (EMD) to decompose the fluctuations or trends of different scales in the signal into a series of data sequences with different characteristic scales, that is, Intrinsic Mode Functions (IMF), and then uses the Hilbert transform to process the decomposed signals to obtain the complete time–frequency distribution of the processed signals. The HHT algorithm is an adaptive signal decomposition method based on local signal characteristics, completely getting rid of linear and stationary constraints, suitable for analyzing non-linear non-stationary signals; HHT can adaptively generate IMF, with full adaptability; it is not restricted by the Heisenberg uncertainty principle, and can achieve high accuracy in time and frequency at the same time, suitable for analyzing abrupt signals, with a high signal-to-noise ratio. The HHT algorithm has developed rapidly in fields such as fault diagnosis, but it also has shortcomings such as end effects and modal aliasing [8]. Lv Yong and others [9] used EMD to spatially reconstruct the impact pulse signal and then used singular value decomposition, and performed envelope analysis on the decomposed principal components, which can extract weak impact features in strong background signals and noise. The Ensemble Empirical Mode Decomposition (EEMD) adds noise assistance based on EMD, which to some extent solves the mixing problem of EMD [10, 11], but the end effect problem still cannot be solved. In order to optimize or solve the deficiencies of the above algorithms, Wang and others [12] proposed a solution to suppress the EMD end effect by optimizing the algorithm itself, and at the same time integrated the EMD and Smoothed Pseudo Wigner-ville Distribution (SPWVD) algorithms to achieve higher time–frequency resolution. In addition, when using empirical mode decomposition signals, the overall average empirical mode decomposition [13, 14], local mean decomposition [15] and other algorithms have improved the overshoot, undershoot and not smooth enough and modal aliasing problems that occur when fitting the envelope line in the Hilbert-Huang transform. The hybrid learning paradigm composed of EEMD and least squares support vector machine regression LSSVR has been proven to be helpful for predicting high volatility time series [10]. These methods can accurately extract multi-scale signals with time–frequency characteristics in some aspects, but they have limitations when dealing with signals affected by stress. Changes in environmental stresses and operating conditions such as temperature, vibration, and input control signals will affect the degradation path, making the degradation model more complex. Although Petri nets, Piecewise Deterministic Markov Process (PDMP), and Particle Filtering (PF) have been introduced to model the evolution process of equipment performance characteristics under various working conditions, they are not suitable for solving problems where the degradation rate highly depends on the working conditions.

The signal processing of multi-dimensional data (such as images) is very similar to waveform signal processing, but it is more complex than waveform signal processing because it has an extra dimension. In practice, for fault diagnosis problems where fault detection information cannot be directly obtained, image processing technology has a very strong ability to extract features from complex original images. When the original image can provide sufficient and clear information to identify fault modes and detect faults, image processing seems unnecessary. In this case, image processing can still perform automatic fault detection by extracting features. In addition to obtaining original images through data collection, some waveform signal processing techniques such as wavelet transform, adaptive time–frequency analysis, etc., also produce images [16], in this case, combining image processing with waveform processing can achieve better results. However, there is less research on the application of image processing in fault diagnosis and prediction.

Numerical data includes the original data obtained through data collection, as well as the feature values extracted from the original signal through signal processing. Numerical data seems simpler than waveform and image data, but when the number of variables is large, the complexity lies in the correlation structure between the variables. Multivariate analysis techniques, such as Principal Component Analysis (PCA) and Independent Component Analysis (ICA), are very useful in dealing with data with complex correlation structures. In addition, trend analysis techniques are also commonly used to analyze numerical data, such as regression analysis and time series models, etc.

In summary, traditional time–frequency analysis methods such as Fourier transform cannot handle non-stationary non-linear random signals. Methods such as wavelet change, Wavelet Packet Decomposition (WPD), and Empirical Mode Decomposition (EMD) can handle these high-frequency dispersion signals well, and represent the local features of the signal in the time–frequency domain, with better time and frequency resolution, and have been widely used in signal denoising, feature extraction, signal decomposition, and fault identification, etc. WT and WPD, as automatic decomposition methods, have the problem of insufficient threshold sensitivity. The EMD decomposition result has a modal aliasing problem. EEMD has good adaptability, adopts data series decomposition combined with multi-scale features, and overcomes the shortcomings of EMD and WPD. In order to further accurately extract fault features, it is not enough to only analyze multi-scale feature signals using the above decomposition methods. Mixed decomposition methods have been applied in recent years. In addition, by converting non-stationary acoustic emission and vibration signals into image data through wavelet transform or EMD method, combining image processing with waveform processing to achieve feature extraction of fault signals can also achieve better results.

7.2.3 Feature Extraction Method Combining Image Processing and Waveform Processing

In the diagnosis process of mechanical systems, it is necessary to select appropriate signal processing tools in conjunction with case-related knowledge to achieve feature extraction of fault signals. In the various non-stationary non-linear multi-modal signals shown in Table 7.1, the non-linear non-stationary impact pulse slow-changing signal can be converted into an image signal containing time–frequency features through empirical mode decomposition, and the non-stationary acoustic emission signal and vibration signal can be converted into image signals through wavelet transform. The time–frequency feature extraction method based on signal-image conversion shown in Fig. 7.1 can convert various non-stationary non-linear waveform signals detected into image information that is easy for the deep learning framework to process.

In the feature extraction process, the original vibration, acoustic emission, impact pulse and other waveform signals in the MRO data are compressed into grayscale images after being converted into spectrum coefficient matrices through CWT/EMD. The specific process is: ① Randomly extract 1024 consecutive sampling points from the original signal; ② The 1024-point signal is converted through CWT or EMD, and converted into a $1024 \times S$ coefficient matrix. In practical applications, the scale factor of S ranges from 1 to S, as long as the value of S is large enough, sufficient information of the original signal can be obtained, and the value of S can be selected as 1024 during the experiment; ③ The CNN model will produce quite large computational

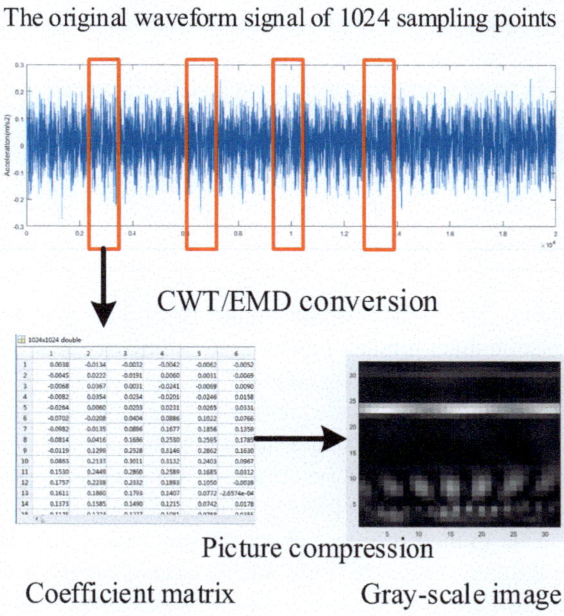

Fig. 7.1 Time–frequency feature extraction method based on signal-image conversion

complexity when dealing with the ultra-large scale coefficient matrix of 1024×1024, and some image compression methods, such as based on bicubic interpolation, can be used to reduce the size and computational complexity of the coefficient matrix.

(1) Continuous Wavelet Transform

The wavelet replaces the infinite long triangular function base of the Fourier transform with a finite long decaying wavelet base, which can not only obtain the frequency, but also locate the time, using a normalized function $\psi(\mathrm{t}) \in L^2(\mathrm{R})$:

$$\overline{\psi}(\mathrm{t}) = \int_{-\infty}^{+\infty} \psi(\mathrm{t})dt = 0$$

$$\|\psi(\mathrm{t})\|^2 = \int_{-\infty}^{+\infty} |\psi(\mathrm{t})|\mathrm{d}\, t = 1$$

The basic wavelet function $\psi(\mathrm{t})$ is often referred to as the wavelet basis function. By scaling and translating $\psi(\mathrm{t})$, let a and b represent the scaling factor (scale factor) and translation factor respectively, and let the function after scaling and translation be $\psi_{\mathrm{a,\,b}}(\mathrm{t})$, then we have:

$$\psi_{\mathrm{a,b}}(\mathrm{t}) = |a|^{-\frac{1}{2}}\psi\left(\frac{t-b}{a}\right), \mathrm{a, b} \in \mathrm{R}, \mathrm{a} > 0$$

$\psi_{\mathrm{a,b}}(\mathrm{t})$ is a wavelet basis function that depends on the parameters a and b, also known as a continuous wavelet basis function. The scale factor a can stretch or compress the wavelet function to change the oscillation frequency (corresponding to frequency), and the translation factor b changes the position of the time window (corresponding to time).

For any signal function $\mathrm{f(t)} \in \mathrm{L}^2(\mathrm{R})$, its continuous wavelet transform is defined as:

$$\mathrm{CWT_f(a, b)} = \mathrm{a,b(t)} >= |a|^{-\frac{1}{2}} \int_{-\infty}^{+\infty} f(t)\overline{\psi}\left(\frac{t-b}{a}\right)\mathrm{d}t$$

where, $\overline{\psi}(t)$ is the complex conjugate of the wavelet basis function $\psi(t)$; $CWT_f(a, b)$ is the inner product reflecting the similarity between the signal function $f(t)$ and the wavelet function $\psi_{a,\,b}(t)$.

By changing the scale factor a and the time index corresponding to the translation factor b (number of scales and sampling points), a continuous wavelet transform coefficient matrix can be obtained, which reflects the amplitude of the continuous wavelet transform coefficient with the change of scale and time.

Due to their different characteristics, many wavelet basis functions are applied to different research fields, such as Haar, Meyer, Coiflet, Symlet, Gabor, Morlet, etc. Among these functions, Morlet wavelet has been proven to be superior to other wavelet basis functions in the analysis of non-stationary vibration signals, and can be selected as the wavelet basis function for fault diagnosis [17].

Morlet wavelet is a complex wavelet, defined in the time domain as:

$$\psi(t) = \exp(-\beta^2 t^2/2)\cos(\pi t),$$

where β is the only parameter controlling the shape of the fundamental wave. The continuous wavelet transform coefficient of Morlet wavelet is also a complex number, and the square of its modulus is defined as the wavelet power spectrum coefficient matrix, the number of rows and columns of the matrix correspond to the number of scales and the number of sampling points in the time direction.

As the value of β increases, the time domain resolution increases and the frequency domain resolution decreases. With an appropriate β value, the continuous wavelet transform using Morlet wavelet can obtain the local features of the signal and good resolution in the time and frequency domains. Therefore, the project proposes to use the continuous Morlet wavelet transform to convert one-dimensional vibration or acoustic emission signals into a two-dimensional coefficient matrix with rich state information in the form of a grayscale image, and to select the scale value to adjust the image size according to the actual situation.

(2) Empirical Mode Decomposition

Unlike Fourier decomposition and wavelet decomposition methods based on a priori harmonic basis functions and wavelet basis functions, Empirical Mode Decomposition (EMD) decomposes signals based on the time scale characteristics of the data itself, without the need to pre-set any basis functions. It has a very obvious advantage in dealing with non-stationary and non-linear signals, and is considered a major breakthrough in linear and steady-state spectral analysis based on Fourier transform since 2000.

Since most of the data to be analyzed are not Intrinsic Mode Functions (IMF), at any point in time, the data may contain multiple oscillation modes. Simple Hilbert transform cannot fully characterize the frequency characteristics of general data, and it is necessary to perform EMD decomposition on the original data to obtain the IMF. The EMD method decomposes complex signals into a finite number of IMFs, each IMF component contains the local feature signals of the original signal at different time scales (stationarization of non-stationary data), and then obtains the time–frequency spectrum through Hilbert transform to obtain physically meaningful frequencies. The basis functions of EMD are obtained directly and adaptively based on the local characteristics of the signal sequence time scale.

EMD assumes that non-stationary non-linear complex signals are composed of a finite number of multi-element IMFs of different scales, and IMFs contain local features of signals at different time scales. IMF simultaneously satisfies the assumed conditions: ① The number of local extreme points and zero-crossing points are equal or differ by at most one; ② At any time, the average value of the upper envelope formed by local maxima and the lower envelope formed by local minima must be zero. Different IMFs do not have the same frequency in the same time period, they are orthogonal to each other.

The original signal produces m Intrinsic Mode Functions (IMF) after EMD decomposition, and IMF1 is defined as $(a_1^1, a_2^1, \ldots, a_n^1)$, the superscript 1 represents IMF component 1, IMF p is represented as $(a_1^p, a_2^p, \ldots, a_n^p)$, the power spectrum coefficients of each sampling point of IMF p E_i^p are $|a_i^p|^2$, n is the number of sampling points, then the cyclic matrix of IMF $a[m][n]$ and its power spectrum coefficient matrix can be obtained as $E[m][n]$.

7.3 Fault Diagnosis Method Based on Convolutional Neural Network and Ensemble Learning

7.3.1 CNN Model with Multi-layer Multi-scale Feature Optimal Configuration

The function of the CNN convolution layer is to detect the local features of the previous layer, while the pooling layer is to semantically merge similar features. The pooling layer performs statistical calculations on the local spatial region of the feature map obtained by convolution, calculating the average or maximum value of the local spatial region of the feature map. Adjacent pooling units read data from local areas by moving a few rows or columns, reducing the dimensionality of features, thereby reducing the possibility of network overfitting and achieving translation invariance of the input image. The last layer of the standard CNN model is more invariant and robust than the previous layers, and contains global information, but it will lose the detailed information of the previous layers, limiting the transmission of network information, while the previous layers contain more accurate detailed information. Combining the two will allow for more accurate classification. Therefore, the outputs of multiple convolution layers and maximum pooling layers can be used as inputs to the fully connected layer, constructing a CNN model based on multi-layer multi-scale feature optimal configuration for fault feature learning and image classification. The training dataset updates network parameters by backpropagating errors. After training, the CNN model can automatically extract representative features, classify images, and be applied to fault diagnosis. Figure 7.2 shows a deep CNN learning model based on LeNet-5 with 6 layers.

To reduce computational complexity, the size of the input image should be as small as possible while ensuring that the image contains sufficient fault information, such as the 32×32 pixel image shown in Fig. 7.2. By convolving 6 5×5 convolution kernels (also known as filters or feature detectors) with the input image and applying the ReLU activation function, 6 28×28 feature maps are obtained in the C1 convolution layer. The S2 layer is a pooling layer, containing 6 14×14 feature maps. Each feature map of C1 is sampled into the corresponding feature map in the S2 layer. Layers C3 and S4 are formed in a similar way. The FC5 layer is a fully connected layer with

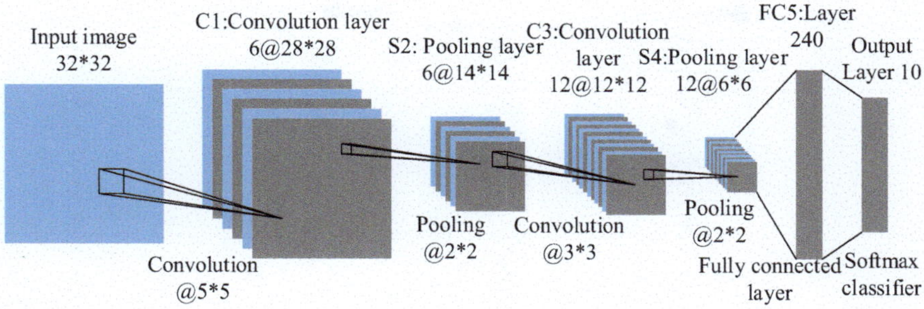

Fig. 7.2 Deep CNN learning model based on LeNet-5

240 feature maps, each of size 1 × 1, each pixel is connected to the 6 × 6 adjacent area of the feature map in S4. The classification result is finally output through the Softmax classifier. The FC5 layer feature map $\{ x^{(i)} \}_{i=1}^{m}$. The input to the Softmax classifier, the probability distribution of each input sample belonging to each class:

$$p(y^{(i)} = j|x^{(i)}; \theta) = \frac{\exp(\theta_j^T x^{(i)})}{\sum_{l=1}^{k} \exp(\theta_l^T x^{(i)})}, y = \text{argmax}_j p(y^{(i)} = j|x^{(i)}; \theta)$$

In the formula, $j = 1, 2, \ldots, k$. k is the dimension of the output layer. In addition, θ represents the parameters of the Softmax classifier (weight matrix and bias vector), and the cost function is defined as:

$$J(\theta) = -\frac{1}{m}\left[\sum_{i=1}^{m}\sum_{j=1}^{k} 1\{y^{(i)} = j\}\right] \log p(y^{(i)} = j|x^{(i)}; \theta)$$

In the formula, $1\{\cdot\}$ is the indicator function. The optimization algorithm can be applied to solve the minimum value of $J(\theta)$ to optimize the CNN parameters.

7.3.2 *Ensemble Learning Method Based on Multi-layer Multi-scale Deep CNN and RFs*

As previously mentioned, only the features extracted from the FC5 fully connected layer in the standard CNN model are used for fault classification. However, other lower-level features extracted, such as S2 and S4, are more capable of accurately describing local features and may contain some important sensitive information that does not exist in the FC5 layer, which can be utilized to improve classification accuracy. Therefore, the multi-scale features of the CNN model can be integrated through multiple random forest classifiers to construct an ensemble learning model with optimal configuration features for diagnosis. As shown in Fig. 7.3, the extracted

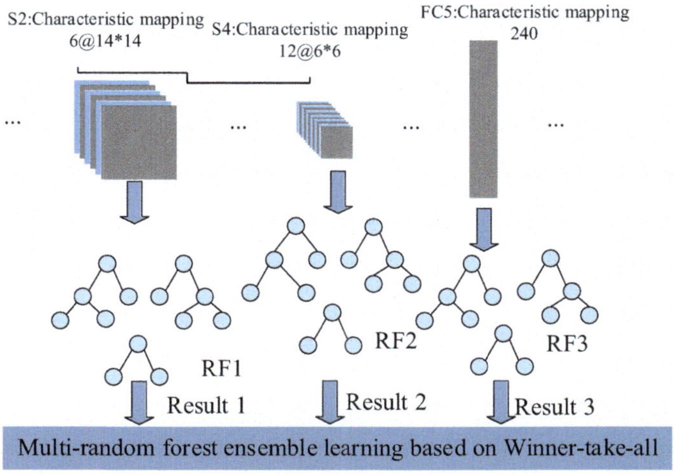

Fig. 7.3 Ensemble learning model based on multi-layer multi-scale feature CNN and multiple random forest classifiers

features of S2, S4, and FC5 are input into three random forest classifiers respectively. The trained CNN serves as a trainable feature extractor, and RF serves as the basic classifier. Each RF classifier is trained in parallel using different levels of feature mapping in the CNN. Once the training is completed, the outputs of multiple random forest classifiers can be integrated through some decision strategies, such as the winner-takes-all strategy, to achieve ensemble learning.

7.3.3 Experiment Analysis

To evaluate the effectiveness of the proposed fault diagnosis method, the proposed model and algorithm are simulated in MATLAB. After achieving the expected results, they are integrated into the MRO maintenance service scheduling platform of the rail transit operation enterprise for application verification with actual data.

In addition, for the multi-layer multi-scale CNN and RF ensemble learning method proposed in this project, the project team used the public bearing vibration dataset provided by the Case Western Reserve University Bearing Data Center [18] and the existing dataset in the laboratory for preliminary exploration and algorithm verification, and achieved preliminary results.

Two datasets I and II were generated in the experiment. The original vibration signal data is transformed into a coefficient matrix by CWT with a scale factor of 1024. The scatter plot of the multi-layer features of the transformed CNN is shown in Fig. 7.4. The multi-layer features under the same load conditions are concentrated, while the features under different load conditions are separated. This shows that the

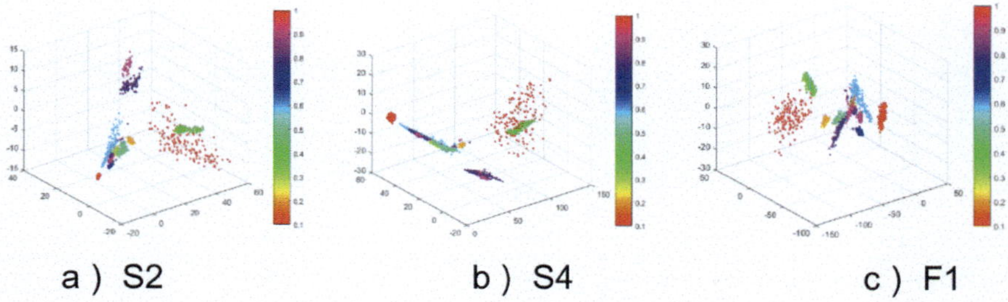

Fig. 7.4 Scatter plot of CNN multi-layer features

CNN model has a strong ability to extract representative features, and the transformation features at different levels are different. Fully utilizing multi-layer features can help the diagnosis results.

The feature mappings extracted from S2, S4, and F1 (the first layer of the fully connected layer FC5) are input into three RF classifiers respectively (each RF classifier consists of 50 decision trees), and the training error curves of datasets I and II are shown in Fig. 7.5.

The training error of the three RF classifiers is close to 0, which indicates that the feature mappings of the previous layers also contain important information that helps the diagnosis results. Based on multi-layer features, Table 7.2 gives the accuracy results of Softmax, RF1, RF2, RF3, and the ensemble learning classifier. The RF2 classifier used in the S4 layer has the highest accuracy for dataset I, which is 99.45%, and the accuracy for dataset II is 99.2%. For dataset II, the accuracy of using feature mappings of other layers is between 94.8% and 96.6%, which is lower than the proposed ensemble method. The training and testing samples of dataset II are selected under different loads, and it can be concluded that the introduction of the multi-classifier ensemble learning method helps to improve the robustness and generalization ability of fault classification.

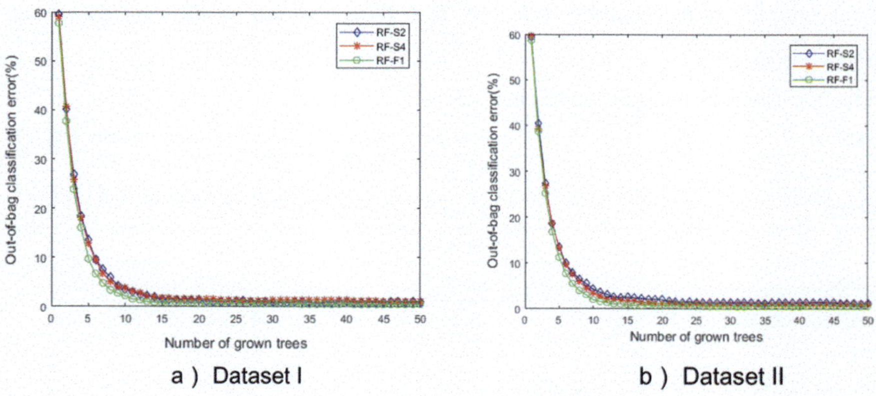

Fig. 7.5 Training error of 3 RF classifiers

Table 7.2 Accuracy comparison

Method	Accuracy (Dataset I) (%)	Accuracy (Dataset II) (%)
CNN + Softmax	99.05	94.8
CNN + RF1	99.4	96.6
CNN + RF2	99.45	99.2
CNN + RF3	91.5	94.8
Multi-layer multi-scale CNN + multiple random forests	99.45	99.2

A preliminary comparison of several fault diagnosis results based on traditional machine learning models (BPNN, SVM, KNN) and deep learning models (CNN, DBN, SAE) was made. All experiments were conducted on datasets I and II, and the results are shown in Table 7.3.

The diagnostic accuracy of BPNN, SVM, and KNN for Dataset I are 78.1%, 82.43%, and 76.9% respectively, and for Dataset II are 69.8%, 74.5%, and 65.4% respectively. Due to the shallow architecture of traditional machine learning methods, they cannot explore the complex relationship between health conditions and signal data, and the manually extracted features cannot represent the original data, resulting in much poorer diagnostic results. The classification accuracy of the proposed method is significantly better than other deep learning-based methods, especially for Dataset II, which shows a huge improvement, thus preliminarily verifying that the proposed method has strong feature learning ability and generalization ability of the integrated classifier.

The proposed method was also validated using bearing condition monitoring data stored in the Baosteel MRO management system. For detailed information, please refer to "http://imss.tongji.edu.cn:9080/". The vibration signal was collected from a mill with a frequency of 10 kHz, and the accelerometer was placed on the bracket. There are four different health conditions, namely normal condition (No), rolling ball defect (RD), inner ring defect (ID), and outer ring defect (OD). All 1024 data point samples were randomly selected from the condition monitoring data, with

Table 7.3 Accuracy comparison with other methods

Method	Accuracy (dataset I) (%)	Accuracy (dataset II) (%)
BPNN	78.1	69.8
SVM	82.43	74.5
KNN	76.9	65.4
DBN	89.45	86.53
SAE	93.1	85.4
CNN	99.05	94.8
Multi-layer multi-scale CNN + multiple random forests	99.45	99.2

2000 training samples and 500 test samples under each load condition. The signal image conversion results based on CWT are shown in Fig. 7.6. There are no obvious differences between the vibration signals in the time domain of the first three health states, but differences between the grayscale images containing fault features can be easily found in the time–frequency domain, demonstrating the effectiveness of the CWT conversion.

As the number of training samples increases, the parameters of the CNN model also change. After multiple tests, the best parameters for the experimental case are

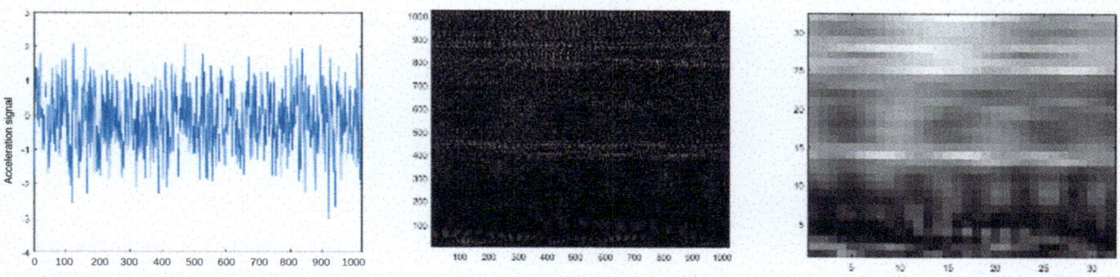

(a) Normal condition

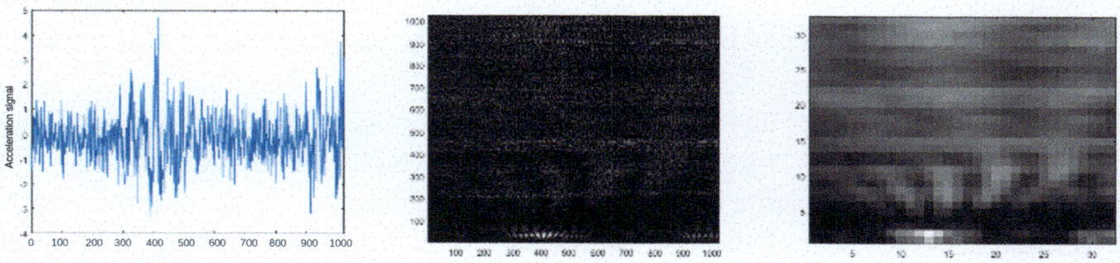

(b) Rolling ball defect condition

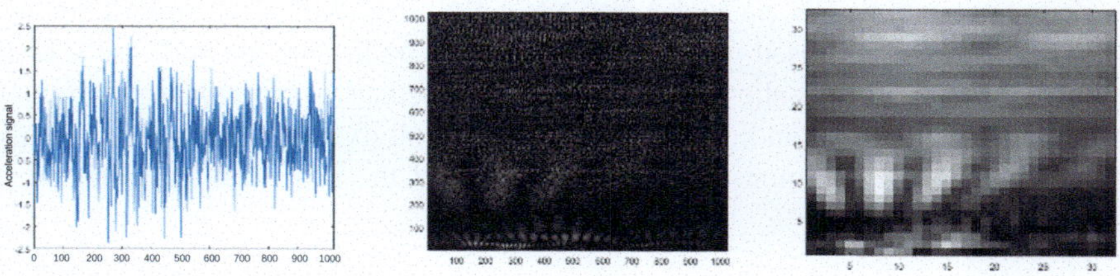

(c) Inner race defect condition

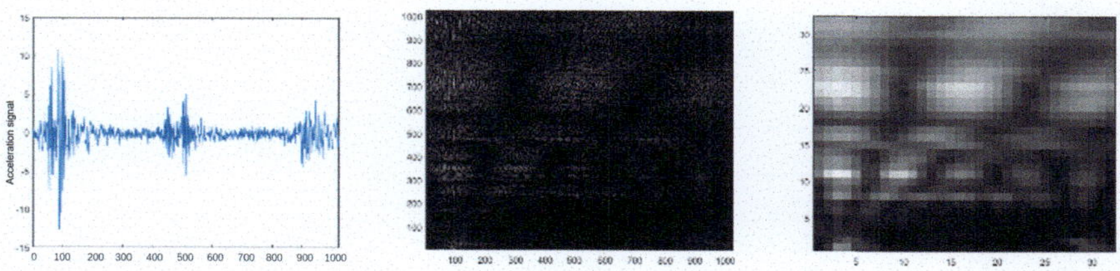

(d) Outer race defect condition

Fig. 7.6 Conversion results of the original vibration signal waveform based on CWT

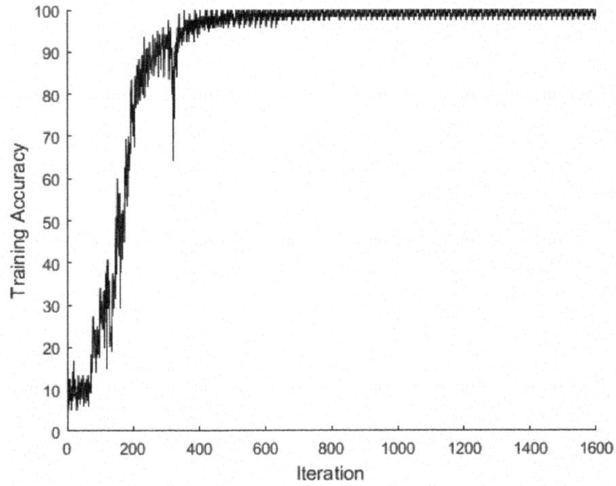

Fig. 7.7 Accuracy curve of training

as follows: batch size is set to 120, the number of epochs is set to 100, and the initial learning rate is set to 0.01. The CNN model can achieve a convergence speed of 99.5% through weighted and bias adjustments, and the accuracy curve of training is shown in Fig. 7.7.

Extracting multi-layer features of CNN for training different classifiers, the training accuracy results are shown in Table 7.4, which is less accurate than the previous case studies. Despite trying to adjust the architecture of the proposed CNN model, the accuracy results did not improve significantly. Upon final inspection of the original data, it was found that there were some "dirty data" in the samples, including redundant data and lost data.

Compared with BPNN, SVM, KNN, CNN, DBN, and SAE, the accuracy comparison results of the proposed method with the above traditional methods are shown in Table 7.5. The results show that the proposed method is superior to traditional machine learning-based and deep learning-based fault diagnosis methods in terms of average accuracy, demonstrating the potential of the proposed fault diagnosis method.

Table 7.4 Comparison of accuracy of the proposed method

Method	Accuracy (%)
CNN + Softmax	91.1
CNN + RF1	95.66
CNN + RF2	94.49
CNN + RF3	92.71
Multi-layer multi-scale CNN + multiple random forests	95.66

Table 7.5 Comparison of accuracy with other methods

Method	Accuracy (%)
BPNN	74.3
SVM	72.41
KNN	66.9
DBN	82.41
SAE	85.1
CNN	91.1
Multi-layer multi-scale CNN + multiple random forests	95.66

7.4 Fault Diagnosis Method Based on Convolutional Neural Networks and Transfer Learning

In many machine learning algorithms, there is a common assumption that the training dataset and the test dataset should come from the same feature space and maintain the same distribution [19, 20]. However, in many real-world problems, the distribution of the training dataset often differs from the test dataset, and the model trained using the training dataset sometimes cannot perform well in the test dataset. To solve this problem, researchers have proposed transfer learning methods, the goal of which is to apply the knowledge learned from previous tasks to new tasks. In fault diagnosis, methods based on transfer learning and deep convolutional networks can reuse parameters trained on the source domain dataset, improving diagnostic performance and increasing convergence speed.

7.4.1 Concept of Transfer Learning

The industrial application of machine learning is divided into the following two categories [21].

① The maturity of the training model allows these models to be deployed on a large scale to millions of user machines and has been widely adopted. Over the past few years, people have gained the ability to train increasingly accurate models. For many tasks, the most advanced models have reached a level of performance so good that they no longer pose a barrier to users. For example, the latest residual network on ImageNet has surpassed humans in object recognition, Google's smart reply can automatically handle 10% of all replies, the error rate of speech recognition has gradually dropped to below the error rate of human typing, machine learning models can diagnose skin cancer as accurately as dermatologists, Google's neural machine translation system has been applied to more than ten languages, Baidu can generate realistic voices in real time, and there are many more similar examples.

② The accuracy of the model is extremely dependent on data, and the improvement of model performance requires a large amount of labeled data. In some tasks and domains, these data are available and have been carefully organized over many years.

At the same time, when applying machine learning models in natural environments, the model will face many situations it has never encountered before and does not know how to handle. For example, each user has their own performance indicators and has or generates data different from the training dataset. Moreover, each model needs to handle tasks that are similar but not exactly the same as those during training. In all cases, even models that perform at or beyond human level on the training set will see a significant drop in performance, and may even become completely unusable in some scenarios. Transfer Learning (TL) can help handle these scenarios, especially when applied to domains where labeled data is not abundant.

According to the review by Pan and Yang [19], transfer learning involves the concepts of domain and task. A domain D consists of a feature space X and a marginal probability distribution $P(X)$, where $X = x_1, \ldots, x_n \in X$. For classification documents with many bag-of-words representations, X is the space of all document representations, x_i is the binary feature of the ith word, X is a specific document. Given a domain $D = \{X, P(X)\}$, a task T consists of a label space y and a conditional probability distribution $P(Y|X)$, which is usually learned from training data consisting of feature-label pairs $x_i \in X, y_i \in Y$. In the example of document classification, Y is the set of all labels, i.e., true or false, y_i is either true or false.

Given a source domain D_s, a corresponding source task T_s, as well as a target domain D_t, and a target task T_t, the purpose of transfer learning is: when $D_s \neq D_t, T_s \neq T_t$, with information coming from D_s and T_s, to learn the conditional probability distribution $P(Y_t|X_t)$ in the target domain D_t. In most cases, it is assumed that the number of labeled target samples that can be obtained is limited, and the number of labeled target samples is far less than the source samples.

Since both the domain D and task T are defined as tuples, given the source domain and target domain D_s and D_t, where, $D = \{X, P(X)\}$, and given the source task and target task T_s and T_t, where $T = \{Y, P(Y|X)\}$. The source and target situations can vary in four scenarios:

① $X_s \neq X_t$, the feature spaces of the source domain and target domain are different. For example, documents are written in two different languages, in the context of natural language processing, this scenario is often referred to as cross-language adaptation.

② $P(X_s) \neq P(X_t)$, the marginal probability distributions of the source domain and target domain are different. For example, two documents have different topics, this scenario is often referred to as domain adaptation.

③ $Y_s \neq Y_t$, the label spaces of the two tasks are different. For example, in the target task, documents need to be assigned different labels. In fact, this scenario often occurs in scenario 4, because different tasks have different label spaces, but have the same conditional probability distribution, which is extremely rare.

④ $P(Y_s|X_s) \neq P(Y_t|X_t)$, the conditional probability distributions of the source task and target task are different. For example, the source and target documents are unbalanced in categories, this scenario is quite common in practice, methods such as oversampling, undersampling, and SMOTE [22] are widely used.

7.4.2 Fault Diagnosis Model Based on CNN and TL

(1) CNN Model

A complete CNN model mainly includes three types of network layer structures: convolutional layer, pooling layer, and fully connected layer. Literature [23] provides the definition and calculation process of CNN. First, the input feature map always convolves with different convolution kernels, sometimes the convolution kernel is also called the convolution filter, and then the bias is added to each convolution result. Then, the output of the previous operation is used as the input of the activation function to obtain the output feature map. Finally, the complete process can be defined as follows:

$$x_i^l = f\left(\sum\nolimits_{j \in M_i} x_j^{l-1} k_{ji}^l + b_i^l\right) \tag{7.1}$$

In the formula, x_i^l is the ith output feature map of the lth layer; M_i is the selection range for input feature mapping; k_{ji}^l is the convolution filter, connecting the j th output feature map of layer $l-1$ and the ith input feature map of layer l; b_i^l is the bias attached to the ith input feature map of layer l; $f(\cdot)$ is the activation function used to enhance the non-linear expression ability of the model. Usually, each output map can correspond to multiple input maps. After the convolution layer, a subsampling layer is always used to reduce the dimension of the output feature map while keeping the number of feature maps unchanged. Here, the max pooling layer is used as the subsampling layer, defined as follows:

$$y_{\max} = \max(y_{i,j});\ a \le i \le b, c \le j \le d \tag{7.2}$$

In the formula, $y_{i,j}$ defines the subsampling subregion. i and j represent the height and width of the sampling window, respectively. The max pooling layer reduces the number of model parameters and avoids overfitting to a certain extent. Finally, the features extracted by stacking convolution layers and pooling layers are used as inputs to the fully connected layer, and then a Softmax classifier [24] is added to recognize the category of the object. The probability that input sample $x^{(n)}$ belongs to class m can be obtained as follows:

$$p(y^{(n)} = m|x^{(n)}; \theta) = \frac{e^{\theta_m^T x^{(n)}}}{\sum_{i=1}^{k} e^{\theta_i^T x^{(n)}}} \tag{7.3}$$

$$y_{\max} = \operatorname{argmax} p(y^{(n)} = m|x^{(n)};\theta) \tag{7.4}$$

In the formula, $n \in [1, l]$; l represents the number of training samples; $m \in [1, k]$; k represents the number of categories. In addition, θ represents the list of model parameters, including weights and biases. CNN minimizes the loss function through the backpropagation algorithm to obtain the optimal parameters. Generally, the loss function can be defined as follows:

$$L(\theta) = -\frac{1}{2}\left[\sum_{n=1}^{l}\sum_{m=1}^{k} I\{y^{(n)} = m\} \log p(y^{(n)}) = m|x^{(n)};\theta)\right] \tag{7.5}$$

In the formula, I {True} = 1.

(2) *CNN* structure design

In deep convolutional neural networks, lower-level networks learn low-level features and have better generalization capabilities, while higher-level networks learn high-level features and are more suitable for specific tasks [25]. A well-designed CNN structure can improve classification accuracy.

Figure 7.8 shows an 8-layer CNN model: 3 convolutional layers (C1 ~ C3), 2 sampling pooling layers (S1, S2), 1 input layer, 1 fully connected layer (FC), and 1 classification layer (Softmax). The training samples are grayscale images, and the input layer only has one channel (multiple channels can achieve multi-sensor data fusion, for example, 3 channels can solve the RGB three-color problem of images), the size is 32 × 32.

Generally speaking, using a smaller convolution kernel can reduce model parameters and alleviate computational burden. In the experiment, a 3 × 3 convolution kernel is applied in all convolution layers. In the C1 layer, there are 8 convolution kernels, corresponding to 8 output feature maps. Then, the maximum pooling technique is used to sample the output feature maps from the C1 layer. The stride of the sampling window is set to 2, and the size is 2 × 2. The size of the feature map in the S1 layer is half of that in the C1 layer, but it has the same number of feature maps. The C2 layer, S2 layer, and C3 layer are constructed in a similar way. The FC layer is a fully connected layer, used to combine all output feature maps in the C3

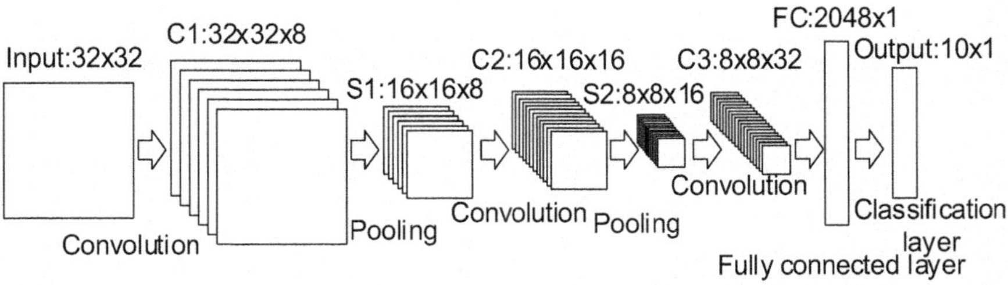

Fig. 7.8 CNN structure diagram

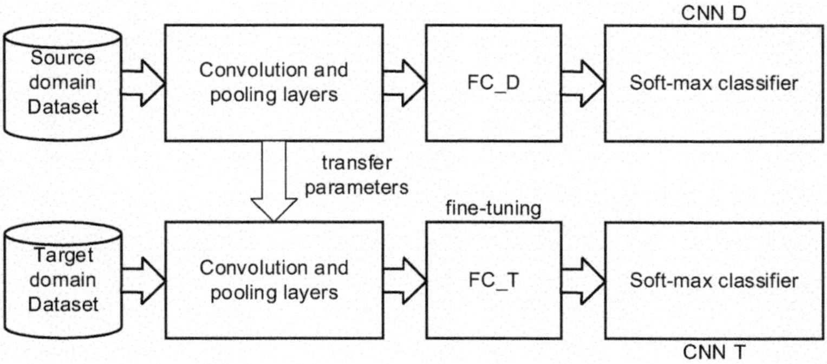

Fig. 7.9 Transfer learning based on CNN

layer. The Softmax classifier classifies all training samples into corresponding categories. In addition, during the training process, zero-padding technology is applied to keep the output feature map and the input feature map the same size, and the batch normalization method is used to accelerate CNN training.

(3) Fine-tuning of *CNN* model based on transfer learning

Fine-tuning is to transfer some parameters of the well-trained CNN model to the target CNN with a similar structure. As shown in Fig. 7.9, the fine-tuning CNN model based on transfer learning is divided into three stages: ① Use source domain samples to train CNN D to obtain the best structure model, making it have outstanding feature extraction and pattern recognition capabilities; ② Use the parameters of the lower network of CNN D to initialize CNN T; ③ Use target domain samples to fine-tune the fully connected layer of CNN T to better adapt to the new task requirements.

7.4.3 Experimental Analysis

First, train the best CNN model on the source domain as the original model, then use a small dataset on the target domain to fine-tune the model, and finally verify the effectiveness of the proposed bearing fault diagnosis method through two experiments: ① Compare the results of the fine-tuned model and the original model on the target domain; ② Retrain the model on the target domain and compare it with the fine-tuned model.

(1) Experimental dataset

The public bearing dataset used in the experiment is provided by CWRU [18]. Data is sampled at a frequency of 12 kHz under four different motor loads, with load ranges from 0 to 3. The bearing has three fault locations, including ball fault (BF), inner ring fault (IF), and outer ring fault (OF). In addition, each location has four

Table 7.6 Source domain dataset

Data sequence	Data Type	Motor load	Fault size
1	Normal	0	/
2	BF	0	0.007
3	IF	0	0.007
4	OF	0	0.007
5	BF	0	0.014
6	IF	0	0.014
7	OF	0	0.014
8	BF	0	0.021
9	IF	0	0.021
10	OF	0	0.021

fault sizes, respectively 0.007in^2, 0.014in, 0.021in, and 0.028in. This book only uses the normal data from the drive end and the fault data from the first three fault sizes.

The source domain dataset is created using the normal and fault data under 0 motor load. In the experiment, there are a total of 10 types of faults, and more related details are listed in Table 7.6. Similarly, the target domain dataset can be created, and Table 7.7 lists more details. It should be noted that the motor load and feature space between the source domain dataset and the three target domain datasets are different.

There are 1000 training samples in the source domain and 200 training samples in the target domain. However, both domains have 1000 test samples and the number of samples for each type of fault is the same.

(2) Results and Discussion

The original signal data was converted into a coefficient matrix using the CWT method, with a scale factor set to 1024 to obtain a 1024 $\times$ 1024 grayscale image.

Table 7.7 Target domain dataset

Data sequence	Data type	Motor load	Fault size
1	Normal	1/2/3	/
2	BF	1/2/3	0.007
3	IF	1/2/3	0.007
4	OF	1/2/3	0.007
5	BF	1/2/3	0.014
6	IF	1/2/3	0.014
7	OF	1/2/3	0.014
8	BF	1/2/3	0.021
9	IF	1/2/3	0.021
10	OF	1/2/3	0.021

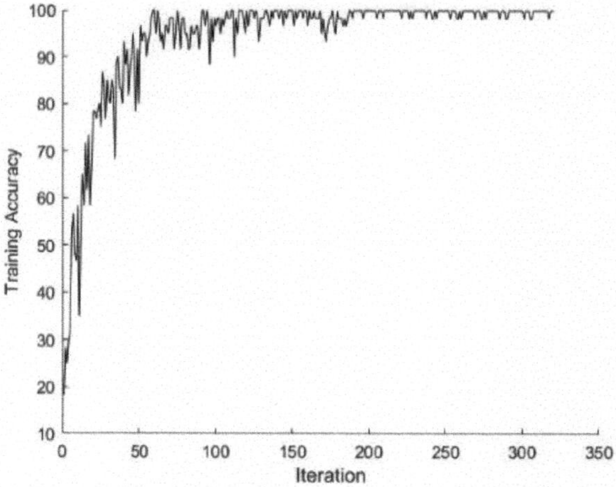

Fig. 7.10 Original model

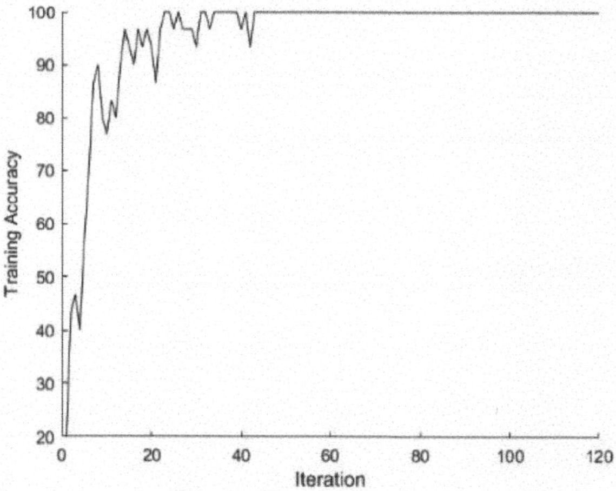

Fig. 7.11 Fine-tuned model

The image was compressed to a size of 32 × 32 using bicubic interpolation. The training accuracy curve in the source domain is shown in Fig. 7.10, with the training accuracy nearly reaching 100%. This process was repeated 10 times, and the final average accuracy of the test samples under 0 motor load was 99.18%. However, using the original model, the accuracy of the test samples under 1 motor load was only 78.2%. Then, the original model was fine-tuned using a small training sample under 1 motor load, and the accuracy of the test samples reached 99.3%. In addition, another experiment was conducted to prove that the convergence speed of the fine-tuned model (the training accuracy curve is shown in Fig. 7.11) is faster than training a model from scratch in the target domain (the training accuracy curve is shown in Fig. 7.12). The experimental results are listed in Table 7.8, and the fine-tuned model performs better.

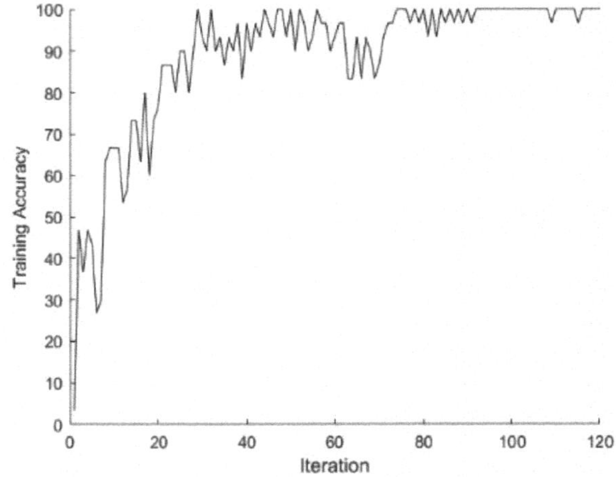

Fig. 7.12 Retrained model

Table 7.8 Experimental results

Accuracy (%)	Model without fine-tuning	Fine-tuning	Model training in target domain
Target domain 1	78.2	99.3	98.6
Target domain 2	77.5	99.1	98.9
Target domain 3	73.7	99.1	99.1

References

1. SUN Y, WANG X, TANG X. Deep learning face representation from predicting 10000 classes. IEEE Int Conf on Computer Vision and Pattern Recognition, 2014: 1891–1898.
2. DING X, HE Q. Energy-fluctuated multiscale feature learning with deep ConvNet for intelligent spindle bearing fault diagnosis. IEEE Transactions on Instrumentation & Measurement, 2017, 66(8): 1926–1935.
3. LEE J, NAM J. Multi-level and multi-Scale feature aggregation using pre-trained convolutional neural networks for music auto-tagging. IEEE Signal Processing Letters, 2017, 24(8): 1208–1212.
4. WANG Z, LAI C, CHEN X, et al. Flood hazard risk assessment model based on random forest. Journal of Hydrology, 2015, 527: 1130–1141.
5. ALLEN J B. Short time spectral analysis, synthesis, and modification by discrete fourier transform. IEEE Transactions on Acoustics, Speech, and Signal Processing, 1977, 25(3): 235–238.
6. Zhou Yufeng, Cheng Jingquan. Wavelet transform and its application. Physics, 2008, 37(1): 24–32.
7. HUANG N E, SHEN S S P. Hilbert-Huang transform and its applications. London: World Scientific, 2005.
8. HU J, WANG J, ZENG G. A hybrid forecasting approach applied to wind speed time series. Renew Energy, 2013(60): 185–194.
9. Lv Yong, Li Yourong, Wang Zhigang. Fault diagnosis of rolling mill main drive reducer based on empirical mode decomposition. Vibration, Testing and Diagnosis, 2007, 27(2): 112–115.
10. TANG L, YU L, WANG S, et al. A novel hybrid ensemble learning paradigm for nuclear energy consumption forecasting. Appl Energy, 2012(93): 432–443.

11. NGUYEN D N, DIEULLE L, GRALL A. Remaining useful lifetime prognosis of controlled systems: a case of stochastically deteriorating actuator, Math Probl Eng, 2015(2015).
12. WANG T, ZHANG M C, YU Q H, et al. Comparing the applications of EMD and EEMD on time-frequency analysis of seismic signal. Journal of Applied Geophysics, 2012, 83(1): 29–34.
13. IMAOUCHEN Y, KEDADOUCHE M, ALKAMA R, et al. A frequency-weighted energy operator and complementary ensemble empirical mode decomposition for bearing fault detection. Mechanical Systems and Signal Processing. 2017, 82(1): 103–116.
14. Luo Lei, Huang Boyan, Sun Jinwei, Wen Liang. Research on active noise control system based on overall average empirical mode decomposition. Acta Automatica Sinica.2016, 42(9): 1432–1439.
15. YU J, LV J. Weak fault feature extraction of rolling bearings using local mean decomposition-based multi-layer hybrid denoising. IEEE Transactions on Instrumentation and Measurement.2017, 66(12): 3148–3159.
16. JARDINE A, LIN D, BANJEVIC D. A review on machinery diagnostics and prognostics implementing condition-based maintenance. Mechanical Systems and Signal Processing, 2006, 20: 1483–1510.
17. LIN J, QU LS. Feature extraction based on morlet wavelet and its application for mechanical fault diagnosis. Journal of Sound & Vibration, 2000, 234(1): 135–148.
18. LOPARO K. Case western reserve university bearing data centre website. Available: http://csegroups.case.edu/bearingdatacenter/pages/download-data-file, 2012.
19. PAN S.J, YANG Q. A survey on transfer learning. IEEE Transactions on Knowledge and Data Engineering, 2010, 22(10): 1345–1359.
20. WEISS K, KHOSHGOFTAAR T M, WANG D D. A survey of transfer learning. Journal of Big Data, 2016, 3(1): 9.
21. Baijiahao. Depth|A comprehensive overview of transfer learning: from basic concepts to related research. http://baijiahao.baidu.com/s?id=1563010399699279&wfr=spider&for=pc, 2017.
22. ISHII T, ASHIHARA M, ABE S. Kernel discriminant analysis based feature selection. Neurocomputing, 2008, 71(13-15): 2544–2552.
23. BOUVRIE J. Notes on convolutional neural networks. Neural Nets, 2006.
24. JIANG Z, LIU M, XU G. Fault diagnosis method based on transfer learning and deep convNet. 2nd International Workshop on Structural Health Monitoring for Railway System (IWSHM-RS 2018), Qingdao, China, 2018.
25. ZEILER M D, FERGUS R. Visualizing and understanding convolutional network. European Conference on Computer Vision, 2014.

Chapter 8
Data-Driven Fault Prediction Model and Methods

In recent years, intelligent algorithms have received attention. Facing the uncertainty and dynamics of equipment failure and maintenance needs in the MRO operating network environment, combined with models based on Markov stochastic process models and neural network fusion models, wavelet decomposition, etc., this chapter mainly discusses the fault prediction model based on quantum multi-agent neural networks, fault prediction models based on machine learning, and fault prediction models based on deep learning, and other data-driven theories and methods of equipment fault prediction.

8.1 Fault Prediction Model Based on Quantum Multi-agent Neural Network

In recent decades, for the fault prediction problem of large complex equipment with non-linear, uncertain, and dynamic time-varying characteristics, the industry and scholars have conducted a large amount of research on Kalman filtering, hidden Markov models, expert systems, gray models, autoregressive moving average models, and support vectors [1–3]. In the information processing mode of simulating the human brain's neural system, the neural network (Neural Network, NN) is used to process the prediction of complex systems, which has the characteristics of reliability, scalability, and good robustness. The existing neural network models mainly include Back Propagation Neural Network (BPNN), Radial Basis Function Neural Network (RBFNN), Wavelet Neural Network (WNN), Elman Neural Network (ENN), and Generalized Regression Neural Network (GRNN) and other models. Among them, BPNN is currently the most widely used, with the advantages of simple structure and the ability to approximate non-linear continuous functions with any precision, and has been successfully applied to engineering prediction and other fields. Specifically, after extracting a large amount of feature information from the equipment monitoring

M. Liu et al., *Intelligent Predictive Maintenance*, Advanced and Intelligent Manufacturing in China, https://doi.org/10.1007/978-981-97-2677-6_8

signal, the BPNN framework can predict the deterioration state of the equipment by processing the time series data, thereby predicting the moment of fault occurrence.

Neural networks have realized the precise mapping between input and output, and have been proven to be a powerful prediction algorithm. In order to make the life prediction of the equipment fast, and at the same time have the same accuracy and reliability as the model-based method, Ghafir and others proposed a new type of multilayer feedforward BPNN, gave three types of neural network prediction models based on range, function, and sensor structure, and verified them through a single-axis gas turbine engine model [4]. AI-Garni and others proposed a framework based on artificial neural networks for reliability analysis of Boeing 737 tires based on flight operating time and landing times [5].

BPNN has problems such as slow convergence speed and easy to fall into local minimum, and the important parameters of the initial weight and threshold BPNN framework are closely related to the prediction accuracy and speed. The intelligent agent (Agent) has superior environmental perception and response capabilities. In order to achieve autonomous computing, multi-agent technology has flourished in recent years. Quantum computing applies the theory of quantum mechanics, such as quantum bits and state superposition, can process a large number of quantum states in parallel at the same time, and has stimulated in-depth research on quantum algorithms in the past few decades, such as quantum genetic algorithms, quantum ant colony algorithms, and quantum particle swarm optimization algorithms.

Based on the theory of multi-agent and quantum computing, by introducing the concepts of quantum computing and multi-agent systems, a Quantum Multi-agent based Neural Network (QMA-BPNN) model is established [6], which improves the initial weight and threshold selection method of the BPNN algorithm, the error propagation algorithm, and the selection of the number of hidden layer nodes, obtains the optimal initial weight and threshold, improves the learning rate and prediction performance of the BPNN, and solves the problem of complex equipment fault prediction. Through experimental analysis of public bearing datasets, it is found that under different numbers of hidden layer neurons, the prediction results of QMA-BPNN are more accurate and converge faster than those of BPNN based on genetic algorithms (GA-BPNN).

8.1.1 Neural Network Fault Prediction Model

Fault trend analysis through monitoring the abnormal values of key component sensors, such as pressure, temperature, fluid height and flow speed, vibration signals, etc., is very useful for equipment health status prediction. For most mechanical equipment, different degrees of performance decline can be reflected in the continuous data changes of state parameters. For data-driven methods, a prediction model based on monitoring signal feature values is constructed, and faults are identified through certain fault diagnosis techniques. Figure 8.1 is a three-layer BP neural network structure used for predicting feature values, which can approximate any

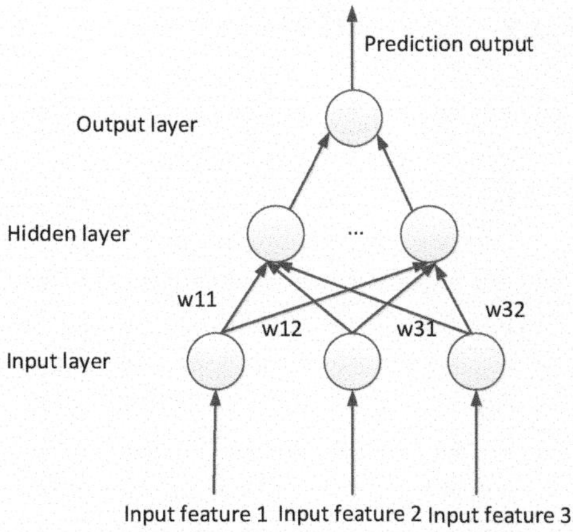

Fig. 8.1 Three-layer BP neural network structure for predicting feature values

continuous function with any precision. For the BP neural network, the nature of the mapping relationship between the input and output vectors is defined by the weight and threshold, and the difference between the network output and the actual response vector is reduced by the gradient descent algorithm in the direction. The training set vectors repeatedly adjust the weights and thresholds on the network until the error measure reaches an acceptable value or meets the convergence conditions of the inherent algorithm.

The BP neural network is prone to local minimum errors, and the network weights and thresholds are adjusted using the error gradient descent algorithm, which can easily lead to the network training results falling into the minimum. As the most important factors in the network, the initial weights and thresholds play a decisive role in the final result of fault prediction. Generally, they are random within a certain time interval, which may lead to the training results gradually decreasing along the error function to reach the minimum. That is to say, once the initial values are defined, the convergence direction of the network is determined. Moreover, the convergence speed of network training is also related to the selection of initial weights and thresholds. The initial weights and thresholds can be optimized through the quantum multi-agent algorithm to avoid falling into local extremes, speed up the convergence of BPNN, and improve the convergence accuracy. In addition, there are some optimization measures: ① The number of nodes in the hidden layer is an important parameter of BPNN. If the number of nodes is too small, the network cannot establish complex judgment boundaries; if the number is too large, the network will lose its ability to induce and judge; ② Due to the fluctuation of eigenvalues, normalization and denormalization operations are used before training and after testing to prevent small value information from being weakened by large data; ③ Tansig and logsig are usually used as the transfer functions of the first two layers, and the transfer functions of the last two layers of the neural network are selected from the network toolbox in MATLAB software, and the output of the network is between [0, 1].

8.1.2 QMA-BPNN Algorithm

(1) The concept of Agent Although there are various debates about the definition of Agent in the field of distributed artificial intelligence research, people believe that autonomy is the main feature of the agent, that is, it can complete tasks and achieve goals without people or other aids. Especially in the multi-agent system (Multi-agent System, MAS), the agent has attributes and methods, just like object-oriented programming, it has inherent beliefs, ideas, and purposes related to its environment. The agents in MAS interact with the environment to achieve information dissemination and search for the global optimal solution. In this section, the agent α represents a candidate solution for the initial weights and thresholds of the BP neural network, α = (α1, α2, …, αM), M is the total number of weights and thresholds.

Definition 8.1 *Multi-agent Grid (MAG).* All agents are in a square grid with a size of N × N as shown in Fig. 8.2, which is called a multi-agent grid or MAG. The size of the grid should be appropriate. Too complex grid will lead to large calculations and low efficiency, and too simple grid is not conducive to the search for the optimal solution, N is usually an even number. Each agent in MAG represents the weights between the input layer and the hidden layer, the thresholds of the hidden layer neurons, the weights between the hidden layer and the output layer, and the thresholds of the output layer neurons. The entire agent system is divided into independent organizations as shown in Fig. 8.2. Let αij represent the agent in the i-th row and j-th column of MAG, then the definitions of other members in the organization are as follows: ① If i ≤ N − 1, j ≤ N − 1, αi + 1, j, αi, j + 1, αi + 1, j + 1 are interconnected. ② If i = N, j ≤ N − 1, α1, j, αi, j + 1, αi − 1, 1 are interconnected. ③ If i ≤ N − 1, j = N, αi, 1, αi, j + 1, αi + 1, 1 are interconnected. ④ If i = N, j = N, α1, 1, α1, N are interconnected., $\alpha_{N,1}$ are interconnected.

Definition 8.2 *Elite Intelligent Agent.* The most powerful intelligent agent in the Multi-Agent Grid (MAG) (for example, the agent with the highest fitness value), including Group Elite Agent (GEA), Current-generation Elite Agent (CGEA), and Multi-generation Elite Agent (MGEA).

GEA represents the strongest intelligent agent in the organization, CGEA represents the strongest intelligent agent of the current generation, and MGEA represents the strongest intelligent agent over multiple generations. The QMA-BPNN algorithm uses the final optimization result as MGEA.

(2) Concept of Quantum Intelligent Agent (QAgent)

In quantum computing, the basic unit of information is described by quantum bits, which can be represented as follows:

$$|\varphi>=\alpha|0>+\beta|1> \tag{8.1}$$

where α and β satisfy the following normalization condition:

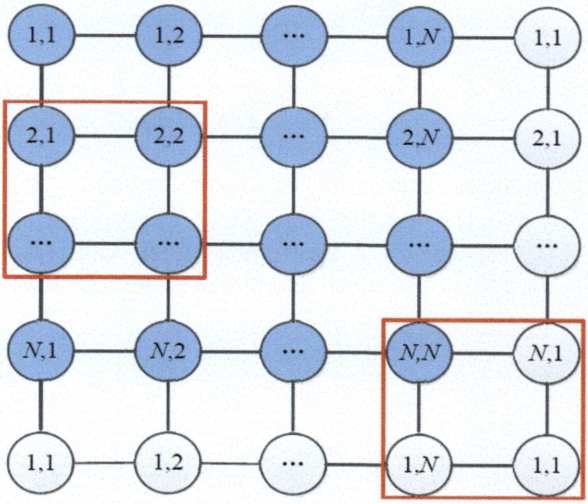

Fig. 8.2 Structure of MAG

$$|\alpha|^2 + |\beta|^2 = 1 \tag{8.2}$$

In Eqs. (8.1) and (8.2), the complex numbers α and β represent the probability amplitude of the quantum bit. Therefore, a quantum bit can also be defined as $[\alpha\ \beta]^{\mathrm{T}}$. According to the properties of probability, a quantum bit can be represented as shown in Fig. 8.3.

Therefore, $\alpha = \cos\theta$ and $\beta = \sin\theta$; then, the quantum bit can also be represented as follows:

$$|\varphi >= [\cos\theta \sin\theta]^{\mathrm{T}} \tag{8.3}$$

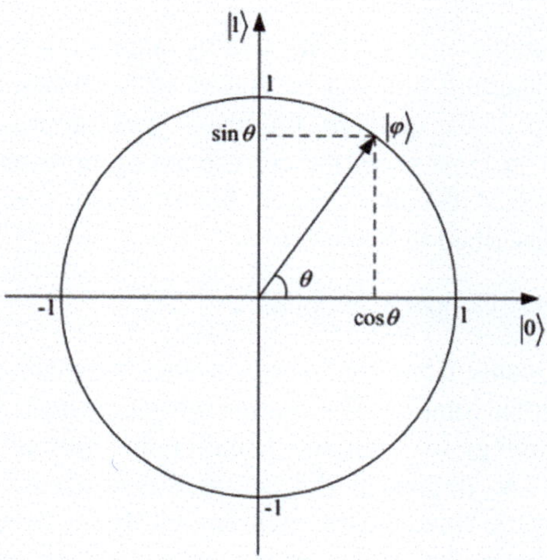

Fig. 8.3 Quantum bit

Specifically, a system of three quantum bits

$$\left\langle \begin{array}{c|c|c} \sqrt{3}/2 & 1/\sqrt{2} & -1/\sqrt{5} \\ 1/2 & -1/\sqrt{2} & 2/\sqrt{5} \end{array} \right\rangle \tag{8.4}$$

Can be represented by the state probabilities 3/40, 3/10, 3/40, 3/10, 1/40, 1/10, 1/40, and 1/10 for the states $|000\rangle$, $|001\rangle$, $|010\rangle$, $|011\rangle$, $|100\rangle$, $|101\rangle$, $|110\rangle$, and $|111\rangle$, respectively. A three-qubit system can represent eight information states, demonstrating that quantum bits can better represent population diversity.

$$\begin{aligned} &-\frac{\sqrt{3}}{2\sqrt{10}}|000\rangle + \frac{\sqrt{3}}{\sqrt{10}}|001\rangle + \frac{\sqrt{3}}{2\sqrt{10}}|010\rangle - \frac{\sqrt{3}}{\sqrt{10}}|011\rangle \\ &-\frac{\sqrt{1}}{2\sqrt{10}}|100\rangle + \frac{\sqrt{1}}{\sqrt{10}}|101\rangle + \frac{\sqrt{1}}{2\sqrt{10}}|110\rangle - \frac{\sqrt{1}}{\sqrt{10}}|111\rangle \end{aligned} \tag{8.5}$$

Definition 8.3 *Quantum Intelligent Agent (QAgent).* QAgent is an intelligent agent represented by quantum bits α_{ij}, as shown in Eq. (8.6).

$$\alpha_{ij} = \left[\begin{bmatrix} \alpha_{1,1}\ \alpha_{1,2}\ \cdots\ \alpha_{1,k_1} \\ \beta_{1,1}\ \beta_{1,2}\ \cdots\ \beta_{1,k_1} \end{bmatrix} \cdots \begin{bmatrix} \alpha_{m,1}\ \alpha_{m,2}\ \cdots\ \alpha_{m,k_m} \\ \beta_{m,1}\ \beta_{m,2}\ \cdots\ \beta_{m,k_m} \end{bmatrix} \cdots \begin{bmatrix} \alpha_{M,1}\ \alpha_{M,2}\ \cdots\ \alpha_{M,k_M} \\ \beta_{M,1}\ \beta_{M,2}\ \cdots\ \beta_{M,k_M} \end{bmatrix}\right] \tag{8.6}$$

where, M is the dimension of the problem; $\begin{vmatrix} \alpha_{m,1}\ \alpha_{m,2}\ \cdots\ \alpha_{m,k_m} \\ \beta_{m,1}\ \beta_{m,1}\ \cdots\ \beta_{m,k_m} \end{vmatrix}$ is the solution for the m th dimension, $m = (1,2, \ldots, M)$.

(3) Fitness evaluation function

Intelligent agents encoded by quantum bits cannot be directly evaluated for fitness. Quantum bits can be converted to binary for evaluation. Specifically, for the ith bit of the quantum bit string, generate a random number ε in the interval [0, 1]. If $|\alpha_i|^2 < \varepsilon$, set the corresponding position of the binary string to 1; otherwise, it is 0. After obtaining the binary string, complete the fitness (Fitness) evaluation as described in Table 8.1.

(4) Cooperative operator

Quantum bits obtain better values by updating individual quantum bits through directional rotation. This section introduces a quantum rotation gate based on a cooperative strategy to obtain the evolution direction of the Quantum Agent (QAgent).

As defined in 8.1, an intelligent agent can sense the environment and respond to the surrounding environment. In the process of evolution, only quantum intelligent agents with higher fitness can produce offspring, forming the next generation of multi-agent grid MAG, and ultimately producing the optimal individual. The

Table 8.1 Fitness evaluation process

Step	Contents
Step 1	For each quantum agent (QAgent) α in MAG, $\alpha = \left[\begin{bmatrix}\alpha_{1,1}\ \alpha_{1,2}\ \dots\ \alpha_{1,k_1}\\ \beta_{1,1}\ \beta_{1,2}\ \dots\ \beta_{1,k_1}\end{bmatrix} \vert\cdots\vert \begin{bmatrix}\alpha_{m,1}\ \alpha_{m,2}\ \dots\ \alpha_{m,k_m}\\ \beta_{m,1}\ \beta_{m,2}\ \dots\ \beta_{m,k_m}\end{bmatrix} \cdots \begin{bmatrix}\alpha_{M,1}\ \alpha_{M,2}\ \dots\ \alpha_{M,k_M}\\ \beta_{M,1}\ \beta_{M,2}\ \dots\ \beta_{M,k_M}\end{bmatrix}\right]$, set: Bin(α) = [(Bin$_{1,1}$ Bin$_{1,2}$ … Bin_{1,k_1}), …, (Bin$_{m,1}$ Bin$_{m,2}$ … Bin_{m,k_m}), …, (Bin$_{M,1}$ Bin$_{M,2}$ … Bin_{M,k_M})], $m = 1$
Step 2	When $m \le M$, set: $i = 1, x_m = 0$
Step 3	When i $\le k_m$, if random[0, 1] > $\vert\alpha_{m,i}\vert^2$, then set: Bin$_{m,j}$ = 1, otherwise Bin$_{m,j}$ = 0
Step 4	$i = i + 1$, end
Step 5	For x_m, perform binary encoding conversion Bin$_{m,1}$ Bin$_{m,2}$ … Bin_{m,k_m}
Step 6	$m = m + 1$, end
Step 7	Output $(x_1, x_2, \dots, x_M)$
Step 8	Evaluate fitness using the root mean square error of the neural network

cooperative strategy adopted is divided into two steps, which are related to the organizational elite intelligent agent GEA and the contemporary CGEA elite intelligent agent respectively.

First, the cooperation with the organizational elite intelligent agent GEA is as follows. Let $R(i, j)$ be a set of interacting quantum intelligent agents, $\alpha^*_{p,k} \in R\,(i, j)$, $\forall\, \alpha_{p,\,k} \in R\,(i, j)$, the fitness $\alpha^*_{p,k} \ge \alpha_{p,\,k}$.

If the fitness $\alpha_{(i,\,j)} \ge \alpha^*_{p,k}$, $\alpha_{(i,\,j)}$ remains unchanged; if the fitness $\alpha_{(i,\,j)} < \alpha^*_{p,k}$, $\alpha_{(i,\,j)}$ will die and be replaced by offspring, denoted as $\alpha'_{i,j}$, $\alpha_{(i,\,j)}$ is based on quantum rotation, the formula is as formula (8.7).

$$\begin{bmatrix}\alpha,_{m,k}\\ \beta,_{m,k}\end{bmatrix} \begin{matrix}\cos(\theta_{m,k}) - \sin(\theta_{m,k})\ \alpha_{m,k}\\ \sin(\theta_{m,k}) - \cos(m,k)\ \beta_{m,k}\end{matrix} \tag{8.7}$$

where, $[\alpha, \|_{m,k}\beta\|,_{m,k}]^T$ and $[\alpha_{m,k}\beta_{m,k}]^T$ respectively represent a quantum bit of $\alpha'_{i,j}$ and $\alpha_{(i,\,j)}$; $\theta_{m,\,k}$ is defined as in Eq. (8.8).

$$\theta_{m,k} = \Delta\theta_{m,k} * \mathrm{sign}(\alpha_{m,k} * \beta_{m,k}) \tag{8.8}$$

$\theta_{m,k}$ is the rotation angle of the quantum bit to the state “0” or “1”, for the search of $\theta_{m,k}$, as shown in Table 8.2.

(5) Crossover operation operator

To avoid falling into local minima, a crossover operation is introduced. Specifically, single-point crossover is used, a crossover point is randomly determined, and the quantum bits at the crossover point of the two quantum intelligent bodies are exchanged to increase the diversity of the population. If the ith position is chosen as a split point, the crossover operation is as shown in formula (8.9).

Table 8.2 Rotation angle of the QMA-BPNN method

Bin($\alpha_{i,j}$ [m, k])	Bin ($\alpha^*_{p,k}$ [m, k])	Fitness ($\alpha_{i,j}$) < Fitness ($\alpha^*_{p,k}$)	$\Delta\,\theta_{m,k}$	Sign($\alpha_{m,k} * \beta_{m,k}$)			
				$\alpha_{m,k} * \beta_{m,k} > 0$	$\alpha_{m,k} * \beta_{m,k} < 0$	$\alpha_{m,k} = 0$	$\beta_{m,k} = 0$
0	0	False	0	0	0	0	0
0	0	True	0	0	0	0	0
0	1	False	0	0	0	0	0
0	1	True	0.05π	− 1	+ 1	± 1	0
1	0	False	0.01π	− 1	+ 1	± 1	0
1	0	True	0.025π	+ 1	− 1	0	± 1
1	1	False	0.005π	+ 1	− 1	0	± 1
1	1	True	0.025π	+ 1	− 1	0	± 1

$$\begin{bmatrix} \alpha_{1,1} & \ldots & \alpha_{1,i} & \alpha_{1,i+1} & \ldots & \alpha_{1,n} \\ \beta_{1,1} & \ldots & \beta_{1,i} & \beta_{1,i+1} & \ldots & \beta_{1,n} \end{bmatrix}$$
$$\begin{bmatrix} \alpha_{1,1} & \ldots & \alpha_{1,i} & \alpha_{2,i+1} & \ldots & \alpha_{2,1} & \ldots & \alpha_{2,i} & \alpha_{2,i+1} & \ldots & \alpha_{2,n} \\ \beta_{2,1} & \ldots & \beta_{2,i} & \beta_{2,i+1} & \ldots & \beta_{2,n} \end{bmatrix}$$
$$\begin{bmatrix} \alpha_{2,1} & \ldots & \alpha_{2,i} & \alpha_{1,i+1} & \ldots & \alpha_{1,n} \\ \beta_{2,1} & \ldots & \beta_{2,i} & \beta_{1,i+1} & \ldots & \beta_{1,n} \end{bmatrix} \tag{8.9}$$

Similarly, cooperation with contemporary elite intelligent entities CGEA can be achieved by replacing the individual of CGEA with $\alpha^*_{p,k}$.

(6) Mutation operator

In addition to interacting with other nearby quantum intelligent entities, quantum intelligent entities can also change their own information under certain conditions, which is called mutation. Inspired by the quantum NOT gate and considering the quantum bit representation of quantum intelligent entities, the mutation of the QMA-BPNN model is achieved by randomly swapping states "0" and "1", for example, swapping α and the corresponding quantum bit β.

For example, for the following four quantum bit system, if the second position is selected for mutation, the mutation process is as shown in formula (8.10).

$$\begin{matrix} \frac{1}{\sqrt{2}} & \frac{1}{\sqrt{2}} & \frac{1}{2} & \frac{\sqrt{3}}{2} \\ \frac{1}{\sqrt{2}} & -\frac{1}{\sqrt{2}} & \frac{\sqrt{3}}{2} & -\frac{1}{2} \end{matrix} \rightarrow \begin{matrix} \frac{1}{\sqrt{2}} & -\frac{1}{\sqrt{2}} & \frac{1}{2} & \frac{\sqrt{3}}{2} \\ \frac{1}{\sqrt{2}} & \frac{1}{\sqrt{2}} & \frac{\sqrt{3}}{2} & -\frac{1}{2} \end{matrix} \tag{8.10}$$

This effect can be considered as a phase change of the quantum bit, changing from the original angle to the opposite angle to increase population diversity.

8.1.3 Prediction Algorithm Based on Quantum Multi-agent Neural Network

The computational framework of QMA-BPNN is shown in Fig. 8.4.

First, some parameter definitions are given.

① t—Evolutionary generations.
② MAG^t—The multi-agent grid MAG of the tth generation.
③ $\alpha_{i,j}^t$—The intelligent agent in the ith row and jth column of the multi-agent grid MAG represented by quantum bits.
④ α_{bet}^t—The contemporary elite intelligent agent CGEA of the t th generation.
⑤ $BEST(t)$—Stores the tth element in the collection of the elite intelligent agents CGEA of each generation.

The computational process of the Quantum Multi-Agent Back Propagation Neural Network (QMA-BPNN) can be summarized in the following steps.

Step 1: Determine the structure of BPNN.

Step 2: Let $t = 0$, generate the initial MAG^t, for example, $MAG^t = (\alpha_{i,j}^t)_{N \times N}$.

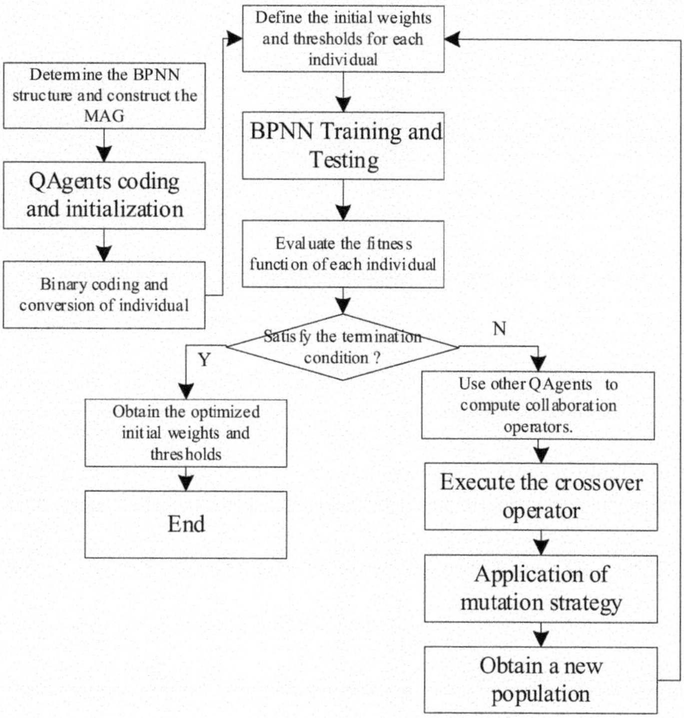

Fig. 8.4 Computational framework of QMA-BPNN

Step 3: Use $\alpha_{i,j}^{t}$ as the initial weights and thresholds of the BPNN, and carry out BPNN training and testing.
Step 4: Evaluate the fitness of $\alpha_{i,j}^{t}$. Select the contemporary elite intelligent agent CGEA from the multi-agent grid MAG, α_{best}^{t}, and store it in $BEST(t)$.
Step 5: If the termination condition is met, output the best solution in $BEST(t)$, i.e., the multi-generation elite intelligent agent MGEA, and retrain and test the network for fault prediction; otherwise, go to step 6.
Step 6: Cooperate $\alpha_{p,k}^{*}$ and $BEST(t)$, cross and mutate $\alpha_{i,j}^{t}$, to form the next generation MAG^{t+1} of MAG^{t}.
Step 7: Let $t = t + 1$, return to step 3.

8.1.4 Algorithm Analysis

The predictive performance of the QMA-BPNN model is tested through numerical approximation experiments designed with test functions and rolling bearing fault prediction. All algorithms are developed on the MATLAB R2012b simulation platform.

(1) Numerical Experiment

The QMA-BPNN method is compared with the standard BPNN and the BPNN based on the Genetic Algorithm (GA-BPNN) to determine the adaptability of applying quantum computing and multi-agent algorithm optimization to neural networks, in order to solve approximation problems including the gamma function, complementary error function, and right-tail χ^{2} distribution function. The training and testing datasets for these functions are provided by the Mathematica software package in MATLAB.

The gamma function has one parameter, defined by the following integral.

$$\Gamma(z) = \int_{0}^{\infty} t^{z-1} e^{-t} dt \tag{8.11}$$

The approximation of the gamma function uses a 1-20-1 configuration neural network, where the input nodes correspond to the z values in the range of $0 < z < 20$, and the output nodes correspond to $\ln\Gamma(z)$.

The complementary error function is defined as follows.

$$erfc(x) = \frac{2}{\sqrt{\pi}} \int_{0}^{\infty} e^{-t^{2}} dt \tag{8.12}$$

Since there is a direct relationship between the complementary error function and the cumulative integral of the Gaussian probability distribution function (also known as the normal distribution function), the latter is chosen as the second special function for the approximation ability test, defined as follows.

$$Q(x) = \frac{1}{\sqrt{2\pi}} \int_0^{\infty} e^{-t^2/2} dt = \frac{1}{2} erfc\left(\frac{x}{\sqrt{2}}\right) \tag{8.13}$$

The approximation of this function uses a 1-20-1 configuration neural network, where the input nodes are set to values in the range of 0–20, and the output nodes take the logarithm of the result.

The third is the cumulative distribution function of the χ^2 probability distribution, defined as Eq. (8.14).

$$Q(\chi^2|\nu) = \frac{1}{2^{\nu/2}\Gamma(\nu/2)} \int_{\chi^2}^{\infty} e^{-t/2} t^{\nu/2-1} dt, \nu > 0, \chi^2 \geq 0 \tag{8.14}$$

where, ν is the degree of freedom. The network structure is configured as 2-20-1, where the two input nodes correspond to χ^2 and ν values, where $1 \leq \chi^2 \leq 100$, $1 \leq \nu \leq 100$.

During network training, the Root Mean Square Error (RMSE) is used to compare function performance. RMSE is defined as formula (8.15).

$$\text{RMSE} = \frac{1}{PI} \sqrt{\sum_{p=1}^{P} \sum_{i=1}^{I} (D_{pi} - O_{pi})^2} \tag{8.15}$$

where, D_{pi} and O_{pi} are the expected and measured values respectively; P is the number of training times under different parameter values (such as the size of the learning rate); I is the number of this output node. Meanwhile, the population size of the Genetic Algorithm Optimized BP Neural Network (GA-BPNN) is set to 36, which is also the size of the QMA-BPNN grid. Each test is run independently 30 times. Table 8.3 lists the simulation results of BPNN, GA-BPNN, and QMA-BPNN. It can be seen that QMA-BPNN can better approximate these functions.

(2) Validation using bearing dataset

Table 8.3 Mean root mean square error (mean RMSE) of various neural network algorithms for approximation experiments of three different distribution functions

Neural network algorithm	Gamma function	Right-tailed normal distribution function	Right-tailed χ^2 distribution function
BPNN	1.25×10^{-6}	2.8×10^{-9}	1.72×10^{-5}
GA-BPNN	9.86×10^{-7}	1.52×10^{-9}	8.96×10^{-7}
QMA-BPNN	5.01×10^{-7}	7.33×10^{-11}	6.42×10^{-8}

Predicting the deterioration trend of mechanical components can provide enough time for maintenance engineers and replace components before actual failures occur. Bearings are crucial to all forms of rotating machinery and are one of the most common mechanical components. Unwarned bearing failures can have disastrous consequences in many cases, such as in helicopters, transport vehicles, continuous casting machines, and wind turbines. This section describes the bearing health status prediction based on QMA-BPNN, aiming to improve the performance of rotating machinery by reducing bearing failures. The simulation uses bearing data provided by the Intelligent Maintenance Systems Center at the University of Cincinnati.

To obtain real data, four force-lubricated bearings were installed on the shaft. A PCB 353b33 high-sensitivity quartz ICP accelerometer was installed on the bearing body. The speed was always kept at 2000 r/min, and a radial load of 6000 lb (1 lb = 0.45359237 kg) was placed on the bearing through a spring mechanism. The second dataset used contains accelerometer measurements, including vibration data sampled every ten minutes at a rate of 20 kHz over 7 days by the NI-DAQ 6062E data acquisition card. At the end of the fault test experiment, bearing 1 had an outer ring fault. The predictive ability of the QMA-BPNN detection algorithm on the dataset. In machine vibration monitoring, peak values are extracted from the original acceleration signal, and the acceleration peak value is selected as the feature value to evaluate the status of the machine or component. To find better results, experiments with different numbers of hidden layer neurons were conducted. Finally, the results of QMA-BPNN and GA-BPNN were compared, and QMA-BPNN had better generalization performance and learning speed. In both cases, bearing fault prediction is divided into two stages: the best search for initial weights and thresholds and BPNN learning. For the first stage, the maximum number of iterations is 20; the corresponding parameters for the second stage are set to 300, and the learning rate and acceptable error are 0.1 and 0.01, respectively.

In the experiment, the number of hidden layer neurons was 2 and 4, respectively. The GA-BPNN and QMA-BPNN algorithms and the bearing dataset 2 from the Intelligent Maintenance Systems Center at the University of Cincinnati were used. The sample $t + 1$ was predicted using samples t and $t - 1$. Figure 8.5 shows the peak after 10 min. The peak signal is represented by discrete points, and the predicted value is represented by a continuous line. As can be seen from Fig. 8.5, the vibration signal has a large jump before the fault, which is an important feature that truly reflects the bearing degradation trend. For the above number of hidden layer neurons, the results of this method are more accurate, especially at the moment of sharp rise adjacent to the time of fault occurrence, with smaller errors.

Figure 8.6 shows the error evolution curves of the GA-BPNN and QMA-BPNN algorithms, indicating that QMA-BPNN is superior to GA-BPNN in terms of evolution speed and accuracy. As the number of hidden layer neurons increases, the curve error decreases, confirming that appropriately increasing the number of hidden layer neurons can improve prediction accuracy. The Mean Square Error (MSE) was used to evaluate the predictive performance of 20 independent experiments, and the evaluation results are shown in Table 8.4.

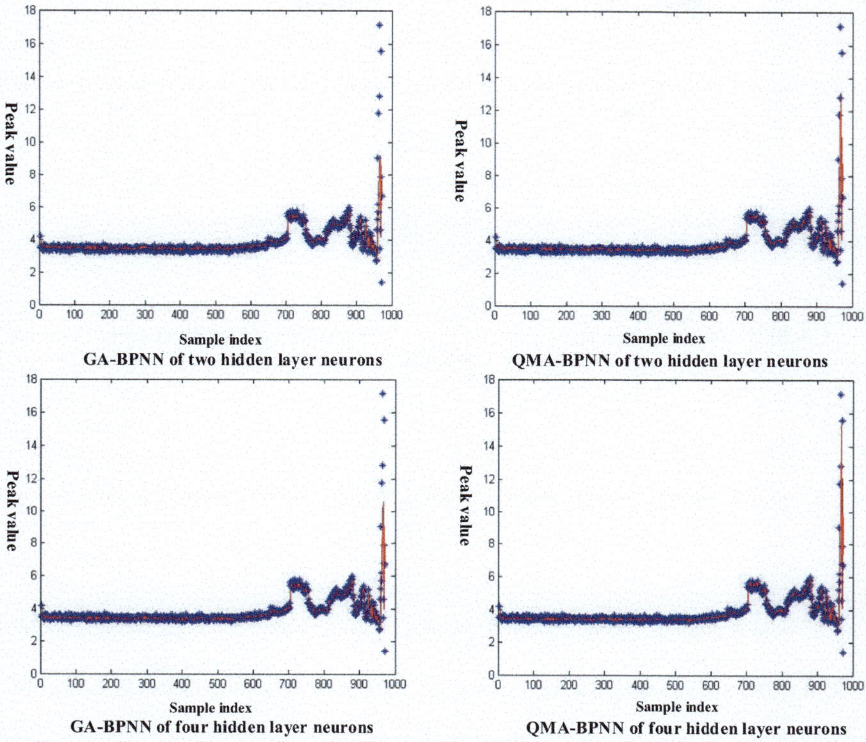

Fig. 8.5 Bearing vibration data prediction results based on GA-BPNN and QMA-BPNN (electronic version)

MSE, the error norm of the training sample (Error Norm) N_{tse}, and the error norm of the test sample N_{cse} are given by Eqs. (8.16)–(8.18). *Pre_out* and *tar_out* are the predicted and target output values, n_{ts} is the size of the training sample, n_{cs} is the number of test samples. QMA-BPNN uses the 2-2-1, 2-4-1, and 2-8-1 frameworks, and has a small error in both the training set and the test set.

$$MSE = \frac{1}{n_{ts}} \sum_{i=1}^{n_{ts}} (Pre\_out - tar\_out)^2 \tag{8.16}$$

$$N_{\text{tse}} = \sqrt{\sum_{i=1}^{n_{ts}} (Pre\_\ out - tar\_\ out)^2} \tag{8.17}$$

$$N_{\text{cse}} = \sqrt{\sum_{i=1}^{n_{cs}} (Pre\_\ out - tar\_\ out)^2} \tag{8.18}$$

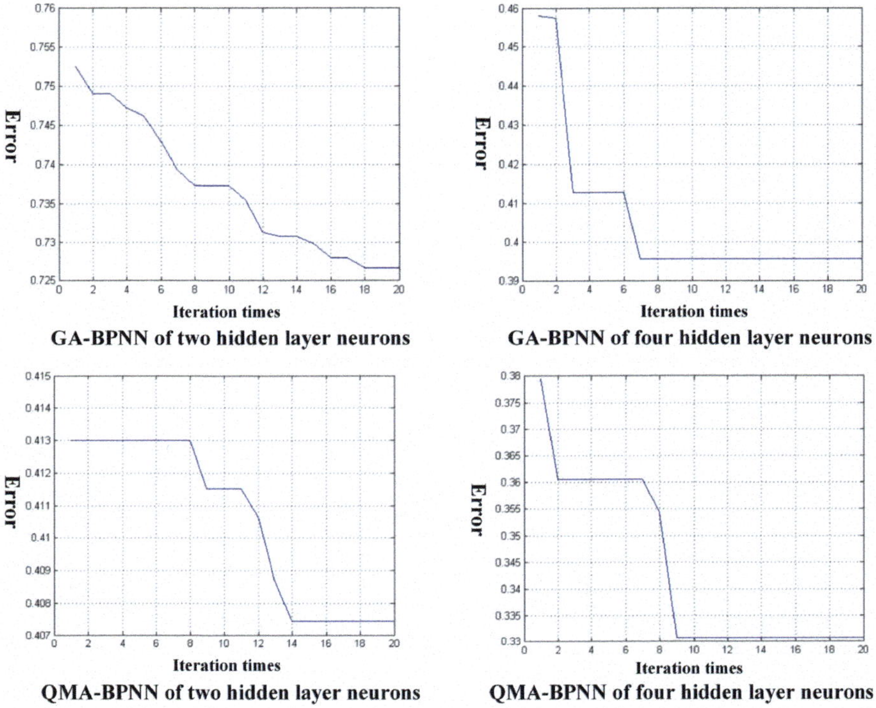

Fig. 8.6 Error evolution curve of GA-BPNN and QMA-BPNN

Table 8.4 Comparison of QMA-BPNN and GA-BPNN performance

BPNN Framework	GA-BPNN			QMA-BPNN		
	N_{cse}	N_{tse}	*MSE*	N_{cse}	N_{tse}	*MSE*
2-2-1	0.7021	1.2456	0.0021	0.5664	1.1894	0.0018
2-4-1	0.5459	1.284	0.0019	0.503	1.1782	0.0016
2-8-1	0.5192	1.1634	0.002	0.4357	1.1612	0.0015

8.2 Machine Learning-Based Fault Prediction Model

Fault prediction is a very rich field, and business problems can be approached from many different modeling perspectives. In the following sections, the process of fault prediction modeling is explained through methods such as binary classification and multi-classification machine learning [7].

8.2.1 *Fault Prediction Modeling Techniques*

(1) Fault Prediction Modeling Based on Binary Classification

Fault prediction modeling based on binary classification mainly predicts the probability of equipment failure in the future time period.

The division of time periods is determined by business rules and available data. Some common time periods include the minimum lead time required to purchase spare parts to replace potentially damaged components, or the time required to deploy maintenance resources to perform maintenance routines to fix potential problems within that time period, known as the future range time period "X".

In the discussion of binary classification used for fault prediction modeling, it is necessary to label both positive and negative types of situations (Fig. 8.7), where the positive type represents a fault (Label = 1), and the negative type represents normal operation (Label = 0). The labels belong to type categories, and the goal is to find a model that identifies whether each new example will fail or work normally in the future "X" units of time. Each example is a record of a unit of time belonging to an asset, and the operating conditions of this unit of time are conceptually described and extracted using feature engineering of historical and other data sources previously described.

Construct data labels as shown in Fig. 8.8. To create a predictive model to answer the question "What is the probability of an asset failing in the next 'X' units of time", take 'X' records before the asset fails and mark them as "about to fail" (Label = 1), and mark all other records as "normal" (Label = 0). The label is a categorical variable.

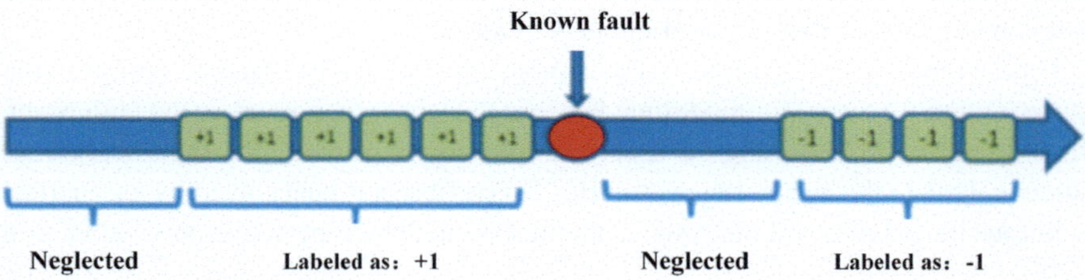

Fig. 8.7 Labeling of faults

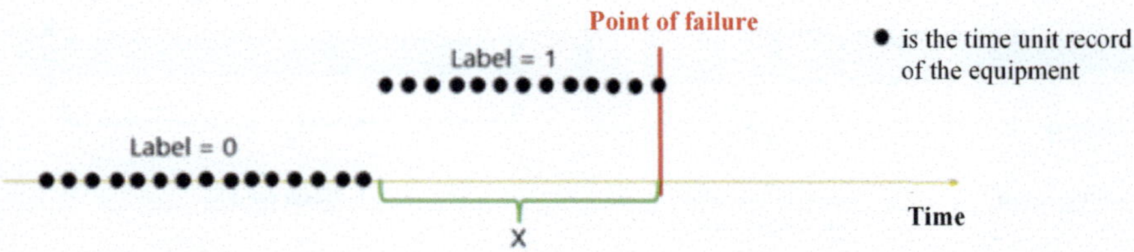

Fig. 8.8 Data labels for binary classification

For flight delays and cancellations, choose 'X' as one day to predict delays within the next 24 h. Mark all flights within 24 h before the fault occurs as 1 s. For ATM cash withdrawal faults, two binary classification models were established to predict the fault probability of transactions in the next 10 min, and also to predict the fault probability in the next 100 cash withdrawals. The first model marks all transaction faults that occurred in the last 10 min as 1. The second model marks all faults that occurred in the last 100 cash withdrawals as 1. For circuit breaker faults, the task is to predict the probability of the next circuit breaker command fault, in which case 'X' is chosen as a future command. For train door faults, a binary classification model is established to predict faults within the next 7 days. For wind turbine faults, 'X' is chosen as 3 months.

(2) Regression Analysis of Predictive Modeling

Regression analysis in predictive modeling is mainly used to calculate the remaining useful life (RUL) of an asset, which is defined as the amount of time the asset will run before the next fault occurs.

Like binary classification, each example is a record of a unit of time 'Y' belonging to an asset. However, the goal of regression analysis is to find a regression model to calculate the remaining useful life of each new instance, i.e., the remaining normal operating time before a fault. This time period is referred to as a multiple of 'Y'. Each test case has a remaining useful life, which can be calculated by measuring the remaining time before the next fault for that example.

The construction of labels for regression analysis is shown in Fig. 8.9. Considering the question "How long is the remaining useful life of the equipment", labels for the regression model can be constructed by taking each record before a fault and marking them by calculating how many units of time remain before the next fault. In this method, the label is a continuous variable.

Unlike binary classification, assets without any faults in the dataset for regression analysis cannot be used for modeling, because the label is set based on the fault point. If we do not know how long the asset has been running before the fault occurs, we cannot calculate the remaining useful life. This problem is best solved by a statistical technique called survival analysis. This method is not discussed in this book, as it

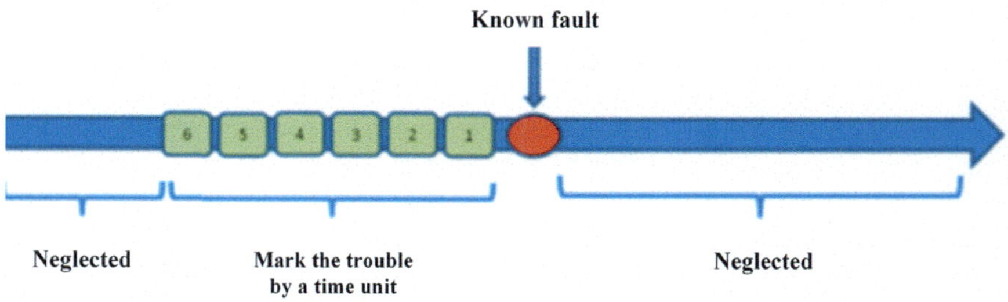

Fig. 8.9 Data labels for regression analysis

may introduce new problems in predictive maintenance involving frequent interval time-varying data.

(3) Multi-classification Model of Predictive Modeling

The multi-classification methods of predictive modeling can be used to predict two variables. The first is by allocating assets to one of several possible time periods to provide a time range for each asset failure. The second is to determine the likelihood of a failure in the future due to one of several failure modes. This allows maintenance personnel with relevant knowledge to address these issues in advance. Multi-classification modeling techniques focus on determining the most likely root cause of a given failure, which can provide targeted suggestions for repairing the failure, so as to take top-level maintenance operations to solve the failure. By listing the root causes of the failure and the related repair operations, technicians can more effectively complete the first repair operation after the failure.

Label construction: Considering two issues, namely "What is the probability of the device failing in the next aZ time units, where a is the number of periods", and "What is the probability of asset failure caused by failure P_i in the next X unit time, where i is the number of possible failure modes", these techniques are marked in the following ways.

For the first question, mark it by taking an aZ record before the device fails and using time buckets (3 Z, 2 Z, Z) as its label, and mark all other records as "normal" (Label = 0). The label is a categorical variable (as shown in Fig. 8.10).

For the second question, mark it by obtaining X records before the device fails and marking them as "about to fail due to problem P_i " (label = P_i), and mark other records as normal (label = 0). Here, the label is a categorical variable (as shown in Fig. 8.11).

The model can assign failure probabilities and non-failure probabilities based on each P_i. These probabilities can be sorted by metric to allow prediction of the most likely problems to occur in the future. The use case of aircraft component failure can be constructed as a multi-classification problem, which can predict the possibility of two different pressure valve components causing failure in the next month.

In order to better recommend maintenance operations after failure, there is no need to label data within the future time range. This is because this model does not predict future failures, but only predicts the most likely root cause once a failure

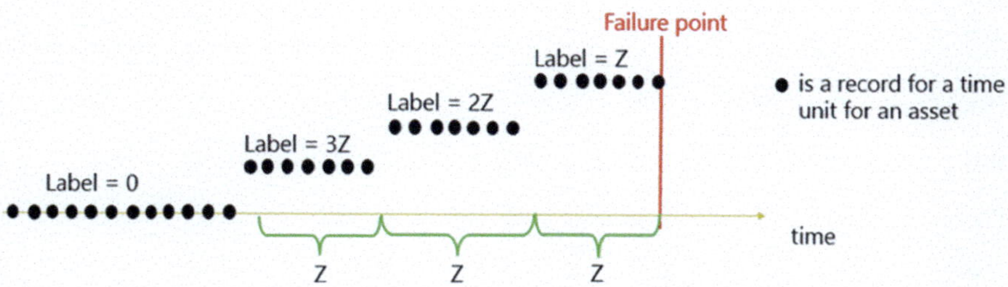

Fig. 8.10 Multi-class classification labels for failure time prediction

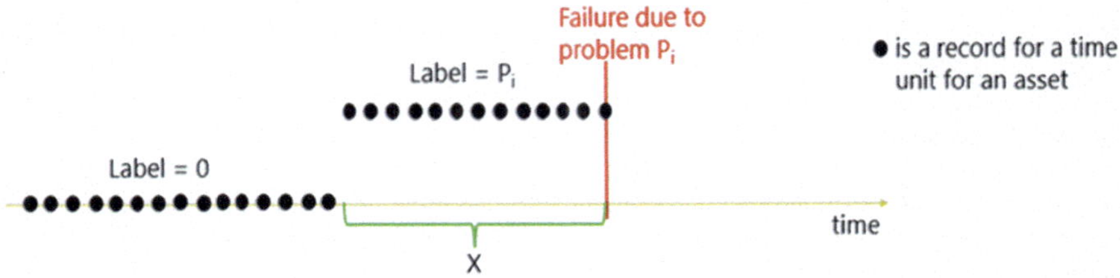

Fig. 8.11 Multi-class classification labels for root cause prediction

has occurred. The elevator door failure belongs to the third case, and its goal is to predict the cause of the failure based on the historical data of operating conditions. This model is suitable for predicting the most likely root cause after a failure occurs. A key advantage of this mode is that it will help inexperienced technicians easily diagnose and repair problems that would normally require years of experience to solve.

8.2.2 Training, Simulation, and Evaluation

By training and evaluating four regression models: Decision Forest Regression (DFR), Boosted Decision Tree (BDT), Poisson Regression (PR), and Neural Network Regression (NNR), Fig. 8.12 shows the comparison results of the four models on parameters such as Negative Log Likelihood (NLL), Mean Absolute Error (MAE), Root Mean Squared Error (RMSE), Relative Absolute Error (RAE), Relative Squared Error (RSE), Coefficient of Determination (CoD), etc. Among them, Decision Forest Regression and Boosted Decision Tree Regression perform best on the two main indicators MAE and RMSE.

In this step, binary classification modeling is explained from two aspects. First, four binary classification models: Logistic Regression (LR), Boosted Decision Tree (BDT), Decision Forest (DF), and Neural Network (NN) are trained and evaluated; second, Fig. 8.13 compares the results of the four models to determine the best

Algorithms	NLL	MAE	RMSE	RAE	RSE	CoD
DFR	462.724302192819	21.141062	30.147662	0.574998	0.526317	0.473683
BDT	Infinity	21.283579	29.615866	0.578874	0.507913	0.492087
PR	NA	23.243022	29.973356	0.632167	0.520249	0.479751
NNR	NA	75.505023	86.185055	2.053597	4.301346	-3.301346

Fig. 8.12 Training and evaluation of four typical regression models

Algorithms	Accuracy	Precision	Recall	F-Score
LR	0.92	0.947368	0.72	0.818182
BDT	0.91	0.9	0.72	0.8
DF	0.92	0.947368	0.72	0.818182
NN	0.94	0.952381	0.8	0.869565

Fig. 8.13 Comparison of indicators of four binary classification models

model. The two-layer neural network algorithm performs best in terms of accuracy, precision, recall, and F-measure.

In this step, two multi-classification models are trained and evaluated: multi-class logistic regression and multi-class neural network. Figure 8.14 compares the results of “multi-class logistic regression” and “multi-class neural network”, the latter performs better in terms of overall accuracy, average accuracy, micro-average precision, macro-average precision, micro-average recall, and macro-average recall.

Multiclass Logistic Regression

Metrics

Overall accuracy	0.88
Average accuracy	0.92
Micro-averaged precision	0.88
Macro-averaged precision	0.792622
Micro-averaged recall	0.88
Macro-averaged recall	0.74

Confusion Matrix

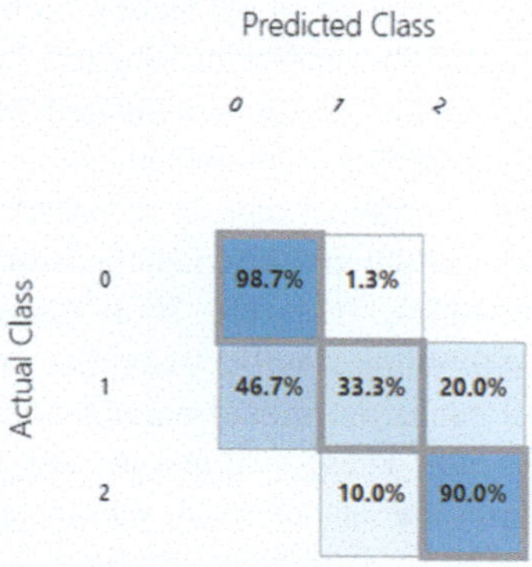

Multiclass Neural Network

Metrics

Overall accuracy	0.92
Average accuracy	0.946667
Micro-averaged precision	0.92
Macro-averaged precision	0.868946
Micro-averaged recall	0.92
Macro-averaged recall	0.84

Confusion Matrix

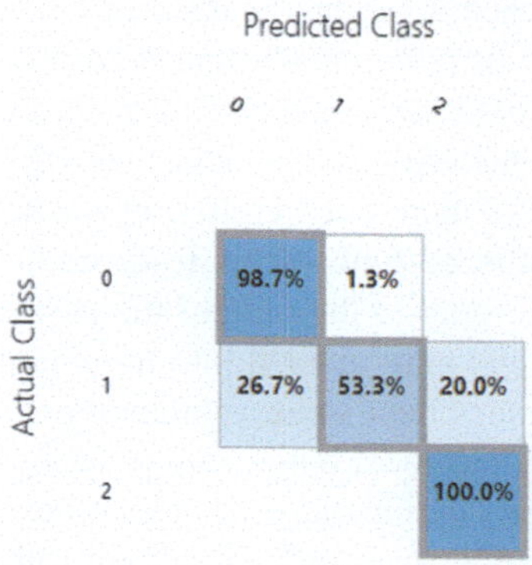

Fig. 8.14 Comparison of algorithms and indicators of 2 multi-classification models

MZOE	MAE	RMSE
0.09	0.09	0.3
0.07	0.07	0.264575

Fig. 8.15 Comparison of training results of 2 models

When evaluating ordinal regression models, the results are similar to those of regression models. Figure 8.15 shows the comparison results between the models trained in Fig. 8.14 in terms of parameters such as Mean Zero One Error (MZOE), MAE, RMSE, etc.

8.2.3 Verification and Testing Methods

In predictive modeling, like any other solution that includes timestamp data, typical training and testing need to consider aspects of time variation to better generalize unseen future data.

(1) Cross-validation

Many machine learning algorithms rely on many hyperparameters that can significantly change model performance. The optimal values of these hyperparameters are not automatically calculated when training the model, but should be specified by data scientists.

There are several methods to find the optimal values of hyperparameters. The most common is "k-fold cross-validation", which randomly divides instances into "k" folds. For each set of hyperparameter values, the learning algorithm runs k times. In each iteration, the instances in the current iteration are used as the validation set, and the remaining instances are used as the training set. At the end of the loop for each set of hyperparameter values, the average of k performance metrics is calculated, and the hyperparameter values with the best average performance are selected.

As mentioned earlier, in predictive maintenance problems, data is recorded as events of time series from multiple data sources. These records can be sorted according to the time of the labeled records or examples. Therefore, if the dataset is randomly divided into training and validation sets, some training examples will be later than some validation examples. This will cause the model parameters to fall into local optimization during the training process. These evaluations may be overly optimistic, especially in cases where time series are unstable and change their behavior over time. The end result is that the selected hyperparameter values may not be ideal.

A better way to find the optimal values of hyperparameters is to divide the examples into training and validation sets in a time-dependent manner so that all validation examples are later than all training examples. Then, for each set of hyperparameter values, train the algorithm on the training set, measure the model's performance on the same validation set, and select the hyperparameter values that show the best performance. When time series data is not stationary and evolves over time, the performance of the model formed by the hyperparameter values selected through the training/validation split is better than that randomly selected through cross-validation. The best hyperparameter values found using training/validation splitting or cross-validation are used to generate the final model by training the learning algorithm on the entire data.

After establishing the model, it is necessary to evaluate its performance on a new dataset. The simplest way might be to evaluate the model's performance on the training data. However, this evaluation result is overly optimistic because the model is built based on the dataset used to evaluate performance. A better evaluation method could be the performance metric of the hyperparameter values calculated on the validation set or the average performance metric calculated from cross-validation. But for the same reasons mentioned earlier, these evaluations are still overly optimistic.

One method is to randomly divide the data into training, validation, and test sets. The training and validation sets are used to select the values of the hyperparameters and train the model with them. The model's performance is measured on the test set. Another method related to predictive maintenance is to decompose the examples into training, validation, and test sets in a time-based manner, so that the test examples lag behind all training and validation examples in time. After the split, model generation and performance measurement are the same as previously described. When the time series is stationary and easy to predict, both methods will produce similar evaluations of future performance. However, when the time series is unstable and/or difficult to predict, the second method will have better predictive performance than the first method.

(2) Time-based division

As a best practice, a time-dependent bidirectional split between the training set and the test set is described. However, for the training and validation sets, the same logic should be applied to the time-dependent split.

Suppose there is a stream of timestamp events, such as measurements from various sensors. The features of training and testing examples are defined together with their labels through a time range containing multiple events. For example, for binary classification, as described in the feature engineering and modeling techniques section, features are created based on past time, and labels are created based on future events within "X" time units in the future. Therefore, the time range of the labels of the examples is later than the time range of their features. For time-dependent splitting, choose a point in time, at which a model with optimized hyperparameters is trained using historical data up to that point. To prevent future labels beyond the training cutoff from leaking into the training data, choose the latest time range to mark the training examples as X units before the training cutoff date. In Fig. 8.16, each solid

Fig. 8.16 Time division for training and testing (electronic version)

circle represents a row in the final feature dataset calculated according to the above method. Suppose, the figure shows the records that should enter the training and testing sets when a time-dependent split is implemented for X = 2 and W = 3.

The green squares represent records available for training that belong to time units. As previously explained, each training example in the final feature table is generated by looking up the past 3 periods for features and looking up 2 future periods for labels before the training day cutoff. When any part of the 2 future periods used for this example has exceeded the training cutoff, the example is not used in the training set, because it is assumed that visibility beyond the training cutoff is not available. Due to this constraint, the black examples represent records of the final labeled dataset that should not be used in the training dataset. These records will also not be used in the test data because they are before the training cutoff, and their label time range partly depends on the training time range, which should not be the case, as it is desired to label training and testing completely independently to prevent label information leakage.

This technology allows for the overlap of data generated by features between training and testing examples near the training cut-off. Depending on data availability, separation is achieved by not using any examples within the *W* time units of the training cut-off in the test set.

In practice, regression models used to predict remaining useful life are more severely affected by leakage issues, and using random splits can lead to extreme overfitting. Similarly, in regression problems, the split should allow for records of assets that had failures before the training cut-off to be used for the training set, and assets that had failures after the cut-off to be used for the test set.

As a general method, another important best practice for splitting data for training and testing is to use splits divided by asset ID, ensuring that no assets used for training are used for testing, as the idea of testing is to ensure that the model provides accurate results when predicting with new assets.

(3) Dealing with imbalanced data

In classification problems, if one class has more examples than the others, the data is considered imbalanced. Ideally, you want to have enough representation of each class in the training data to distinguish between different classes. If a class is less than 10% of the data, it can be said that the data is imbalanced, and the underrepresented data set is referred to as the minority class. Broadly, in many cases, it is found that one class in an imbalanced data set is severely underrepresented compared to other classes, such as only accounting for 0.001% of data points. Class imbalance is a problem in many fields, including fraud detection, network intrusion, and predictive maintenance, where failures rarely occur during the generation period of assets that make up the minority class examples.

In the case of class imbalance, the performance of most standard learning algorithms is compromised because they aim to minimize the overall error rate. For example, for a dataset with 99% negative class examples and 1% positive class examples, 99% accuracy can be achieved by marking all instances as negative. However, this misclassifies all positive examples, so despite the very high accuracy measure, this algorithm is useless. Therefore, overall accuracy measures like error rate are insufficient in imbalanced learning situations. Other measures, such as precision, recall, F1 score, and cost-adjusted ROC curve, are used for evaluation in imbalanced datasets.

There are some methods that can help remedy the class imbalance problem, such as sampling methods and cost-sensitive learning.

(4) Sampling Methods

Sampling methods are used in imbalanced learning (including modifying the dataset through some mechanism) to provide a balanced dataset. Although there are many different sampling techniques, the most straightforward are random oversampling and undersampling.

Simply put, random oversampling is the selection of random samples from the minority class, duplicating these samples and adding them to the training dataset. This increases the total number of samples in the minority class, ultimately balancing the number of samples from different classes. One danger of oversampling is that multiple instances of certain samples may make the classifier overly specific, leading to overfitting. This results in high training accuracy, but the performance on unseen test data may be very poor. Conversely, random undersampling selects random samples from the majority class and removes these samples from the training dataset. However, removing samples from the majority class may cause the classifier to miss important concepts related to the majority class. Mixed sampling, oversampling the minority class while undersampling the majority class, is another viable method. There are many other more complex sampling techniques available, and effective sampling methods for class imbalance is a popular research area that receives ongoing attention and contributions from various sources. Different techniques are used to determine the most effective method, usually determined by data scientists through research and experimentation, and it highly depends on data attributes. Moreover, it is especially important to ensure that sampling methods are only applied to the training set and not the test set.

(5) Cost-Sensitive Learning

In predictive maintenance, failures that make up the minority class are of more concern than regular examples. This is often referred to as unequal loss or asymmetric cost for misclassifying elements of different classes, where predicting a positive incorrectly as a negative costs more, and vice versa. The desired classifier should be able to give high-accuracy predictions on the minority class without affecting the accuracy of the majority class.

There are several ways to achieve this. By assigning high costs to misclassifications of the minority class and trying to minimize the overall cost, the problem of unequal loss can be effectively handled. Some machine learning algorithms inherently use this concept, such as SVM (Support Vector Machine), which incorporates the cost of positive and negative examples into itself during training. Similarly, using boosting methods has shown good performance in the case of imbalanced data (like boosted decision tree algorithms).

(6) Evaluation Metrics

As mentioned earlier, class imbalance leads to poor performance because algorithms tend to classify majority class examples better in the case of minority classes, as the overall misclassification problem can be greatly improved when the majority class is correctly labeled. This results in low recall and becomes a bigger problem when the cost of false alarms to the business is extremely high. Accuracy is the most popular measure used to describe classifier performance. However, as mentioned above, because it is very sensitive to data distribution, accuracy is ineffective and does not reflect the true performance of the classifier's capabilities. Instead, other evaluation metrics should be used to evaluate imbalanced learning problems. In this case, precision, recall, and F1 score should be looked at as initial indicators when evaluating the performance of predictive maintenance models. In predictive maintenance, recall represents how many failures in the test set were correctly identified by the model. A higher recall means the model has successfully captured the true failures. Precision measures involve the false alarm rate, where a lower precision corresponds to a higher false alarm. The F1 score considers both precision and recall, with the best value being 1 and the worst value being 0.

In addition, for binary classification, decile tables and lift charts can provide rich information for evaluation. They only focus on the positive class (failures) and provide a more complex picture of algorithm performance than what can be seen by just looking at a fixed operating point on the ROC (Receiver Operating Characteristic) curve. By sorting the test examples by the failure probability predicted by the model before the threshold determines the final label, a decile table is obtained. Then, the sorted examples are grouped by decile (e.g., the 10%, 20%, 30% of examples with the highest probabilities, etc.). The algorithm's performance change in each decile can be estimated by calculating the ratio between the true positive rate of each decile and its random baseline (i.e., 0.1, 0.2…). A lift chart can be used to plot the decile values by plotting pairs of true positive rates and random true rates for all deciles. Typically, the first decile is the focus of the results, as the greatest gains are seen here. When used for predictive maintenance, the first decile can also be seen as a representative of those "at risk."

8.3 Deep Learning-Based Fault Prediction Model

The characteristic of deep learning is its deep network architecture, which is stacked in multiple layers in the network to fully extract representative information from the original input data [8]. With the help of complex deep structures, it can model high-level abstractions of data more effectively than shallow networks, thus extracting features more effectively. Deep learning methods have gained great interest and achievements in various fields such as image recognition and speech recognition. Due to the high similarity between the original data of equipment health monitoring and the methods of image processing research, deep learning architecture has great potential in PHM and RUL estimation.

Ren et al. proposed a comprehensive deep learning method for multi-bearing remaining useful life (RUL) collaborative prediction combining time-domain and frequency-domain features [9], and verified the effectiveness and superiority of the method through numerical experiments on actual datasets. Liao et al. used a new regularization term and an unsupervised self-organizing map (SOM) algorithm to propose a Restricted Boltzmann Machine (RBM) for representation learning [10], to predict the RUL of equipment. Zhang et al. proposed a multi-objective deep belief network (DBN) method [11], evolving multiple DBNs simultaneously, achieving better predictive performance through NASA's turbofan engine degradation problem.

In the deep learning architecture, deep convolutional neural networks (DCNN) designed specifically for variable and complex signals can also be utilized. In the past few years, DCNN has achieved significant success in various applications. DCNN was first proposed by LeCun et al. for image processing, and it has the ability to preserve data information without considering scale, displacement, and distortion invariance [12]. A large amount of research in computer vision, speech processing, etc., benefits from the local receptive field, shared weights, and spatial subsampling characteristics of DCNN. DCNN models include one-dimensional convolutional neural networks, two-dimensional convolutional neural networks, and three-dimensional convolutional neural networks. One-dimensional convolutional neural networks are often used for sequence data processing, two-dimensional convolutional neural networks are often used for image text recognition, and three-dimensional convolutional neural networks are mainly used for medical image and video data recognition. Babu et al. [13] built a two-dimensional deep convolutional neural network to predict the system RUL using normalized variable time series of sensor signals, where one dimension of the two-dimensional input is the number of sensors. Their work uses average pooling and places a linear regression layer at the top. Although the deep CNN architecture shows strong ability in feature extraction, its application research in equipment remaining life prediction is very limited. By adopting the DCNN architecture and extracting local data features through the deep learning network, better predictions can be obtained.

8.3.1 Convolutional Neural Network Model

Convolutional Neural Networks (CNN) were first proposed by LeCun for image processing. They have two features: spatially shared weights and spatial pooling, and have achieved significant success in various research and industry fields such as computer vision, natural language processing, and speech recognition. The convolutional layer applies multiple filters (convolution kernels) to the original input data to generate features, and the pooling layer then extracts important local characteristics. The input data of CNN is usually two-dimensional data (as shown in Fig. 8.17), and abstract spatial features are learned through interleaved and stacked convolution kernels and pooling operations.

The image on the far left of the CNN model is the input layer composed of several matrices, the same as the Deep Neural Network (DNN). The Convolution Layer is unique to CNN, and the activation function of the convolution layer uses ReLU (x) = max (0, x). The next is the Pooling Layer, also unique to CNN, which does not have an activation function. The combination of Convolution Layer + Pooling Layer can appear many times in the hidden layer, in fact, this number is according to the needs of the model. Of course, the combination of Convolution Layer + Convolution Layer, or Convolution Layer + Convolution Layer + Pooling Layer can also be flexibly used when building the model, there are no restrictions on these. However, the more common deep CNN is a combination of several convolution layers + pooling layers, followed by a Fully Connected Layer (FC), and the output layer uses the Softmax activation function for image recognition classification. In recent years, there has been a clear development trend in the image field: deeper and deeper convolutional neural network levels, from the initial 8 layers, to 19 layers, 22 layers, and even 152 layers of network structure. As the network structure deepens, the error rate of the ImageNet Large Scale Visual Recognition Challenge (ILSVRC) has gradually dropped from 16.4% in 2012 to 3.57% (Fig. 8.18, where ILSVRC is the Large Scale Visual Recognition Challenge).

Deep learning algorithms adopt the concept of Channels. In the general deep learning framework (such as tensorflow, mxnet, etc.), the channel is a required parameter in the conv2d function. Tensorflow believes that the number of channels for general RGB images in the input samples is 3 (Red, Green, Blue), and the number of channels for monochrome images is 1, while mxnet believes that the meaning of the channel is the number of convolution kernels in each convolution layer. From

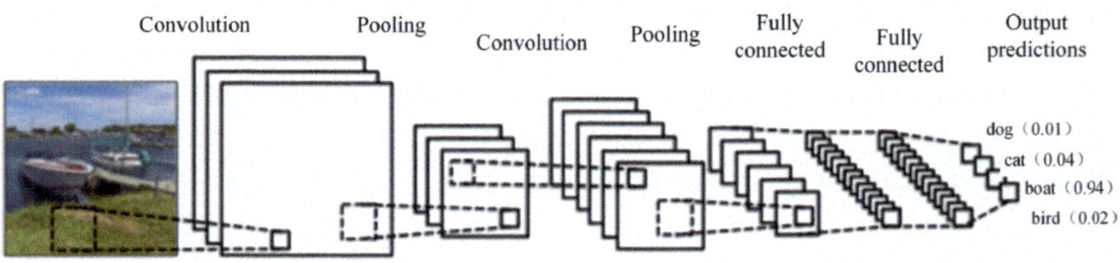

Fig. 8.17 Convolutional neural network model

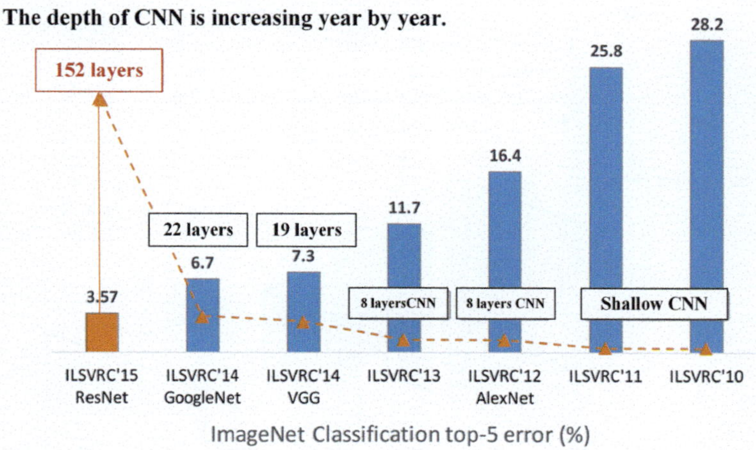

Fig. 8.18 The performance of increasingly deep CNNs in the ILSVRC competition continues to improve

this, it can be seen that tensorflow and mxnet respectively give the definitions of the input channel (in_channels) and output channel (out_channels) of the convolution kernel. The input channel is the number of channels of the input image, and the output channel is the number of channels after convolution, which is consistent with the number of convolution kernels.

According to the number of channels and convolution kernels, the convolution process of the CNN model can be divided into the following 4 situations.

(1) Single-channel single convolution kernel convolution calculation

For a monochrome (single-channel) 7×7 image, if a 3×3 convolution kernel (also known as a filter) is used for convolution calculation, a 5×5 feature map can be obtained, as shown in Fig. 8.19.

(2) Single-channel multi-convolution kernel convolution calculation

The features extracted by a single convolution kernel are insufficient, so multiple convolution kernels can be added. For example, 32 convolution kernels can learn 32 features. For the monochrome (single-channel 6×6) image shown in Fig. 8.20, if 32 3×3 convolution kernels (also known as filters) are used for convolution calculation, 32 4×4 feature maps can be obtained.

(3) Multi-channel single convolution kernel convolution calculation

Suppose there is a $4 \times 4 \times 3$ RGB image sample (Fig. 8.21). If a $2 \times 2 \times 3$ convolution kernel is used for convolution operation, the number of input channels of the image is 3, and the number of input channels of the convolution kernel is the same as the number of channels of the data to be convolved, which is 3. Since there is only one convolution kernel, the final result is $3 \times 3 \times 1$, and the number of output

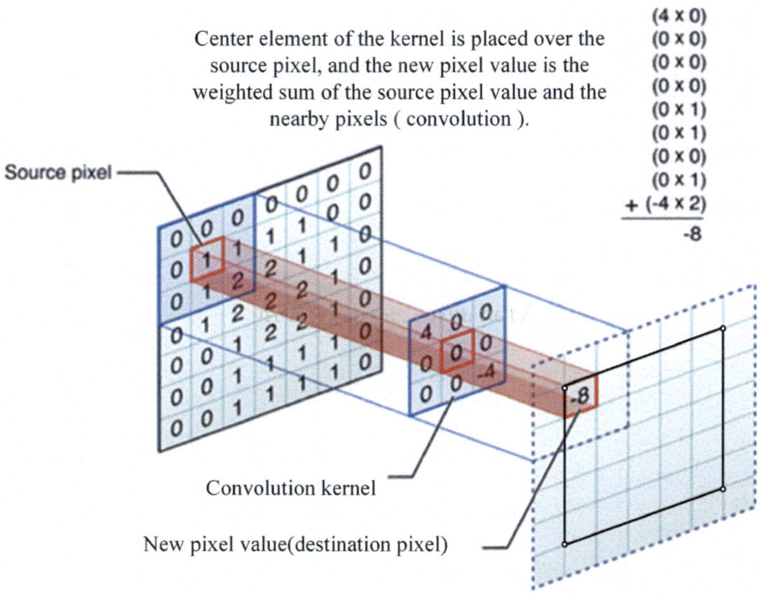

Fig. 8.19 Single-channel single convolution kernel convolution calculation

Fig. 8.20 Single-channel multi-convolution kernel convolution calculation

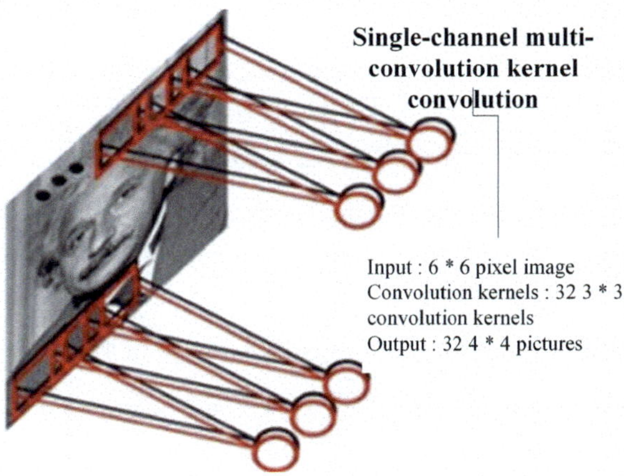

channels is 1. In actual applications, multiple convolution kernels will be used. If another convolution kernel is added, the convolution calculation will be performed according to the multi-channel multi-convolution kernel, and the result will be $3 \times 3 \times 2$, with the number of output channels being 2.

(4) Multi-channel multi-convolution kernel convolution calculation

Suppose the width of the image is W, the height is H, and the number of channels is D (usually RGB three-channel $D = 3$ represents color images). For generality, the

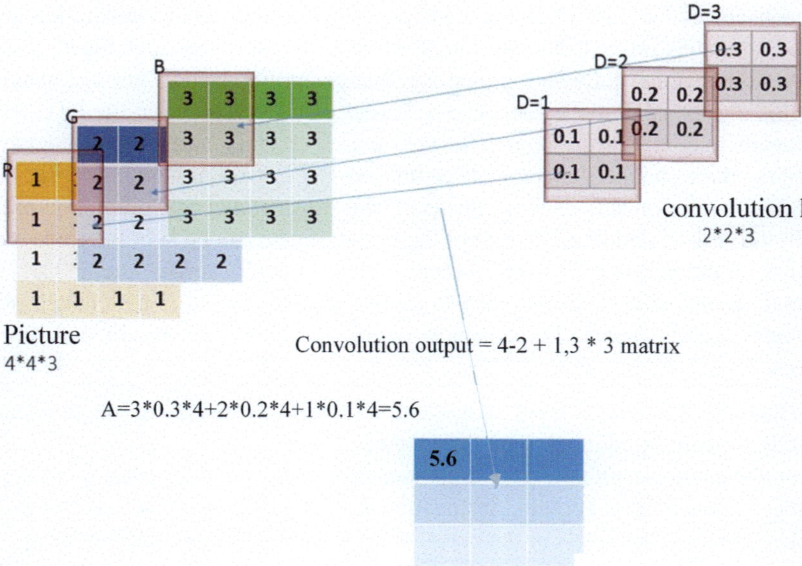

Fig. 8.21 Multi-channel single convolution kernel convolution calculation

number of channels is represented by D. In Fig. 8.22, W and H are both 6. Since the image being processed is D channels, the size of the convolution kernel is kernel_size × kernel_size × D (in the figure it is 3 × 3 × 3). Given the number of convolution kernels M (in the figure $M = 2$), the actual size of the convolution kernel is $M \times$ kernel_size × kernel_size × D.

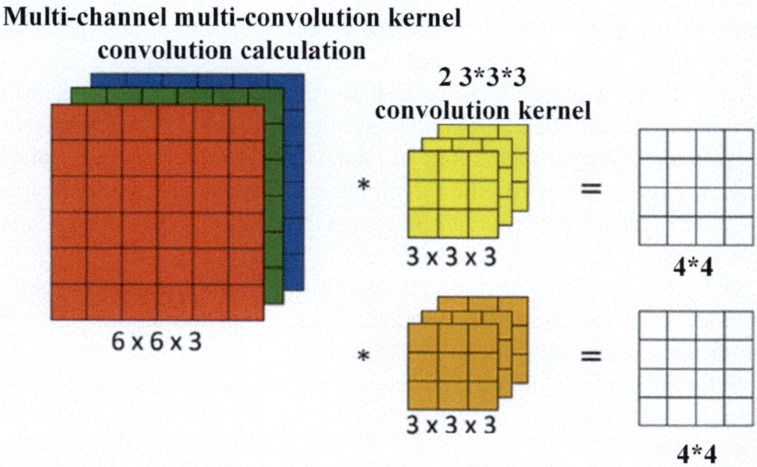

Fig. 8.22 Multi-channel multi-convolution kernel convolution calculation

For each channel of the D channel image, a two-dimensional convolution is performed separately on each channel, and then the D channels are added together to get the two-dimensional convolution output at that position. For RGB three-channel images, this means using the kernel_size × kernel_size size kernel on the R, G, B three channels to convolve the $W \times H$ image on each channel, and then adding the outputs of the three channels to get the two-dimensional convolution output result. If there are M convolution kernels, you can get M two-dimensional convolution output results. In the case of padding, the size of the output image can be kept the same as the original, so the output is $M \times (W$ -kernel_size + 1) × (H -kernel_size + 1).

The main relationship between each layer in the CNN model is the change of height, width, and channels. The channels mentioned above can be divided into three types:

① The channels of the initial input image sample depend on the type of image, such as RGB 3 channels, monochrome is 1 channel;
② The output channels after the convolution operation, out_channels, depend on the number of convolution kernels. At this time, out_channels will also serve as the input channels in_channels of the convolution kernel for the next convolution;
③ The input channels in_channels in the convolution kernel are the output channels out_channels of the previous convolution. If it is the first convolution, it is the channels of the 1-channel sample image.

8.3.2 *Various CNN Prediction Models for Multi-variate Time Series Data*

When dealing with multi-variate time series data, CNN models with different channels and numbers of convolution kernels are used for classification. The input end of the model is generally divided into two types: ① input features processed by traditional signal processing, processed by different filters, and then expanded by left–right or frame skipping; ② directly input the original spectrum, treating the spectrum diagram as an image for processing. In the practice of speech recognition, Baidu's speech recognition technology has gone through various iterative algorithm models shown in Fig. 8.23 [14]. The early input end used the first type of deep neural network (DNN) model, and the current deep CNN model is based on the speech spectrum after time–frequency analysis.

The first mode is to use various artificially designed filter groups for feature extraction after the speech signal undergoes Fourier transformation, which causes information loss in the frequency domain, especially in the high-frequency area. Moreover, for the sake of computational considerations, a very large frame shift must be adopted, causing information loss in the time domain, which is more prominent when the speech rate is fast. The second mode treats the time–frequency spectrum obtained from the entire speech signal time–frequency analysis as an image, and uses the DCNN (Deep Convolutional Neural Network) model widely used in image processing for recognition, overcoming the diversity problem of speech signals.

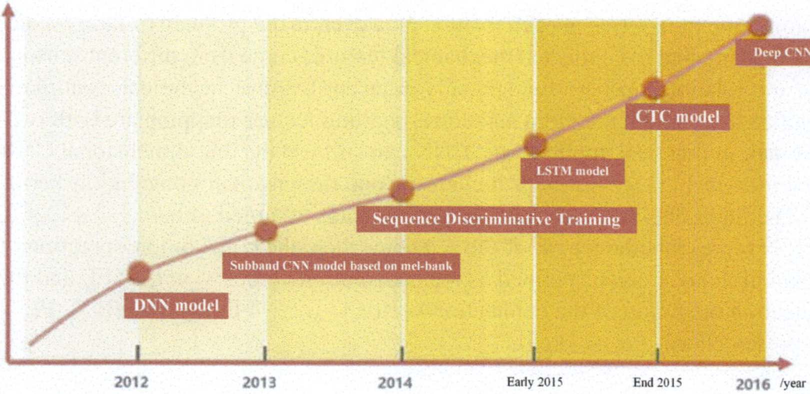

Fig. 8.23 Iterative algorithm models of Baidu's speech recognition technology in recent years

Through a large number of comparative experiments, it was found that the DCNN architecture not only significantly improves the performance of the HMM (Hidden Markov Model) based speech recognition system, but also improves the performance of the CTC (Connectionist Temporal Classification) based speech recognition system. Compared with the traditional speech recognition framework with traditional speech features as input, directly converting the time–frequency analysis into a speech spectrum as DCNN input improves the accuracy rate, and IBM, Microsoft, and iFlytek have successively launched their own DCNN models. In summary, according to the processing method of the input end multi-element time series data, such as direct signal input or processed signal input, one-dimensional, two-dimensional, and three-dimensional CNN models can be used to model the prediction problem of multi-element time series data. (1) One-dimensional CNN architecture model for RUL prediction In order to better extract CNN features, the time window method is used for sampling preparation. The standardized original sensor measurement data can be directly input into the CNN network as model input, without the need for prior expertise in prediction and signal processing, which can promote the industrial application scope of the proposed method. The deep CNN framework can successfully extract high-level abstract features of the signal and estimate the relevant RUL values based on the learned knowledge. Compared with traditional machine learning methods, this method can achieve higher prediction accuracy by using time windows, data normalization, and deep CNN architecture. In recent years, the development of modern aviation technology has led to the complexity of aircraft systems, requiring high reliability, high quality, and high safety in harsh environments. The engine is a key component of the aircraft, and there is an urgent need to develop new methods to better assess engine performance degradation and estimate the remaining useful life. Literature [15] made predictions based on the RUL of the aero engine, and used NASA's C-MAPSS dataset to verify the effectiveness of this method, and showed the superiority of this structure. In the original collected data, the input data is in a two-dimensional format: one dimension is the number of features, and the other dimension is the time series of each feature. However, in the problem of fault prediction, considering that the collected mechanical features come from different sensors, because the relationship between spatially adjacent features in the data sample is not significant, although the input and corresponding feature mapping are both two-dimensional, in practical applications, CNN can still use the one-dimensional CNN model shown in Fig. 8.24, which includes a one-dimensional convolution kernel (filter). The input one-dimensional time series data is assumed to be x = {×1, × 2, ..., xN}, N represents the length of the sequence data, the convolution operation of the convolution layer can be defined as the convolution kernel w, w ∈ R^FL and the multiplication operation of the connection vector x_{i:i + FL − 1}, x_{i:i + FL − 1} is represented as:

$$x_{i:i+F_L-1} = x_i \oplus x_{i+1} \oplus \cdots \oplus x_{i:i+F_L-1}$$

In the formula, $x_{i:i+F_L-1}$ is a time series signal (time scale) of length F_L starting from the ith point x_i. Each data sample is connected into a longer nesting, and the final convolution operation is defined as:

$$z_i = \varphi\left(w^T x_{i:i+F_L-1} + b\right)$$

In the formula, w^{T} is the transpose of the weight matrix w; b and φ are the bias term and the nonlinear activation function, respectively. The output z_i can be regarded as the learned feature of the convolution kernel w on the corresponding subsequence

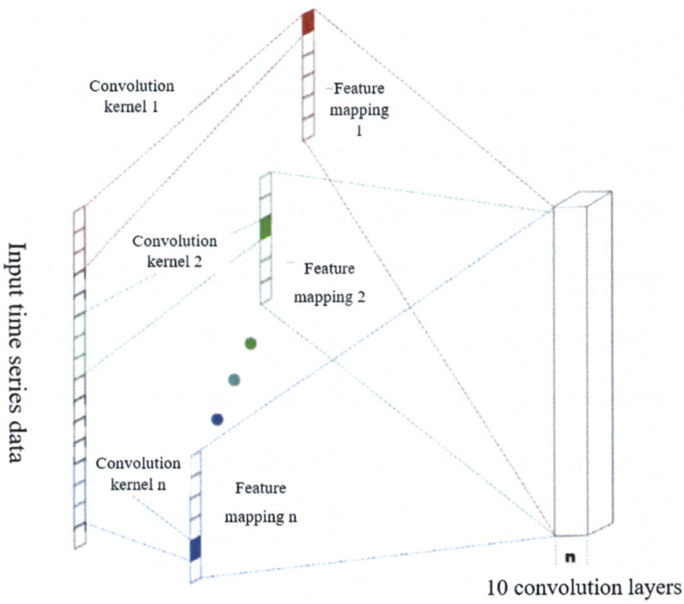

Fig. 8.24 One-dimensional CNN model

$x_{i:i+F_L-1}$. By moving the filter window from the first point to the last point of the sample data, the feature map of the jth filter can be obtained, denoted as:

$$z_i = \{z_j^1, z_j^2, \ldots z_j^{N-F_L+1}\}$$

In the formula, z_j is the convolution kernel of the jth filter. In CNNs, multiple convolution kernels can be applied in the convolution layer with different filter lengths F_L, and the number of filters F_N (displacement factor) and length have an impact on network performance.

In practical applications, F_N and F_L The determination depends on the specific task. According to the understanding of literature [13], larger filter sizes and numbers usually lead to higher prediction accuracy. However, the computational load also increases. A balance must be struck in actual case studies. In fault diagnosis research, the median of the two parameters is better, with the default $F_N = 10$ and $F_L = 10$.

The pooling layer is applied to the feature maps generated by the convolutional layer. On the one hand, pooling can extract the most important local information in each feature map. On the other hand, this operation can significantly reduce the dimension of features, i.e., the number of model parameters. Therefore, pooling is very suitable for high-dimensional problems like image processing. However, while this operation can improve computational efficiency, it also filters out a large amount of useful information to some extent. Therefore, despite the widespread use of pooling in convolutional neural networks, it is not recommended to use pooling operations in this prediction problem with relatively low original feature dimensions.

Dropout is a technique that can help reduce data overfitting when training neural networks, especially in small training datasets. Overfitting of training data usually results in good network performance on the training dataset, but poor network performance on the test dataset. Dropout provides a simple and effective way to solve this problem, preventing complex co-adaptation of training data and avoiding repeated extraction of the same features. In practice, dropout can be implemented by setting the activation output of some hidden neurons to zero, so that the neurons are not included in the forward propagation training process. However, during the testing process, turning off the dropout switch means that all hidden neurons participate in the test. In this way, the robustness of the network can be enhanced. Dropout is also a simple method of model ensemble within the network, which helps to improve the feature extraction ability of the network.

Deep neural networks can adaptively capture the representation information of the original input signal through multiple nonlinear transformations and approximate complex nonlinear functions. Figure 8.25 shows the system architecture of a multi-layer convolutional network and fully connected layer CNN model for RUL estimation, which consists of 5 layers of convolutional neural networks and 1 layer of fully connected substructures, to achieve regression prediction.

First, the input data sample is in a two-dimensional format. The dimensions of the input are $N_{\text{tw}} \times N_{\text{ft}}$, where N_{tw} represents the time series, and N_{ft} is the number of selected features. The original features are usually measured by multiple sensors. Next, 4 convolutional layers are stacked in the network for feature extraction, all 4

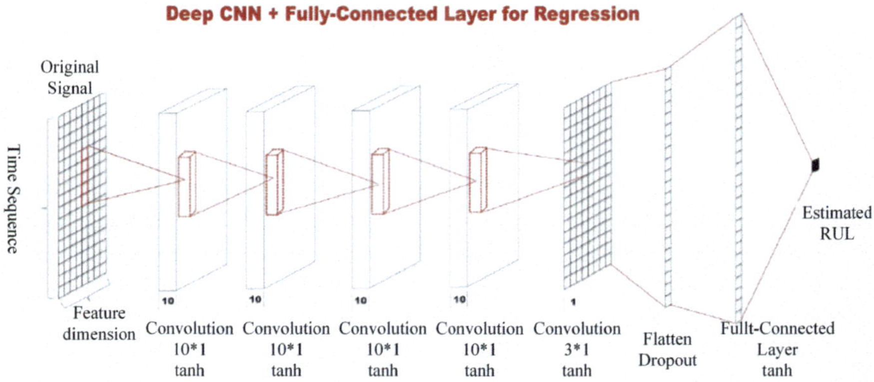

Fig. 8.25 CNN model based on multi-layer convolutional network and fully connected layer

layers have the same configuration: using F_N filters and the size of the filter is $F_L \times 1$. Zero padding operation keeps the dimension of the feature map unchanged. So far, the output obtained is F_N feature maps of size $N_{tw} \times N_{ft}$, the same as the original input sample. Using another convolutional layer with one filter, the previous feature maps are combined into a unique feature map, the size of the convolution kernel (filter) is 3×1. In this way, a high-level representation of each original feature is obtained. Then, the two-dimensional feature map is flattened and connected with the fully connected layer. Dropout technique is used on the last feature map to alleviate the overfitting problem. Finally, a neuron is attached at the end of the network for RUL estimation.

All layers use tanh as the activation function and use the Xavier initializer for weight initialization. To further improve prediction performance, the backpropagation algorithm is used for fine-tuning, updating the model parameters to minimize training error, and the Adam algorithm is used for optimization [16].

When using a two-dimensional convolutional neural network for feature extraction, the convolution operation is actually performed in one dimension, that is, the time series dimension of each feature. The goal of stacking multiple convolutional layers is to learn the high-level representation of each original feature separately, while the fully connected layer uses all learned feature representations for final regression prediction. Compared with existing deep CNN prediction methods, this method is more suitable for extracting features from different sensor measurement data, with the aim of understanding the spatial relationship of different features at the beginning and further extracting information from multi-layer abstract representations.

(2) Two-dimensional *CNN* architecture model for *RUL* prediction [17, 18]

In Fig. 8.26, the two-dimensional *CNN* architecture modifies the traditional *CNN* and applies it to the regression prediction of multivariate (multivariable) time series signals: feature learning is performed on each segment of the multivariate time series, and after feature learning, a regular multi-layer perceptron (*Multi-layer Perceptron, MLP*) is connected for *RUL* estimation.

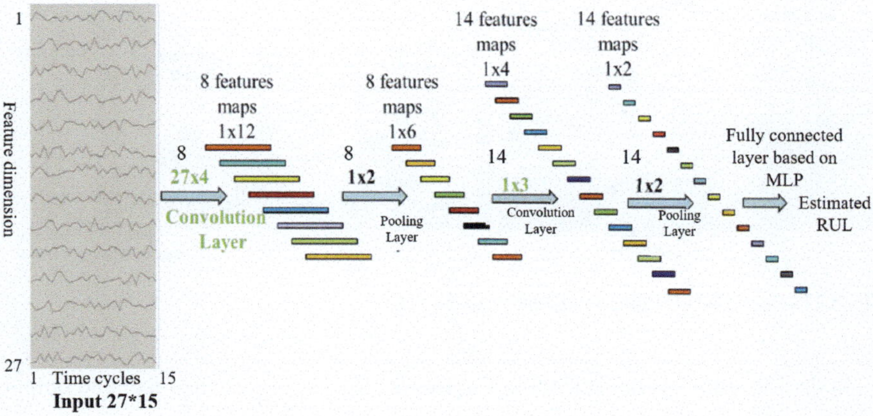

Fig. 8.26 Two-dimensional CNN architecture model

A sliding window strategy is used to divide the time series signal into a set of short signals. Specifically, *CNN* An example used is a data sample containing r samples, each sample containing D attributes (in the case of a single operating state sub-dataset, D attributes serve as d original sensor signals, and in the case of multiple operating state sub-datasets, D attributes included original sensor signals and features extracted related to operating state historical data). Here, r is chosen as the sampling rate (15 is used in the experiment because a test engine trajectory only has 15 time period data samples), and the step length of the sliding window is chosen as 1. A larger step length can be chosen to reduce the number of instances, thereby reducing computational cost. For training data, the true *RUL* of the matrix instance is determined by the last recorded true *RUL*. Specifically, in this work, 2 pairs of convolutional layers and pooling layers, as well as a regular fully connected layer, were used. The model includes D channel inputs (in the figure, D = 27, from *C-MAPSS* dataset's 21 sensors and *PHM* 2008 data challenge dataset's 6 sensor data), each input length is 15. This segmented multivariate time series (D × 15) is input into 2 stages of convolution and pooling layers. Then, all the feature maps of the final layer are connected into a vector, serving as the *RUL* estimate's *MLP* input. The training phase uses the standard backpropagation algorithm to estimate *CNN* parameters, and the stochastic gradient descent method is used to optimize the objective function, i.e., the cumulative square error of the *CNN* model. In the convolutional layer, the feature map of the previous layer is convolved with several convolution kernels, and the output of the convolution operator with added bias (to be learned) and the feature map of the next layer are calculated through the activation function. The calculation of the output feature map of the convolutional layer is as follows:

$$x_j^l = sigm(z_j^l), z_j^l = \sum_i x_j^{l-1} * k_{ij}^l + b_j^l$$

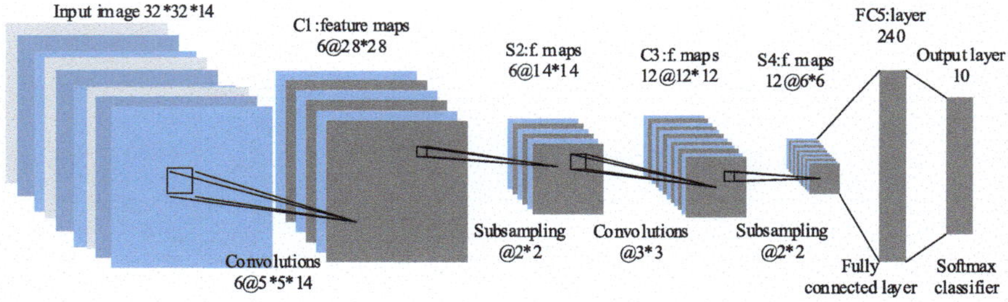

Fig. 8.27 Deep CNN learning model based on LeNet-5

In the formula, * is the convolution operator; x_j^{l-1} and x_j^l are the input and output of the convolution filter; *sigm*() is the *sigmoid* function; z_j^l is the input of the nonlinear *sigmoid* function. The *sigmoid* function is used for its simplicity. The first convolutional layer applies 8 convolution kernels (filters) of size 27 × 4, and the second convolutional layer uses 14 convolution kernels of size 1 × 3. In the pooling layer, the input features are subsampled by an appropriate factor, thereby reducing the resolution of the feature map and increasing the invariance of the features to input distortion. *Average* pooling is used at each stage, with no overlap. The input feature map is divided into a set of non-overlapping regions through average pooling, and the output of each sub-region is the mean. The calculation of the output feature map of the pooling layer is as follows:

$$x_j^{l+1} = down(x_j^l)$$

In the formula, x_j^l is the input; x_j^{l+1} is the output of the pooling layer; *down*() is the subsampling function of average pooling, and the size of the pool filter used by the pooling layer is 1 × 2.

(3) Three-dimensional CNN architecture model for RUL prediction

If the signals from 14 sensors are transformed into 32 × 32 × 14 three-dimensional image data through time–frequency analysis signal processing methods such as wavelet transform or empirical mode decomposition (32 × 32 is the result of image compression), and then combined with the deep *CNN* model based on *LeNet*-5, it forms the three-dimensional *CNN* prediction model shown in Fig. 8.27.

8.3.3 Algorithm Analysis

(1) Experimental dataset: C-MAPSS dataset

The *NASA* turbofan engine degradation problem dataset contains simulated data generated by a model-based simulation program, namely the *NASA* developed

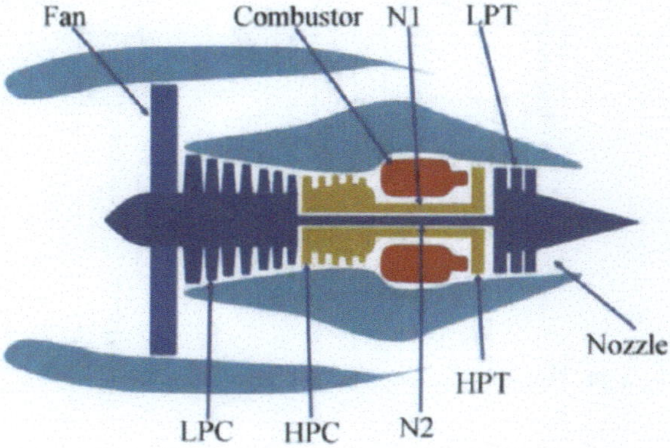

Fig. 8.28 Simplified process of the simulated engine in the C-MAPSS toolkit

Commercial Modular Aero-Propulsion System Simulation (*NASA C-MAPSS*) dataset. Figure 8.28 shows the main elements of this engine model.

C-MAPSS is a software tool that simulates a large commercial turbofan engine. It is written in the *MATLAB* and *Simulink* environment and includes many editable input parameters, allowing users to choose to input some specific parameter values about operating conditions, closed-loop controllers, environmental conditions, etc. *C-MAPSS* simulates a 90,000 lb thrust engine model. The software package includes an atmospheric model that can simulate the following operations: ① from sea level to an altitude of 40,000 ft (1 ft = 0. 3048 m); ② Mach number from 0 to 0. 90; ③ sea level temperature from – 60 °C to 103 °C. It also includes a power management system that allows the engine to operate at a wide range of thrust levels under all-round flight conditions. In addition, the built-in control system includes a fan speed controller and a set of regulators and limiters. The latter includes three high limit regulators to prevent the engine from exceeding its core speed, engine pressure ratio, and high-pressure turbine (*High-Pressure Turbine*, *HPT*) outlet temperature design limits. The limit regulator prevents the high-pressure compressor (*High-Pressure Compressor*, *HPC*) outlet static pressure from being too low, and a plus-minus speed limiter for the core speed. A comprehensive logical structure integrates these control system components in a way similar to the way used in actual engine controllers. In addition, all gains of the fan speed controller and four limit regulators are planned so that the controller and regulator can work normally in all ranges of flight conditions and power levels. Figure 8.29 shows how various components are assembled in the simulation.

C-MAPSS has 14 inputs as shown in Table 8.5, which can generate multiple output data as shown in Table 8.6. The inputs include fuel flow and a set of 13 health parameter inputs, allowing users to simulate the effects of faults and degradation in any of the 5 rotating components of the engine (fan, *LPC*, *HPC*, *HPT* and *LPT*). The study uses 21 variables out of 58 different outputs in the model. *C-MAPSS* provides a set of graphical user interfaces to simplify input and output control for various possible uses, including open-loop analysis, controller design, and simulation of engine and control system responses under various conditions.

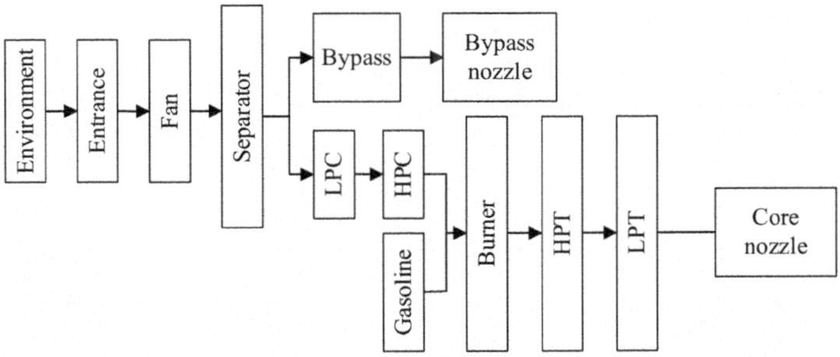

Fig. 8.29 Different components and their connections

Table 8.5 C-MAPSS inputs for simulating the 5 rotating components of the engine

Name	Parameter
Fuel flow	Wf
Fan efficiency modifier	Fan eff mod
Fan flow modifier	Fan flow mod
Fan pressure-ratio modifier	Fan PR mod
LPC efficiency modifier	LPC eff mod
LPC flow modifier	LPC flow mod
LPC pressure-ratio modifier	LPC PR mod
HPC efficiency modifier	HPC_eff_mod
HPC flow modifier	HPC_flow_mod
HPC pressure-ratio modifier	HPC PR mod
HPT efficiency modifier	HPT eff mod
HPT flow modifier	HPT flow mod
LPT efficiency modifier	LPT eff mod
LPT flow modifier	LPT flow mod

The C-MAPSS dataset includes 4 sub-datasets, which are composed of multivariate time series data obtained from 21 sensors. Each sub-dataset includes a training set and a test set, as shown in Table 8.7.

The training dataset includes sensor records of multiple aero-engines operating under different operating conditions (Regime ID in Table 8.8) and fault modes (Table 8.7).

Each engine unit has varying degrees of initial wear and unknown manufacturing variations, but is still healthy. Over time, the engine units begin to degrade until they reach system failure, i.e., the last data entry corresponds to the time period when the engine unit is declared unhealthy. In addition, the sensor records in the test dataset terminate at some time before system failure, and the goal of this task is to estimate the remaining useful life (RUL) of each engine in the test dataset. For verification, the actual RUL values of the test engine units are also provided. A comprehensive

Table 8.6 C-MAPSS outputs

Parameter	Description	Unit
21 Available parameters		
T2	Total temperature at fan inlet	°R
T24	Total temperature at LPC outlet	°R
T30	Total temperature at HPC outlet	°R
T50	Total temperature at LPT outlet	°R
P2	Pressure at fan inlet	psia①
P15	Total pressure in bypass-duct	psia
P30	Total pressure at HPC outlet	psia
Nf	Physical fan speed	r/min
Nc	Physical core speed	r/min
epr	Engine pressure ratio (P50/P2)	—
Ps30	Static pressure at HPC outlet	psia
phi	Ratio of fuel flow to Ps30	pps/psi
NRf	Corrected fan speed	r/min
NRc	Corrected core speed	r/min
BPR	Bypass Ratio	—
farB	Burner fuel–air ratio	—
htBleed	Bleed Enthalpy	—
Nf_dmd	Demanded fan speed	r/m
PCNfR_dmd	Demanded corrected fan speed	r/m
W31	HPT coolant bleed	lb·m/s
W32	LPT coolant bleed	lb·m/s
Parameters for calculating health index		
T48(EGT)	Total temperature at HPT outlet	°R
SmFan	Fan stall margin	—
SmLPC	LPC stall margin	—
SmHPC	HPC stall margin	—

Table 8.7 Training and test sets under different operating conditions (fault modes in the C-MAPSS dataset)

Dataset	C-MAPSS				PHM 2008
	FD001	FD002	FD003	FD004	
Engine units for training	100	260	100	249	218
Engine units for testing	100	259	100	248	218
Operating Conditions	1	6	1	6	6
Fault Conditions	1	1	2	2	2
Training samples(default)	17,731	48,819	21,820	57,522	
Testing samples	100	259	100	248	

Table 8.8 6 different operating conditions

Engine ID	Operational parameter 1	Operational parameter 2	Operational parameter 3
1	0	0	100
2	20	0.25	20
3	20	0.7	0
4	25	0.62	80
5	35	0.84	60
6	42	0.84	40

evaluation of the four sub-datasets was conducted, and the tail information of the C-MAPSS dataset is shown in Table 8.7, with the four sub-datasets denoted as FD001, FD002, FD003, and FD004.

The training process uses all available engine measurement data points as training samples, each data point is labeled with its RUL as the target, and a piecewise linear degradation model is used to obtain the RUL label for each training sample [19]. The testing process typically uses a single data point corresponding to the last recorded cycle on each engine unit as a test sample and provides the actual RUL of the test sample.

The multivariate time data in the C-MAPSS dataset contains engine unit measurement data from 21 sensors [20]. However, some sensor readings have constant output throughout the life of the engine and cannot provide valuable information for RUL estimation. Therefore, 14 measurements from the 21 sensors are used as original input features, which are 2, 3, 4, 7, 8, 9, 11, 12, 13, 14, 15, 17, 20, and 21, and the min–max normalization method is used to normalize the measurement data collected by each sensor to values within (− 1, 1).

Unlike common regression problems, it is difficult to determine the expected output value of the input data for RUL prediction. This is because in many industrial applications, if there is no precise prediction model based on physical mechanisms, it is impossible to evaluate the precise health status of the system at each stage and estimate the RUL of the system. For this popular dataset, a piecewise linear degradation model has been verified to be appropriate and effective [19]: generally, the engine units work normally in the early stage and then degrade linearly. According to the research in reference [19], it is assumed that there is a fixed RUL label: R_{early}, as the target label for early data points. It is worth noting that, R_{early} has a significant impact on the prediction of the dataset.

In RUL estimation and other multivariate time series problems, compared with single time step multivariate data point sampling, time series data can usually obtain more information and have better predictive performance. To make full use of multivariate time information, a time window is used for data preparation. Let N_{tw} represent the size of the time window. At each time step, all past sensor data within the time window are collected to form a high-dimensional feature vector, which is used as the input to the CNN network. Figure 8.30 shows the normalized data samples from 14 selected sensors related to a single engine in the training sub-dataset FD001,

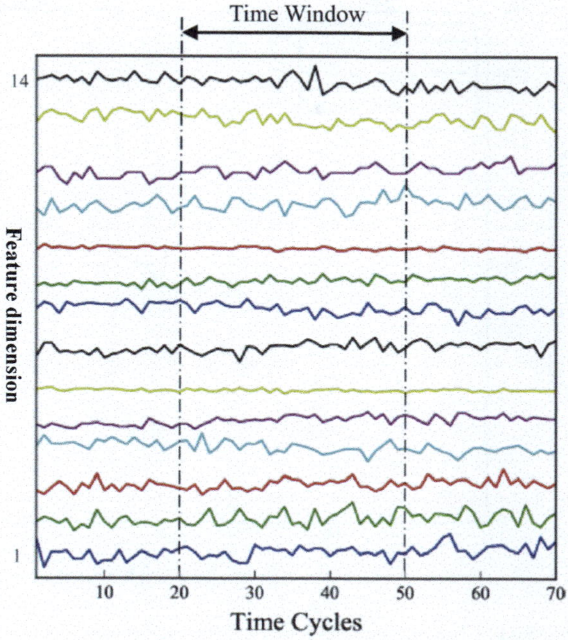

Fig. 8.30 Training samples from 14 selected features (time window of 30)

within a time window of size 30. The shape of the data sample corresponds to the input size of the CNN network.

(2) Experimental result analysis of the one-dimensional CNN model

Figure 8.31 shows the RUL prediction results for the last recorded data points of the FD001 test engine units. The test engine units are classified from small to large by label for better observation and analysis. It can be seen that the predicted RUL values are generally close to the actual values. Especially in areas where the RUL value is small, the accuracy after prediction is often higher. This is because when engine components are close to failure, the failure characteristics are enhanced and can be captured by the proposed network for better prediction.

In addition, Fig. 8.32 shows the RUL estimation of the life of the test engine components before the last recorded cycle in FD001. Among the 100 test engine units, the 21st (#21), 24th (#24), 34th (#34), and 8th (#) test engine units were selected for testing. In this example, the RUL estimation of the last part of the engine unit's life is not shown. This is because in the test dataset, the last part of the sensor measurement data is not provided to test the performance of the prediction algorithm. The actual RUL value of the last recorded cycle is given in the dataset, and the corresponding RUL label of the previous life cycle is obtained accordingly.

In this work, two main measurement methods were used, namely the scoring function and the Root Mean Square Error (RMSE), which are introduced as follows.

① Scoring function The scoring function uses the scoring function in the PHM 2008 Data Challenge, as shown in Eq. (8.19), where N is the number of engines in the

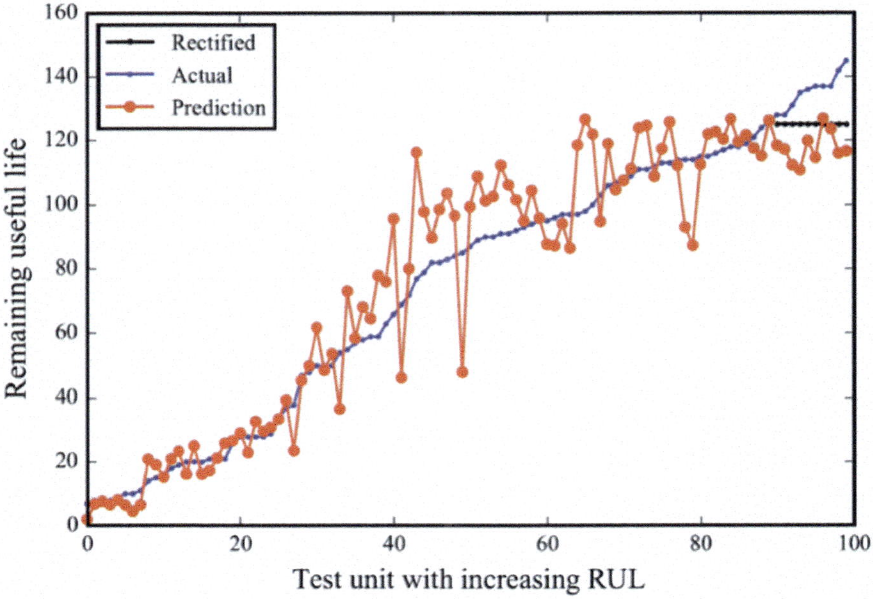

Fig. 8.31 Classification prediction of 100 test engine units in FD001 (electronic version)

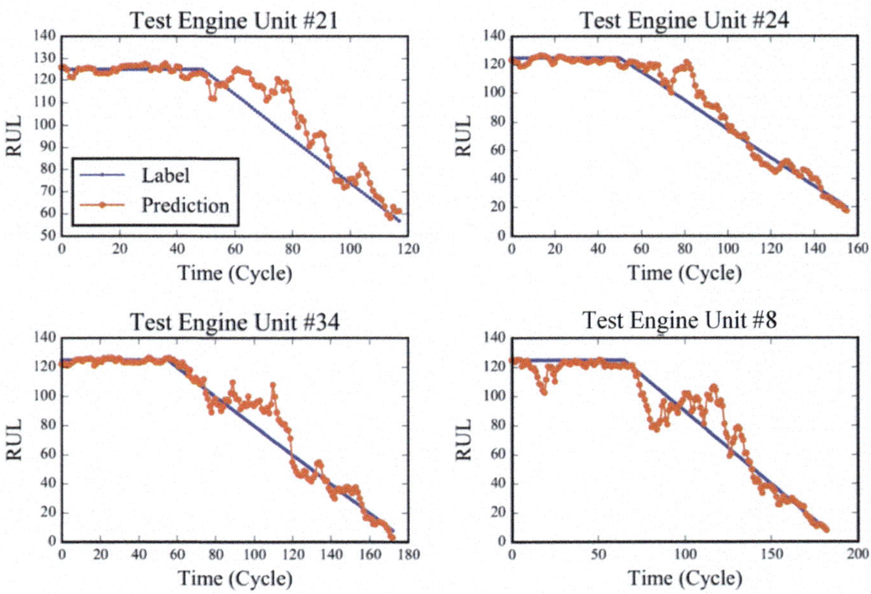

Fig. 8.32 RUL prediction of test engine units in FD001 (electronic version)

test set, S is the calculated score, h_i = estimated RUL—actual RUL.

$$S = \begin{cases} \sum_{i=1}^{N} (e^{-\frac{h_i}{13}} - 1) \; h_i < 0 \\ \sum_{i=1}^{N} (e^{\frac{h_i}{10}} - 1) \;\; h_i \geq 0 \end{cases}, \tag{8.19}$$

The scoring function penalizes late predictions (maintenance performed too late) more than early predictions (which may waste maintenance resources, but will not cause too much harm), which is consistent with the risk reversal in the aerospace industry. However, this function has several drawbacks: a. A single outlier (with a later prediction) will dominate the overall performance score (refer to the exponential growth on the right side of Figure 8.33), thereby masking the true overall accuracy of the algorithm; b. It does not consider the prediction range of the algorithm, the time before the prediction horizon evaluates failure, the algorithm can accurately estimate the RUL value with a certain confidence level; c. The scoring function supports artificially lowering the score by underestimating the RUL. Despite these drawbacks, the scoring function is still used for comparison with the methods in other literature.

② Root Mean Square Error In addition to the scoring function, the Root Mean Square Error (RMSE) of estimating RUL is also used as a performance measure of the model. The reason for choosing *RMSE* is that it gives equal weight to early and late predictions. Using *RMSE* in combination with the scoring function will avoid this problem: artificially lowering the score by underestimating the score will result in a higher *RMSE*. *RMSE* is defined as follows:

$$RMSE = \sqrt{\frac{1}{N} \sum_{RMSE=i=1}^{N} h_i^2} \tag{8.20}$$

The comparison chart between these two evaluation indicators is shown in Fig. 8.34. It can be seen that when the absolute error value is low, the result of the scoring function is lower than that of the *RMSE*, and the relative characteristics of these two evaluation indicators will help in the discussion of the experimental results.

The comprehensive comparison results of different prediction methods are shown in Table 8.9 and Fig. 8.34. The effectiveness of the method was verified through the default experimental settings of four sub-datasets. It can be seen that the proposed deep learning method achieved the best results in all cases.

In the experiment, NN is the basic Neural Network, also known as Multi-layer Perceptron (MLP), which includes a hidden layer with 500 neurons. This is a reasonable number in methods based on neural networks, and a dropout rate of 0.5 is used to improve the generalization ability of the neural network; DNN is a Deep Neural Network with 4 hidden layers, the number of neurons in the hidden layers are 500,

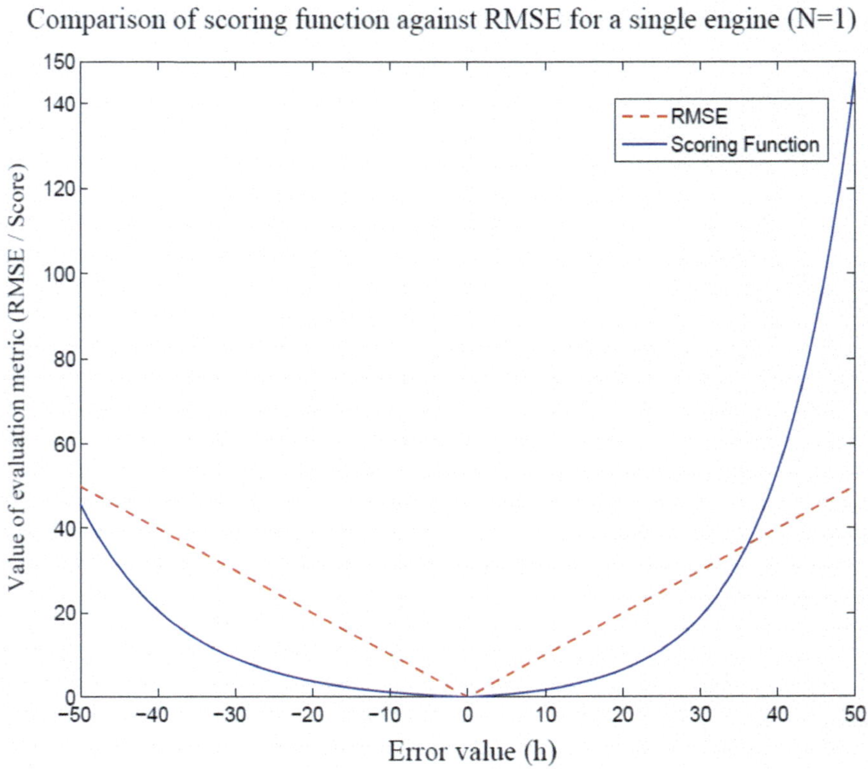

Fig. 8.33 Comparison of evaluation metrics

400, 300, and 100 respectively. DNN has a similar depth to all CNN networks, providing a relatively fair comparison, considering that deeper architectures are often difficult to train effectively, each hidden layer also uses dropout technology; Recurrent Neural Network (RNN) has the ability to process dynamic information due to its feedback connections from the hidden layer or output layer to the previous layer [21], it is a more effective model for time series data, 5 recursive layers are used for comparison in the RNN network, the method has the same depth and similar computational burden; due to the problem of gradient vanishing in the backward propagation process of model training, traditional RNN may not be able to capture long-term dependencies, as a variant of RNN, Long Short Term Memory (LSTM) can prevent errors from disappearing or exploding in backward propagation [22], allowing each recurrent unit to adaptively capture dependencies at different time scales, 5 LSTM layers and 1 fully connected layer are used for comparison.

The experimental results show that the proposed Deep Convolutional Neural Network (DCNN) structure is suitable for prediction problems. The stacked convolutional layers contribute to the learning ability of the network. The RNN structure is the second best structure utilizing cyclic information flow. Although LSTM is a

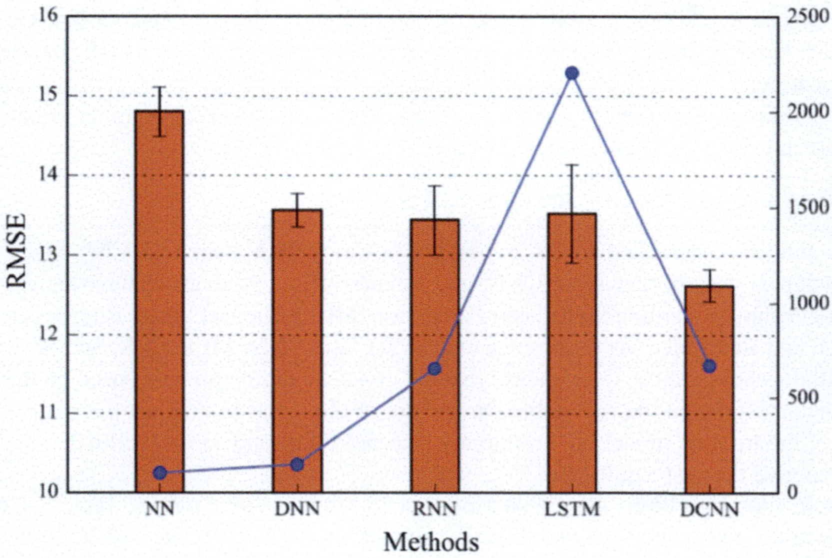

Fig. 8.34 Predictive performance of different methods

Table 8.9 Comparison of different methods on the C-MAPSS dataset

Project		NN		DNN		RNN		LSTM		DCNN	
		Mean	STD	Mean	STD	Mean	STD	Mean	STD	Mean	STD
FD001	*RMSE*	14.80	0.31	13.56	0.21	13.44	0.43	13.52	0.61	12.61	0.19
	Score	496.3	14.3	348.3	17.5	339.2	29.0	431.7	42.4	273.7	24.1
FD002	*RMSE*	25.64	0.34	24.61	0.33	24.03	0.26	24.42	0.45	22.36	0.32
	Score	18,255	1402	15,622	872	14,245	622	14,459	815	10,412	544
FD003	*RMSE*	15.22	0.34	13.93	0.34	13.36	0.38	13.54	0.29	12.64	0.14
	Score	522.3	17.1	364.3	19.3	315.7	24.2	347.3	28.0	284.1	26.5
FD004	*RMSE*	25.80	0.44	24.31	0.24	24.02	0.41	24.21	0.36	23.31	0.39
	Score	20,422	1231	16,223	895	13,931	1102	14,322	1043	12,466	85

Note Mean refers to the average value, and STD refers to the standard deviation

more advanced variant of RNN, its performance is not as good as RNN in this case study. Basic neural networks and deep neural networks are also competitive. These algorithms show that the use of original feature selection, data preprocessing, and time window application for sampling preparation is an effective method for further feature extraction. Traditionally, DNN with 4 hidden layers has overfitting problems. Good prediction results can be achieved through regularization techniques, such as dropout.

In addition, due to the increased difficulty of the sub-datasets, the RMSE and scores of FD002 and FD004 are relatively high. On the one hand, the increase in operating states and failure modes makes the prediction problem more complex. On the other hand, with the increase in the number of engine parts, there is a high chance of overfitting.

In summary, the above comparison results show that the DCNN method can provide reliable RUL predictions under different circumstances.

(3) Data analysis of the two-dimensional CNN model

In two publicly available datasets, a comparison was made between the CNN-based regression model (abbreviated as CNN) and the currently most advanced three regression algorithms, including Multi-layer Perceptron (MLP), Support Vector Regression (SVR), and Relevance Vector Regression (RVR). The CNN, MLP, SVR, and RVR algorithms choose to use the standard fivefold cross-validation process based on the training set to obtain the adjustable parameters of the algorithm, adjust the parameters of the training model on 4 randomly selected folds, and select the final value that can give the best results.

These four algorithms were tested on four C-MAPSS sub-datasets. Table 8.10 shows that CNN achieves lower RMSE values than MLP, SVR, and RVR on all sub-datasets, regardless of operating conditions, indicating that the proposed deep learning method can discover more informative features than shallow features and naive MLP networks. Among the four methods, MLP has higher RMSE values on all four sub-datasets than the other methods, indicating that naive deep models can even affect performance, further verifying the necessity of exploring modern deep learning techniques. SVR has lower RMSE values than MLP and RVR on single operating condition datasets (i.e., the first and third sub-datasets). In addition, on multiple operating condition datasets (i.e., the second and fourth sub-datasets), RVR achieves lower RMSE values than MLP and SVR. This shows that none of the existing traditional methods can consistently outperform the other methods, while the CNN method consistently achieves significantly better results in multiple datasets.

Similarly, in the C-MAPSS dataset, Table 8.11 describes the comparison results of four methods from the perspective of the evaluation function. Compared with MLP, SVR, and RVR, CNN obtained lower (better) score values in the dataset under multiple operating states, namely the 2nd and 4th sub-datasets, as well as the dataset under a single operating state, namely the 1st sub-dataset. Among the four methods, regardless of the operating state, the MLP method obtained the highest score value (worst result). The score of CNN on the dataset under a single operating state (namely

Table 8.10 Comparison of RMSE of algorithms

Algorithm	C-MAPSS dataset			
	FD001	FD002	FD003	FD004
MLP	37.5629	80.0301	37.3853	77.3688
SVR	20.9640	41.9963	21.0480	45.3475
RVR	23.7985	31.2956	22.3678	34.3403
CNN	18.4480	30.2944	19.8174	29.1568

Table 8.11 Comparison of score function values of various algorithms

Algorithm	C-MAPSS dataset			
	FD001	FD002	FD003	FD004
MLP	1.7972×10^4	7.8028×10^6	1.7409×10^4	5.6166×10^6
SVR	1.3815×10^3	5.8990×10^5	1.5983×10^3	3.7114×10^5
RVR	1.5029×10^3	1.7423×10^4	1.4316×10^3	2.6509×10^4
CNN	1.2867×10^3	1.3570×10^4	1.5962×10^3	7.8864×10^3

the 3rd sub-dataset) is slightly higher than that of RVR, although the *RMSE* value is lower. Combined with the characteristics of each evaluation indicator, it means that some outliers may cause slightly high scores when predicting RUL. Based on these observations, the performance of the RUL estimation method also depends on the operating state.

References

1. BENKEDJOUH T, et al. Remaining useful life estimation based on nonlinear feature reduction and support vector regression. Eng Appl Artif Intell, 2013, 26(7):1751–1760.
2. TIAN Z, WONG L, SAFAEI N. A neural network approach for remaining useful life prediction utilizing both failure and suspension histories. Mech Syst Signal Process, 2010;24(5):1542–1555.
3. MOGHADDASS R, ZUO M J. An integrated framework for online diagnostic and prognostic health monitoring using a multistate deterioration process. Reliab Eng Syst Saf, 2014:92–104.
4. GHAFIR M F A, LI Y G, WANG L. Creep life prediction for aero gas turbine hot section component using artificial neural networks. Journal of Engineering for Gas Turbines and Power, 2014, 136(3):1504–1513.
5. AI-GARNI A Z, JARMAL A. Artificial neural network application of modeling failure rate for Boeing 737 tires. Quality and Reliability Engineering International, 2011, 27(2);209–219.
6. WU W, LIU M, LIU Q, et al. A quantum multi-agent based neural network model for failure prediction. Journal of Systems Science and Systems Engineering, 2016, 25(2):210–228.
7. Microsoft.https://azure.microsoft.com/en-us/documentation/articles/machine-learning-algorithm-choice/.
8. HINTON G E. Reducing the dimensionality of data with neural networks. Science. 2006, 313(5786):504–507.
9. REN L, CUI J, SUN Y, et al. Multi-bearing remaining useful life collaborative prediction:a deep learning approach. J Manuf Syst, 2017, 43(2):248–256.
10. LIAO L,JIN W,PAVEL R. Enhanced restricted boltzmann machine with prognosability regularization for prognostics and health assessment. IEEE Trans Ind Electron, 2016, 63(11):7076–7083.
11. ZHANG C,LIM P,QIN A K,et al. Multiobjective deep belief networks ensemble for remaining useful life estimation in prognostics. IEEE Trans Neural Netw Learn Syst, 2017, 28(10): 2306–2318.
12. LECUN Y, BOTTOU LBENGIO Y, et al. Gradient-based learning applied to document recognition. Proceedings of the IEEE, 1998, 86(11): 2278–2324.
13. BABU G S,ZHAO P,LI X L. Deep convolutional neural network based regression approach for estimation of remaining useful life. Database systems for advanced applications:21st international conference. Cham:Springer International Publishing; 2016: 214–228.

14. ITBEAR. Baidu voice recognition technology breakthrough, 2016. https://www.csdn.net/article/a/2016-11-03/15833064.
15. LI X,DING Q,SUN J Q. Remaining useful life estimation in prognostics using deep convolution neural networks. Reliability Engineering and System Safety, 2018 (172):1–11.
16. KINGMA D, BA J. Adam:a method for stochastic optimization. arXiv preprint arXiv: 14126980, 2014.
17. SZEGEDY C,LIU W,YANGQING J,et al. Going deeper with convolutions. Proceedings of IEEE conference on computer vision and pattern recognition, 2015; 1–9.
18. REN W,LIU S,ZHANG H,et al. Single image dehazing via multi-scale convolutional neural networks. European conference on computer vision. Cham:Springer International Publishing, 2016: 154–169.
19. RAMASSO E. Investigating computational geometry for failure prognostics. Int J Prognostics Health Manage 2014; 5(1):005.
20. SAXENA A,GOEBEL K,SIMON D. Damage propagation modeling for aircraft engine run-to-failure simulation. Proceedings of International Conference on Prognostics and Health Management. Denver, CO, 2008: 1–9.
21. MALHI A, YAN R, GAO R X. Prognosis of defect propagation based on recurrent neural networks. IEEE Trans Instrum Meas, 2011, 60(3): 703–711.
22. GUO L, LI N, JIA F, et al. A recurrent neural network based health indicator for remaining useful life prediction of bearings. Neurocomputing, 2017, 240: 98–109.

Chapter 9
Maintenance Optimization Scheduling and Decision Making in Intelligent Factories

This chapter explores the optimization scheduling and decision-making model based on fault prediction for equipment maintenance in smart factories, considering the maintenance service demands generated by various maintenance strategies and the inventory control strategies adopted. It covers aspects such as types of maintenance demands and strategies, joint optimization strategies for maintenance and parts inventory, and predictive maintenance and parts inventory joint optimization models.

9.1 Joint Optimization of Maintenance and Inventory

9.1.1 Maintenance Demand

As shown in Fig. 9.1, the deterministic, uncertain, and predictive maintenance service demands generated by maintenance strategies such as fault maintenance, preventive maintenance, and predictive maintenance are explored. The optimization scheduling and decision-making model based on fault prediction for equipment maintenance within a single enterprise or smart factory is given, along with the joint optimization scheduling and decision-making strategy for predictive maintenance and spare parts inventory in smart factories.

① Fault Maintenance: Due to the occurrence of random/sudden faults, the uncertain (random) maintenance service demand is determined through fault diagnosis.

② Preventive Maintenance: Combining age-based preventive maintenance for single-component systems, block-based periodic (regular) preventive maintenance, sequential preventive maintenance, fault limit maintenance, repair limit maintenance, etc., as well as group maintenance and opportunistic maintenance for multi-component systems. The lifetime distribution of equipment and

M. Liu et al., *Intelligent Predictive Maintenance*, Advanced and Intelligent Manufacturing in China, https://doi.org/10.1007/978-981-97-2677-6_9

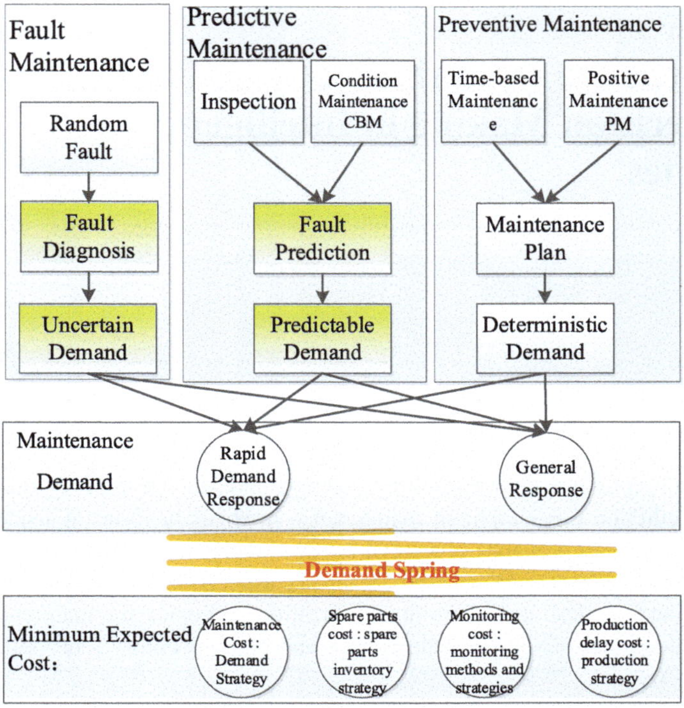

Fig. 9.1 Uncertainty demand management model

core components is estimated to provide maintenance demand and preventive maintenance plans.

③ Predictive Maintenance: Based on the status data of the equipment (from real-time monitoring of the equipment) and environmental operation data (from spot checks, status detection), a MRO data-driven fault prediction model is built on the basis of fault feature extraction. The remaining useful life (RUL) and loss of functionality of the equipment and core components are predicted in real time to provide predictive maintenance demand and plans.

According to the response time of the demand, the predictive demand, deterministic demand, and uncertain demand generated by the above maintenance strategies are divided into rapid response demand and ordinary demand. According to the type of demand, corresponding maintenance and inventory joint optimization strategies are formulated.

9.1.2 *Classification of Maintenance Strategies and Inventory Control Models*

Complex equipment system maintenance strategies involve single-component and multi-component equipment systems, inventory control strategies, maintenance features, maintenance delays, multi-level maintenance networks, business optimization objective functions, and optimization simulation technologies [1–3]. Based on existing joint optimization models, Fig. 9.2 presents a feature classification framework for a complex equipment maintenance and spare parts inventory joint optimization model.

(1) Optimization Criteria

Under the same maintenance strategy, different maintenance cost structures or different maintenance recovery degrees (minimum, imperfect, perfect) will be classified into the same maintenance strategy. Generally, the optimal system maintenance strategy [4, 5] mainly includes the following contents shown in Fig. 9.3.

① Minimize the system maintenance cost rate to the maximum extent;
② Maximize system reliability measures;
③ Maximize the reduction of the system maintenance cost rate while meeting the system reliability requirements;
④ When meeting the system maintenance cost requirements, maximize the system reliability measures.

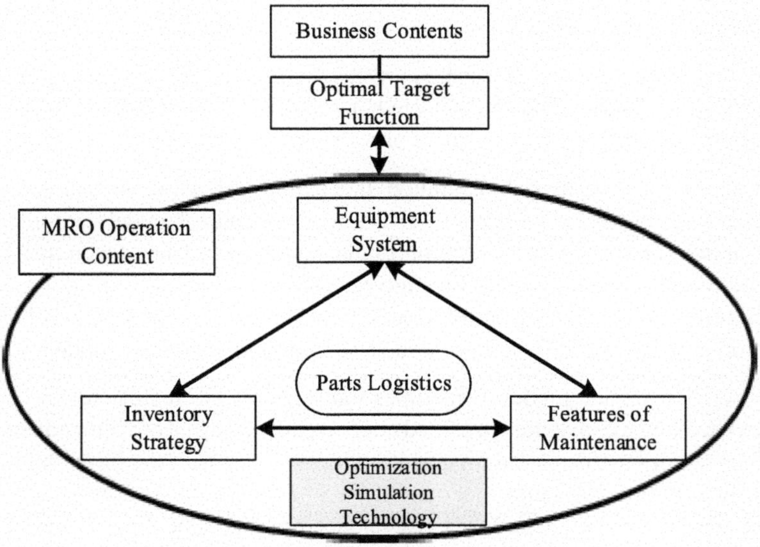

Fig. 9.2 Joint decision on maintenance and spare parts inventory in complex equipment system preventive maintenance

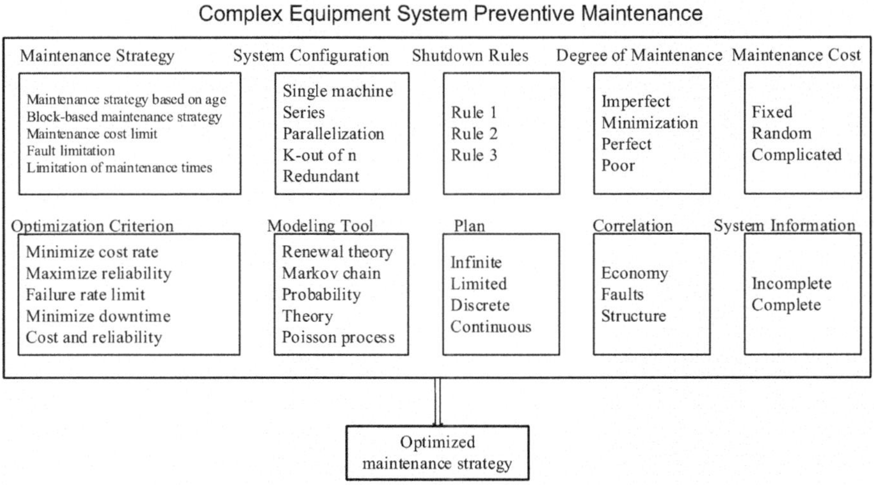

Fig. 9.3 Optimization problem of system or equipment preventive maintenance

(2) Single-component and Multi-component Systems

A single-component system is defined as a system composed of one component, and equipment can also be regarded as a single-component system as a whole. A multi-component system is a system composed of multiple components, which may be the same or different, and sometimes multiple pieces of equipment can also be regarded as a multi-component system.

(3) Inventory Control Model and Spare Parts Inventory Strategy

Inventory control models and spare parts inventory strategies (Inventory Policies) have been widely studied. Basten and van Houtum [6], van Houtum and Kranenburg [7] provided a research review of spare parts inventory strategies.

① Inventory Control Model: Common independent demand inventory control models can be divided into deterministic inventory control models and stochastic inventory control models according to their main parameters (such as whether the demand quantity and lead time are certain).

a. Deterministic Inventory Control Model. Deterministic inventory control models are divided into periodic review models and continuous review models.

The periodic review model (sometimes translated as the regular review model) involves checking the inventory at a fixed interval and placing an order to replenish the existing inventory to the maximum inventory level S. That is, if the inventory level at the time of inspection is I, the order quantity is $S-I$. There are six scenarios, divided into not allowing stockouts, allowing stockouts, and implementing replenishment, each of which is divided into instant arrival and delayed arrival. The most commonly used periodic review model is the one that does not allow stockouts and has instant arrival. Its optimal order cycle is: $T = \sqrt{2C_R/HD}$ Where C_R is the unit order cost, in yuan; H is the unit product inventory maintenance cost, in yuan/(piece·year); D is the demand rate (annual demand), in pieces/year. Maximum inventory level: $S = TD$. The continuous review model is an inventory control model where both the reorder point and the order quantity are fixed. It is suitable for situations with large demand, high stockout costs, and high demand volatility. The continuous inventory review model needs to determine two parameters: the reorder point and the order quantity, that is, to solve the two parameter setting problems of the (s, Q) strategy. There are six types of continuous inventory review models: not allowing stockouts, instant arrival; not allowing stockouts, delayed arrival; allowing stockouts, instant arrival; allowing stockouts, delayed arrival; implementing replenishment, instant arrival; implementing replenishment, delayed arrival. The most common continuous review model is the one that does not allow stockouts and has instant arrival. The most classic Economic Order Quantity (EOQ) model is of this type. Optimal order quantity: $Q^* = \sqrt{2DC_R/H}$ Where C_R is the unit order cost, in yuan; H is the unit product inventory maintenance cost, in yuan/(piece·year); D is the demand rate (annual demand), in pieces/year. b. Stochastic inventory control model. The problems to be solved by the stochastic inventory control model are: determining the economic order quantity or economic order period, determining the safety stock level, and determining the reorder point and the maximum inventory level after ordering. The stochastic inventory control model also divides into continuous review and periodic review. When both demand and lead time are random variables, the inventory model is more complex. The inventory analysis and control discussed above have relatively mature theories and methods. Interested readers can refer to relevant materials and research literature, and no further introduction is given here. ② Spare parts inventory strategy. Because the independent demand inventory control uses the reorder point control strategy, there are many inventory replenishment strategies for the reorder point method. The most basic spare parts inventory strategies are four: continuous review of fixed order quantity, fixed reorder point strategy, that is, (s, Q) strategy; continuous review of fixed reorder point, maximum inventory strategy, that is, (s, S) strategy; periodic review strategy, that is, (t, S) strategy; comprehensive inventory strategy, that is, (t, s, S) strategy. a. (s, Q) strategy. The (s, Q) strategy shown in Fig. 9.4 continuously checks the inventory. When the inventory drops to the reorder point level s, an order is placed. The spare parts order quantity remains unchanged each time, all being the fixed value Q. This strategy is suitable for situations with large demand, high stockout costs, and high demand volatility.

b. (s, S) strategy. The (s, S) strategy and the (s, Q)) The strategy is the same as the continuous check type strategy, which means that the inventory status needs to be checked at all times. When the inventory is found to be reduced to the order point level s, start ordering, and after ordering, the maximum inventory remains unchanged, that is, the constant S. If the inventory is I when the order is issued, then the order quantity is $S-I$. The difference between this strategy and the (s, Q) strategy is that its order quantity is determined by the actual inventory, so the order quantity is variable.

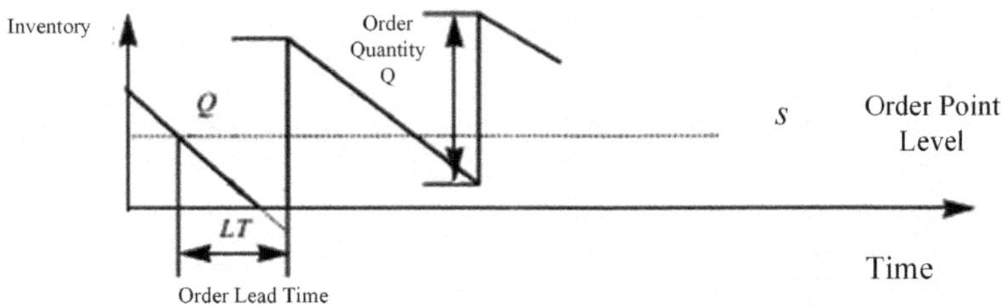

Fig. 9.4 (s, Q) strategy

c. (*t*, *S*) strategy. As shown in Fig. 9.5, the (*t*, *S*) strategy checks the inventory at regular intervals and places an order to replenish the existing inventory to the maximum inventory level *S*. If the inventory is *I* at the time of inspection, the order quantity is $S-I$. After a fixed inspection period *t*, an order is placed, and at this time the inventory is I_1, and the order quantity is $S-I_1$; after a certain time *LT* (order lead time, which can be a random variable), the inventory is replenished $S-I_1$, and the inventory reaches point *A*; after another fixed inspection period *t*, another order is placed, the order quantity is $S-I_2$, and after a certain time, the inventory reaches a new height *B*. In this way, the inventory is checked periodically and replenished continuously. The (*t*, *S*) strategy does not set an order point, only sets a fixed inspection cycle and maximum inventory, and is suitable for some unimportant or small-quantity materials.
d. (*t*, *s*, *S*) strategy. The (*t*, *s*, *S*) strategy is a combination of the (*t*, *S*) strategy and the (*s*, *S*) strategy, as shown in Fig. 9.6. This replenishment strategy has a fixed inspection cycle *t*, maximum inventory *S*, and fixed order point level *s*. After a certain inspection cycle *t*, if the inventory is below the order point, an order is placed, otherwise no order is placed. The size of the order quantity is equal to the maximum inventory minus the inventory at the time of inspection. When the fixed inspection period reaches point *A*, the inventory has dropped below the order point level *s*, so an order should be placed, and the order quantity is equal to the difference between the maximum inventory *S* and the current inventory I_1, that is, $S-I_1$.

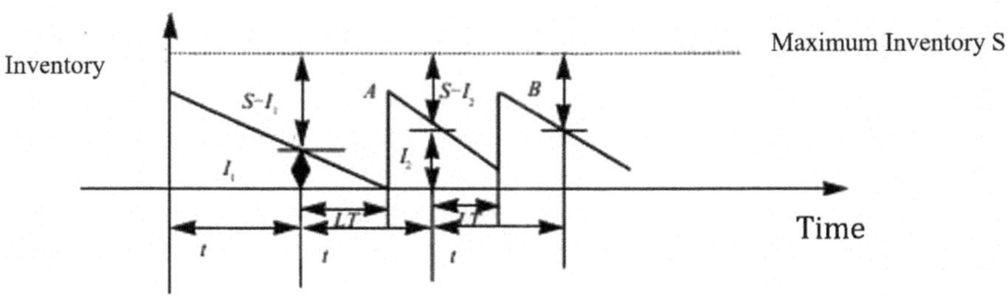

Fig. 9.5 (t, S) strategy

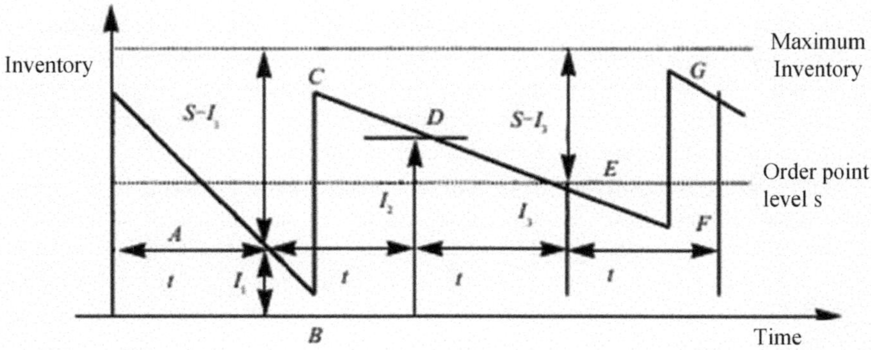

Fig. 9.6 (t, s, S) strategy

After a certain order lead time, the order arrives at point *B*, the inventory is replenished to point *C*, and the inventory position at the second inspection period is at point *D*, which is higher than the order point level position line, so there is no need to order. When the third inspection period arrives, the inventory point is at *E*, which is equal to the order point, and another order is placed, and the order quantity is $S{-}I_3$. In this way, the cycle continues, achieving periodic inventory replenishment.

Based on the above 4 basic inventory strategies, many spare parts inventory strategies can be extended. Moreover, in the usual literature, spare parts inventory is usually handled according to the given demand, so the underlying maintenance plan is ignored, and most maintenance research assumes that there are an infinite number of spare parts in the inventory, and there is relatively little literature on the joint optimization of maintenance and inventory.

(4) Maintenance strategies and characteristics

Many types of maintenance strategies have been used in maintenance decision practice and have been extensively studied in different situations. Maintenance strategies can be divided into three major categories: failure maintenance, preventive maintenance, and predictive maintenance.

① Failure maintenance (Corrective Maintenance, CM) strategy is sometimes referred to as failure-based maintenance (Failure-based Maintenance, FBM), which is a passive maintenance strategy. Once a failure occurs, parts are replaced or repaired as much as possible. If spare parts are not available, maintenance will be delayed or a large-scale downtime will occur.

② Preventive Maintenance (PM) strategies are divided into single-component systems and multi-component systems. The preventive maintenance strategies for single-component systems include: age or time-based preventive maintenance strategies, including age replacement strategies, random age replacement strategies, etc.; block replacement strategies based on the number of component uses *N*, periodic/regular preventive maintenance strategies, etc.; sequential preventive maintenance strategies based on the longer the component is used, the older the part, the shorter the maintenance cycle; failure limit strategies for replacing parts when the failure rate or reliability reaches a predetermined level, as well as time limit for repair strategies, number of repairs limit strategies, repair cost limit strategies, reference time strategies, mixed age strategies, preventive maintenance strategies, etc. For the preventive maintenance strategies of multi-component systems, if the components are independent of each other, the maintenance strategies of single-component systems can be adopted for each component. If there are economic, failure-related, or structural relationships between the components, the relationships between the components need to be considered comprehensively, and group maintenance or opportunistic maintenance strategies should be adopted.

a. Group maintenance strategy is a maintenance strategy based on the idea that if one part is maintained, the entire component should also be maintained (maintenance projects are grouped or clustered according to logically related fields). In many practical situations, the system is composed of several groups of identical components. By replacing failed components in groups, rather than replacing failed components individually, costs can be reduced. This cost saving comes from economies of scale, mainly due to quantity discounts or reduced maintenance installation costs per component. In the group maintenance model, Okumoto and Elsayed [8] proposed a group maintenance strategy for independent identical components, that is, group maintenance is carried out within a specified time interval. Gertsbakh [9] found that group replacement is the best strategy when the number of failed components exceeds a specified number. Ritchken and Wilson [10] proposed a group maintenance strategy with two decision variables, group maintenance is carried out when a fixed time interval expires or a fixed number of units fail (whichever comes first).
b. Opportunistic maintenance strategy is a maintenance strategy based on the idea that when a subsystem fails, other similar parts should also be replaced, etc. When a component in the system fails, the opportunity to maintain the failed component is usually used to carry out preventive maintenance in advance for other components in the system that need maintenance in the short term. This early preventive maintenance is often referred to as opportunistic maintenance, which can greatly reduce the proportion of unplanned maintenance in the system, reduce downtime and maintenance costs. The control conditions for opportunistic maintenance are mainly reflected in: ⓐusing working time as the control condition for opportunistic maintenance, that is, when a component is maintained, if the working time of another component exceeds a given threshold, opportunistic maintenance is initiated; ⓑadopting a reliability threshold control condition, that is, when a component is maintained, if the failure rate of another component at this time is higher than a given threshold, opportunistic maintenance is initiated; ⓒadopting a risk control condition, that is, when a component is maintained, if the risk level of another component at this time is higher than a given threshold, opportunistic

maintenance is initiated. Although these opportunistic maintenance strategies with control conditions can guarantee the preventive maintenance benefits of the system to a certain extent, they often ignore the optimal preventive maintenance information of the component itself.

In practice, the information of component opportunistic preventive maintenance and planned preventive maintenance can be combined, and the optimization model of system preventive maintenance can be established with the failure timing in the system as the optimization variable and the minimum component maintenance cost rate as the goal, so as to realize the complementary advantages of the two preventive maintenance methods and effectively reduce the preventive maintenance cost of the system.

③ Predictive Maintenance (PdM) strategy, also known as Condition based Maintenance, the maintenance department monitors different equipment characteristics, such as vibration, temperature and other signal characteristics, measures the condition of the equipment, and when the measured value reaches a certain threshold or the normal function of the system is affected, the part is replaced.

Compared with other maintenance strategies, predictive maintenance strategy is based on the actual system condition to carry out maintenance operations, which can be more efficient, reduce the number of failures (thus reducing downtime), minimize maintenance costs (but will increase detection costs), and improve operational safety. Ahmad [11], Bousdekis [12], Hong [13], Li [14], Rasmekomen [15] and others have studied Condition based Maintenance strategies, and Olde Keizer et al. [16] developed a state strategy for multi-component systems that does not consider inventory decisions.

In the maintenance process, for non-repairable components, a maintenance strategy of complete replacement after failure is adopted. For repairable components, the degree of maintenance can be distinguished according to the degree of spare parts repair or restoration to normal state. The degree of maintenance can be divided into the following situations.

- Perfect Maintenance/Repair: It's as good as new after maintenance, for example, the replacement of a failed system.
- Minimal Maintenance/Repair: The failure rate is the same as the original system after maintenance.
- Imperfect Maintenance/Repair: Maintenance between perfect maintenance/repair and minimal maintenance/repair. The main reasons for imperfect maintenance/repair are: the failed parts have been repaired; some failed parts have been partially repaired; some failed parts have been partially repaired, but other parts have been damaged; the condition of the unit/component has not been correctly assessed and inspected; the repair activities are not timely, etc.
- Worse Maintenance/Repair: The repair activity increases the failure rate of the system, but there is no downtime. The main reasons are: hidden failures have not been detected; human errors have caused further damage; wrong parts have been replaced.

- Worst Maintenance/Repair: Maintenance or repair that causes the system to go down.

(5) Maintenance Delay

Maintenance delay, also known as maintenance lead time/LT (time from order to delivery), may reduce these delays during the maintenance process, such as diagnostic time, replenishment lead time, technician response time, and startup time.

(6) Multi-level Maintenance Network

To ensure responsiveness and maintain low inventory costs, warehouses within a country or region can be organized and utilized through a multi-level network. For example, a two-tier, multi-retailer, continuous inspection inventory system can be adopted.

(7) Objective optimization function

The optimization objective focuses on cost, downtime, service level, environmental impact, safety, etc., with cost minimization being the most commonly applied objective function for these types of problems. Maintenance costs are incurred by the corresponding maintenance activities, including inventory costs, replacement costs, inspection costs, safety costs, economic losses, age-related production costs, etc. Inventory costs consist of holding costs, spare parts ordering costs, and stockout costs. Replacement costs include spare parts procurement costs, labor costs, system downtime costs, damage costs, etc., while inspection costs are constituted by the cost of the inspection system.

(8) Optimization, simulation technology

Relevant simulation models (such as Monte Carlo simulation and discrete event simulation), meta-heuristic algorithms (such as simulated annealing algorithm, genetic algorithm, and scatter search algorithm), and full enumeration algorithms can all be applied in the simulation process of maintenance optimization problems, helping to find (near) optimal problem solutions.

9.2 Joint Optimization Strategies of Maintenance and Spare Parts Inventory

The interconnection of maintenance and spare parts inventory management often troubles managers and researchers. The content and standards of maintenance characteristics and spare parts inventory control strategies determine the joint optimization problem of maintenance and spare parts inventory.

Spare parts inventory is for the needs of maintenance and replacement of parts. The management of spare parts inventory is different from the inventory management of other work-in-progress or finished products. The level of spare parts inventory

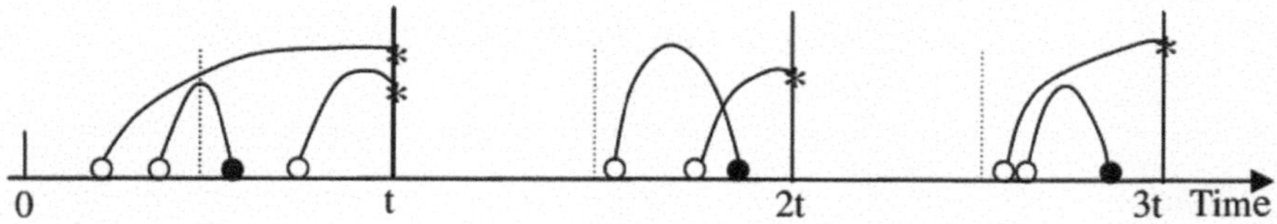

Fig. 9.7 Replacement strategy of spare parts during maintenance o Defect point; • Failure point; * Replacement point of defective parts in preventive maintenance cycle t; o The connecting arc between • represents the delay time; o The connecting arc between * represents the delay time inspected due to preventive replacement

largely depends on how the equipment is used and maintained. The failure process of parts generally follows a two-stage failure process. The first stage is the time interval between new parts and early defect identification, and the second stage is from this defect identification point "o" to the final complete failure (Major failure/State F) " • ", this interval is also known as the failure delay time [1]. During the maintenance process, although the parts have not failed, they are replaced (preventive replacement) at times t, 2t, 3t, etc. after inspection, as shown in Fig. 9.7, and the level of spare parts inventory decreases with preventive replacement.

Spare parts demand is usually generated by predictive, preventive, or failure maintenance needs, which are difficult to predict based on historical data of past spare parts usage. Therefore, the optimal spare parts inventory control strategy is also difficult to obtain. However, maintenance costs are related to the availability of spare parts and the penalty costs of unavailable spare parts, such as the extended downtime costs for waiting for spare parts and the emergency inspection costs for purchasing spare parts. Moreover, appropriate planned maintenance interventions can reduce the number of part failures and related costs, but their performance depends on the availability of spare parts. Wang [1] proposed a joint optimization method for spare parts inventory control and preventive maintenance inspection intervals, with the decision variables being the order interval, PM interval, and order quantity. Due to the randomness of failures, stochastic dynamic programming can be used to solve the joint optimal solution of spare parts inventory and maintenance stochastic costs within a finite time, using the delay time shown in Fig. 9.7 to construct the probability of the number of failures and the number of defective products determined within a PM cycle.

Therefore, based on maintenance characteristics (such as block-based, age-based, and state-based maintenance strategies) and continuous, periodic/regular or just-in-time (JIT) inventory control strategies of spare parts, considering the production plan output rate, order delay cost, and state monitoring cost, the following five joint optimization models and decision strategies of maintenance and spare parts inventory within a single enterprise are given.

9.2.1 Block-Based Maintenance Strategy and Regular Inspection Inventory Strategy

Acharya et al. [17] developed an optimization model for a joint block replacement and regular spare parts replenishment strategy, which analyzes a system with n independent and identically distributed parts. It is assumed that the block replacement and inventory ordering intervals are the same in a single-cycle model, and the ordering interval is a multiple of the replacement interval in a multi-cycle model. In addition, the lead time for spare parts procurement is ignored, and an iterative process is used to optimize these overlapping intervals in a single-cycle model or multi-cycle model. This iterative process starts with the selection of interval increments and replacement intervals, with the interval increasing in each iteration. For each iteration, the spare parts order level and total cost are calculated. The spare parts order level is based on the Laplace inverse transform of the failure distribution, with k cycles and n. The ordering requirements of components are approximated by a normal distribution, and the corresponding expected total cost is calculated through mathematical formulas, including holding costs, ordering costs, shortage costs, repair and preventive maintenance costs. When the stop condition is met, the solution with the smallest total cost is selected. The author proves the trade-off between inventory and maintenance strategies through a multi-cycle model. Unlike Acharya, who uses update functions and Laplace transforms to calculate the mean and variance of the number of failures within a time interval, Chelbi and Ait-Kadi [18] use convolution calculation algorithms to calculate the mean and variance of different equipment life distribution functions. Although different distribution functions are assumed, the iterative processes used by these researchers are very similar. In addition, the authors of these two papers assume in their calculations that component demand follows a normal distribution. Yoo [19] based on the total number of failure replacements within a certain time interval for the demand for spare parts, using n high values of statistically identical update processes, they can be formalized into a probability mass function. This function is evaluated using a recursive algorithm, which starts from the expression of zero operating units and then calculates to the expression of N operating units. Once this expression is calculated, the cost function can be minimized to find the optimal values for replacement time and spare parts inventory level. Brezavscek and Hudoklin [20] tested a joint block replacement and regular inspection model on electric locomotives on Slovenian railways. They extended Acharya's paper through an uncertain lead time, and the reordering interval equals the replacement interval minus the lead time. Huang et al. [21] not only generalized the model of Brezavscek and Hudoklin [20] by introducing a random lead time, but also proved the existence and uniqueness of the minimum value in the model when the order level in the objective function is the only decision variable. By repeating this process for each replacement time, the minimum cost for a specific problem setting can be derived, and the model was tested in the same example proposed by Brezavscek and Hudoklin [20].

9.2.2 Block-Based Maintenance Strategy and Continuous Inventory Inspection Strategy

Sarker and Haque [22] developed a joint optimization model considering block-based maintenance strategies and continuous inventory inspection strategies. The authors used a gamma distribution of maintenance time and introduced a working unit composed of several statistically independent components into the model of Kabir and Al-Olayan [23]. A simulation model was established and optimized through finite enumeration. Ilgin and Tunali [24] applied a simulation optimization model based on genetic algorithms to a motor block production line in an automobile factory. They studied block-based replacement strategies and allowed a variable lead time and multiple types of spare parts. In addition, single-component replacements and multi-component replacements were performed. Cheap spare parts are ordered according to the two-bin system (also known as the replenishment method or two-bin system, using two bins to control inventory and ordering operations. When the materials in the first bin are exhausted, it is the time to start ordering. There will be no shortage due to the exhaustion of materials in the first bin, because there is a second bin for backup, and the materials in the second bin can maintain supply until the order quantity of the first bin arrives), while expensive spare parts are ordered according to the continuous inspection strategy. The focus of their paper is the development of genetic algorithms, which can reduce costs by 53% compared to the current parameter values of specific automobile factories. Nguyen et al. [25, 26] focused on studying factories of chemical processes, different failure modes allocate different costs and repair times, and there is only one spare part for each part in the failure maintenance inventory. Due to this assumption, decisions related to inventory are limited to choosing whether to keep a unit of inventory. Although they used some simplified assumptions in the model, they were the first to add an extra labor dimension to this type of model. The cost function includes inventory costs, maintenance costs, economic losses, and labor-related costs. In addition, maintenance is not only limited by the number of spare parts but also by the number of employees. Under binding constraints, priority rules are used to schedule maintenance, and the model is analyzed using Monte Carlo simulation. In literature [25], they considered an enumeration method to find the optimal values of maintenance strategies, inventory strategies, and the number of employees. In literature [26], they considered the inventory of multiple component spare parts, they changed the preventive maintenance strategy from regular replacement to age-based sequential preventive maintenance, allowed imperfect preventive maintenance, and added employee skills to the model, which was optimized through genetic algorithms. Sergent et al. [27] extended the work of Nguyen et al. by allowing different types of spare parts to have different replacement rates.

9.2.3 Age-Based Maintenance Strategy and Regular Inventory Inspection Strategy

Armstrong and Atkins [28] studied the age replacement and regular inventory inspection strategy of a single-component system. In this system, the best combination of replacement and ordering time is found by searching to minimize costs. They considered the cost of damage or corrective replacement, making their model a joint optimization model. Based on their assumptions, a joint cost function was established, proving that it is unimodal and pseudo-convex in each dimension of replacement and ordering time. Pseudo-convex functions are very similar to convex functions that seek local minima. Generally, the global optimal combination of replacement and ordering time can be found at the KKT (Kuhn-Tucker) point that satisfies the Kuhn-Tucker conditions, which are necessary conditions for solving the optimal solution of nonlinear programming. Here, the ordering time plus the lead time is less than the placement time. In their case, the cost of separate optimization increased by 3% compared to joint optimization. Armstrong and Atkins extended this model in 1998. They considered an age replacement strategy and regular inventory inspection strategy, in which major and minor failures are distinguished. A major failure is resolved by replacing the machine (perfect replacement), and a minor failure can be resolved by minimal repair. Despite the addition of age-related replacement costs, non-decreasing operating costs, and service constraints, pseudo-convexity still exists. They were able to prove that the same is true when adding deterministic lead times for planned orders and emergency orders, but when one or two (for example, planned orders and emergency orders) order delivery times are allowed to be random, they did not successfully prove or deny these results.

9.2.4 Age-Based Maintenance Strategy and Continuous Inventory Inspection Strategy

Kabir et al. [23] differentiated between emergency and regular ordering costs, and incorporated random lead times based on the Weibull distribution into their model. Furthermore, their model dealt with the joint optimization of age replacement and continuous inventory inspection strategies (t, s, S), which no longer allowed simultaneous replacement and ordering times. They combined discrete event simulation and finite enumeration to determine the optimal values of this strategy. Their first model included no more than one operating component, which was extended to multiple machine systems in other papers. The main conclusion that can be drawn from their findings is that sequential optimization does not guarantee global optimality. Sarker and Haque [22] did not add any elements to overcome the problem of local optimal decline in the simulation optimization process. Genetic algorithms can replace finite enumeration to further optimize (t, s, S) strategies, and manage to reduce the cost of the objective function and the total computation time. Chen et al. [29] analyzed a multi-level network system, their supply chain consisted of multiple suppliers, a distributor, and different user groups. These users were assigned certain priorities. The researchers studied the joint optimization of inventory configuration, continuous inspection, and age-based preventive replacement. The scheme based on the joint strategy was compared with three other schemes that did not include inventory configuration and/or preventive replacement. The model was optimized using a simulation optimization method based on scatter search, and Arena and Opt Quest software was used to perform these calculations. The results showed that the joint strategy has certain advantages from the perspective of the entire supply chain, but this is not necessarily applicable to all agents of the supply chain.

The interconnection of maintenance and spare parts inventory management often troubles managers and researchers. The deterioration of inventory affects decisions and increases losses. Block replacement and regular inventory inspection strategies are used to evaluate the joint optimization problem of multi-component systems under inventory deterioration. Jiang et al. [30] used the Deterministic Deteriorating Inventory (DDI) model to describe the inventory deterioration when inventory data is available, and the Stochastic Deteriorating Inventory (SDI) model to describe the inventory deterioration when inventory data is not available. A joint optimization model was established with the preventive replacement interval and the maximum inventory level as decision variables to minimize the total cost. This work proved the existence of the optimal inventory level and provided the uniqueness condition for the DDI model. Through numerical analysis, not only was the validity of the model verified, but it was also shown that the total cost rate is sensitive to the maximum inventory level. Moreover, it reduced the optimal preventive replacement interval and increased the optimal maximum inventory level to balance the impact of inventory deterioration. Monte Carlo experiments showed that the proposed strategy is better than the strategy that does not consider inventory deterioration.

9.2.5 Condition Based Maintenance Strategy and Regular and Continuous Inventory Inspection Strategy

When measurement techniques are used to estimate the state of a component, a more realistic remaining life distribution replaces the general life distribution used in preventive maintenance models based on the entire component population. More accurate information can be obtained to set replacement and spare parts ordering times by dynamically updating the life distribution of the component after each inspection. The updated life distribution was integrated into the model of Armstrong and Atkins [28], and the same nonlinear programming solution method was relied on to solve the problem of Condition based Maintenance strategy, regular and continuous inventory inspection strategy. Inspection costs were added to this model as part of maintenance-related costs in order to make accurate comparisons with preventive maintenance cases.

With the addition of inspection costs to the cost formula, many scholars have studied predictive maintenance and continuous inventory inspection strategies. This model uses a combination of simulation and genetic algorithms to solve the joint strategy. Numerical examples show that compared with separate optimization, the cost of joint optimization is reduced by an average of 3.78%. Wang et al. [31, 32] wrote several papers on the optimization of state replacement and spare parts strategies. First, they considered a single-component system, which allowed them to develop a mathematical model. The calculation process was based on their analytical mathematical model, some equations were calculated through an iterative process, some equations were calculated accurately, and the derivation of the optimal decision parameters (ordering time and replacement time) was based on the solution of the genetic algorithm. Secondly, they extended to systems with multiple identical components. The degradation of the system was modeled through a Markov chain and a Monte Carlo simulation process, combined with an enumeration algorithm, to solve the optimal value of the decision variables. On this basis, simulation optimization based on genetic algorithms was used to determine the optimal decision variables of the continuous inventory inspection (s, S) and Condition based Maintenance strategy. The failure rate in the traditional preventive maintenance model is usually a time function, and the failure rate of this model is a function of the degree of system degradation, which is called a state-based failure rate. The above model was tested with oil monitoring data from a tractor engine, and it was found that the solution method was beneficial.

Jiang [30] proposed a preventive maintenance and inventory control strategy for multi-component systems. The deterioration of spare parts affects decisions and increases losses. In the case of spare parts deterioration, block replacement and periodic inventory inspection strategies are used to evaluate the joint optimization problem of multi-unit systems. When inventory data is available, the Deterministic Deteriorating Inventory (DDI) model is used to describe the deteriorating inventory, and when inventory data is not available, the Stochastic Deteriorating Inventory (SDI) model is used. An analytical joint optimization model is established, with the preventive replacement interval and the maximum inventory level as decision variables, to minimize the total system cost. This work proved the existence of the optimal inventory level and provided the uniqueness condition under the DDI model. Numerical experiments were conducted on Slovenian railway locomotives, verifying the effectiveness of the model. The results show that the total cost rate is sensitive to the maximum inventory level, so it is necessary to study this work. It reduced the optimal preventive replacement interval and increased the optimal inventory level to balance the impact of inventory deterioration. Monte Carlo experiments show that the proposed strategy is better than the strategy that does not consider inventory deterioration.

Olde Keizer et al. [3] In 2017, the issue of joint optimization of multi-component system status maintenance and spare parts planning was formalized as a Markov decision process, with the goal of minimizing long-term average cost per unit time. Numerical results show that for systems composed of a small number of components, the (s, S) inventory strategy is far from optimal. Significant savings can be achieved through state-based maintenance decisions and spare parts ordering time selection strategies. For multi-component systems, opportunistic maintenance is a useful strategy. In this approach, multiple maintenance activities are performed simultaneously, which can not only reduce maintenance costs but also affect the availability of equipment. Research shows that the modeling of opportunistic maintenance optimization for multi-component systems has received considerable attention. In recent years, some studies have considered both CBM and opportunistic maintenance strategies in the maintenance modeling of multi-component systems. Alaswad et al. [33] proposed a CBM strategy for a multi-component system with continuous random degradation behavior while considering economic relevance. Koochaki et al. [34] examined the effectiveness of opportunistic maintenance on CBM by simulating a small system composed of three components. The combination of timely maintenance using CBM strategies and opportunistic maintenance to reduce maintenance costs has formed Condition-based Opportunistic Preventive Maintenance (CBOM) [35]. For multi-component systems adopting the CBOM strategy, a degradation state space division method is proposed, which models opportunistic maintenance for continuous degradation systems with the same and different components. During the modeling process, the representation of all possible maintenance requirements and their corresponding probabilities at each maintenance decision point is derived. However, when considering the joint optimization of maintenance and spare parts supply for multi-component systems, the spare parts inventory limits the satisfaction of repair needs; therefore, the maintenance activities and spare parts ordering and holding activities that can be carried out can be jointly determined by the observation state of the system and the spare parts inventory. Therefore, the joint modeling of preventive maintenance optimization and spare parts supply for multi-component systems should combine system-level spare parts inventory and maintenance requirements for analysis. A joint optimization of state opportunistic maintenance and spare parts supply strategy for a general multi-component system with a known number of identical components is proposed. According to the proposed strategy, maintenance activities and spare parts orders are determined based on the joint state composed of system observation states and spare parts inventory. First, at each maintenance decision point, all possible maintenance requirements and their corresponding probabilities are derived from the degradation state space division method. Then, possible scenarios of maintenance activities for general maintenance requirements are analyzed, and restrictions on inventory status, spare parts ordering, and holding are made. Further, based on the stationary law of degradation states and the probability of spare parts inventory, the probabilities of these activities and their numerical solutions are derived. Based on these probabilities, an expected long-term cost rate model is established using semi-regenerative process theory to jointly determine the optimal maintenance and spare parts supply strategy. Currently, no research has considered the joint optimization of maintenance, spare parts inventory, and buffer inventory [36]. Maintenance activities have an interaction between the intermediate buffer area and the spare parts inventory, as shown in Fig. 9.8. The system cost is related to maintenance decisions, buffer size, and spare parts inventory level. Literature [36] analyzed an intermediate buffer of a dual-machine inventory system

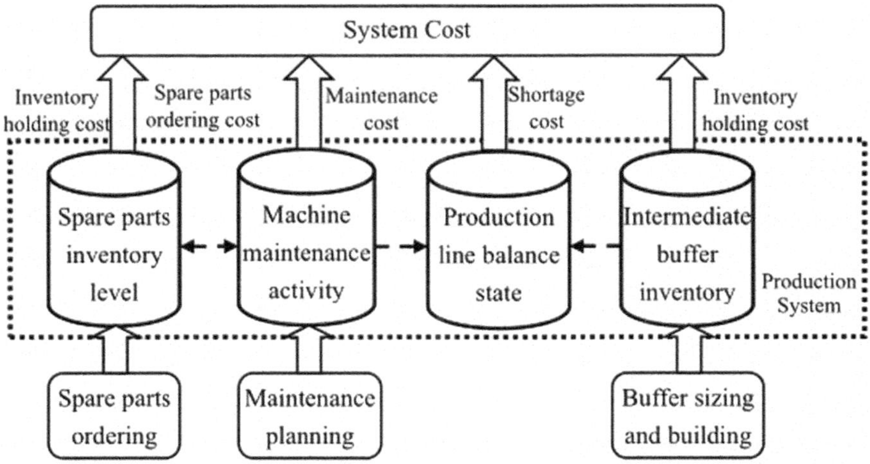

Fig. 9.8 The relationship between maintenance, spare parts, and intermediate buffering and their impact on system costs

involving maintenance and spare parts orders and derived the expected maintenance cost related to buffer inventory through mathematical analysis.

The demand for factory spare parts is, to some extent, driven by maintenance needs. Therefore, the most appropriate maintenance and replenishment strategies must be used to jointly optimize planned maintenance and related spare parts inventory. A simulation-based method [37] solved the problem of random failure and replacement of defective parts at the time of periodic (T) inspection. The inspection uses the concept of delay time modeling, and strategies for several periodic inspections and continuous inspection supplements were compared. A real environment was provided by a paper mill. Practitioners who cooperate with such factories were surveyed to collect real data. Simulation results show that in the factory environment studied, the periodic inspection strategy where the number of spare parts orders is twice the number of inspections is the most cost-effective. For comparison, the characteristics of various considered strategies were also discussed. Using a modular approach, a simulation model ProModel was developed for inventory control strategies for periodic and continuous inspections. A process-based discrete event simulation language was used to model the process, modeling a system with a single device as a continuous production system considering all major assumptions and cost data situations. The operation and maintenance process can fully utilize the proactive concept of event-driven, and the event-driven information system processes sensor data and provides proactive suggestions. The proactive event-driven decision model [38] is used for the joint optimization of predictive maintenance and spare parts inventory, solving the decision process in the "detection-prediction-decision-action" model, which can be embedded in the Event Driven Architecture (EDA) for real-time processing under the e-maintenance conceptual framework. This method was tested in a real production scenario in the automotive lighting equipment industry. By changing the company's maintenance strategy from a time-based maintenance strategy to a Condition-based Maintenance (CBM) strategy, maintenance and inventory costs can be significantly reduced.

9.3 Predictive Maintenance and Spare Parts Inventory Joint Optimization Model

In the joint CBM and inventory strategy of a single-component system, the monotonic strategy structure has been proven to be optimal. In such a strategy, the degradation threshold is used to decide when to order a spare part, the arrival time of the spare part, and when to replace the component. Some scholars have already studied the sequential or joint optimization problem of CBM and spare parts inventory in single-component systems. In practice, systems often contain multiple components, and applying single-component system strategies to a multi-component system often cannot achieve optimization. The main reasons are [39]: ① Multi-component systems have various types of relevance such as economic, structural, or failure-related, and maintenance and inventory optimization decisions depend on the complete system state rather than a single component; ② Multi-component systems can share a set of identical spare parts, and the same spare parts in different subsystems operate under different states, with different failure rates. The number of shared spare parts should be determined at the system level, and the decomposition scheme at the component level will lead to higher inventory levels and costs [40].

Many scholars have conducted research on the CBM and inventory integration methods of multi-component systems. Van Horenbeek and others have considered the economic and structural relevance of components as well as the sequential optimization of maintenance and inventory [41]. However, separate or sequential maintenance and inventory optimization may not necessarily produce a globally optimal strategy, and their joint optimization strategy needs to be studied. The aforementioned paper considered the (s, S) inventory strategy, which, as mentioned before, has been applied in most inventory control software packages. However, in the maintenance process, spare parts only need to be replaced when they are close to failure. When to purchase spare parts is determined by the component status information used to schedule maintenance activities. There is no research on the procurement problem of the multi-component system status, which has obvious cost-saving potential. Olde Keizer [3] studied the joint optimization problem of Condition based Maintenance and inventory decisions of multi-component systems with a shared spare parts pool, and formalized it as a Markov Decision Process (MDP) [42], while Wang and others [31, 32] formalized it as a simulation-based method. Some Markov maintenance process models have been applied to single-component degradation processes.

9.3.1 Simulation-Based Joint Optimization Model

The simulation-based joint optimization model consists of n completely identical independent components, with the general assumptions for each component as follows.

① The degradation level of each component can be described by the random variable X_t, which increases monotonically within the component's lifespan. For convenience, when the component is considered to be in a new state, it can be defined as $t = 0$: $X_0 = 0$. The larger X_t, the closer the system is to failure.
② When component failure occurs, the degradation level X_f varies in different life cycles. Compared to time-induced failure, X_f is also referred to as degradation failure. In summary, X_f is random and uncertain, and it can be assumed that X_f follows a certain failure distribution, such as Poisson distribution, gamma distribution, etc.
③ The degradation level of the component can be determined by instantaneous and non-destructive checks at discrete time points $t_k = k^* \Delta t$ (which is the unit length of time for inspection). To save money, it may not be necessary to check the unit at every t_k moment.
④ The increment of X_t between two consecutive times t_{k-1} and t_k is: $\Delta X_k = X_k - X_{k-1}$.

It is assumed that these increments ΔX_k are negative, stationary, and statistically independent, and follow the same distribution as X_f. In these studies, the system only considers one or two components, each of which is affected by a continuous degradation process. In many cases, a system may consist of several identical components. Moreover, in industrial maintenance practice, it is common for these identical components to use one or more identical spare parts together, which can save the cost of ordering and holding inventory for spare parts.

In addition, it is worth mentioning that the multi-component system considered is not the common multi-component system in the literature. They should also consider the interaction between components in maintenance decisions, including economic dependence, random dependence, and structural dependence.

The Condition based Maintenance strategy is a Control Limit Policy (CLP) with a preventive replacement threshold L_p (component degradation level), and the inventory strategy is (s, S), where S is the maximum inventory level, and s is the reorder level. Kabir and Al-Olayan [23] were the first to propose a continuous inspection (s, S) strategy for a single-component system, using an age-based replacement strategy. Kabir et al. [23] extended the integrated age replacement and spare parts supply strategy for multi-component systems, in which each component is also the same and independent. To reduce the cost of ordering and holding inventory for spare parts, the (s, S) inventory strategy is applicable to a system with multiple identical components, which may require the same spare parts. Therefore, the (s, S) inventory strategy is used in the proposed maintenance strategy.

Below are some relevant assumptions and possible decisions based on state-based replacement and spare parts orders.

① Each component at the discrete time $t_{kT} = kT$, $k \in N$ for regular inspections, where, $T = k_T * \Delta t$ is the fixed inspection interval, and the degradation level of component j at each inspection time is denoted as X_j. The unit cost of each inspection for each component is C_i, and the inspection time can be ignored. Without loss of generality, Δt can be taken as 1, and for simplicity, k_T can be directly represented as T in the formula. Moreover, although all components in the system have the same inspection interval, they may not be inspected at the same time due to differences in the starting points of their lifecycles.

② The available maintenance operations for each component are preventive replacement before failure and corrective replacement after failure, which can only be implemented at discrete time t_k. Replacement can be a complete component replacement or a major overhaul, and the component is as good as new after replacement. It is assumed that maintenance is instantaneous. The cost of preventive replacement is C_p, and the cost of corrective replacement is C_c. In most practical situations, it can be considered that $C_\text{c} > C_\text{p}$.

③ The spare parts inventory adopts the (s, S) inventory strategy. Under this strategy, the inventory level is initially set to S, and it can be known at the discrete t_k moment without any cost. Corrective replacement and preventive replacement deplete the spare parts inventory, and when the inventory drops to s, an $S-s$ purchase order is placed. If an order has been placed but the ordered spare parts have not been delivered, no new order needs to be generated, i.e., only one order is allowed at the same time. The order delivery time t_s is considered to be a constant and independent of the order quantity, where $t_\text{s} = k_\text{s} * \Delta t$, for simplicity, k_s is directly marked as t_s. In addition, after the delivery of the ordered spare parts, some spare parts can be used immediately for replacement, and the remaining spare parts should be stored. Moreover, it is also assumed that the spare parts in the inventory will not degrade (i.e., they always remain in good condition). The inventory holding cost per unit time is C_h, and the order cost per unit spare part is C_s.

④ Based on the availability of spare parts and the observed degradation of component j at each inspection time t_{kT}, decisions should be made on component replacement and spare parts ordering. These possible decisions are as follows.

a. If $x_j < L_\text{p}$, unit j remains as is.
b. If it is found that $x_j \geq L_\text{p}$, and there is at least one spare part in the inventory, then the operating component j is replaced in time.
c. If it is found that $x_j \geq L_\text{p}$, and there are no available spare parts in the inventory (in this case, a spare parts purchase order must be placed), then component j continues to operate. In addition, after this inspection, component j will no longer be inspected. In this case, if the ordered spare parts arrive and the original component j is still working, preventive replacement (also known as delayed preventive replacement) will be performed immediately; if the original component j fails before the spare parts arrive, the original component j remains in a failed state while waiting for the required

spare parts to arrive, and once the spare parts arrive, corrective replacement is immediately performed on component *j*.

d. Without inspection, the failure of each operating component can be immediately detected. At the discrete moment t_k after the failure occurs, if there is at least one spare part in the inventory, the corresponding unit will be correctly replaced without any delay. Otherwise, the failed component will wait for the ordered spare parts, and once the ordered spare parts arrive, corrective replacement will be performed. C_d is the cost loss generated by the component per unit time in the failed state.
e. After each corrective/preventive replacement of component *j*, it must be decided whether to immediately follow the (*s*, *S*) Inventory policy orders $S-s$ spare parts.

⑤ After replacement, another new life cycle of the duplicate part *j* begins.

In addition, assuming the following events are instantaneous, they occur continuously at each discrete point t_k according to the recommended order replacement strategy.

① Check the inventory level to see if there are ordered spare parts before arrival.
② Put the newly arrived spare parts into the warehouse.
③ Consider the first unit to see if it fails. If it runs well, make inspection and maintenance decisions according to the rules suggested by the above strategy.
④ Make spare parts ordering decisions based on the above strategy suggestions.
⑤ Repeat steps ③ and ④ for the 2nd, 3rd, …, *n* units in turn.

Figure 9.9 gives an example of the evolution of a component degradation with two operating parts systems and the corresponding spare parts unit situation.

Based on the degree of degradation, state-based maintenance strategy, and (s, S) inventory strategy, the choice of the following decision variables significantly affects the performance of the proposed strategy: inspection interval T, preventive threshold of degradation degree Lp, reorder quantity s, and maximum inventory level S.

For a slowly degrading component, it is unnecessary and expensive to inspect it too frequently; on the contrary, when the inspection interval T is too large, the probability of downtime before the next inspection increases. Similarly, a low preventive threshold Lp will lead to frequent preventive replacements and cannot fully utilize the remaining life of aging but still working components. Considering spare parts orders, inventory holding costs may increase with high s and S, but low s and S will lead to high unavailability of spare parts, increasing the risk of lack of spare parts when deciding to replace them. The optimal inventory level may be greater than the total number of units n in the system. When the holding cost Ch is much lower than the ordering cost Cs and the cost loss Cd generated by the component per unit time, it may be more economical to store more than n spare parts in the inventory, and the unavailability of the component will lead to profit loss caused by Cd. Therefore, to balance replacement/ordering/holding costs and component unavailability, that is, to reduce total costs, maintenance managers and engineers must choose appropriate values for T, S, s, and Lp. When the cost rate is minimal, the state replacement and spare parts supply strategy with the optimal decision variables corresponding to the

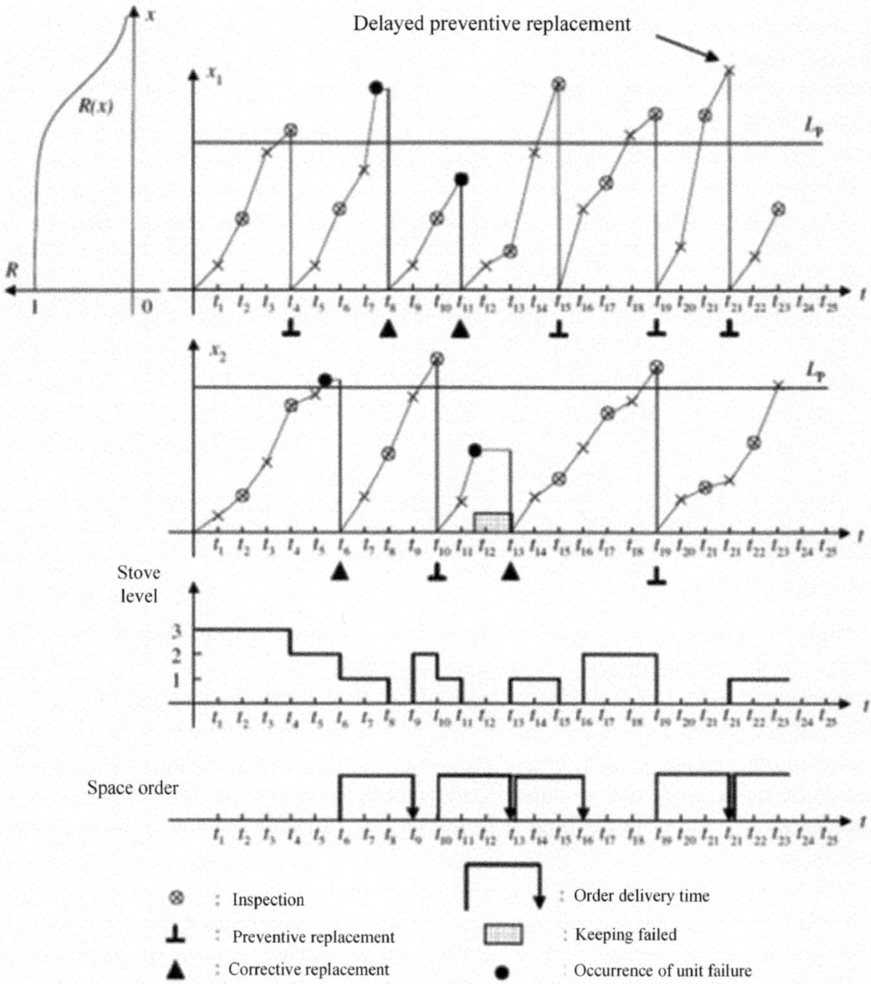

Fig. 9.9 Relationship between preventive replacement, inventory level, and spare parts order

minimum cost rate is called the optimal strategy. Also, if the average cost rate EC∞ can be represented as a function fEC(T, S, s, Lp) of T, S, s, and Lp, the optimization strategy can be defined as a constrained optimization problem, namely:

$$
\begin{aligned}
\min \quad & EC_{\infty} = \min f_{EC}\left(T, S, s, L_p\right) \\
\textit{subject to} \quad & \\
& T = 1, 2, 3 \ldots \\
& S = 1, 2, 3 \ldots \\
& s < S, s = 0, 1, 2, 3 \ldots \\
& L_p > 0
\end{aligned}
$$

Due to the complexity of the order-replacement strategy, it is difficult to derive the analytical formula of the function f_{EC} (T, S, s, L_p). However, over an infinite time span, the average cost per unit time (i.e., the average cost rate) can also be simply expressed as:

$$EC_{\infty} = \lim_{t_m \to \infty} \frac{C_i{*}N_i + C_p{*}N_p + C_c{*}N_c + C_s{*}N_s + C_d{*}\sum_{k=1}^{m} N_d(t_k) + C_h{*}\sum_{k=1}^{m} N_h(t_k)}{t_m{*}n}$$

In the formula, $N_{\rm i}$ is the total number of inspections over the time span $t_{\rm m}$; $N_{\rm p}$ is the total number of preventive replacements within the time $t_{\rm m}$; $N_{\rm c}$ is the total number of corrective replacements within the time span $t_{\rm m}$; $N_{\rm s}$ is the total number of spare parts orders within the time span $t_{\rm m}$; $N_{\rm d}$ (t_k) is the number of failed units before the corrective replacement at t_k; $N_{\rm h}$ (t_k) is the number of spare parts in stock before the corrective and preventive replacements at t_k.

The average cost rate EC_{∞} can be evaluated by simulation methods, and the time value of money is not considered in the study.

(1) Simulation model

There are mainly two methods to evaluate the related costs (or reliability or availability) of the system under the maintenance strategy: the first one is to establish a mathematical model of the system cost rate, so as to get the exact value of the cost rate, and use mathematical methods to minimize the cost rate. However, many maintenance strategies are not easy to analyze, so simulation must be used to analyze these strategies; the second one is to optimize the maintenance strategy through simulation, often using algorithms to optimize deterministic optimization problems. When using simulation methods, only the approximate value of the cost rate can be obtained, and the "optimized" strategy is the closest to the optimal solution.

Under the proposed state replacement and spare supply strategy, the maintenance activities of each component in the system are carried out according to the degradation situation of the corresponding component and the inventory level. The inventory level is determined by the spare supply activities and the spare good use generated by the component replacement. It is difficult to establish a mathematical model to calculate the cost rate. The Monte Carlo (MC) simulation method can be used to evaluate the performance of the strategy—the cost rate. By simulating a large amount of system life cycle history data, the MC method can estimate the overall average value of the related quantity, which is used to calculate the performance standard of the strategy. Each simulated history can be regarded as a virtual experiment, in which the unit degradation process is tracked throughout the life cycle.

① Modeling the degradation and replacement of components First, this process needs to model the degradation of components:

$$x_j = x_j^{last} + \Delta x_j$$

In the formula, Δx_j is a random increment of the degradation degree of component j within a unit time, which is randomly generated according to the given function $g_x(x)$.

Next, determine whether component j fails in the current life cycle:

$$F_j = \begin{cases} 0, \; x_j < X_{fj} \\ 1, \; x_j \geq X_{fj} \end{cases}$$

where, X_{fj} is randomly generated at the beginning of the current life cycle based on the known failure function $F_x(x)$.

According to the rules of the proposed order replacement strategy, the following formula represents the decision of immediate or delayed preventive replacement:

$$IPR_j = \begin{cases} 1, \; if \; (F_j = 0 \; and \; x_j \geq L_p \; and \; I_j = 0 \; and \; SL > 0) \\ or \; (F_j = 0 \; and \; x_j \geq L_p \; and \; IN_j = 0 \; and \; SL > 0) \\ 0, \; otherwise \end{cases}$$

where, SL represents the number of available spare parts; I_j represents the components that need to be inspected.

If unit j fails and at least one spare part is available, corrective replacement is determined, the formula is:

$$ICR_j = \begin{cases} 1, \; if \; F_j = 1 \; and \; SL > 0 \\ 0, \; otherwise \end{cases}$$

So, we can set $IR_j = \begin{cases} 1, \; IPR_j = 1 \text{ or } ICR_j = 1 \\ 0, \text{ others} \end{cases}$

x_j^+ should be obtained, to be used as the next discrete time's x_j^{last}:

$$x_j^+ = \begin{cases} 1, \; IPR_j = 1 \text{ or } ICR_j = 1 \\ x_j, \qquad \text{others} \end{cases}$$

ST_j should be set to represent the start time of the current life cycle of part j:

$$ST_j = \begin{cases} t_k, \; IR_j = 1 \\ ST_j, \; \text{Other} \end{cases}$$

② Detection of modeling components I_j represents the component j that needs to be detected, which should be set before any replacement decision:

$$I_j = \begin{cases} 1, \; t_k - ST_j \neq 0, (t_k - ST_j) \bmod T = 0, IN_j = 1 \\ 0, \qquad Other \end{cases}$$

Under the proposed strategy, if it is found that $x_j \geq L_p$ and no spare parts are available, then component j continues to run and no further detection is carried out after this detection, as follows:

$$IN_j = \begin{cases} 0, & ST_j \neq t_k, x_j \geq L_p, SL = 0, I_j = 1 \\ 1, & ST_j = t_k \\ IN_j, & \text{other} \end{cases}$$

③ Modeling spare parts orders. Before making a decision at the current discrete time, if an order has been placed before, please check whether the ordered spare parts are delivered at this time:

$$NA = \begin{cases} S - s, & t_k - LastOT = t, OC = 1 \\ 0, & other \end{cases}$$

where, *LastOT* represents the last order time.

After setting *NA*, immediately update the available spare parts quantity:

$$SL = SL^- + NA$$

where, SL^- can be derived based on the number of spare parts remaining at the last discrete time. After each part replacement, *SL*, *OC* and *LastOT* need to be reset as follows:

$$SL = \begin{cases} SL - 1, & IR_j = 1 \\ SL, & IR_j = 0 \end{cases}$$

$$OC = \begin{cases} 1, & SL \leq s, OC = 0 \\ 0, & (t_k - LastOT = t, OC = 1) or t_k = 0 \\ OC, & other \end{cases}$$

$$LastOT = \begin{cases} t_k, & SL \leq s, OC = 0 \\ LastOT, & other \end{cases}$$

④ Modeling the average cost rate After all replacement activities are completed at the current discrete time, accumulate N_i, N_p, N_c and record N_s, expressed as:

$$N_i = N_i + \sum_{j=1}^{n} I_j$$

$$N_p = N_p + \sum_{j=1}^{n} IPR_j$$

$$N_c = N_c + \sum_{j=1}^{n} ICR_j$$

$N_s = N_s + NA$

$$N_d(t_k) = \sum_{j=1}^{n} F_j$$

$N_h(t_k) = N_c + SL^-$

After simulating all events within the time span t_m, the average cost rate can be obtained.

To further describe the MC simulation, Fig. 9.10 provides a flowchart of the proposed order replacement strategy simulation program.

The first step "Initialize strategy parameters" is to input system parameters C_i, C_p, C_c, C_h, C_s, C_d, t_s, w_x, w_y, L_p, T, S, s, n, t_m, and set $F_j = 0$, $I_j = 0$, $ICR_j = 0$, $IN_j = 0$, $IPR_j = 0$, $IR_j = 0$, $LastOT = 0$, $OC = 0$, $SL = S$, $ST_j = 0$, $t = 0$, $x_j = 0$, the simulation time span t_m should be set large enough. When simulating the proposed order-replacement strategy, the maintenance decision and corresponding cost can also be determined according to Fig. 9.11.

Due to the discrete time grid used for modeling the order-replacement strategy, it is impossible to accurately determine the time of the failure unit status. According to the aforementioned simulation model, when a unit fails at time t (where $t_{k-1} < t \leq t_k$), the time length of the failure status until t_k is considered as one unit, while the exact result should be $t_k - t$ (<1) unit time. Therefore, the export cost generated by the unit under failure status is an approximate result.

(2) Technical Advantage

The advantages of the state-based replacement and spare supply strategy over PHM can be easily identified, thereby optimizing the maintenance decision of the transport truck engine. Firstly, this strategy can optimize the detection frequency of the motor, and the PHM method cannot obtain the optimal detection interval. Secondly, the state-based replacement-spare supply strategy is to jointly optimize motor maintenance and spare management, PHM does not consider spare availability, and spares are important for maintenance activities. Finally, the new replacement-spare supply strategy can optimize the maintenance and spare ordering activities of all motors in the entire transport fleet, while PHM only focuses on one motor.

① The state-based reliability theory can characterize uncertain failure degradation. Based on this theory, the reliability of the system can be evaluated according to the actual degradation of the system, which is an advantage compared to the traditional time-based reliability that only involves the working age of the system.

② In a multi-unit system, by optimizing maintenance and spare supply decisions and applying the proposed order-replacement strategy, it can be found that when the lead time greatly increases, the increase in the cost rate may be small. This conclusion indicates that the joint optimization of state-based maintenance and spare ordering activities is necessary and beneficial, and the evaluation of cost parameters is more or less important for optimizing state-based replacement and spare supply strategies.

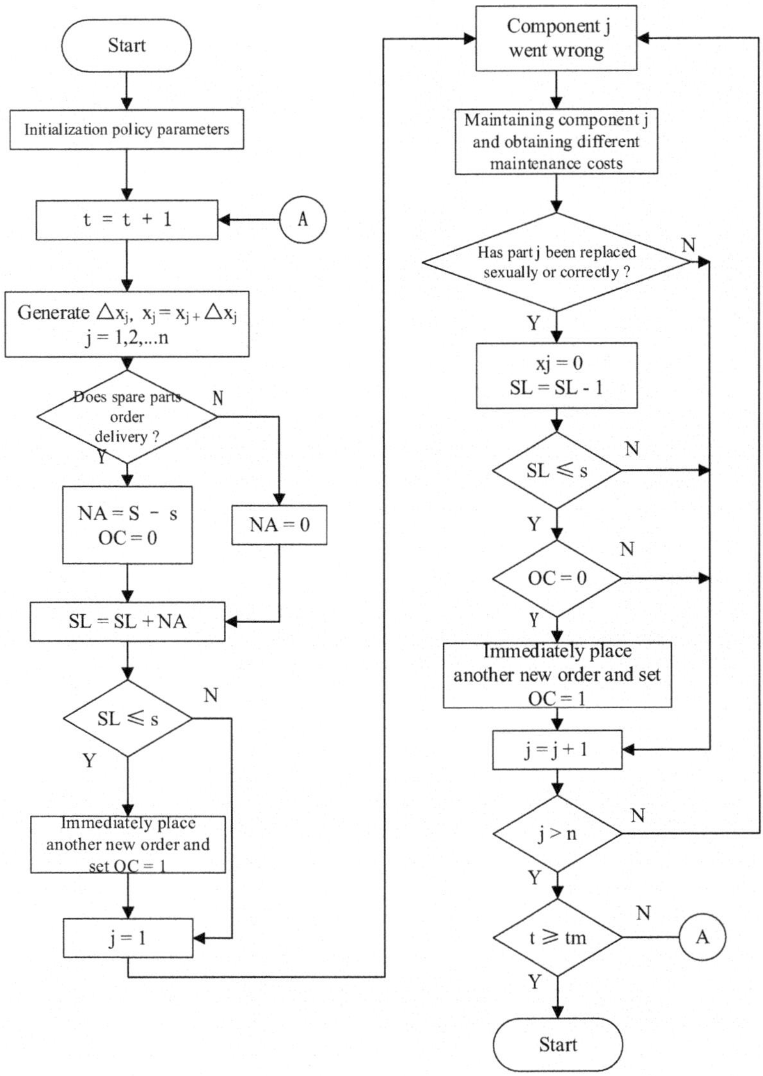

Fig. 9.10 Simulation program flow

③ In the case where spare motors are always available, it can be found that compared with the existing mine motor maintenance strategy, the implementation of the proposed strategy can save funds. Practice has proved that the application of the maintenance model can effectively reduce maintenance costs.

Suggestions for further work are as follows;

① The relationship between the component degradation process and the degradation failure distribution needs further research;

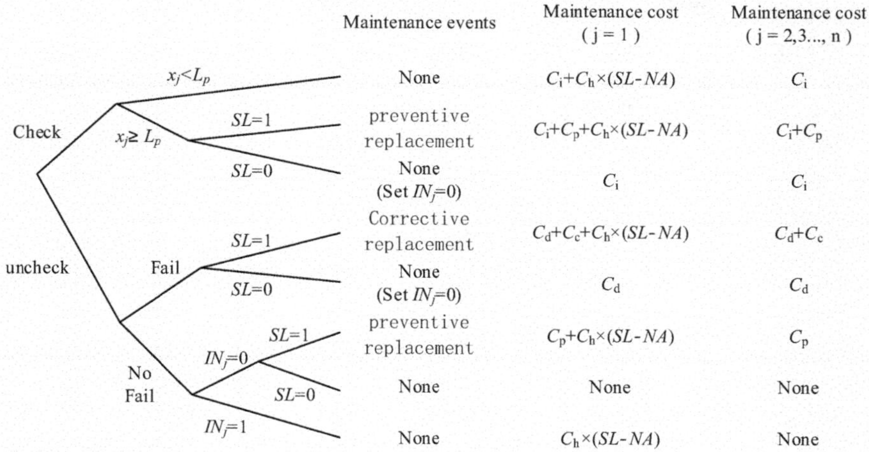

Fig. 9.11 Simulation process of order-replacement strategy

② The state-based regular inspection replacement and spare supply strategy can be extended to the order replacement strategy of non-periodic inspections;

③ Optimize the state-based maintenance and spare supply strategy to adapt to the deterioration system of multiple different components, and should also include economic, random, structural dependencies between components;

④ Minimum repair or maintenance can also be included in the order-replacement policy;

⑤ Extending the constant delivery period and negligible repair/inspection time to random delivery period and random repair/inspection time seems to be another challenging task.

9.3.2 Markov Decision Process Based Joint Optimization Model

Consider a discrete-time system composed of N functionally and degradation independent components, a component j (where $j = 1,2,\ldots,N$) state model can be modeled using $L_j + 1$ different states $(0,1,\ldots, L_j)$. Here, state 0 means the construction is as good as new, and state L_j indicates that the component has failed. Components share a spare pool, although the components are the same, they may be included in different subsystems, these subsystems may operate under different states, which may lead to different failure rates. If a component is replaced, the old component will be discarded, and the new component state is state 0. Compared with the expected life of the component (years) and the lead time of the spare parts (months), the repair time is usually shorter (days), therefore, the replacement process is considered instantaneous, but it can only be arranged when the required spare parts are in the warehouse. Spares can be ordered in any quantity and arrive after a fixed T time unit lead time (usually on the order of months). After performing possible maintenance and inventory operations, component j may degrade. It is assumed that the component state degrades rather than improves, and it is assumed that the degradation of component j follows a Poisson distribution with parameter μ_j.

The order of events is as follows: at the beginning of each time unit, a spare order arrives (the order has been placed T time units ago). Next, operating costs are generated for each component, which depend on the state of the component. Let $C_o^j = (O_0^j, O_1^j, \ldots, O_{Lj}^j)$ represents the operating cost vector of component j. If component j is in state u, then the operating cost for each time is O_u^j. In this way, not only can the downtime cost of failed components be included (by choosing a high value for O_{Lj}^j), but also the revenue loss cost caused by component degradation. After the operating cost is incurred, the component can be replaced as long as there are enough spare parts. In practice, failure replacement is sometimes more expensive than preventive replacement because component failure can also cause damage to other parts of the system. Bouvard [43], Park [44], de Jonge [45] and others have considered this difference. Define the replacement cost vector related to the state of component j as $R^j = (R_0^j, R_1^j, \ldots, R_{L_j}^j)$, including the purchase price of new parts. For $u > v$, $R_u^j \geq R_v^j$. After possible maintenance operations are performed, spare parts can be ordered, and a fixed cost F is incurred for each order, which is independent of the number of spare parts ordered. For each spare part still on hand at the end of the time unit, an inventory holding cost H is incurred.

(1) Model

The Markov Decision Process (MDP) model consists of a set of possible initial states τ. The set of executable activities at state i (where $i \in \tau$) is A_i. The transition probability from state i (where $i \in \tau$) to state $\bar{i}$($\bar{i}$ (where $\bar{i} \in \tau$) after executing activity α (where $\alpha \in A_i$) is defined as p^{α} $(i$: $\bar{i})$. The expected cost function of the system after executing activity α (where $\alpha \in A_i$) at state i (where $i \in \tau$) is $c^{\alpha}(i)$. To limit the size of the state space, it is assumed that the inventory level (the number of existing spare parts plus the number of ordered spare parts) cannot exceed the maximum $\overline{S}$.

① State Space: The state space tracks the status of each component $(x_1, x_2, \ldots, x_N)$ and the status of each order $(s_1, s_2, \ldots, s_{T1})$, s_l represents the number of spare parts ordered $l(l = 1,2,\ldots, T-1)$ time units ago, s_h represents the number of existing spare parts. $N + T$ dimensional state space:

$$\tau = \{(x_1, x_2, \ldots, x_N, s_1, s_2, \ldots, s_{T-1}, s_h)\}$$

where, $j = 1,2,\ldots, N$, $l = 1,2,\ldots, T-1$; $x_j \in \{0,1,\ldots, L_j\}$ is the state of component j; $s_l \in \{0,1,\ldots, \overline{S}\}$ is the number of spare parts ordered l time units ago; $s_h \in \{0,1,\ldots, \overline{S}\}$ is the number of existing spare parts.

The size of the state space is limited by logical constraints, the inventory level cannot exceed the maximum value $\overline{S}$, so $s_1 + s_2 + \ldots + s_{T-1} + s_h \leq \overline{S}$.

② Activity Space: At the start of each activity, it is necessary to decide whether to replace some components and the number of spare parts to order. Therefore, the $N+1$ dimensional activity space can be represented as:

$$A = \{(\delta_1, \delta_2, \ldots, \delta_N, \omega)\}$$

where, $j = 1,2,\ldots, N$, $\delta_j = \begin{cases} 1, & \text{if component } j \text{ replaced} \\ 0, & \text{other} \end{cases}$, the number of spare parts ordered $\omega \in \{0,1,\ldots, \overline{S}\}$.

For any possible $x_j \in \{0,1,\ldots, L_j\}$, $s_l \in \{0,1,\ldots, \overline{S}\}$ and s_h, the following set of activities is allowed for different states of the state space:

$$A_{\{(x_1,x_2,\ldots,x_N,s_1,s_2,\ldots,s_{T-1},s_h)\}}$$
$$= \left\{(\delta_1, \delta_2, \ldots, \delta_N, \omega) : \sum_{j=1}^{N} \delta_j \le s_h, \omega \le \bar{S} - s_h - \sum_{l=1}^{T-1} s_l + \sum_{j=1}^{N} \delta_j \right\}$$

③ Transition probability: If no maintenance is performed, the probability of component j transitioning from state x_j to state $\overline{x_j}$ is $p^j_{x_j\overline{x_j}}$. If component j is replaced, its state will change to the new state 0, and the probability of it transitioning from state 0 to $\overline{x_j}$ is $p^j_{0\overline{x_j}}$. Therefore, we can define the transition probability from state x_j to state $\overline{x_j}$ under the effect of maintenance activity $p^j_{x_j\overline{x_j}}(\delta_j)$:

$$p^j_{x_j\overline{x_j}}(\delta_j) = \begin{cases} p^j_{x_j\overline{x_j}}, & \text{if } \delta_j = 0 \\ p^j_{0\overline{x_j}}, & \text{if } \delta_j = 1 \end{cases}$$

In numerical experiments, it is assumed that the degradation increment (the amount of degradation within two time units) of component j follows a Poisson distribution with parameter μ_j ($j = 1, 2, \ldots, N$). To illustrate this, let x_j be a Poisson distribution with mean μ_j, the transition probability of component j is:

$$p^j_{x_j\overline{x_j}} = \begin{cases} p(X_j = \overline{x_j} - x_j), & \text{if } \overline{x_j} < L_j \\ p(X_j \ge \overline{x_j} - x_j), & \text{if } \overline{x_j} = L_j \end{cases}$$

After the activity $(\delta_1, \delta_2,\ldots, \delta_N,\omega)$ is executed, the system transitions from the state $(x_1, x_2,\ldots, x_N, s_1, s_2,\ldots, s_{T-1}, s_h)$ to the state $(\overline{x_1}, \overline{x_2},\ldots, \overline{x_N}, \overline{s_1}, \overline{s_2},\ldots, \overline{s_{T-1}}, \overline{s_h})$ with the transition probability:

$$p^{j(\delta_1,\delta_2,\ldots,\delta_N,\omega)}_{(x_1,x_2,\ldots,x_N,s_1,s_2,\ldots,s_{T-1},s_h)(\overline{x_1},\overline{x_2},\ldots,\overline{x_N},\overline{s_1},\overline{s_2},\ldots,\overline{s_{T-1}},\overline{s_h})}$$
$$= \begin{cases} \prod_{j=1}^{N} p^j_{x_j\overline{x_j}}(\delta_j), \ if\, \overline{s_1} = \omega, \overline{s_h} = s_h - \sum_{j=1}^{N} \delta_j + s_{T-1}, \overline{s_l} = s_{l-1} (l \in \{0, 1, \ldots, T-1\}) \\ 0, \qquad other \end{cases}$$

④ Expected cost: In each time unit, there will be costs. The cost of performing activity (x_1, x_2,..., x_N, s_1, s_2,..., s_{T-1}, s_h) with (δ_1, δ_2,..., δ_N,ω) is composed of operating costs, part replacement costs (if replacement occurs), order costs (if spare parts orders are generated), and holding costs (if there are spare parts in stock), for example:

$$c^{(\delta_1,\delta_2,\ldots,\delta_N,\omega)}(x_1, x_2, \ldots, x_N, s_1, s_2, \ldots, s_{T-1}, s_h)$$
$$= \sum_{j=1}^{N} O^j_{x_l} + \sum_{j=1}^{N} \delta_j R^j_{x_l} + FI_{(\omega>0)} + \left(s_h - \sum_{j=1}^{N} \delta_j\right)H$$

In the formula, $I_{(\omega>0)}$ indicates that if the expression in the brackets is true, the indicator function is 1, otherwise it is 0. Note that this MDP formula is constructed for the case where the lead time is $T > 1$. In the case where $T = 1$, there is no need to track the time of placing orders, because the order has already arrived before making the next decision. Therefore, the state space can be reduced to $N + 1$ dimensional space, only tracking the state of each component and the quantity of existing spare parts. The action space and transition probabilities are also simplified to $T = 1$.

⑤ (s, S) inventory strategy The inventory level (represented by IP) includes available spare parts $\left(s_h - \sum_{j=1}^{N} \delta_j\right)$ and the quantity of ordered parts $\sum_{l=1}^{T-1} s_l$. As long as the inventory level is less than or equal to s, the (s, S) inventory strategy issues an S–IP spare parts order. It can be seen as a special case of the MDP, by limiting the set of activities allowed in each state, for any x_j(j = 0,1,..., N), s_l (l = 0,1,...,T−1) and s_h:

$$A_{\{(x_1,x_2,\ldots,x_N,s_1,s_2,\ldots,s_{T-1},s_h)\}} = \{(\delta_1, \delta_2, \ldots, \delta_N, \omega) : \sum_{j=1}^{N} \delta_j \le s_h,$$
$$\omega = I_{\{s_h+\sum_{l=1}^{T-1} s_l-\sum_{j=1}^{N} \delta_j \le s\}}\left(S - s_h - \sum_{l=1}^{T-1} s_l + \sum_{j=1}^{N} \delta_j\right)\}$$

(2) Algorithm flow

As a performance standard, enterprises care about the long-term average unit time cost. The inventory level cannot exceed the maximum value of $\overline{S}$. The assumption that the state space and action space are both finite and the cost function is defined. If the model is a single chain, there exists a stationary average optimal policy, and the value iteration algorithm can be used to find this policy. However, the proposed MDP model contains multiple loop states. For example, when N = 2 and T = 3, for all states i ∈ τ, consider the fixed policy f(i) = (0,0,0) (no maintenance, no need to order spare parts). Under this policy, in the long run, each component will be in the failure state Lj, and the initial number of spare parts in stock will remain unchanged. Thus, the generated transition matrix will contain $\overline{S} + 1$ loop states, and the model to be dealt with is multi-chain rather than single-chain. However, the realistic assumption is: any optimal policy will order a new spare part for each replaced component (at some point in time), and as long as a spare part is available, the faulty component will be replaced, because the operating cost of the faulty component is high enough (e.g., downtime cost). Therefore, the minimum average cost per time unit is independent of the initial state and can be effectively determined by applying the value iteration algorithm. v_n represents the value function obtained in the nth iteration.

Step 0: Initialization.

For all i ∈ τ, choose an ε > 0 and $v_0(i)\left(0 \le v_0(i) \le \min_{\alpha\in A_{\{i\}}} c^\alpha(i\right)$, set n = 1;

Step 1: For all i ∈ τ, calculate the value function $v_n(i)$:

$$v_n(i) : \min_{\alpha\in A_{\{i\}}}\left\{c^\alpha(i) + \sum_{i\in\tau} p^\alpha\left(i : \bar{i}\right)v_{n-1}\left(\bar{i}\right)\right\}$$

And choose a minimizer v_n static strategy of $(i)f_n$:

$$f_n(i) \in \arg\min_{\alpha\in A_{\{i\}}}\left\{c^\alpha(i) + \sum_{i\in\tau} p^\alpha(i{:}\bar{i})v_{n-1}(\bar{i})\right\}$$

Step 2: Let $M_n\cdot\cdot = \max_{i\in\tau}\{v_n(i) - v_{n-1}(i)\}$ and $m_n\cdot = \min_{i\in\tau}\{v_n(i) - v_{n-1}(i)\}$, if $0 \le M_n - m_n \le \varepsilon m_n$, the algorithm stops, otherwise, set $n\cdot = n + 1$, continue steps 1 and 2.

(3) Numerical analysis

Consider a system composed of two components with the same failure rate and independent degradation. By considering the two components, the optimal maintenance and inventory decisions can be represented in two dimensions. This system basically illustrates some characteristics of the optimal strategy and the impact of limiting possible inventory operations to the (s, s) inventory strategy.

Consider a fixed failure level $L = 4$, each component can be in one of five different states. Assume that there are 4 inspections per year, with a time unit length of 3 months. Although current technology allows the system to be continuously monitored by placing sensors, a mixed inventory inspection strategy of periodic and continuous has been applied in practice. In the process industry, for safety reasons, the use of Wi-Fi sensors can be prohibited, and installing sensors on some other systems (such as large systems like railway networks) may not be cost-effective. In addition, effective Condition based Maintenance requires a measurable parameter that is closely related to the occurrence of faults, which is often problematic. For example, natural gas distribution companies lack suitable models to link the internal fouling of heat exchangers with useful process data (such as temperature, flow rate) [46], so regular (periodic) visual inspections become the only appropriate method to properly determine the condition of the asset. Therefore, regular and continuous inventory inspections will continue to coexist in the near future.

Parameters	Explanation	Value
L	Failure level	4
O	Operating costs of states 0, 1, 2, 3, 4	(0, 0, 0, 0, 100)
R	Replacement costs of states 0, 1, 2, 3, 4	(5, 5, 5, 5, 5)
F	Fixed cost per order	0
H	Inventory holding cost per spare part unit time	0.5
T	Fixed lead time	3
M	Degradation parameter	0.2

The expected life E is equivalent to choosing the degradation parameter μ within the unit time, $\mu = \frac{1}{E}$. Assume that the degradation increment per unit time of each component follows a Poisson distribution with parameter $\mu = 0.\ 2$. This means that, without maintenance, the transition probability of each component from state u to state v within a unit time follows the upper triangular transition matrix P:

$$P = \begin{bmatrix} 0.82 & 0.16 & 0.02 & 0.00 & 0.00 \\ 0.00 & 0.82 & 0.16 & 0.02 & 0.00 \\ 0 & 0 & 0.82 & 0.16 & 0.02 \\ 0 & 0 & 0 & 0.82 & 0.18 \\ 0 & 0 & 0 & 0 & 1 \end{bmatrix}$$

The value iteration algorithm is used to calculate the long-term average operating cost per unit time based on the joint Condition based Maintenance and inventory strategy. Using $\varepsilon = 0.0005$ as the stopping criterion, this means that the final average cost differs from the actual average cost by at most 0.05%. In addition, for all states $i \in \tau$, use $v_0\ (i) = 0$ as the initial value. The results show that the maximum inventory $\overline{S} = 2$ is a sufficiently high level, and increasing this value will not change the optimal strategy and the corresponding average cost. All experiments were processed using Python 3.4.3 on a computer with a 3.30GHz quad-core processor and 16.0GB RAM. For this example, the state space consists of 250 states, and the value iteration algorithm converges after 24 iterations, taking approximately 0.04s. Figure 9.12 shows the convergence sequence for $\{M_n, n \geq 1\}$ and $\{m_n, n \geq 1\}$. These sequences perform well and converge relatively quickly.

Figure 9.13 shows the optimal maintenance and inventory strategy obtained in the basic case. In this figure, the best maintenance and inventory activities show any combination of parts 1 and 2 states x_1, x_2, respectively giving the number of spare parts required l time units before s_l ($l = 0$, 1) and the number of spare parts s_h.

Note that, $\overline{S}$ selection determines the number of implementations (s_1, s_2, s_h). The average unit time cost of the optimal strategy is equal to 1.57 units of time. It can be seen that due to sharing a spare parts pool, the optimal maintenance decision of a component depends on the states of the two components. When there are enough spare parts on hand $s_h = 2$, the optimal replacement decision for component j ($j = 1, 2$) depends solely on its own degradation level x_j (if $x_j \geq 2$) to replace. However,

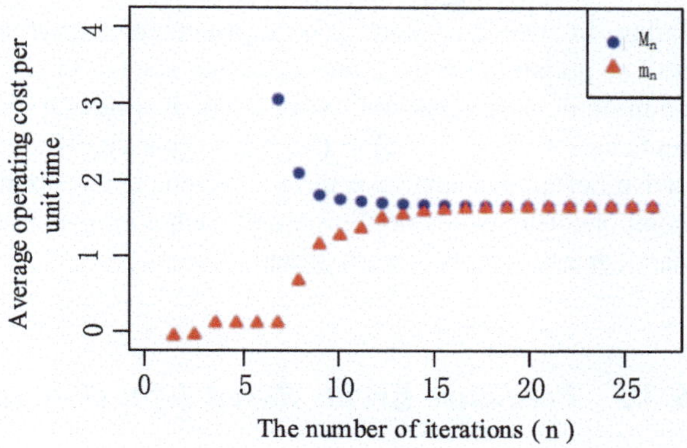

Fig. 9.12 Numerical iteration process

$(s_1, s_2, s_h) = (0, 0, 0)$

x_1 \ x_2	0	1	2	3	4
0	00	00	00	00	00
1	00	00	00	00	00
2	00	00	00	00	00
3	00	00	00	00	00
4	00	00	00	00	00

$(s_1, s_2, s_h) = (0, 0, 0)$, and $(0, 1, 0)$

x_1 \ x_2	0	1	2	3	4
0	00	00	00	00	00
1	00	00	00	00	00
2	00	00	00	00	00
3	00	00	00	00	00
4	00	00	00	00	00

$(s_1, s_2, s_h) = (2, 0, 0)$, $(0, 2, 0)$, and $(1, 1, 0)$

x_1 \ x_2	0	1	2	3	4
0	00	00	00	00	00
1	00	00	00	00	00
2	00	00	00	00	00
3	00	00	00	00	00
4	00	00	00	00	00

$(s_1, s_2, s_h) = (0, 0, 1)$

x_1 \ x_2	0	1	2	3	4
0	00	00	01	01	01
1	00	00	01	01	01
2	10	10	00	01	01
3	10	10	10	00	01
4	10	10	10	10	10

$(s_1, s_2, s_h) = (1, 0, 1)$

x_1 \ x_2	0	1	2	3	4
0	00	00	01	01	01
1	00	00	01	01	01
2	10	10	00	01	01
3	10	10	10	01	01
4	10	10	10	10	10

$(s_1, s_2, s_h) = (0, 1, 1)$

x_1 \ x_2	0	1	2	3	4
0	00	00	01	01	01
1	00	00	01	01	01
2	10	10	10	01	01
3	10	10	10	10	01
4	10	10	10	10	10

$(s_1, s_2, s_h) = (0, 0, 2)$

x_1 \ x_2	0	1	2	3	4
0	00	00	01	01	01
1	00	00	01	01	01
2	10	10	11	11	11
3	10	10	11	11	11
4	10	10	11	11	11

00	Nonsubstitution
10	Replacement part 1
01	Replacement part 2
11	Replace all

(white)	No order required
(grey)	Part 1 Standby
(black)	Part 2 Standby

Fig. 9.13 Numerical case

when there is only one spare part on hand, this decision depends on the complete system state. For example, $(s_1, s_2, s_h) = (0, 0, 1)$ indicates that only one spare part remains, with no outstanding purchase orders. In principle, the component with the highest degree of degradation will be replaced, as long as the component is at least in a state of degradation 2. However, when components 1 and 2 are both close to failure in state 2 or state 3, no replacement is performed. Instead, the spare part will be prepared for the component that is most likely to fail in the near future, and additional spare parts will be ordered. If a new spare part was ordered two time units ago, i.e., $(s_1, s_2, s_h) = (0,1,1)$, no spare parts are kept, as new spare parts will arrive before the next inspection. Therefore, the optimal replacement decision for components depends on the state of each component, the number of spare parts on hand, and the status of each order.

9.3.3 Predictive Events Based Joint Optimization Model

Equipment failure in the manufacturing process poses significant problems for human safety, environmental impact, and reliable industrial processes. The unexpected failure of manufacturing equipment is uncertain, and its degradation process is also random, leading to a high degree of uncertainty in the maintenance decision-making process [47]. Therefore, to reduce unexpected equipment failures, eliminate unplanned downtime, and reduce maintenance-related costs, appropriate maintenance management policies and related information systems are needed [38]. Since maintenance and spare parts inventory management are closely linked [2], accurate and reliable assessment of spare parts is the basis for reliable maintenance modeling and spare parts inventory planning decisions [2, 38, 47].

Predictive maintenance decisions for equipment need to strike a balance between premature replacement costs and unexpected failure costs. Moreover, the ordering time and inventory quantity of spare parts should be planned to minimize holding costs and avoid stockouts. Due to the development of the latest technology and information systems, predictive methods and decision models for joint optimization of maintenance and inventory based on real-time status data from equipment (collected through sensors) and predictive information (such as Remaining Useful Life RUL, remaining life distribution) have begun to emerge. The most representative work, Elwany [47], improved Armstrong's [28] decision model, updating the RUL estimate in real-time with each data collection.

Due to the availability of large amounts of data generated by high-frequency events from various sources, the path has been paved for coupling predictive decision methods with sensor event-driven frameworks. Event-driven architectures can support effective event handling and improve scalability, while also having the ability to handle probability distribution functions rather than parameters (such as RUL). Bousdekis et al. [38] considered the framework of Condition-Based Maintenance (CBM), the concept of e-maintenance, and proactive event-driven principles, using an Event-Driven Architecture (EDA) to develop a joint optimization decision model for predictive event-based maintenance and spare parts inventory. The model can be deployed in a sensor-based real-time industrial big data environment as a decision function in the "detect-predict-decide-execute" cycle.

Companies maintain spare parts inventory for use during maintenance. The number of spare parts in inventory depends on maintenance needs, i.e., failure maintenance, preventive maintenance, or predictive maintenance operations that require related spare parts. Therefore, the maintenance process and spare parts inventory management are closely linked, and both aspects should be considered when optimizing company operations [2]. Most research on joint optimization decisions for maintenance and inventory relies on failure arrival times or reliability distributions (these parameters or distributions come from experimental setups or manufacturing company specifications), rather than real-time data. Therefore, it is not possible to update maintenance recommendations based on actual and/or predicted equipment health status. Although many papers on real-time prediction have been published in recent years, no literature has explored joint decision models for maintenance and spare parts inventory based on real-time prediction due to the lack of effective decision steps in the CBM framework. Moreover, in the field of joint optimization decisions, almost all published papers are for CBM strategies based on the actual degradation level of parts, not for future degradation, future failure, or other predictive information. Since predictive information is available, predictive maintenance and Just-In-Time (JIT) ordering of spare parts can be recommended [2]. Joint optimization decisions for predictive maintenance and inventory have great potential to validate predictive maintenance practices and the impact of predictive information on inventory costs.

(1) Model

Combining cost-risk analysis and reliability analysis, the joint optimization decision model for predictive event-based maintenance and spare parts inventory aims to provide timely and reliable recommendations for the optimal time for maintenance and the optimal time for ordering spare parts. To provide the probability distribution function of failure occurrence and its parameters, the decision model is triggered by predictive events (predictive events are developed based on complex event patterns), including machine learning, statistical, and data mining algorithms. Degradation models usually follow exponential, gamma distribution, or Weibull distribution [47]. In some cases, cumulative damage does not significantly affect the degradation rate, and a linear degradation model can be used directly [47], without the need for a probability distribution function belonging to the exponential family.

Each factor in the long-term maintenance and inventory cost equation in the decision model represents the cost-risk based on real-time predictive event input. At each time period, there are different associated costs represented as a function of the execution time of maintenance activities, the duration of which may be unknown or random, and there is a cost for each time unit. In addition, at $t = 0$, a predictive event is received, and a recommendation is provided. The long-term maintenance cost as a function of time is given by Eq. (9.1), and the long-term inventory cost as a function of time is given by Eq. (9.2). In addition, Table 9.1 provides an explanation for each variable.

$$Cm(t) = cf(t)P^{\varepsilon}(0,t) + (cf(t) + cp(t))P^{\varepsilon}_{\alpha}(t,T) + cp(t)\overline{P}^{\varepsilon}(0,T) \tag{9.1}$$

$$C_o(t) = c_s(t)P^{\varepsilon}(0, t+L) + \big(c_s(t) + c_p(t)\big)P^{\varepsilon}_{\alpha}(t+L, T) + c_p(t)\overline{P}^{\varepsilon}(0, T) \quad (9.2)$$

According to the terms of reliability analysis, the density function of an event ε, $g^{\varepsilon}(t)$ represents the probability of ε occurring at time t, and the cumulative distribution function of $g^{\varepsilon}(t)$ is represented as $G^{\varepsilon}(t)$, known as the life distribution function of ε. $G^{\varepsilon}(t)$ represents the probability of the event ε occurring between time 0 and t [38]. When an activity α is applied to reduce the probability of an unwelcome event, α is associated with a new event density function $g^{\varepsilon}_{\alpha}(t)$, which indicates the probability of ε occurring within time t, even though α has already occurred before time t, because the activity α cannot prevent the occurrence of ε. Therefore, the probability distribution calculation formula is shown in formulas (9.3) and (9.4).

$$P^{\varepsilon}(t_1, t_2) = \frac{G^{\varepsilon}(t_2) - G^{\varepsilon}(t_1)}{1 - G^{\varepsilon}(t_1)} \quad (9.3)$$

$$P^{\varepsilon}_{\alpha}(t_1, t_2) = \frac{G^{\varepsilon}_{\alpha}(t_2) - G^{\varepsilon}_{\alpha}(t_1)}{1 - G^{\varepsilon}(t_1)} \quad (9.4)$$

Minimizing formula [Formula (9.1)]: $C_m(t)$ can yield the optimal maintenance time t_m. In this way, when the long-term replacement cost is the lowest, time-based maintenance can be transformed into Condition based Maintenance by applying the same predetermined activities. This formula consists of three factors representing cost risk.

① The cost incurred due to the probability of failure before the execution of maintenance activities. This factor indicates that the longer the wait to perform a certain maintenance, the greater the possibility of a failure occurring.

Table 9.1 Explanation of variables

Variable	Explanation
$P^{\varepsilon}(t_1, t_2)$	The probability distribution function of failure ε occurring in the time interval (t_1, t_2) under the condition that no failure has occurred at time t_1
$P^{\varepsilon}_{\alpha}(t_1, t_2)$	The probability distribution function of failure ε occurring in the time interval (t_1, t_2) under the condition that no failure has occurred at time t_1 and action α has been implemented at time t_1
$\overline{P}^{\varepsilon}(t_1, t_2)$	Under the condition that no failure has occurred at time t_1, the failure ε The probability distribution function that does not occur within the time interval (t_1, t_2)
$c_f(t)$	The failure and update cost of implementation time
$c_p(t)$	The planned maintenance cost of implementation time
$c_s(t)$	The stockout cost of the time function
L	Lead time for ordering
T	Time to the next planned maintenance

② The cost incurred due to the probability of failure despite taking mitigation measures. This factor indicates that the probability of failure gradually decreases over time.
③ The cost of performing maintenance activities at the end of the next planned maintenance. This factor is considered mainly because, despite predicting a failure, no failure has occurred.

Minimizing formula [Formula (9.2)]: C_o (t) can yield the optimal time t_o for ordering spare parts. In this way, spare parts can be ordered in a JIT manner, reducing long-term inventory costs to a minimum. This formula considers the issue of spare parts scrapping that affects inventory costs and also includes three cost risk factors:

① The cost incurred due to the probability of failure before the moment of spare parts ordering plus the lead time;
② The cost incurred due to the probability of failure despite taking action, but due to the lack of more spare parts;
③ The cost of performing maintenance activities at the end of the next planned maintenance. Considering this factor, despite having predicted a failure, no failure occurred within the decision cycle, and the ordered spare parts are still left in the warehouse until the next planned maintenance.

(2) Case Study

The model was validated in the production scenario of the automotive lighting equipment industry. The production process includes the production of car headlight assemblies and automated transport assembly, embedding sensors and measuring equipment into quality assessment equipment to collect data from each production stage. Due to the large volume of the production process, the complex parts production equipment is expensive, and improvements in detecting, predicting, and eliminating or mitigating failures can be measured in tens of thousands of euros. One component of the headlight is the cover lens, the production process of which includes two steps: forming and painting. The forming process ensures the geometric shape of the lens, and the painting is for the external vehicle environment. The failure to be reduced is a 25% scrap rate. The company implements planned maintenance, dusting and cleaning the injection molding machine every Monday and Thursday at 9:00. The goal is to convert planned maintenance into predictive maintenance to reduce maintenance costs, while ordering spare parts in a JIT manner under insufficient inventory conditions.

Sensors measure the dust level in the workshop, environmental factors such as temperature and humidity, and other known factors affecting the function of the forming machine, as well as the scrap rate of the cover lens. In the detection (diagnosis) stage, a Complex Event Processing (CEP) engine (Fig. 9.14) detects a complex pattern, indicating the start of abnormal behavior and degradation process of the equipment. It sends an event to the prediction stage and triggers an online predictive analysis service, which uses statistical/machine learning methods to provide a prediction of a scrap rate exceeding 25%. This predictive event triggers the decision stage, which formulates online and gives recommendations for the best time to clean

Fig. 9.14 Event-driven decision model

and the best time to order related spare parts. The execution stage deals with the configuration and continuous monitoring of key performance indicators, as well as the adaptability of all stages in the "detection-prediction-decision-execution" process cycle, which can lead to continuous improvement of business performance. The joint optimization decision model solves the problem of the decision stage.

In the case study, the planned maintenance cost is 325 euros, lasting 1h, while the failure cost, i.e., the scrap rate (including the cost of corrective measures), is 85 euros per hour. Therefore, the fixed planned maintenance cost equals 325 euros, and the linearly increasing failure and corrective cost equals 85 t. The shortage cost is 140 euros per hour, the holding cost is 65 euros per hour, and the delivery period L equals 2h. The next planned maintenance (mold cleaning) is within 10h. After 5h, the sensor measures a high dust content and detects an anomaly, with an expected failure time of 4h ($\lambda = 0.25$) to predict the distribution of usable life, the prediction process triggers the decision-making process. Therefore, by combining Eqs. (9.1), (9.2) with Eqs. (9.3), (9.4), we obtained the decision model for expected maintenance costs, as shown in Eq. (9.5), and Eq. (9.6) for the expected spare parts ordering costs.

$$C_{\mathrm{m}}(t) = (85t)\left(1 - \mathrm{e}^{-0.25t}\right)(85t + 325) \\ \left(1 - \mathrm{e}^{1-0.25\,(s-t)}\right) + 325\left(\mathrm{e}^{-0.25t} + \mathrm{e}^{1-0.25\,(s-t)} - 1\right) \tag{9.5}$$

$$C_{\mathrm{o}}(t) = (140t)\left(1 - \mathrm{e}^{-0.25\,(t+2)}\right) + (140t) \\ \left(1 - \mathrm{e}^{1-0.25\,(s-t-2)}\right) + 75t\left(\mathrm{e}^{-0.25\,(t+2)} + \mathrm{e}^{1-0.25\,(s-t-2)} - 1\right) \tag{9.6}$$

By optimizing Eq. (9.5), the recommended optimal maintenance time is 3.54h, with a cost of 348.8 euros. By optimizing Eq. (9.6), the recommended optimal time to order spare parts is 1.32h, with a cost of 616.6 euros. The results are shown in Fig. 9.15.

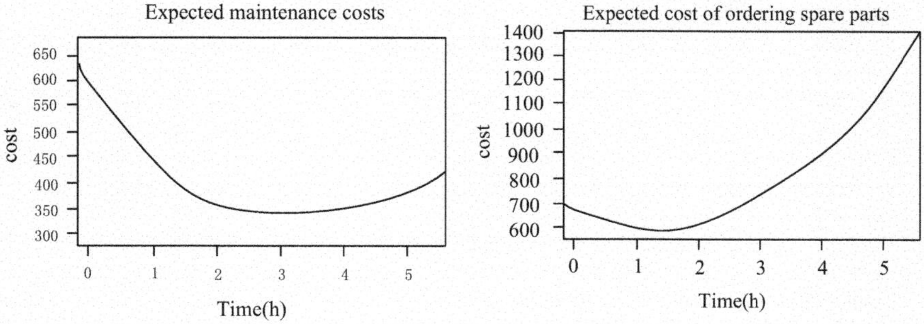

Fig. 9.15 Numerical simulation case

References

1. WANG Wenbin. A stochastic model for joint spare parts inventory and planned maintenance optimisation. European Journal of Operational Research, 2012 (216): 127-139.
2. HORENBEEK A V, BURE J, CATTRYSSE D, et al. Joint maintenance and inventory optimization systems: A review. Int.J.Production Economics, 143 (2013): 499-508.
3. OLDE KEIZER M, TEUNTER R, VELDMAN J. Joint condition-based maintenance and inventory optimization for systems with multiple components. European Journal of Operational Research, 2017 (257): 209–222.
4. WANG Hongzhou. A survey of maintenance policies of deteriorating systems. European Journal of Operational Research, 2002 (139): 469-489.
5. WU S, ZUO M J. Linear and nonlinear preventive maintenance. IEEE Transactions on Reliability, 2010, 59 (1): 242-249.
6. BASTEN R, VAN HOUTUM G. System-oriented inventory models for spare parts. Surveys in Operations Research and Management Science, 2014, 19 (1), 34-55.
7. VAN HOUTUM G, KRANENBURG B. Spare parts inventory control under system availability constraints. International Series in Operations Research and Management Science, 2015, 227.
8. OKUMOTO K, ELSAYED E. An optimal group replacement policy. Naval Research Logistics Quarterly, 1983, 30: 667-674.
9. GERTSBAKH, I B. Optimal group maintenance of a system with observable state parameter. Advances in Applied Probability, 1984, 16: 623–625.
10. RITCHKEN P, WILSON J. (m, T) Group maintenance policies. Management Science, 1990, 36: 632-639.
11. AHMAD R, KAMARUDDIN S. An overview of time-based and condition-based maintenance in industrial application. Computers and Industrial Engineering, 2012, 63 (1): 135-149.
12. BOUSDEKIS A, MAGOUTAS B, APOSTOLOU D, et al. A proactive decision making framework for condition-based maintenance. Industrial Management and Data Systems, 2015, 115 (7): 1225-1250.
13. HONG H, ZHOU W, ZHANG S, et al. Optimal condition-based maintenance decisions for systems with dependent stochastic degradation of components. Reliability Engineering and System Safety, 2014, 121: 276-288.
14. LI H, DELOUX E, DIEULLE L. A condition-based maintenance policy for multi-component systems with Lévy copulas dependence. Reliability Engineering and System Safety, 2016, 149: 44-55.
15. RASMEKOMEN N, PARLIKAD A. Condition-based maintenance of multi-component systems with degradation state-rate interactions. Reliability Engineering and System Safety, 2016, 148: 1-10.

16. OLDE KEIZER M, TEUNTER R, VELDMAN J. Clustering condition-based maintenance for systems with redundancy and economic dependencies. European Journal of Operational Research, 2016, 251 (2): 531–540.
17. ACHARYA D, NAGABHUSHANAM G, ALAM S. Jointly optimal block-replacement and spare provisioning policy. IEEE Transactions on Reliability, 1986, 35 (4): 447-451.
18. CHELBI A, AIT-KADI D. Spare provisioning strategy for preventively replaced systems subjected to random failure. International Journal of Production Economics, 2001, 74 (1-3): 183-189.
19. YOO, Y K, KIM K J, SEO J. Optimal joint spare stocking and block replacement policy (cost modelling of spare stocking and block replacement). International Journal of Advanced Manufacturing Technology, 2001, 18 (12): 906–909.
20. BREZAVSCEK A, HUDOKLIN, A. Joint optimization of block-replacement and periodic-review spare-provisioning policy. IEEE Transactions on Reliability, 2003, 52 (1): 112–117.
21. HUANG R, et al. Modeling and analyzing a joint optimization policy of block-replacement and spare inventory with random-leadtime. IEEE Transactions on Reliability, 2008, 57 (1): 113-124.
22. SARKER R, HAQUE A. Optimization of maintenance and spare provisioning policy using simulation. Applied Mathematical Modelling, 2000, 24(10): 751-760.
23. KABIR A B M Z, AL-OLAYAN A S. A stocking policy for spare part provisioning under age based preventive replacement. European Journal of Operational Research, 1996, 90(1): 171-181.
24. ILGIN M A, TUNALI S. Joint optimization of spare parts inventory and maintenance policies using genetic algorithms. The International Journal of Advanced Manufacturing Technology, 2007, 34(5-6): 594-604.
25. NGUYEN D Q, BRAMMER C, BAGAJEWICZ M. New tool for the evaluation of the scheduling of preventive maintenance for chemical process plants. Industrial and Engineering Chemistry Research, 2008, 47(6): 1910-1924.
26. NGUYEN D Q, BAGAJEWICZ, M. Optimization of preventive maintenance scheduling in processing plants. Proceedings of the 18th European Symposium on Computer Aided Process Engineering. Elsevier Science Ltd, 2008: 319–324.
27. SERGENT L, SORENSON J, BAGAJEWICZ, M. Optimal preventive maintenance scheduling in process plants. Working Paper, 2008.
28. ARMSTRONG M J, ATKINS, D A. Joint optimization of maintenance and inventory policies for a simple system. IIE transactions, 1996, 28: 415–424.
29. CHEN, M C, HSU C M, CHEN S W. Optimizing joint maintenance and stock provisioning policy for a multi-echelon spare part logistics network. Journal of the Chinese Institute of Industrial Engineers, 2006, 23(4): 289–302.
30. JIANG Y, CHEN M, Zhou D H. Joint optimization of preventive maintenance and inventory policies for multi-unit systems subject to deteriorating spare part inventory. Journal of Manufacturing Systems, 2015(35): 191-205.
31. WANG L, CHU J, MAO W. A condition-based order-replacement policy for a single-unit system. Applied Mathematical Modelling, 2008, 32(11): 2274-2289.
32. WANG L, CHU J, MAO W. A condition-based replacement and spare provisioning policy for deteriorating systems with uncertain deterioration to failure. European Journal of Operational Research, 2009, 194(1): 184-205.
33. ALASWAD S, XIANG Y. A review on condition-based maintenance optimization models for stochastically deteriorating system. Reliability Engineering & System Safety, 2016, 157: 54-63.
34. KOOCHAKI J, BOKHORST J A, WORTMANN H, et al. Condition based maintenance in the context of opportunistic maintenance. International Journal of Production Research, 2012, 50(23): 6918-6929.
35. ZHANG X, ZENG J. Joint optimization of condition-based opportunistic maintenance and spare parts provisioning policy in multiunit systems. European Journal of Operational Research, 2017(262): 479-498.

36. GAN S, ZHANG Z, ZHOU Y, et al. Joint optimization of maintenance, buffer, and spare parts for a production system. Applied Mathematical Modelling, 2015(39): 6032-6042.
37. ZAHEDI-HOSSEINI F, SCARF P, SYNTETOS A. Joint optimisation of inspection maintenance and spare parts provisioning: a comparative study of inventory policies using simulation and survey data. Reliability Engineering and System Safety, 2017(168): 306-316.
38. BOUSDEKIS A, PAPAGEORGIOU N, MAGOUTAS B, et al. A proactive event-driven decision model for joint equipment predictive maintenance and spare parts inventory optimization. Procedia CIRP, 2017(59): 184-189.
39. CHO D, PARLAR M. A survey of maintenance models for multi-unit systems. European Journal of Operational Research, 1991, 51(1): 1-23.
40. KARSTEN F, BASTEN R. Pooling of spare parts between multiple users: How to share the benefits. European Journal of Operational Research, 2014, 233(1): 94-104.
41. VAN HORENBEEK A, PINTELON L. A joint predictive maintenance and inventory policy. / /P W TSE, J MATHEW, K Wong, et al. Engineering asset management systems, professional practices and certification. In Lecture notes in mechanical engineering, 2015: 387–399.
42. BORRERO J, AKHAVAN-TABATABAEI R. Time and inventory dependent optimal maintenance policies for single machine workstations: An MDP approach. European Journal of Operational Research, 2013, 228(3): 545-555.
43. BOUVARD K, ARTUS S, BERENGUER C, et al. Condition-based dynamic maintenance operations planning & grouping. Application to commercial heavy vehicles. Reliability Engineering and System Safety, 2011, 96(6): 601–610.
44. PARK M, PHAM H. A generalized block replacement policy for a k-out-of-n system with respect to threshold number of failed components and risk costs. IEEE Transactions on Systems, Man and Cybernetics, Part A: Systems and Humans, 2012, 42(2): 453-463.
45. DE JONGE B, KLINGENBERG W, TEUNTER R, et al. Optimum maintenance strategy under uncertainty in the lifetime distribution. Reliability Engineering and System Safety, 2015, 133(0): 59-67.
46. VELDMAN J, WORTMANN H, KLINGENBERG W. Typology of condition based maintenance. Journal of Quality in Maintenance Engineering, 2011, 17(2): 183-202.
47. ElWANY A, GEBRAEEL N. Sensor-driven prognostic models for equipment replacement and spare parts inventory. IIE Transactions, 2008, 40(7): 629–639.

Chapter 10
Large-Scale Maintenance Service Forecasting and Optimization Configuration

This chapter introduces the service demand prediction and optimization of MRO operation network, demand prediction and service provision model, service provider management, service spare parts prediction and management model, and service resource configuration, etc., in the context of equipment MRO operation network. It presents a wide range of intelligent predictive maintenance service prediction and optimization configuration models and methods.

10.1 Service Demand Prediction and Optimization in MRO Operation Network

Due to the high randomness of production system failures, the MRO industry is characterized by multiple participants, high uncertainty of maintenance demand, and rapid response. How to accurately predict maintenance demand in the face of multiple uncertain collaborative entities (equipment manufacturers, spare parts suppliers, repairers, equipment users, and domain maintenance experts with equipment maintenance expertise for specific application areas) and highly random spare parts demand in the MRO network (the problem of spare parts supply and demand asymmetry), minimize safety stock (even "zero inventory"), while meeting the supply demand of spare parts; how to maximize the response to maintenance demand under limited service resources, and provide matching MRO maintenance services (the problem of difficult service resource matching), and establish long-term, comprehensive cooperation with customers. These issues have become the common demands of spare parts suppliers, equipment manufacturers, repairers, and professionals in the MRO network. In order to dynamically predict maintenance demand, reasonably dispatch service resources, and deploy maintenance services in real time, Smart Predictive With the uncertainty of maintenance requirements in the Service Predictive Maintenance (SPdM) network as the background, this research investigates the

M. Liu et al., *Intelligent Predictive Maintenance*, Advanced and Intelligent Manufacturing in China, https://doi.org/10.1007/978-981-97-2677-6_10

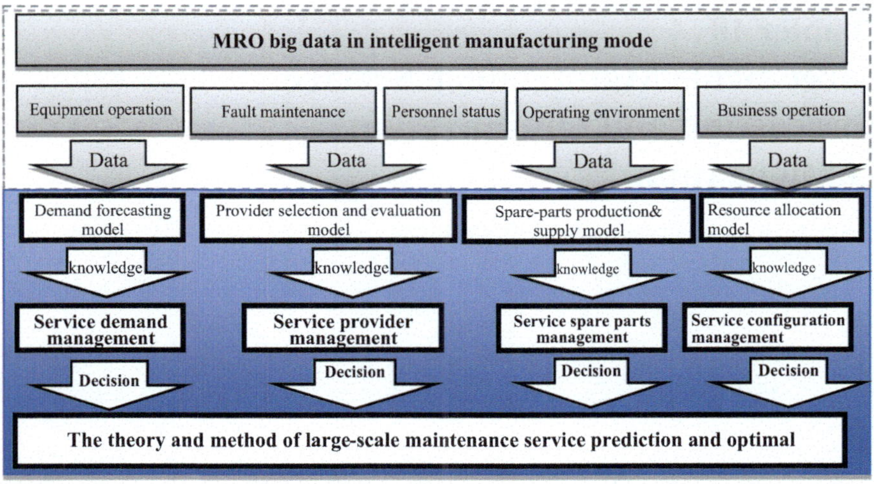

Fig. 10.1 MRO network maintenance service collaborative scheduling and optimization decision

management and prediction models of uncertain service requirements in a dynamic network environment, the selection and evaluation of service providers, the prediction and management of service spare parts based on improved stochastic planning, and the configuration of service resources based on fuzzy stochastic planning and profit sharing models. It establishes the theory and methods of large-scale maintenance service prediction and optimization configuration for the intelligent PdM network as shown in Fig. 10.1.

(1) In terms of service demand management

Demand management is one of the core contents of MRO operation services, and all product service provision activities are based on scientific and reasonable demand plans and related operation strategies. Currently, agile and efficient service supply chains should be demand-driven. Passively responding to user demands leads to frequent movement of service resources such as manpower and equipment, which inevitably reduces the utilization rate of service resources and increases service costs. Only by accurately predicting uncertain demands can the passive service provision process be automated. However, traditional service demand management research mostly focuses on the characteristics that affect the demand itself, such as service price, service level, inventory, etc., and rarely considers the root cause of uncertain demand. Therefore, existing service demand management strategies are difficult to gain market and user recognition.

For manufacturing enterprises, the downtime caused by equipment failure can cause huge losses to the enterprise. Therefore, proactive preventive maintenance of equipment is more meaningful than passive corrective maintenance. The goal of MRO service chain demand management is to actively provide MRO services to users by predicting uncertain demands, maximize equipment availability and reliability, and reduce the probability of failure. However, how to accurately predict service demand is a topic of common research for all members of the MRO service chain. For MRO services aimed at industrial products, the uncertainty of demand mainly comes from the randomness of the product degradation process. Therefore, quantifying the product degradation process is a necessary prerequisite for predicting service demand. From this point of view, choosing an appropriate stochastic process model to describe product degradation and predicting demand based on degradation status is of great significance for optimizing service demand management.

(2) In terms of service provider management

The MRO service chain is a service network with the product's main manufacturer at the core and multiple partners participating. These partners include spare parts suppliers, third-party professional service providers, etc. The participation of multiple parties with different interest demands makes the capabilities of the partners uneven. They have significant differences in service scale, capability, quality, etc., which invisibly brings great challenges to the management of service providers. In addition, MRO service providers are very different from traditional parts suppliers. Choosing the right service provider must fully consider factors such as the availability of services to users, service costs, and service quality. Under the MRO service chain environment, how to objectively and dynamically select and evaluate providers is a problem that urgently needs to be solved in current service provider management.

Although the existing service provider management theories and methods are relatively mature and complete, and have applicability and effectiveness in different fields, they cannot directly apply to the selection and evaluation of MRO service providers. Unlike the provision of tangible products, the provision of MRO services not only needs to evaluate the objective strength of the provider (such as financial reserves), but also needs to evaluate its subjective ability (such as service level). Due to the fuzziness and volatility of subjective abilities, existing static evaluation models are no longer applicable to the dynamic evaluation and selection of service providers. Quantifying the fuzziness and uncertainty of the provider's service ability, and using it as the basis for dynamic evaluation and selection, is an important means for scientific and efficient management of providers.

(3) In terms of service spare parts management

The level of MRO service spare parts management directly affects the level of service capability, and spare parts management has become a research hotspot worldwide. In fact, competition between enterprises, the requirement to improve service levels, the huge profits of service spare parts, and the low cost brought to enterprises by scientific and reasonable management plans all urge enterprises to better manage service spare parts. As manufacturing enterprises outsource MRO services, how to formulate scientific and effective spare parts production and sales plans, while meeting user spare parts availability and reducing the total service cost to the lowest, is the core issue of MRO service chain management.

Unlike general service spare parts, MRO service spare parts are assembled in large-scale manufacturing equipment. Usually, these manufacturing equipment work in extremely harsh environments (high temperature, high pressure, high load), and the requirements for production accuracy and storage environment are more stringent. For example, as one of the main users of MRO services, steel companies are capital-intensive and have large inputs and outputs. The equipment of their production systems generally accounts for more than half of the fixed assets, and the availability of spare parts is an important prerequisite for the normal operation and maintenance of the production system. Continuous casting machines are the main production system equipment of steel companies. Their parts operate in high-temperature and high-vibration environments, which undoubtedly puts extremely high requirements on the high-temperature resistance and impact resistance of related spare parts, and also leads to uncertainty in the production and storage process of spare parts. Therefore, when optimizing the modeling of MRO service spare parts production and sales plans, it is necessary to fully consider the uncertainty in the production and storage process of spare parts in order to formulate more realistic spare parts management strategies.

(4) In terms of service configuration management

The configuration and optimization of MRO service resources is the key to realizing the co-creation of value by all members of the service chain. The core issue is how to use limited service resources to maximize user satisfaction. The resources here not only include tangible products (such as spare parts), but also intangible services (such as expert knowledge). The main problem to be solved by MRO service resource configuration is how to obtain service delivery capabilities, how to ensure the availability of service resources, and how to scientifically and effectively configure resources to achieve optimization. In the current service chain environment, many random factors and uncertainties coexist, such as the randomness of user demand and the uncertainty of service provider capabilities, making it difficult for their service strategies to maximally respond to user demand and provide matching professional services. Therefore, there is still a lot of room for improvement in service resource configuration and optimization management.

Although most service resource configuration models take user demand as a key input variable, they mostly assume that their demand is deterministic. In addition, existing resource configuration models almost never consider the fuzziness of service provider capabilities, but assume that they can "respond to any request" under any circumstances. In fact, the MRO service chain is a service network with the product's main manufacturer at the core and multiple partners participating. The provision of its services is also a process involving multiple parties, so the uneven capabilities of partners will inevitably lead to the fuzziness of service levels. It can be seen that quantifying the uncertainty of demand and the fuzziness of provider capabilities is the key to achieving flexible resource configuration.

10.2 Demand Prediction and Service Provision Model Based on Product Degradation Status

In the MRO service chain, an efficient service provision process should be driven by demand. In fact, the successful implementation of almost all service provision activities depends on a reasonable and accurate demand plan. However, due to the high uncertainty caused by the random degradation process of equipment, the MRO demand for large-scale manufacturing equipment is different from the maintenance demand for ordinary industrial products. In recent years, although there have been many research results on service strategy optimization in the SSC and IPS2 fields, most of these research results focus on passive service provision through the composite characteristics of demand patterns. There are still few attempts to quantify service demand through the root cause of demand generation, and then develop more proactive and effective service provision strategies. Although the relevant research results provide good optimization theories and methods for service and demand strategies, they have a common feature: they model the demand attributes (such as price, service level, inventory level, etc.) in the best way, but they do not ask about the generation process of demand itself, which may not be suitable for the optimization of MRO service provision strategies. Only by modeling the generation process of uncertain demand can we help transform passive service management into active. Taking the MRO service of continuous casting equipment (CC) as an example, CC is the main production equipment of steel companies, with a long service cycle, generally more than eight years. In order to ensure output and profit, MRO service outsourcing is the main means for companies to reduce costs and risks. The production interruption caused by CC failure will bring huge losses to the company. Therefore, proactive MRO services that can guarantee CC reliability (such as preventive maintenance, preventive replacement, etc.) are more important than passive post-failure repairs. Modeling the source of MRO service demand—the degradation process and reliability of equipment is a more effective method to seek the optimal service provision strategy. By formulating the optimal MRO service provision and demand strategy based on the industrial maintenance process of product degradation, we intend to solve two problems: how to model the random process of product degradation? How to use the proposed model to predict service demand and then formulate the optimal service provision plan? For this purpose, the geometric process and the gamma process are combined, and the random monotonicity of the geometric process is used to model the accelerated degradation process of the product. Further, based on the proposed accelerated degradation model, the analytical expression of the two-dimensional joint decision including the deterministic service plan and the random service plan is derived, and the optimal solution is finally obtained using the discrete approximation iterative algorithm.

10.2.1 Accelerated Degradation Model of Production Equipment

In this section, an accelerated degradation model of production equipment will be constructed: ① give the model description and model assumptions; ② propose an accelerated degradation model combining geometric process and gamma process. (1) Model Description and Model Assumptions Consider a production system (a piece of production equipment or a group of equipment) that requires MRO services. To ensure the reliability of this production system, the MRO service provider (such as the original equipment manufacturer) will implement periodic and random MRO services on the production system according to demand. Based on the industrial maintenance characteristics of production equipment, three types of MRO services are considered here: periodic inspection, preventive replacement, and post-failure replacement. Although it is not possible to accurately predict when the system will fail (i.e., when the post-failure replacement service is needed), the occurrence of failures can be delayed through periodic inspections and preventive replacements, and the demand for periodic inspections and preventive replacements can be predicted using system reliability. The decision problem is to find the optimal inspection cycle and preventive replacement time, and then formulate the optimal service provision strategy. As discussed earlier, the service demand for periodic inspection and preventive replacement can be predicted based on system reliability, and system reliability can be quantified by the random process of system degradation. Therefore, modeling the system degradation process is a prerequisite for service strategy optimization. To make model construction more convenient, the following model assumptions are given first.

Assumption 10.1 The production system gradually degrades with use and service life, and its degradation can be characterized by a random process. Let $X(t), t > 0$ represent the degradation state at time t, where $X(0)$ represents the system in a brand new state. As time goes by, the system performance gradually degrades, that is, $X(t)$ is a monotonically increasing random process. The system's failure threshold is set to L, that is, the system fails when $X(t) \geq L$. The system's downtime is T_D, and the downtime loss per unit time is C_D. Through post-failure replacement (Corrective Replacement, CR), the system can be repaired as new, and the cost of post-failure replacement is C_{CR}.

Assumption 10.2 During the working period, the system needs to be inspected regularly, and the inspection cycle is denoted as T, and the cost of each inspection is C_i.

Assumption 10.3 Let $\xi, \xi < L$ be the threshold for preventive maintenance (Preventative Maintenance, PM). When the system is inspected, if $X(t) \in (0, \xi)$, the system does not need PM and continues to operate; if $X(t) \in [\xi, L)$, PM is implemented on the system, and $X(t)$ is set to zero after PM, but the system's degradation will accelerate. The unit time cost of PM is C_{PM}, the time of the i-th PM is Y_i, and

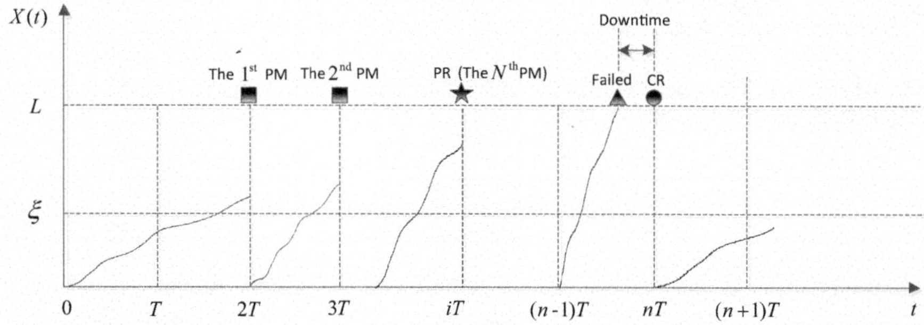

Fig. 10.2 The process of implementing MRO services on the system

Y_i is related to X(t), that is, Y_i is related to X(t). $i = h(X(t))$, where $h(\cdot)$ is a monotonically increasing function.

Assumption 10.4 As the service life and the number of maintenance times increase, the system degradation gradually accelerates. At the N th PM, a preventative replacement (PR) is implemented on the system, after which the system is repaired as new. Assume the cost of PR is C_{PR}.

Based on the above model assumptions, Fig. 10.2 intuitively depicts the process of implementing MRO services on the system.

Regarding the above four model assumptions, the following two points need to be explained.

① In Assumption 10.3, it is assumed that Y_i is a monotonically increasing function of $X(t)$, which indicates that the more severe the system degradation, the longer the time spent on PM.

② In Assumption 10.4, it is assumed that as the service life and the number of maintenance times increase, the system degradation will gradually accelerate, which can also be seen in Fig. 10.2. As time goes by, $X(t)$'s slope gets larger and larger until the system fails. Although there will be no residual damage after PM, the process of degradation will accelerate. This means that degradation is irreversible, that is, degradation is the accumulation of damage.

(2) Mathematical model of accelerated degradation

As mentioned above, system degradation appears in the form of cumulative damage, and the gamma process is very suitable for describing this feature. The gamma process has a monotonically degrading path and has been widely used. In addition, it is necessary to describe the accelerated degradation of the system. Here, accelerated degradation refers to the time t_x from state 0 to state x is randomly decreasing. For this, the typical random monotonic process—geometric process will be used to describe this accelerated degradation. In this section, the basic concepts of the gamma process and geometric process are first given. Then, the gamma process and geometric process are combined to propose a mathematical model of accelerated degradation.

① Basic concepts of gamma process and geometric process

Definition 10.1 The gamma process model $\{X(t), t \geq 0\}$ is a random process, its probability density function is [1]:

$$f_t(x) = \frac{\beta^{\alpha t} x^{\alpha t - 1}}{\Gamma(\alpha t)} \mathrm{e}^{-\beta x}, x \geq 0 \tag{10.1}$$

In the formula, α and β are the shape parameter and scale parameter respectively; $\Gamma(\cdot)$ is the gamma function $\Gamma(s) = \int_0^{+\infty} u^{s-1} e^{-u} du, s > 0$. The expectation and variance of the gamma process are respectively $\frac{\alpha}{\beta}$ and $\frac{\alpha}{\beta^2}$.

Definition 10.2 Given two random variables X and Y, if $P(X > \alpha) \geq P(Y > \alpha), \forall \alpha \in R$, then X is said to be stochastically greater than Y, denoted as $X \geq_{st} Y$ [2]. If for all $n = 1, 2, \ldots$, we have $X_n \geq_{st} X_{n+1}$(or $X_n \leq$ st X_{n+1}), then $\{X_n, n = 1, 2, \ldots\}$ is said to be stochastically decreasing (or stochastically increasing).

Definition 10.3 Suppose $\{\zeta_n, n = 1, 2, \ldots\}$ is a sequence of independent, non-negative random variables. If the distribution function of ζ_n is $U_n(t) = U\left(a^{n-1} t\right), n = 1, 2, \ldots$ where $U(t)$ is the cumulative distribution function of ζ_1, a is a positive constant, then $\{\zeta_n, n = 1, 2, \ldots\}$ is called a geometric process, and a is its scaling factor.

Note that according to Definitions 10.2 and 10.3, if $a > 1$, then $\{\zeta_n, n = 1, 2, \ldots\}$ is stochastically decreasing; if $a < 1$, then $\{\zeta_n, n = 1, 2, \ldots\}$ is stochastically increasing; when $a = 1$, the geometric process $\{\zeta_n, n = 1, 2, \ldots\}$ degenerates into a renewal process.

② Mathematical model of accelerated deterioration: According to Definition 10.1, the distribution function of the time t_x to change from state 0 to state x, denoted as $F_x(t)$, can be expressed as:

$$F_X(t) = P\{X(t) \geq x\} = \frac{\Gamma(\alpha t, x\beta)}{\Gamma(\alpha t)} \tag{10.2}$$

where, $\Gamma(s,b) = \int_b^{+\infty} u^{s-1} e^{-u} du, s > 0, b \geq 0$.

After the $i-1$th MRO service, the system enters the ith service cycle, $i = 1, 2, \ldots, N$. Let the random variable $t_x^{(i)}$ represent the time for the system state 0 to change to state x in the i th service cycle. Based on the previous discussion, $t_x^{(i)}$ randomly decreases, that is, by Definition 10.2: $t_x^{(i-1)} \leq_{st} t_x^{(i-1)}$.

Let the distribution function of $t_x^{(i)}$ be $F_x^{(i)}(t)$, according to Definition 10.3 and formula (10.2), we have:

$$F_x^{(i)}(t) = F_x^{(1)}\left(a^{i-1} t\right) = \frac{\Gamma\left(\alpha a^{i-1} t, x\beta\right)}{\Gamma\left(\alpha a^{i-1} t\right)}, a > 1 \tag{10.3}$$

Therefore, we can obtain the probability density function of the gamma process of the i th service cycle $X_i(t)$:

$$f_t^{(i)}(X) = \frac{\beta^{\alpha a^{i-1}} x^{\alpha a^{i-1} t-1}}{\Gamma\left(\alpha a^{i-1} t\right)} \mathrm{e}^{-\beta x}, X \geq 0 \tag{10.4}$$

Equation (10.4) is the mathematical model of accelerated degradation, where a^{i-1} α and β are the shape parameter and scale parameter, respectively. From this model, it can be seen that the average degradation rate μ_i during the ith service cycle can be expressed as $\mu_i \frac{\alpha a^{i-1}}{\beta} = \frac{\alpha}{\beta}\mu_1$, where μ_1 is the average degradation rate of the first service cycle. Since $a > 1$, it indicates that μ_i increases with the increase of i. This means that as the MRO service cycle increases, the system degradation gradually accelerates. It can be seen that Eq. (10.4) well characterizes the characteristics of system accelerated degradation.

10.2.2 Demand Forecasting and Service Provision Model Based on Product Degradation Status

This section will consider the two-dimensional joint strategy (T, N), where T is the inspection cycle, N is the total number of PMs implemented before PR. The decision goal is to seek the optimal strategy (T^*, N^*) to minimize the system's average cost rate $C(T, N)$. Next, derive the analytical expression of $C(T, N)$, and then use the discrete approximation iterative algorithm to find the optimal numerical solution.

(1) Calculation of the average cost rate $C(T, N)$

Assume that under the strategy (T, N), the moment when the system restarts after PR is τ_n $(n \geq 1)$, then $\{\tau_1, \tau_2 - \tau_1, \ldots, \tau_n - \tau_{n-1}\}$ constitutes an update process. Let $C(t)$ be the total cost generated in the time interval $[0, t]$, then according to the renewal reward theorem, we have:

$$C(T, N) = \lim_{t \to \infty} \frac{C(t)}{t} = \frac{E(C)}{E(\tau)} \tag{10.5}$$

In the formula, $E(C)$ is the expected total cost of an update cycle; $E(\tau)$ is the expected total length of an update cycle.

According to Assumptions 10.1 and 10.4, the system will be replaced after the Nth PM or failure, so we have:

$$E(\tau) = E\left(\sum_{i=1}^{N} K_i T + \sum_{i=1}^{N-1} Y_i\right) \chi\{\text{The system does not fail at the } i\text{th } PM, i = 1, 2, \ldots, N\}$$

$$+\sum_{m=1}^{N} E\left(\sum_{i=1}^{m} K_i T + \sum_{i=1}^{m-1} Y_i\right)\chi\{\text{The system has already failed at the } m\text{th } PM\}$$

$$= \left[T\sum_{i=1}^{N} E(K_i) + \sum_{i=1}^{N-1} E(Y_i)\right] P_N$$

$$+\sum_{m=1}^{N}\left[T\sum_{i=1}^{m} E(K_i) + \sum_{i=1}^{m} E(K_i) + C_{PM}\sum_{i=1}^{m-1} E(Y_i) + C_{PR}\right] P_m \tag{10.6}$$

In the formula, $\chi(\cdot)$ is an indicator function; $P_N = P$ {The system did not fail at the ith PM, $i = 1, 2, \ldots, N$}; $P_m = P$ {The system has already failed at the mth PM}; K_i, $i = 1, 2, \ldots, N$ is the number of inspections in the ith service period.

Similarly, we can get $E(C)$ as follows:

$$E(C) = \left[C_i \cdot \sum_{i=1}^{N} E(K_i) + C_{\text{PM}} \cdot \sum_{i=1}^{N-1} E(Y_i) + C_{\text{PR}}\right] \cdot P_N$$

$$+\sum_{m=1}^{N}\left[C_i \cdot \sum_{i=1}^{m} E(K_i) + C_{\text{PM}} \cdot \sum_{i=1}^{m} E(Y_i) + C_D \cdot E\left(T_D^{(m)}\right) + C_{\text{CR}}\right] \cdot P_m \tag{10.7}$$

In the formula, $T_D^{(m)}$ represents the time from system failure to the m-th PM.

From formulas (10.6) and (10.7), it can be seen that to obtain the analytical expressions of E(τ) and E(C), it is necessary to first calculate E(Ki), E(TD(m)), PN, Pm, and E(Yi).

① Calculate the average number of inspections E(Ki) in the i-th service cycle. Let Xi(t) be the degradation state at time t in the i-th service cycle. Since the degradation will be reset to zero after PM, we have Xi(0) = 0. Therefore:

$$E(K_i) = \sum_{k=1}^{\infty} kP\{\text{The system does not need to perform } PM \text{ in}$$

$$\text{the previous k} - 1 \text{ detections, and needs to perform } PM \text{ in the kth detection}\}$$

$$= \sum_{k=1}^{\infty} kPX_i((\text{k}-1)\text{T}) < \xi, \xi \le X_i(\text{kT}) < L \tag{10.8}$$

Since the gamma process X_i(t) has independent increments of the gamma distribution, Eq. (10.8) can also be written as:

$$E(K_i) = \sum_{k=1}^{\infty} kP\{X_i((\text{k}-1)\text{T}) < \xi, \xi \le X_i((\text{k}-1)\text{T}) + X_i(\text{T}) < L\}$$

$$= \sum_{k=1}^{\infty} k \int_{0}^{\xi} f^{i}_{(k-1)T}(x) \left[\int_{\xi-x}^{L-x} f^{i}_{T}(y)dy \right] dx \tag{10.9}$$

where, $f^{(i)}_{T}(X)$ is the probability density function of the gamma process $X_i(t)$, as shown in Eq. (10.4).

② Calculate the average time E from system failure to the mth PM $T^{(m)}_{D}$ Suppose the time of system failure is t_D, then the distribution function of t_D $H^{(m)}_{t_D}(t)$ can be written as:

$$H^{(m)}_{t_D}(t) = P\{t_D < t\} = P\{X_m(t) > L\}\Big\} = F^{(m)}_{L}(t) = \frac{\Gamma\left(\alpha a^{m-1}t, L\beta\right)}{\Gamma\left(\alpha a^{m-1}t\right)} \tag{10.10}$$

If $(j-1)T < t_D \le jT, j = 1, 2, \ldots$, then the downtime before replacement is $jT - t_D$, so we have:

$$E\left(T^{(m)}_{D}\right) = \sum_{j=1}^{\infty} E\left[(jT - t_D)|(j-1)T < t_D \le jT\right]$$

$$= \sum_{j=1}^{\infty} P\{X_m((j-1)T)) < \xi\} \int_{(j-1)}^{jT} {}_{T}jT - x)dH^{(m)}_{t_D}(x)$$

$$= \sum_{j=1}^{\infty} \int_{0}^{\xi} f^{(m)}_{(j-1)T}(y)\, dy \int_{(j-1)}^{jT} {}_{T}jT - x)dF^{(m)}_{L}(x) \tag{10.11}$$

In the formula, $f^{(m)}_{(j-1)T}(y)$ is in the form shown in formula (10.4), $F^{(m)}_{L}(x)$ is in the form shown in formula (10.10).

③ Calculate the probability values P_N and P_m P_N represents the probability that the system has no failure at the ith PM, so it can be calculated as follows:

$$P_N = \prod_{j=1}^{N} P\{\text{the system has no failure at the } j\text{th } PM\}$$

$$= \prod_{j=1}^{N} \left(\sum_{K_j=1}^{\infty} P\{\text{the system has no failure at the } j\text{th } PM\text{, and the time of implementing the } j\text{th } PM \text{ is } K_jT\}\right)$$

$$= \prod_{j=1}^{N} \left(\sum_{K_j=1}^{\infty} P\{X_j(((K_j - 1) < \xi, \xi \le X_j(K_jT) < L\} \right)$$

$$=\prod_{j=1}^{N}\left\{\sum_{K_j=1}^{\infty}\int_{0}^{\xi} f_{(K_j-1)T}^{(j)}(x)\left[\int_{\xi-x}^{L-x} f_T^{(j)}(y)dy\right]dx\right\} \tag{10.12}$$

Similarly, we can get:

$$\begin{aligned}P_m &= P\{\text{The system fails at the } m\text{th } PM\}\\ &= \prod_{j=1}^{m-1} P\{\text{The system does not fail at the } j\text{th } PM\\ &\quad P\{\text{The system fails at the } m\text{th } PM\}\\ &= \prod_{j=1}^{m-1}\left\{\sum_{K_j=1}^{\infty}\int_{0}^{\xi} f_{(K_j-1)T}^{(j)}(x)\left[\int_{\xi-x}^{L-x} f_T^{(j)}(y)dy\right]dx\right\}\left\{\sum_{K_m=1}^{\infty}\int_{0}^{\xi} f_{(K_m-1)T}^{(j)}(x)\left[\int_{L-x}^{\infty} f_T^{(m)}(y)dy\right]dx\right\}\end{aligned} \tag{10.13}$$

④ Calculate the average time of the ith PM, $E(Y_i)$. According to Assumption 10.3, Y_i is related to$(X)(t)$, and $Y_i = h(X(\mathrm{t}))$, so we have:

$$\begin{aligned}E(Y_i) &= E(h[X_i(t)]) = \sum_{K_i=1}^{\infty} E(h[X_i(K_i t)])\\ &\quad P\{\text{The time of implementing the } i\text{th } PM \text{ is } K_i T\}\\ &= \sum_{K_i=1}^{\infty} E(h[X_i(K_i t)])P\{X_i((K_i-1)T) < \xi, \xi \le X_i((K_i-1)T) + X_i(T) < L\}\\ &= \sum_{K_i=1}^{\infty}\int_{0}^{\xi} f_{(K_i-1)T}^{(i)}(x)\left[\int_{\xi-x}^{L-x} h(x+y) f_T^{(i)}(y)dy\right]dx\end{aligned} \tag{10.14}$$

Based on the above calculations of $E(K_i), E\left(T_D^{(m)}\right), P_N, P_m$ and $E(Y_i)$, we can obtain the following analytical expression for $C(T, N)$:

$$C(T,N) = \frac{\left[C_i\sum_{i=1}^{N}E(K_i) + C_{PM}\sum_{i=1}^{N-1}E(Y_i) + C_{PR}\right]P_N + \sum_{m=1}^{N}\left[C_i\sum_{i=1}^{m}E(K_i) + C_{PM}\sum_{i=1}^{m}E(Y_i) + C_D E\left(T_D^{(m)}\right) + C_{CR}\right]P_m}{\left[T\cdot\sum_{i=1}^{N}E(K_i) + \sum_{i=1}^{N-1}E(Y_i)\right]\cdot P_N + \sum_{m=1}^{N}\left[T\cdot\sum_{i=1}^{m}E(K_i) + \sum_{i=1}^{m-1}E(Y_i)\right]\cdot P_m} \tag{10.15}$$

where, $E(K_i), E\left(T_D^{(m)}\right), P_N, P_m$ and $E(Y_i)$ are expressed as in Eqs. (10.9), (10.11) to (10.14) respectively.

(2) Find the optimal numerical solution that minimizes $C(T, N)$

Table 10.1 Specific steps of the discrete approximation iterative algorithm

Step No.	Content
1	Choose three positive integers: maximum detection period $T_{\max}$, minimum detection period $T_{\min}$, and step size ΔT
2	Initialize the detection period $T_i = T_{\min}$, and the moment of implementing PR $N = 1$
3	Start calculating $C(T_i, N^*)$ according to formula (10.15), assign $C(T_i, N^*) = C(T_i, N)$
4	Assign $N = N + 1$, calculate $C(T_i, N^*)$ again
5	If $C(T_i, N^*) > C(T_i, N)$, assign $C(T_i, N^*) = C(T_i, N)$, and obtain $N^* = N$, then go to step 4; otherwise, record $C(T_i, N^*)$ and N, then go to step 6
6	Assign $T_i = T_i + \Delta T$. If $T_i \le T_{\max}$, go to step 2; otherwise, go to step 7
7	For all $T_i \in [T_{\min}, T_{\max}]$, calculate $C(T_i, N^*)$, return the minimum average cost rate $C(T, N^*) = \min_{T_i \in [T_{\min}, T_{\max}]}, \{C(T_i, N^*)\}$, obtain the optimal joint strategy (T_i, N^*)

The goal of this section is to find the optimal joint strategy (T^*, N^*) that minimizes $C(T, N)$. From the derivation process of $C(T, N)$, it can be seen that the optimization problem considered is very complex and has non-linear characteristics, so it is theoretically difficult to find its analytical optimal solution. Therefore, a discrete approximation iterative algorithm is used to find the optimal numerical solution (T^*, N^*). Table 10.1 gives the specific steps of this algorithm.

10.2.3 Case Study

This section will verify the feasibility and effectiveness of the proposed model through a case study.

(1) Parameter Setting

Assume that the amplitude (or crack) of a certain part of the production equipment can reflect its degree of deterioration, that is, the larger the amplitude, the more severe the equipment deterioration. When the amplitude exceeds a certain threshold, the equipment fails and stops. Table 10.2 provides the amplitude data for different maintenance cycles.

In addition, the failure threshold of the equipment $L = 6$, the maintenance threshold $\xi = 3$. The maximum inspection cycle $T_{\max} = 15$, the minimum cycle $T_{\min} = 1$. Considering the actual preventive maintenance activities, the step size $\Delta T = 1$ Within the system life cycle, the inspection cost $C_{\mathrm{i}} = 150$, the unit time maintenance cost $C_{\mathrm{PM}} = 100$, the preventive replacement cost $C_{\mathrm{PR}} = 3000$, the replacement cost after failure $C_{\mathrm{CR}} = 4000$, the unit time downtime loss $C_{\mathrm{D}} = 40$. The ith maintenance time Y_i is related to the degradation state of the device at the time of detection x, and their functional relationship is set as $Y_i = h \quad (X) = 1 + \sqrt{\frac{x-\xi}{\xi}} = 1 + \sqrt{\frac{x}{4} - 1}$.

Table 10.2 Amplitude data for different maintenance cycles in mm

Cycle	Working hours/day											
	0	30	60	90	120	150	180	210	240	270	300	330
I	1.004	1.193	1.389	1.391	1.405	1.418	1.433	1.442	1.468	1.512	1.611	1.702
II	1.006	1.205	1.411	1.422	1.445	1.469	1.494	1.503	1.549	1.604	1.713	1.814
III	1.011	1.213	1.422	1.437	1.463	1.489	1.519	1.533	1.584	1.642	1.756	1.864
IV	1.023	1.233	1.453	1.48	1.519	1.559	1.601	1.633	1.704	1.772	1.916	2.068
V	1.076	1.315	1.614	1.646	1.691	1.754	1.825	1.896	2.018	2.102	2.263	2.473

(2) Optimal service provision strategy

In order to calculate the average cost rate $C(T, N)$, it is necessary to obtain the probability density function of the gamma process of the ith service cycle $X_i(t)$ $f_t^{(i)}$ $(X) = \frac{\beta^{\alpha a^{i-1}t} x^{\alpha a^{i-1}t-1}}{\Gamma(\alpha a^{i-1}t)} \mathrm{e}^{-\beta x}$. Therefore, the parameters α, β and a in $f_t^{(i)}$ (X) are estimated based on the amplitude data in Table 10.2. Let α_i and β_i be the shape parameter and scale parameter of $X_i = (t)$, respectively, and let t_{ij} and x_{ij} represent the jth detection time and the corresponding amplitude value of the ith service cycle, where $i = 1, 2, \ldots, n; j = 1, 2, \ldots, m$ In this example, $X_{23} = 1.411$,, $n = 5$ and $m = 12$. Since the increments are $\Delta t_{ij} = t_{ij} - t_{i,j-1}$ and $\Delta X_{ij} = X_{ij} - X_{i,j-1}$, the following log-likelihood function is obtained:

$$\ln L(\alpha_i, \beta_i) = \sum_{j=1}^{m} (\alpha_i \Delta t_{ij} - 1)\ln \Delta x_{ij} - \alpha_i t_{im} \ln \beta_i - \sum_{j=1}^{m} \ln \Gamma(\alpha_i \Delta t_{ij}) - \frac{x_{im}}{\beta_i} \tag{10.16}$$

The partial derivatives of $\ln L(\alpha_i, \beta_i)$ with respect to α_i and β_i can be obtained, resulting in the following maximum likelihood equations:

$$\begin{cases} \frac{\partial \ln L(\alpha_i, \beta_i)}{\partial \alpha_i} = \sum_{j=1}^{m} \Delta t_{ij}(\ln x_{ij} - \psi(\alpha_i \cdot \Delta t_{ij}) - \ln \beta_i) = 0 \\ \frac{\partial \ln L(\alpha_i, \beta_i)}{\partial \beta_i} = \frac{x_{im}}{\beta_i^2} - \frac{\alpha_i t_{im}}{\beta_i} = 0 \end{cases} \tag{10.17}$$

where, $\psi(\cdot) = \frac{\Gamma^{\cdot}(\cdot)}{\Gamma(\cdot)}$ is the digamma function.

By solving Eq. (10.17), we can obtain the maximum likelihood estimates of α_i and β_i. Furthermore, based on the average degradation rate $\mu_{\mathrm{i}} = \frac{\alpha_i}{\beta_i}$ and $\mu_{\mathrm{i}} = a^{\mathrm{i}-1}\mu_1$, the estimated value of a can be obtained using the least squares method.

Based on the above maximum likelihood estimation method and the amplitude data in Table 10.2, we can get the estimated values: $\alpha = 0.0266$, $\beta = 0.0795$ and $a = 1.492$, so we have:

$$f_t^{(i)}(x) = \frac{0.0795^{0.0266 \times 1.492^{i-1}t} \cdot x^{0.0266 \times 1.492^{i-1}t-1}}{\Gamma(0.0266 \times 1.492^{i-1}t)} \mathrm{e}^{-0.0795x} \tag{10.18}$$

According to the discrete approximation iterative algorithm in Table 10.1, we can obtain the optimal replacement strategy N^* and the corresponding average cost rate $C(T_i, N^*)$, the results are listed in Table 10.3.

As can be seen from Table 10.3, the optimal joint strategy (T^*, N^*) is (5,6), and the minimum average cost rate $C(T^*, N^*)$ is 229.2. Therefore, in this example, the optimal service provision strategy is to periodically inspect the equipment, with an inspection cycle of 5, and the equipment needs to be replaced at the 6th PM.

(3) Parameter Analysis

In the model, there are three sets of parameters related to the optimal service supply and demand strategy. The first set of parameters reflects the reliability of the product, such as L, α, a etc.; the second set reflects the benefits of the service provider, such as C_{PM}, C_{PR} etc.; the last set reflects the benefits of the service demander, such as C_{D}. Without loss of generality, one is selected from each set of parameters for the following experiment to analyze the impact of these parameters on the optimal strategy (T^*, N^*).

① Set a to different values. In order to analyze the impact of parameter a on T^*, N^* and $C(T^*, N^*)$, let a change from 1.05 to 1.85, with a step size of 0.05. The calculated results are listed in Table 10.4, and Fig. 10.2 more intuitively shows the trend of (T^*, N^*) and $C(T^*, N^*)$ with the change of a.

As can be seen from Table 10.4 and Fig. 10.3, when the value of a is small, the optimal detection period T^* is longer, and the optimal replacement time N^* is larger. This is because the size of a directly reflects the degree of equipment degradation, that is, the smaller the value of a, the lower the degree of system degradation. In this case, the system does not need to be checked and replaced frequently. As a increases, T^* and N^* gradually decrease. This indicates that as the system degradation rate increases, MRO service providers need to shorten the detection cycle to better control the system degradation status, and replace it in time to reduce the probability of system failure. In addition, the average cost rate $C(T, N^*)$ increases with the increase of a, which indicates that the more severe the system degradation, the higher the maintenance cost.

② Set C_{PR} to different values. In order to study the preventive replacement cost C_{PR} on T^*, N^* and $C(T^*, N^*)$'s influence, causing C_{PR} to change from 1000 to 4200, with a step size of 200, the recalculated results are listed in Table 10.5, and Fig. 10.3 more intuitively shows the trend of T^*, N^* and $C(T^*, N^*)$ with the change of C_{PR}.

From Table 10.5 and Fig. 10.4, it can be seen that the optimal detection period T^* decreases as C_{PR} increases, but the optimal replacement time N^* Increases with the increase of C_{PR}. This phenomenon can be explained as, when the cost of PR C_{PR} is high, MRO service providers should shorten the inspection cycle to improve system reliability, thereby delaying the replacement of the system. The increase of N^* indicates that when the cost of PR C_{PR} increases, the PR of the system should

Table 10.3 Optimal replacement strategy N^* and average cost rate $C(T_i, N^*)$ under different detection periods T_i

Item	Detection period T_i														
	1	2	3	4	5	6	7	8	9	10	11	12	13	14	15
N^*	4	4	5	5	6	7	6	5	7	7	8	8	9	10	9
$C(T_i, N^*)$	301.6	292.3	280.5	263.4	229.2	275.7	316.8	340.4	369.0	402.2	457.1	450.4	513.8	607.0	644.2

Table 10.4 The trend of T^*, N^* and $C(T^*, N^*)$ with the change of parameter a

a	T^*	N^*	$C(T^*, N^*)$
1.05	8	10	184.0
1.10	8	9	184.6
1.15	7	9	185.9
1.20	7	9	188.7
1.25	7	8	193.6
1.30	6	8	201.0
1.35	6	8	209.5
1.40	5	7	219.2
1.45	5	7	230.2
1.50	5	6	243.8
1.55	4	6	262.5
1.60	4	5	297.4
1.65	3	5	335.1
1.70	2	4	372.0
1.75	1	3	404.9
1.80	1	3	453.3
1.85	1	3	486.2

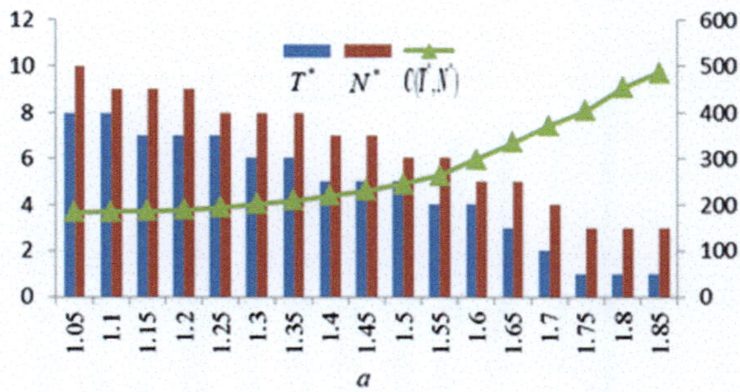

Fig. 10.3 The trend of T*, N* and C(T*, N*) with the change of a (electronic version)

be delayed to reduce the total number of PRs, which is also consistent with the fact that the cost of PR occupies a large part of the total cost. In addition, the slope of the $C(T^*, N^*)$ curve is getting larger and larger, indicating that its increase rate is also increasing with the increase of C_{PR}.

③ Set C_D to different values In order to analyze the impact of downtime loss on T^*, N^* and $C(T^*, N^*)$, let C_D change from 8 to 72, with a step size of 4, the recalculated results are listed in Table 10.6, and Fig. 10.4 more intuitively shows the trends of T^*, N^* and $C(T^*, N^*)$ with C_D.

From Table 10.6 and Fig. 10.5, it can be seen that when the shutdown loss is high, MRO service providers need to not only shorten the detection cycle but also carry

Table 10.5 The trend of T^*, N^* and $C(T^*, N^*)$ with the parameter C_{PR}

C_{PR}	T^*	N^*	$C(T^*, N^*)$
1000	9	3	139.9
1200	8	3	147.8
1400	8	4	153.3
1600	8	4	160.2
1800	7	4	168.9
2000	7	4	174.0
2200	7	5	181.5
2400	6	5	188.4
2600	6	5	193.6
2800	5	6	201.3
3000	5	6	229.2
3200	4	6	248.7
3400	4	7	275.8
3600	3	7	328.0
3800	3	8	360.4
4000	2	8	416.5
4200	2	9	473.6

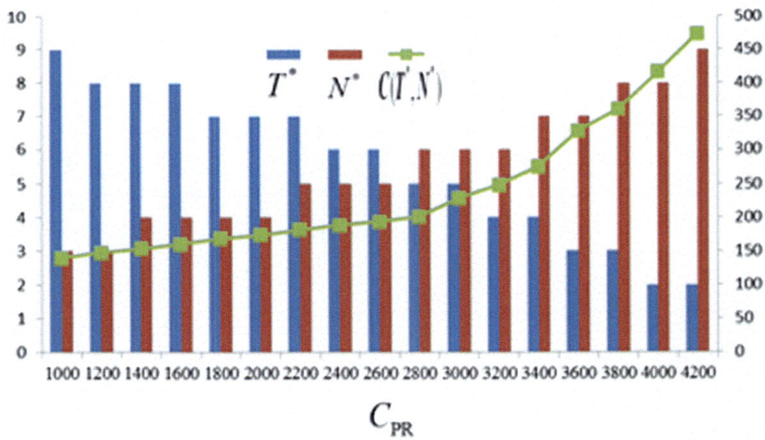

Fig. 10.4 T*, N* and C(T*, N*) trends with C_{PR} (electronic version)

out preventive replacement earlier to reduce the probability of system shutdown due to failure. Another conclusion that can be drawn is that when the detection cycle T^* is short, the average cost rate C(T*, N*) is high, which is due to the system being frequently detected, leading to an increase in costs.

Table 10.6 T^*, N^* and $C(T^*, N^*)$ trends with parameter C_D

C_D	T^*	N^*	$C(T^*, N^*)$
8	8	10	109.4
12	7	9	121.8
16	7	9	135.7
20	6	9	142.9
24	6	8	163.2
28	6	8	187.1
32	6	7	195.0
36	5	6	208.3
40	5	6	229.2
44	4	6	236.9
48	4	5	258.0
52	4	5	285.4
56	3	4	311.6
60	3	4	342.5
64	2	4	389.0
68	2	3	428.4
72	2	3	447.6

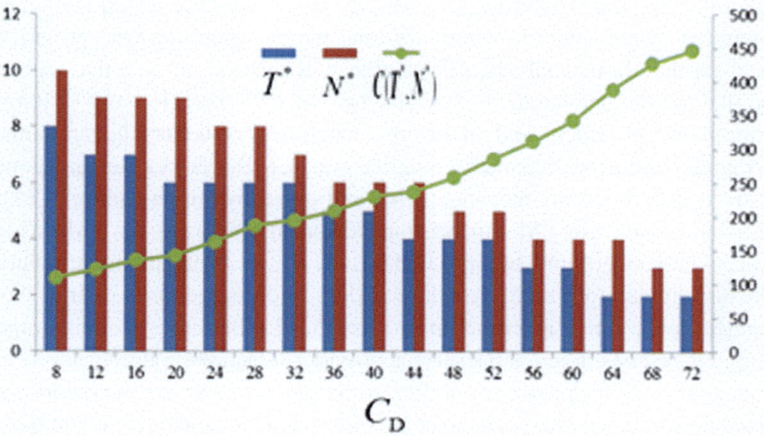

Fig. 10.5 T*, N* and C(T*, N*) trends with changes in C_D (Electronic version)

10.3 Service Provider Management Based on DSSI Theory and FAHP Empowerment Model

MRO service outsourcing is a major means for modern manufacturing enterprises to reduce operating costs and risks. Service provider relationship management is a complex process of managing the interaction between the enterprise and the provider, with the goal of making the service activities between the two more effective. Therefore, service provider selection is a complex and tricky decision problem. At present, many methods and theories about provider evaluation and selection have been proposed in different research fields, such as operations research theory, artificial intelligence, decision analysis theory, etc. Some scholars have pointed out that the most commonly used modeling methods include Analytic Hierarchy Process (AHP), Data Envelopment Analysis (DEA), Multiple Objective Programming (MOP), Fuzzy Logic (FL), Genetic Algorithms (GA), and Artificial Neural Networks (ANN). However, when using these methods, the process of generating relevant criteria and determining their weights is challenging. Due to the limitations of professional knowledge, decision-makers may not be able to determine the evaluation criteria and accurately assign weights to them, and the selection result highly depends on the subjective judgment of the decision-maker. In addition, literature [3, 4] point out that it is difficult for providers to maintain the same service level during different service cycles. On the one hand, they experience fluctuations in their capabilities at different stages, and on the other hand, the market environment is constantly changing. Therefore, an effective service provider selection model must not ignore the dynamism of this issue for long-term cooperation with providers.

Based on the above analysis, this section will provide an effective and simple method that, on the one hand, emphasizes the dynamic characteristics of long-term cooperation with providers, and on the other hand, fully considers the randomness of user demand. Unlike the demand for tangible goods, MRO service demand generally comes from preventive maintenance (PM) and corrective maintenance (CM) after failure. The demand for PM can be planned, but the demand for CM has strong randomness and uncertainty, because CM usually occurs after equipment failure, and equipment degradation is related to many factors, such as age, usage status, working environment, etc., which is a random process [5, 6]. Therefore, the randomness of demand must be considered in the service provider selection model. In addition, due to the uncertainty of the provider's ability to complete the service, the randomness of its service performance also needs to be considered in the model. Based on these two considerations, a novel and universal criterion—Demand Fulfillment Level (DFL)—is introduced based on the Universal Generating Function (UGF) method and Discrete Stress-Strength Interference (DSSI) theory. DFL is defined as the probability that the number of services randomly completed by the provider is greater than the number of services demanded by the user, so the randomness and uncertainty in the selection process are reflected in the model. It is worth mentioning that the implementation process of the selection model is simple, and because the calculation of DFL is based on actual record data rather than subjective judgment, the decision result is more objective and will be updated with the update of data, which is dynamic, helping decision-makers to choose the most suitable service provider for different types of MRO services at different stages.

In order to promote healthy competition among service providers and improve service quality in a targeted manner, a Fuzzy Analytical Hierarchy A multidimensional service quality evaluation mechanism based on Fuzzy Analytic Hierarchy Process (FAHP) is established. Firstly, a fuzzy hierarchy structure containing dual quality attributes (objective attributes and subjective evaluation) is established to fully consider the impact of subjective and objective service quality attributes on the evaluation of providers. Then, an FAHP weighting model is constructed that can dynamically adjust the weight resolution according to the preferences and emphasis of the decision maker. Unlike traditional weighting models, the weight vector under this weighting model is not unique (but the order of weights is unique), but a function containing adjustable parameters. By adjusting the parameters, the candidate providers preferred (or abandoned) by the decision maker can be retained (or excluded), which not only makes the decision maker more rational in choosing providers, but also promotes the service providers to further improve the service quality. Finally, the evaluation results of each candidate service provider are obtained using the combined weight vector and ordinal utility function value.

10.3.1 Theoretical Background

Before establishing the service provider selection and evaluation model, the theoretical knowledge and related concepts needed in the model establishment process are first given, and the focus is placed on the statement of these knowledge and concepts, without delving into their mathematical principles or derivation.

(1) Overview of UGF Method

The concept of UGF was first proposed in literature [7] and later applied to reliability analysis and optimization of multi-state systems.

Suppose X is a discrete random variable, its distribution law can be represented as a vector x and another vector p, vector x contains all possible values of X, and vector p contains the probabilities of taking corresponding values, i.e.,$X = (X_1, X_2, \ldots, X_k)$, $p(p_1, p_2, \ldots, p_k), p_i = p_r(X = X_i), i = 1, 2, \ldots, k$.

Definition 10.4 UGF of Discrete Random Variables.

The UGF of X is defined as a polynomial function of variable z, denoted as $u_X(z)$, and:

$$u_X(z) = p_1 z^{x_1} + p_2 z^{x_2} + \cdots + p_k z^{x_k} = \sum_{i=1}^{k} p_i z^{x_i} \tag{10.19}$$

It should be noted that for a discrete random variable, there is a one-to-one correspondence between its distribution law and UGF.

Definition 10.5 UGF of Discrete Random Variable Functions.

Consider n mutually independent discrete random variables $X_1, X_2, \ldots, X_n$, the UGF of each random variable is respectively $u_{X_1}(Z), u_{X_2}(Z), \ldots, u_{X_n}(Z)$, let $f(X_1, X_2, \ldots, X_n)$ be an arbitrary function of $X_1, X_2, \ldots, X_n$. Introduce the composite operator, then the UGF of the function $f(X_1, X_2, \ldots, X_n)$ is defined as follows:

$$u_f(Z) = \text{composite operator}(u_{X_1}(Z), u_{X_2}(Z), \ldots, u_{X_n}(Z)) \tag{10.20}$$

Definition 10.6 Composite operator.

According to Definition 10.4, we have $u_{X_i}(z) = \sum_{j_i=1}^{k_i} p_{ij_i} z^{x_{ij_i}}, i = 1, 2, \ldots, n,$ where $k_1, k_2, \ldots, k_n$ are the number of possible values for each random variable. To obtain $u_f(z)$, the composite operator is defined as:

$$\begin{aligned} &\otimes\left(\sum_{j_1=1}^{k_1} p_{1j_1} z^{x_{1j_1}}, \sum_{j_2=1}^{k_2} p_{2j_2} z^{x_{2j_2}}, \ldots, \sum_{j_n=1}^{k_n} p_{nj_n} z^{x_{nj_n}}\right) \\ &= \sum_{j_1=1}^{k_1} \sum_{j_2=1}^{k_2} \cdots \sum_{j_n=1}^{k_n} \left(\prod_{i=1}^{n} p_{ij_i} z^{f(x_{1j_1}, x_{2j_2}, \ldots, x_{nj_n})}\right) \end{aligned} \tag{10.21}$$

Property 10.1 UGF operations satisfy the commutative and associative laws, that is:

$$\begin{aligned} u_f(Z) &= \otimes\big(u_{X_1}(Z), u_{X_2}(Z), \cdots, u_{X_i}(Z), u_{X_{i+1}}(Z), \cdots, u_{X_n}(Z)\big) \\ &= \otimes\big(u_{X_1}(Z), u_{X_2}(Z), \cdots, u_{X_{i+1}}(Z), u_{X_i}(Z), \cdots, u_{X_n}(Z)\big), \\ u_f(Z) &= \otimes\big(u_{X_1}(Z), u_{X_2}(Z), \ldots, u_{X_i}(Z), u_{X_{i+1}}(Z), \ldots, u_{X_n}(Z)\big) \\ &= \otimes\Big(\otimes\big(u_{X_1}(Z), u_{X_2}(Z), \ldots, u_{X_{i+1}}(Z)\big), \otimes\big(u_{X_i}(Z), \ldots, u_{X_n}(Z)\big)\Big) \end{aligned}$$

(2) Overview of the DSSI Model

The Stress-Strength Interference (SSI) model has been widely used in the reliability analysis of "components", where the "component" can be an abstract system. Stress-strength analysis is an effective tool in the field of reliability engineering.

Definition 10.7 "Component" reliability.

Let S_1 and S_2 represent the stress acting on a "component" and the stress of the "component" itself, respectively, then the "component" reliability R can be defined as:

$$R = Pr(S_2 > S_1) \tag{10.22}$$

Equation (10.22) is the basic expression of the SSI model, which indicates that the "component" reliability is defined as the probability that the stress is greater than the pressure.

If S_1 and S_2 are two discrete random variables, their distribution laws are as follows:

$$S_1 = \left(S_{11}, S_{12}, \ldots, S_{1k_1}\right), p_1 = \left(p_{11}, p_{12}, \ldots, p_{1k_1}\right) \tag{10.23}$$

$$S_2 = \left(S_{21}, S_{22}, \ldots, S_{2k_2}\right), p_2 = \left(p_{21}, p_{22}, \ldots, p_{2k_2}\right) \tag{10.24}$$

In the formula, k_1 and k_2 are the number of possible values for S_1 and S_2 respectively. Therefore, according to Definition 10.7, the UGFs of S_1 and S_2 are $u_{S_1}(z) = \sum_{j_1=1}^{k_1} p_{1j_1} z^{S_{1j_1}}$ and $u_{S_2}(z) = \sum_{j_2=1}^{k_2} p_{2j_2} z^{S_{2j_2}}$.

If $f(S_1, S_2)$ is a function of S_1 and S_2, based on the previously introduced UGF method, the UGF of the function $f(S_1, S_2)$ can be obtained as

$$u_f(z) = \otimes(u_{S_1}(z), u_{S_2}(z)) = \sum_{j_1=1}^{k_1} \sum_{j_2=1}^{k_2} \left(\prod_{i=1}^{2} p_{ij_i} z^{f(S_{1j_1}, S_{2j_2})} \right) = \sum_{j=1}^{K} P_j z^{f_j} \tag{10.25}$$

where, f_j and P_j, $j = 1,2,\ldots, K$ are respectively the possible values of function $f(S_1, S_2)$ and the probabilities of taking the corresponding values, and $K \leq k_1 k_2$.

Definition 10.8 DSSI model.

If $f(S_1, S_2) = S_2 - S_1$, then the "component" reliability can be written as:

$$R = Pr(f(S_1, S_2) > 0) = \sum_{j=1}^{K} P_j \sigma(f_j) \tag{10.26}$$

Equation (10.26) is known as the DSSI model, where $\sigma(f_j)$ is a binary valued function defined on the possible value set of $f(S_1, S_2)$, and $\sigma(f_j) = \begin{cases} 1, f_j > 0 \\ 0, f_j \leq 0 \end{cases}$.

(3) Overview of FAHP method

AHP is a multi-objective decision-making method established by American operations researcher Satty [8], which combines qualitative and quantitative analysis. It can effectively analyze the non-sequential relationship between the levels of the target evaluation criteria system and effectively integrate the comparison and judgment of decision-makers. The Fuzzy Analytical Hierarchy Process (FAHP) is an improvement on the traditional AHP. It proposes a more reasonable and scientific weight vector calculation method, and can adjust the decision-maker's discrimination ability for weights according to the selection of parameters [9].

Definition 10.9 Consistency matrix.

If a pairwise comparison matrix $A = [a_{ij}]_{n\times n}$ satisfies:

$$a_{ij}a_{jk} = a_{ik},\ \text{for } i, j, k = 1, 2, \ldots, n \tag{10.27}$$

Then, A is called a consistency matrix, or simply a consistent matrix.

Theorem 10.1 If $A[a_{ij}]_{n\times n}$ is a consistent pairwise comparison matrix, then $R = [r_{ij}(\alpha)]_{n\times n}$, $\alpha \geq 81$, is a fuzzy consistent judgment matrix. Here $r_{ij}(\alpha) = \log_\alpha a_{ij} + 0.5$.

Note:

① Theorem 10.1 provides a method to transform a consistent matrix into a fuzzy consistent matrix. Here, $\alpha \geq 81$ is to ensure $0 \leq r_{ij}\ (\alpha) \leq 1$. Without loss of generality, in the calculations of this chapter, we take $\alpha = 243$;

② The size of the fuzzy scale value depends on the decision maker's choice of α. Specifically, $\lim_{\alpha\to\infty} \log \alpha X = 0$, $1/9 \leq x \leq 9$; thus, a consistent matrix $A = [a_{ij}]_{n\times n}$ corresponds to a family of fuzzy consistent matrices $R = [r_{ij}(\alpha)]_{n\times n}$, where $r_{ij}(\alpha) = \log_\alpha a_{ij} + 0.5$, and $\alpha \geq 81$.

Definition 10.10 Fuzzy weight vector.

Let $R = [r_{ij}]_{n\times n}$ be a fuzzy consistent matrix, then its fuzzy weight vector $\omega = (\omega_1, \omega_2, \ldots, \omega_n)^T$ can be determined by the following constrained programming problem:

$$P1: \begin{cases} \min z = \sum_{i=1}^{n}\sum_{j=1}^{n}\left(\log_\beta \omega_i - \log_\beta \omega_j + 0.5 - r_{ij}\right)^2 \\ s.t. \sum_{j=1}^{n} \omega_j = 1 \\ \omega_j > 0, j = 1, 2 \ldots, n \end{cases} \tag{10.28}$$

Here $\beta > 1$.

Definition 10.11 Combined fuzzy weight vector.

Suppose the weight vector of R_P is $\left(\omega_1^{R_P}, \omega_2^{R_P}\right)^T$, R_{Q_1}, the weight vector of $\left(\omega_1^{R_{Q_1}}, \omega_2^{R_{Q_1}}, \ldots, \omega_n^{R_{Q_1}}\right)^T$, R_{Q_2} is $\left(\omega_1^{R_{Q_2}}, \omega_2^{R_{Q_2}}, \ldots, \omega_n^{R_{Q_2}}\right)^T$, then the combined fuzzy weight vector is $\omega^*\left(\omega_1^*, \omega_2^* \ldots, \omega_n^*\right)$, where $\omega_i^* = \omega_i^{R_{Q_1}}\omega_1^{R_P} + \omega_i^{R_{Q_2}}\omega_2^{R_P}$, $i = 1, 2, \ldots n$.

10.3.2 Provider Selection Model Based on DSSI Theory

This section will establish a mathematical model for service provider selection based on DSSI theory. First, the model description and model assumptions are given. Then, based on the DSSI model, a novel evaluation criterion—Demand Fulfillment Level (DFL) is introduced, and the calculation steps of DFL are given. Finally, the dynamic selection rules of providers are formulated.

(1) Model Description and Model Assumptions

In the model, consider a user (decision maker) and multiple MRO service providers. The user outsources different types of MRO services (such as power distribution system maintenance, automation instrument calibration, etc.) to these providers. At each service stage, the decision maker needs to formulate a service plan, including the number of equipment that needs preventive maintenance, and the order quantity of corresponding services. The decision problem of this model is which provider should be chosen to perform which type of MRO service at each service stage. For the need to establish the model, the following model assumptions are given.

Assumption 10.5 The set of candidate providers is represented by $S = \{S_1, S_2,\ldots S_I\}$, where $S_i, i = 1, 2, \ldots, I$ represents the ith provider. At each stage (except the first stage), its decision goal is to select the most suitable supplier from the set S based on the service performance of the provider in the previous stage, and outsource the corresponding type of service to him.

Assumption 10.6 In the first stage, the user (enterprise) randomly selects providers from the set S to sign MRO service outsourcing contracts.

Assumption 10.7 The user has the ability and technology to record information and data of various maintenance activities.

Assumption 10.8 At each time node, the number of services ordered by the user and the number of services completed by the provider can be recorded respectively. The former can be based on preventive maintenance plan information, and the latter can be based on implemented maintenance activity information.

Assumption 10.9 The provider selection model considers multiple periods, each period contains multiple time nodes. Figure 10.6 intuitively depicts the time nodes for recording data and the period nodes for updating outsourcing contracts.

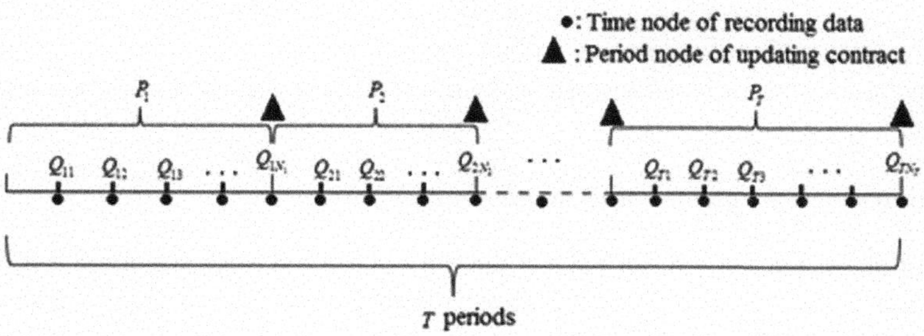

Fig. 10.6 Time nodes for recording data and period nodes for updating outsourcing contracts

Table 10.7 Symbols used in the model and their explanations

i	Provider subscript
j	MRO service subscript
t	Period subscript

For clarity, Table 10.7 lists the symbols used in the model and their corresponding symbol explanations.

Variable X_{ijt}

The quantity of service M_j ordered from provider S_i in the tth stage, it is a random variable

Y_{ijt}

The quantity of service M_j completed by provider S_i in the tth stage, it is a random variable.

DFL_{ijt}

The demand fulfillment level of service M_j by provider S_i in the tth stage.

I The number of providers
J The number of types of maintenance services
T The number of stages
N_t The number of time nodes in the tth stage

(2) Definition of Demand Fulfillment Level (DFL)

This section introduces a novel and universal criterion—DFL, to evaluate the performance of providers. The definition of DFL is mainly based on the previously introduced DSSI theory and UGF method. As introduced earlier, the DSSI model considers two random variables, which are the external pressure applied to the system and the self-stress to maintain the system from failure. Here, the service provider S_i, $i = 1, 2, \ldots, I$ is equivalent to a system, the random number of services ordered by users X_{ijt} is equivalent to pressure, and the random number of services completed by the provider Y_{ijt} is equivalent to stress. Further, DFL is defined as the reliability of the system (provider), that is:

Definition 10.12 Demand Fulfillment Level (DFL).

Assuming the number of services ordered by users X_{ijt} and the number of services completed by the provider Y_{ijt} are random variables, then in the tth stage, the demand fulfillment level DFL_{ijt} of service M_j by provider S_i is defined as the probability that Y_{ijt} is greater than X_{ijt}, that is:

$$DFL_{ijt} = Pr\left(Y_{ijt} > X_{ijt}\right) \tag{10.29}$$

Regarding the definition of DFL, there are several points to note.

① The statistical characteristics of the random variable X_{ijt} are based on a set of observed/recorded data, that is, the recorded parameters O_{ijtn_t}; similarly, the statistical characteristics of the random variable Y_{ijt} are based on another set of observed/recorded data, that is, the recorded parameters F_{ijtn_t}. This indicates that the distribution laws of X_{ijt} and Y_{ijt} can be obtained from their recorded parameters respectively.

② Since the observed/recorded data are objective, DFL_{ijt} will not depend on the subjective judgment of the decision-maker. Moreover, the more observed/recorded data, the more DFL_{ijt} The more evaluations are made of the provider's performance, the more accurate they will be.

③ Although there is only one evaluation criterion in the model DFL_{ijt}, the relationship between other factors affecting the provider's performance (such as price, quality, etc.) and DFL_{ijt} will be covered in the randomness of variables X_{ijt} and Y_{ijt}. In other words, in the long-term operation of MRO services, the impact of these factors will be reflected in DFL_{ijt}. For example, if a provider provides poor quality service at a certain stage, then the user will not order services from him in the next stage, which will lead to a decrease in the data O_{ijtn_t}. If a provider's service capability is weak, it will also lead to a decrease in the data F_{ijtn_t}.

(3) Calculation steps of DFL

Based on the UGF method and DSSI model introduced earlier, the following specific steps for calculating DFL are given.

① Derive the distribution of X_{ijt} and Y_{ijt}. Assume that the recorded data of X_{ijt} are $O_{ijt1}, O_{ijt2}, \ldots, O_{ijtN_t}$, Y_{ijt} 's recorded data are $F_{ijt\,1}, F_{ijt\,2}, \ldots, F_{ijtN_t}$, the histograms of these two sets of recorded data are shown in Fig. 10.7, from which the group interval and the corresponding relative frequency can be directly obtained. Since the distribution of X_{ijt} is determined by two elements, one is the possible value of X_{ijt}, and the other is the corresponding probability of each possible value. Therefore, the midpoint value of each group interval $\left[O_{ijt}^{l_t-1}, O_{ijt}^{l_t}\right]$, $l_t = 2, 3, \ldots, L_t$ is taken as the possible value of X_{ijt}, and the relative frequency of each group interval is taken as the corresponding probability value. Therefore, the following distribution of X_{ijt} can be obtained:

$$X_{ijt}\left(\frac{O_{ijt}^2 - O_{ijt}^1}{2}, \frac{O_{ijt}^3 - O_{ijt}^2}{2}, \ldots, \frac{O_{ijt}^{L_t} - O_{ijt}^{L_t-1}}{2}\right) \otimes \left(X_{ijt}^1, X_{ijt}^2, \ldots, X_{ijt}^{L_t}\right) \tag{10.30}$$

$$p_{ijt} = \left(p_{ijt}^1, p_{ijt}^2, \ldots, p_{ijt}^{L_t}\right) \tag{10.31}$$

where,

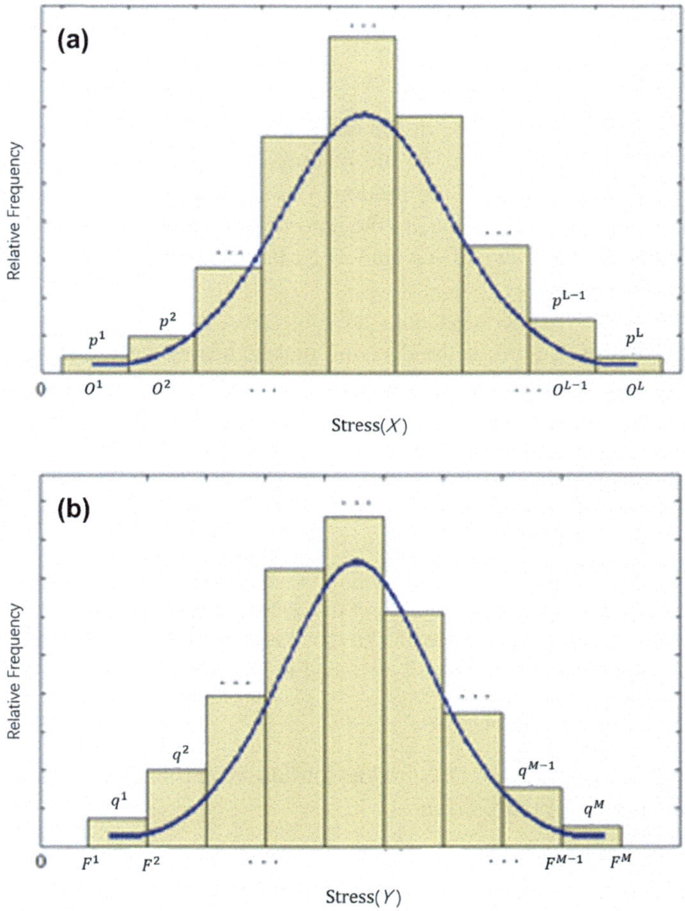

Fig. 10.7 Histogram of pressure and stress data

$$X_{ijt}^{l_t} = \frac{O_{ijt}^{l_t} - O_{ijt}^{l_t-1}}{2}, l_t = 1, 2, \ldots, L_t$$

Similarly, take the midpoint value of each group interval $\left[F_{ijt}^{h_t-1}, F_{ijt}^{h_t}\right]$, $h_t = 2, 3, \ldots, H_t$ as the possible values of Y_{ijt}, and the relative frequency of each group interval as the corresponding probability value. Therefore, the following distribution of Y_{ijt} can be obtained:

$$Y_{ijt} = \left(\frac{F_{ijt}^2 - F_{ijt}^1}{2}, \frac{F_{ijt}^3 - F_{ijt}^2}{2}, \ldots, \frac{F_{ijt}^{H_t} - F_{ijt}^{H_t-1}}{2}\right) \otimes \left(Y_{ijt}^1, Y_{ijt}^2, \ldots, Y_{ijt}^{H_t}\right) \tag{10.32}$$

$$q_{ijt} = \left(q_{ijt}^{1}, q_{ijt}^{2}, \ldots, q_{ijt}^{H_t}\right) \tag{10.33}$$

In the formula, $Y_{ijt}^{h_t} = \frac{F_{ijt}^{h_t} - F_{ijt}^{h_t-1}}{2}$, $h_t = 1, 2, \ldots, H_t$.

② Derive X_{ijt}'s UGF, Y_{ijt}'s UGF and $f(X_{ijt}, Y_{ijt})$'s UGF. According to Definition 10.4, the UGF of a discrete random variable corresponds one-to-one with its distribution, and in the previous step, we have obtained the distributions of X_{ijt} and Y_{ijt}, therefore, their UGFs can be written as:

$$u_{X_{ijt}}(z) = p_{ijt}^{1} z^{X_{ijt}^{1}} + p_{ijt}^{2} z^{X_{ijt}^{2}} + \cdots + p_{ijt}^{L_t} z^{X_{ijt} L_t} = \sum_{l_t=1}^{L_t} p_{ijt}^{l_t} z^{X_{ijt} l_t} \tag{10.34}$$

$$u_{Y_{ijt}}(z) = q_{ijt}^{1} z^{Y_{ijt}^{1}} + q_{ijt}^{2} z^{Y_{ijt}^{2}} + \cdots + q_{ijt}^{H_t} z^{Y_{ijt} H_t} = \sum_{h_t=1}^{H_t} q_{ijt}^{h_t} z^{Y_{ijt} h_t} \tag{10.35}$$

Since $f(X_{ijt}, Y_{ijt})$ is a function of X_{ijt} and Y_{ijt}, according to Definition 10.5 and Definition 10.6, the UGF of $f(X_{ijt}, Y_{ijt})$ can be obtained as follows:

$$\begin{aligned} u_f(z) &= \otimes\left(u_{X_{ijt}}(z), u_{Y_{ijt}}(z)\right) = \otimes\left(\sum_{l_t=1}^{L_t} p_{ijt}^{L_t} z^{Xijt^{l_t}}, \sum_{h_t=1}^{H_t} q_{ijt}^{h_t} z^{Y_{ijt}^{h_t}}\right) \\ &= \sum_{l_t=1}^{L_t} \sum_{h_t=1}^{H_t} \left(p_{ijt}^{L_t} z^{f\left(X_{ijt}^{l_t}, Y_{ijt}^{h_t}\right)} q_{ijt}^{h_t} z^{f\left(X_{ijt}^{l_t}, Y_{ijt}^{h_t}\right)}\right) = \sum_{a=1}^{A} \lambda_a z^{f_a} \end{aligned} \tag{10.36}$$

In the formula, f_a and λ_a, $a = 1, 2, \ldots, A$ are respectively the possible values of the function $f(X_{ijt}, Y_{ijt})$ and the corresponding probabilities, and $A \leq L_t H_t$.

③ Calculate DFL_{ijt} according to the DSSI model. Assume $f(X_{ijt}, Y_{ijt}) = Y_{ijt} - X_{ijt}$, according to Definition 10.5, DFL_{ijt} can be calculated as follows:

$$DFL_{ijt} = \Pr\left(f\left(X_{ijt}, Y_{ijt}\right)\right) > 0 = \sum_{a=1}^{A} \lambda_a \sigma(f_a) \tag{10.37}$$

In the formula, $\sigma(f_a)$ is a binary value function defined on the possible value set of the function $Y_{ijt} - X_{ijt}$, that is, $\sigma(f_a)\begin{cases} 1, f_a > 0 \\ 0, f_a \leq 0 \end{cases}$. It can be seen that Eq. (10.37) transforms the probability in the definition of DFL into a function expression of the sum of the multiplication of λ_a and $\sigma(f_a)$, which is more conducive to computer processing and decision-making.

(4) Dynamic decision rules for provider selection

Without loss of generality, the selection and update process of the provider for service M_j is given. According to the DFL calculation method given in the previous section, the DFL of each provider at different stages for service M_j can be calculated as follows:

$$\begin{aligned} &DFL_{1j1}, DFL_{2j1}, \ldots, DFL_{Ij1}, \\ &DFL_{1j2}, DFL_{2j2}, \ldots, DFL_{Ij2}, \text{ and} \\ &\ldots \\ &DFL_{1jT}, DFL_{2jT}, \ldots, DFL_{IjT}, \end{aligned}$$

$$\begin{aligned} DFL_{i_1j1} &= \max_{1 \le i \le I}\{DFL_{ij1}\}, 1 \le i_1 \le I \\ DFL_{i_2j2} &= \max_{1 \le i \le 1}\{DFL_{ij2}\}, 1 \le i_2 \le I \\ &\ldots \\ DFL_{i_TjT} &= \max_{1 \le i \le I}\{DFL_{ijT}\}, 1 \le i_T \le I \end{aligned}$$

Based on the above formulas, the most suitable service provider (with the highest DFL value) can be dynamically selected for service M_j at different stages. For example, in the first stage, service M_j is outsourced to provider S_{i1}, but in the second stage, when $i_1 \neq i_2$, the outsourcing contract needs to be updated, and service M_j is outsourced to provider S_{i2}. It can be seen that the decision result is not affected by the size of each provider's DFL value, but by the order of these values. Therefore, under different scales of record data sets, although the DFL value of the same provider will be different (because the accuracy is different), the order of DFL values of all providers is not much different, because once the record data set is determined, the accuracy of DFL values of all providers is the same.

In the real decision-making process, if it is necessary to select more providers for the corresponding MRO service according to the actual situation, the decision-maker only needs to sort DFL_{ijt}, $t = 1,2,\ldots,T$; $i = 1,2,\ldots,I$, then set a threshold to select the DFL_{ijt} that are ranked higher, and thus select the corresponding multiple providers.

(5) Case Analysis

This section will verify the feasibility and effectiveness of the proposed model through a case study. Without loss of generality, assume that users can outsource three types of MRO services (i.e., $J = 3$) to four service providers (i.e., $I = 4$). The service process includes four stages (i.e., $T = 4$), each stage includes thirty time nodes (i.e., $N_t = 30$, $t == 1, 2, 3, 4$). The decision goal is to select the most suitable provider based on the service performance of the provider at different stages, and to determine what kind of MRO services should be outsourced to him. For clarity, Tables 10.8 and 10.9 list the recorded parameters O_{ijtn_t} and F_{ijtn_t}.

According to the DFL calculation method proposed in the previous section, the DFL values of each provider regarding the corresponding MRO services at different stages can be calculated. As a demonstration of a calculation step, the calculation process of DFL_{111} is given as follows.

Table 10.8 Recorded parameter O_{ijtn_t}

$t = 1$					$t = 2$				
Item	S_1	S_2	S_3	S_4	Item	S_1	S_2	S_3	S_4
M_1	37, 36, 41,	35, 37, 39,	36, 34, 44,	37, 36, 41,	M_1	41, 37, 36,	39, 37, 37,	34, 48, 44,	34, 49, 45,
	29, 34, 26,	30, 33, 27,	31, 32, 28,	26, 29, 34,		26, 33, 29,	39, 43, 49,	43, 38, 39,	47, 38, 40,
	40, 39, 44,	39, 40, 45,	39, 41, 46,	36, 38, 39,		31, 37, 39,	44, 39, 36,	38, 36, 36,	35, 39, 37,
	36, 33, 47,	34, 32, 46,	38, 33, 46,	44, 47, 36,		23, 24, 33,	29, 34, 29,	37, 42, 44,	29, 24, 34,
	32, 36, 37,	34, 37, 39,	31, 37, 35,	36, 32, 44,		38, 36, 44,	33, 39, 37,	32, 39, 39,	38, 38, 37,
	28, 22, 31,	29, 23, 33,	27, 21, 32,	28, 22, 31,		39, 33, 48,	29, 24, 34,	27, 23, 33,	40, 45, 43,
	33, 46, 51,	35, 44, 53,	34, 45, 50,	37, 33, 46,		33, 47, 54,	34, 49, 53,	54, 39, 37,	53, 45, 38,
	45, 37, 38,	41, 39, 37,	38, 36, 39,	51, 38, 43,		43, 37, 39,	43, 37, 39,	29, 34, 28,	29, 34, 34,
	42, 48, 44,	44, 47, 43,	43, 49, 39,	46, 42, 48,		31, 41, 43,	31, 44, 44,	33, 43, 41,	32, 35, 43,
	46, 41, 42	46, 40, 39	45, 40, 41	44, 42, 41		48, 43, 44	47, 43, 42	49, 44, 42	47, 44, 45
$t = 1$					$t = 2$				
Project	S_1	S_2	S_3	S_4	Project	S_1	S_2	S_3	S_4
M_2	35, 37, 39,	39, 37, 42,	38, 37, 44,	36, 36, 37,	M_2	42, 36, 38,	35, 48, 51,	31, 38, 37,	31, 50, 44,
	36, 33, 47,	24, 33, 28,	28, 33, 29,	27, 33, 29,		26, 34, 29,	42, 39, 38,	29, 24, 34,	49, 36, 39,
	41, 37, 36,	39, 37, 43,	36, 38, 42,	36, 36, 41,		33, 36, 38,	38, 39, 37,	33, 49, 42,	31, 37, 40,
	28, 34, 26,	38, 34, 49,	38, 32, 48,	37, 34, 48,		27, 23, 34,	37, 42, 48,	44, 38, 38,	29, 23, 35,
	31, 31, 36,	34, 38, 40,	31, 39, 39,	31, 39, 39,		39, 37, 41,	31, 37, 37,	39, 38, 37,	37, 39, 37,
	36, 28, 22,	29, 24, 33,	26, 24, 34,	29, 24, 33,		37, 34, 49,	27, 24, 33,	38, 43, 43,	39, 44, 41,
	41, 41, 48,	33, 48, 53,	31, 49, 55,	34, 46, 51,		35, 49, 51,	54, 38, 36,	52, 38, 36,	54, 43, 37,
	43, 52, 45,	44, 37, 39,	44, 39, 39,	43, 36, 39,		43, 37, 37,	26, 34, 28,	29, 34, 30,	24, 34, 35,
	33, 47, 51,	44, 47, 42,	31, 46, 41,	31, 41, 42,		32, 41, 44,	32, 43, 44,	30, 33, 42,	35, 35, 45,
	43, 37, 37	49, 44, 44	47, 41, 42	47, 44, 41		49, 43, 42	49, 44, 41	49, 45, 41	49, 41, 41

(continued)

Table 10.8 (continued)

$t = 1$					$t = 2$				
Item	S_1	S_2	S_3	S_4	Item	S_1	S_2	S_3	S_4
M_3	39, 37, 42, 28, 36, 28, 40, 37, 42, 38, 31, 49, 34, 38, 38, 29, 24, 33, 31, 47, 52, 44, 39, 39, 45, 47, 44, 47, 42, 41	38, 38, 44, 27, 32, 29, 37, 36, 41, 39, 31, 49, 34, 38, 36, 29, 24, 34, 31, 49, 52, 43, 39, 39, 41, 53, 41, 49, 42, 44	39, 36, 39, 29, 31, 28, 38, 37, 44, 39, 35, 49, 33, 38, 38, 27, 22, 32, 34, 46, 52, 45, 37, 38, 34, 47, 43, 49, 44, 43	37, 39, 37, 29, 34, 27, 39, 36, 41, 39, 31, 46, 33, 38, 38, 24, 24, 31, 32, 46, 53, 41, 39, 39, 33, 41, 44, 49, 44, 44	M_3	38, 36, 37, 37, 33, 48, 41, 37, 39, 27, 33, 28, 34, 37, 39, 29, 23, 33, 33, 47, 54, 43, 36, 38, 33, 41, 43, 47, 41, 44	33, 49, 54, 44, 36, 37, 39, 39, 39, 36, 42, 41, 33, 36, 37, 29, 23, 34, 53, 38, 38, 28, 34, 29, 31, 43, 42, 47, 44, 44	35, 47, 44, 48, 39, 40, 34, 38, 39, 27, 25, 33, 37, 39, 36, 40, 43, 44, 51, 37, 39, 29, 35, 32, 33, 34, 44, 47, 45, 44	35, 39, 40, 28, 24, 34, 38, 39, 39, 40, 44, 42, 32, 49, 44, 50, 36, 40, 55, 44, 38, 25, 34, 34, 39, 35, 44, 47, 44, 42
$t = 3$					$t = 4$				
Item	S_1	S_2	S_3	S_4	Item	S_1	S_2	S_3	S_4
M_1	31, 49, 45, 49, 36, 39, 32, 37, 42, 28, 29, 35, 36, 40, 37, 38, 45, 41, 53, 44, 36, 24, 34, 35, 40, 35, 42, 49, 44, 43	32, 39, 45, 25, 29, 35, 38, 38, 38, 37, 44, 42, 33, 49, 45, 48, 38, 53, 48, 44, 37, 25, 35, 35, 38, 35, 45, 50, 45, 43	34, 40, 41, 29, 27, 34, 49, 40, 37, 24, 32, 34, 34, 49, 39, 50, 39, 55, 37, 38, 38, 42, 44, 45, 44, 31, 42, 50, 45, 42	40, 39, 37, 37, 43, 45, 50, 41, 41, 24, 32, 34, 32, 44, 43, 50, 39, 52, 31, 39, 41, 27, 27, 35, 36, 34, 41, 37, 43, 42	M_1	37, 37, 44, 27, 35, 29, 38, 39, 39, 36, 34, 46, 31, 37, 37, 28, 25, 33, 31, 47, 51, 44, 37, 39, 43, 48, 45, 46, 44, 44	34, 37, 37, 29, 24, 32, 38, 39, 44, 30, 34, 27, 41, 37, 45, 39, 34, 49, 35, 47, 53, 44, 36, 39, 43, 49, 41, 47, 43, 41	38, 37, 41, 30, 32, 27, 39, 37, 45, 36, 33, 46, 34, 37, 36, 29, 25, 33, 35, 37, 52, 43, 37, 39, 43, 49, 41, 47, 38, 41	37, 36, 41, 36, 44, 49, 38, 38, 37, 28, 45, 27, 33, 49, 38, 28, 29, 33, 35, 46, 47, 44, 38, 37, 33, 37, 44, 31, 42, 33

(continued)

Table 10.8 (continued)

$t = 1$					$t = 2$				
Item	S_1	S_2	S_3	S_4	Item	S_1	S_2	S_3	S_4
M_2	37, 39, 37, 39, 44, 43, 34, 48, 44, 49, 37, 52, 33, 37, 44, 29, 30, 34, 54,42,36, 25,34,34, 39,35,44, 50,44,44	33,50,45, 49,38,51, 34,37,45, 27,27,34, 39,38,39, 37,45,45, 49,45,37, 24,35,34, 39,34,44, 48,43,41	50,45,37, 24,34,33, 33,41,44, 48,37,52, 34,37,43, 27,30,32, 37,37,39, 39,43,45, 36,35,42, 37,43,43	36,40,39, 39,45,41, 49,41,44, 25,31,35, 33,45,44, 50,40,54, 32,37,41, 30,27,34, 39,35,45, 37,45,44	M_2	40,37,38, 37,33,49, 39,37,43, 30,34,28, 33,37,39, 28,24,34, 35,47,53, 45,38,39, 44,50,51, 47,43,44	39,38,43, 26,35,27, 39,37,37, 37,33,46, 33,36,36, 29,25,33, 32,46,53, 44,37,39, 44,49,41, 47,36,45	37,37,39, 28,35,27, 39,39,45, 38,33,49, 31,37,36, 29,24,33, 32,47,53, 44,37,37, 44,49,42, 52,43,44	35,48,37, 28,28,34, 36,37,44, 37,44,48, 36,38,36, 29,44,29, 34,47,47, 45,38,38, 31,36,43, 34,44,35
$t = 3$					$t = 4$				
Project	S_1	S_2	S_3	S_4	Project	S_1	S_2	S_3	S_4
M_3	32,39,45, 25,29,35, 38,38,38, 37,44,42, 33,49,45, 48,38,53, 51,44,37, 25,35,35, 38,35,45, 50,45,43	50,44,39, 21,31,35, 35,47,41, 48, 38, 53, 35, 39, 41, 30, 29, 31, 36, 38, 37, 40, 45, 44, 37, 35, 41, 49, 45, 45	46, 41, 36, 25, 35, 31, 35, 44, 45, 49, 37, 53, 35, 40, 41, 29, 29, 34, 39, 38, 40, 40, 43, 44, 38, 32, 44, 36, 45, 41	35, 44, 43, 49, 40, 55, 39, 40, 36, 39, 44, 34, 39, 43, 42, 23, 32, 33, 35, 38, 41, 29, 27, 35, 40, 32, 41, 36, 41, 45	M_3	34, 36, 39, 29, 25, 33, 39, 37, 41, 30, 34, 27, 38, 39, 45, 38, 33, 48, 35, 47, 54, 45, 37, 39, 45, 48, 45, 47, 41, 44	34, 47, 55, 43, 36, 40, 37, 39, 45, 27, 35, 28, 38, 37, 39, 37, 34, 47, 35, 37, 39, 30, 24, 34, 41, 48, 41, 41, 38, 42	38, 37, 41, 37, 33, 50, 36, 38, 36, 28, 34, 26, 32, 37, 37, 26, 25, 31, 34, 48, 52, 43, 38, 36, 45, 49, 45, 54, 45, 34	37, 39, 38, 24, 45, 27, 34, 42, 36, 24, 28, 35, 37, 38, 45, 36, 43, 49, 32, 49, 46, 44, 39, 39, 32, 37, 41, 34, 42, 31

Table 10.9 Record parameters F_{ijtm_t}

$t = 1$					$t = 2$				
Item	S_1	S_2	S_3	S_4	Item	S_1	S_2	S_3	S_4
M_1	23, 35,26, 31, 29, 41, 39, 54, 51, 49, 38, 28, 33,44, 33, 38, 37, 38, 39, 41, 40, 41, 48, 50, 43, 46, 41, 45, 46, 53	22, 36, 27, 32, 28, 43, 37, 55, 50, 48, 39, 29, 30, 46, 35, 40, 37, 39, 40, 39, 42, 43, 49, 51, 44, 45, 42, 44, 48, 52	24, 35, 27, 33, 30, 39, 41, 53, 50, 48, 39, 29, 34, 45, 32, 39, 36, 37, 40, 42, 39, 42, 47, 51, 42, 45, 39, 44, 47, 54	23, 36, 26, 31, 30, 38, 41, 51, 50, 48, 36, 39, 31, 42, 32, 33, 37, 38, 42, 40, 41, 45, 38, 44, 44, 47, 41, 48, 46, 54	M_1	33, 44, 36, 46, 35, 40, 29, 36, 26, 31, 35, 41, 41, 38, 41, 38, 45, 49, 39, 54, 50, 51, 40, 30, 39, 44, 38, 45, 46, 55	35, 46, 36, 44, 44, 40, 30, 36, 25, 30, 36, 41, 41, 54, 50, 50,41, 31, 40,38,39, 41,46,46, 38,46,41, 45,46,50	41,54,50, 50,41,29, 33,45,36, 45,44,41, 25,35,26, 30,36,40, 44,38,39, 40,39,46, 40,44,40, 44,45,43	42, 47, 39, 46, 44, 45, 40, 56, 51, 51, 43, 36, 34, 46, 35, 46, 47, 42, 27, 36, 28, 30, 42, 37, 50, 41, 38, 44, 36, 46
M_2	28, 39, 54, 51, 49, 28, 38, 36, 25, 31, 29, 41, 50, 39, 41, 40, 41, 45, 36, 33, 44, 34, 39, 38, 52, 43, 46, 41, 45, 45	28, 36, 25, 30, 30, 39, 38, 53, 50, 50, 39, 29, 35, 45, 34, 40, 34, 39, 39, 40, 41, 39, 49, 51, 44, 45, 39, 44, 45, 54	26, 36, 25, 30, 36, 40, 40, 53, 50, 50, 39, 30, 34, 46, 35, 39, 34, 39, 40, 40, 41, 41, 50, 50, 43, 45, 40, 45, 45, 53	29, 36, 25, 30, 36, 40, 39, 53, 48, 51, 40, 30, 33, 44, 35, 46, 34, 41, 41, 40, 41, 39, 48, 48, 43, 45, 38, 46, 44, 55	M_2	30, 35, 25, 31, 36, 40, 35, 43, 35, 46, 36, 41, 38, 53, 51, 51, 41, 31, 40, 39, 41, 39, 45, 45, 38, 43, 41, 44, 45, 55	41, 39,38, 40, 41,45, 36, 44,35, 45, 44,41, 26, 35,26, 31, 35,40, 40, 55,51, 50, 40,30, 39, 45,41, 43, 45, 51	42, 57,51, 50, 40,31, 35,46,36, 46, 44,42, 27, 35,25, 30, 36,41, 46, 38,38, 41, 39,45, 41, 44,39, 44, 46, 47	40, 46,41, 47, 43,46, 41, 54,50, 52, 43,37, 36, 47,36, 46, 44,46, 26, 37,26, 32, 41,36, 51, 41,40, 46, 35, 44

(continued)

Table 10.9 (continued)

$t = 1$					$t = 2$				
Item	S_1	S_2	S_3	S_4	Item	S_1	S_2	S_3	S_4
M_3	37, 35, 40, 28, 35, 25, 39, 44, 43, 35, 35, 47, 32, 34, 37, 29, 22, 30, 35, 45, 50, 44, 37, 39, 42, 49, 43, 46, 40, 42	24, 36, 25, 29, 31, 40, 41, 56, 54, 48, 40, 30, 35, 47, 36, 39, 34, 41, 38, 39, 39, 40, 46, 51, 45, 43, 39, 46, 44, 55	25, 35, 26, 31, 34, 41, 41, 56, 51, 50, 41, 31, 36, 46, 36, 45, 36, 38, 41, 41, 40, 41, 51, 49, 44, 46, 38, 45, 46, 53	34, 45, 35, 46, 36, 41, 30, 36, 25, 32, 36, 40, 40, 39, 41, 39, 49, 49, 40, 55, 49, 51, 40, 31, 41, 45, 38, 46, 44, 55	M_3	36, 45, 35, 45, 35, 41, 31, 36, 26, 31, 35, 39, 40, 55, 49, 50, 40, 29, 41, 41, 40, 40, 46, 45, 39, 45, 39, 43, 46, 51	39, 54, 51, 50, 41, 31, 46, 38, 38, 39, 40, 45, 35, 46, 35, 46, 44, 40, 26, 36, 26, 29, 35, 38, 41, 45, 40, 43, 46, 50	49, 40, 39, 44, 37, 44, 41, 57, 52, 50, 43, 35, 36, 47, 35, 47, 46, 41, 26, 36, 26, 31, 37, 40, 42, 46, 38, 44, 45, 46	37, 47, 35, 47, 46, 45, 41, 44, 42, 47, 45, 47, 40, 55, 51, 52, 40, 39, 29, 36, 26, 31, 41, 35, 49, 41, 39, 46, 37, 45
$t = 3$					$t = 4$				
Project	S_1	S_2	S_3	S_4	Project	S_1	S_2	S_3	S_4
M_1	41, 56, 50, 51, 42, 38, 36, 47, 34, 46, 46, 46, 40, 45, 42, 46, 45, 44, 29, 35, 28, 32, 41, 36, 50, 42, 39, 45, 37, 46	38, 44, 42, 46, 55, 43, 45, 53, 52, 48, 44, 38, 36, 46, 35, 46, 44, 46, 28, 34, 32, 30, 40, 35, 51, 41, 38, 46, 35, 47	40, 45, 42, 45, 55, 47, 48, 42, 39, 45, 39, 46, 49, 45, 51, 49, 43, 42, 37, 46, 34, 44, 45, 45, 32, 34, 31, 30, 41, 35	50, 44, 49, 49, 46, 41, 40, 44, 36, 45, 43, 43, 40, 46, 41, 46, 57, 42, 49, 41, 38, 47, 39, 46, 32, 34, 31, 28, 42, 37	M_1	26, 35, 41, 32, 29, 40, 38, 55, 51, 48, 38, 29, 34, 45, 33, 39, 37, 39, 40, 39, 41, 42, 48, 51, 44, 46, 40, 46, 45, 55	38, 55, 51, 48, 38, 32, 26, 37, 41, 30, 31, 40, 36, 45, 34, 39, 35, 40, 40, 39, 39, 41, 49, 41, 44, 43, 41, 44, 46, 53	41, 39, 39, 40, 48, 39, 39, 54, 52, 49, 38, 31, 27, 35, 41, 30, 29, 39, 34, 45, 35, 39, 36, 41, 44, 44, 26, 45, 45, 43	40, 38, 39, 41, 48, 38, 35, 45, 36, 39, 37, 38, 38, 54, 51, 49, 38, 31, 36, 36, 42, 28, 31, 38, 43, 45, 27, 44, 45, 47

(continued)

Table 10.9 (continued)

$t = 1$					$t = 2$				
Item	S_1	S_2	S_3	S_4	Item	S_1	S_2	S_3	S_4
M_2	39, 57, 49, 51, 41, 40, 34, 46, 35, 45, 46, 47, 41, 44, 42, 45, 54, 46, 28, 36, 30, 31, 41, 35, 49, 41, 38, 47, 37, 47	46, 54, 52, 49, 46, 39, 39, 46, 42, 47, 55, 45, 52, 42, 39, 45, 35, 46, 35, 46, 34, 47, 44, 47, 29, 35, 32, 31, 41, 36	35, 45, 36, 47, 46, 47, 41, 46, 39, 44, 54, 42, 49, 41, 38, 46, 41, 44, 51, 46, 52, 49, 45, 39, 30, 34, 32, 32, 40, 36	48, 39, 39, 46, 38, 45, 49, 45, 49, 51, 45, 42, 38, 44, 37, 43, 43, 45, 41, 45, 42, 46, 55, 38, 25, 36, 30, 29, 39, 34	M_2	41, 38, 40, 41, 49, 50, 25, 37, 41, 31, 28, 38, 39, 55, 52, 48, 39, 30, 35, 45, 36, 38, 37, 41, 45, 47, 41, 45, 46, 47	41, 38, 39, 40, 48, 42, 39, 55, 50, 49, 39, 31, 27, 36, 40, 31, 30, 41, 35, 46, 35, 38, 36, 41, 45, 44, 40, 45, 47, 55	36, 46, 34, 38, 37, 40, 40, 38, 40, 41, 48, 38, 38, 53, 51, 49, 39, 32, 26, 36, 41, 31, 29, 38, 45, 44, 25, 46, 46, 44	39, 40, 38, 40, 49, 39, 33, 46, 37, 38, 37, 54, 39, 53, 50, 49, 39, 32, 35, 37, 40, 29, 25, 41, 44, 43, 26, 43, 46, 44
M_3	43, 56, 50, 49, 42, 39, 35, 45, 35, 47, 45, 44, 38, 43, 41, 46, 54, 45, 29, 35, 31, 29, 40, 34, 50, 42, 39, 44, 34, 46	50, 41, 38, 46, 38, 47, 47, 43, 50, 48, 43, 41, 38, 44, 41, 47, 56, 46, 36, 45, 33, 46, 43, 44, 31, 36, 32, 32, 42, 33	38, 47, 35, 46, 45, 44, 39, 44, 40, 45, 56, 41, 48, 42, 39, 47, 40, 45, 49, 43, 51, 48, 44, 40, 31, 37, 30, 29, 41, 33	41, 46, 36, 43, 46, 47, 49, 41, 38, 44, 38, 45, 50, 45, 48, 50, 47, 40, 39, 46, 45, 45, 54, 39, 27, 35, 30, 31, 38, 36	M_3	39, 56, 52, 49, 39, 31, 25, 36, 42, 31, 30, 41, 35, 46, 34, 38, 36, 41, 41, 38, 40, 40, 48, 52, 46, 45, 39, 45, 47, 54	39, 40, 38, 41, 49, 40, 38, 53, 51, 48, 38, 32, 26, 36, 41, 32, 31, 42, 33, 47, 36, 39, 35, 39, 43, 43, 41, 47, 46, 54	37, 45, 35, 40, 37, 39, 41, 39, 38, 40, 49, 41, 39, 53, 50, 48, 39, 30, 35, 35, 40, 29, 31, 39, 44, 43, 26, 45, 46, 45	34, 36, 25, 32, 29, 40, 38, 54, 50, 35, 38, 29, 34, 46, 33, 39, 37, 40, 41, 38, 41, 41, 46, 49, 45, 44, 39, 43, 45, 42

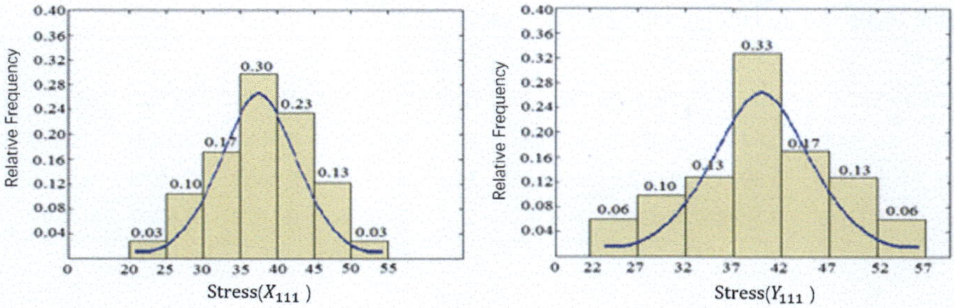

Fig. 10.8 Histogram of X_{111} and Y_{111}

Based on the record parameters in Tables 10.8 and 10.9, two sets of data (O_{1111} and F_{1111}) can be obtained, as shown in Fig. 10.8. Therefore, the distribution laws of X_{111} and Y_{111} are as follows:

$$X_{111} = (22.5, 27.5, 32.5, 37.5, 42.5, 47.5, 52.5),$$
$$p_{111} = (0.03, 0.10, 0.17, 0.30, 0.23, 0.13, 0.03)$$

$$Y_{111} = (24.5, 29.5, 34.5, 39.5, 44.5, 49.5, 54.5),$$
$$q_{111} = (0.06, 0.10, 0.13, 0.33, 0.17, 0.13, 0.06)$$

Therefore, we have:

$$\begin{aligned} DFL_{111} = \Pr(f(X_{111}, Y_{111}) > 0) = \sum_{a=1}^{A} \lambda_a \sigma(f_a) = 0.06 \times 0.03 \\ + 0.10 \times (0.03 + 0.10) + 0.13 \times (0.03 + 0.10 + 0.17) \\ + 0.33 \times (0.03 + 0.10 + 0.17 + 0.30) + 0.17 \\ \times (0.03 + 0.10 + 0.17 + 0.30 + 0.23) + 0.13 \\ \times (0.03 + 0.10 + 0.17 + 0.30 + 0.23 + 0.13) + 0.06 \\ \times (0.03 + 0.10 + 0.17 + 0.30 + 0.23 + 0.13 + 0.03) = 0.59 \end{aligned}$$

Therefore, we can obtain all the DFL_{ijt} (where $i = 1, 2, 3, 4; j = 1, 2, 3; t = 1, 2, 3, 4$), the results are listed in Table 10.10, where the gray area represents the maximum value of each row.

As can be seen from Table 10.10, in the first stage, M_1 should be outsourced to S_2, M_2 and M_3 should be outsourced to S_4; however, in the second stage, users need to update the outsourcing contract and should outsource M_1, M_2 and M_3 all to S_1; similarly, in the third stage, M_1 and M_2 should be outsourced to S_3, M_3 outsourced to S_4; in the final stage, ${}_1$ is outsourced to S_1, M_2 and M_3 are respectively outsourced to S_4 and S_2.

Table 10.10 All the DFL_{ijt} (where $i = 1, 2, 3, 4; j = 1, 2, 3; t = 1, 2, 3, 4$)

$t = 1$					$t = 2$				
Item	S_1	S_2	S_3	S_4	Item	S_1	S_2	S_3	S_4
M_1	0.5900	0.6244	0.5322	0.5456	M_1	0.6189	0.5889	0.5900	0.5922
M_2	0.6044	0.5944	0.6122	0.6389	M_2	0.6144	0.5689	0.6078	0.5856
M_3	0.6078	0.5822	0.6289	0.6333	M_3	0.6244	0.5822	0.6000	0.5767
$t = 3$					$t = 4$				
Item	S_1	S_2	S_3	S_4	Item	S_1	S_2	S_3	S_4
M_1	0.5778	0.5811	0.6011	0.5944	M_1	0.6100	0.5867	0.5844	0.6089
M_2	0.5622	0.5800	0.5956	0.5911	M_2	0.5933	0.6111	0.5700	0.6156
M_3	0.5711	0.5744	0.6067	0.6344	M_3	0.5989	0.6167	0.5967	0.5978

This result indicates that at different stages, it is necessary to dynamically evaluate the service level of each provider. Even for the same service, the outsourcing contract should be formulated according to its corresponding objective DFL value, rather than outsourcing a certain service to a certain provider for a long time. In addition, the data in Table 10.10 also shows the fluctuation of the service level of different providers at different stages. Figure 10.9 intuitively shows the average DFL value of different providers for the corresponding services. From a long-term overall perspective, S_1 is the optimal provider for M_1, and the optimal providers for M_2 and M_3 are S_4. In addition, S_3 and S_4 have a large difference in the average DFL value for different services, especially S_3, which also reflects its large fluctuation in service level; in contrast, S_1 and S_2 have a small difference in the average DFL value for different services, and their service level is relatively stable.

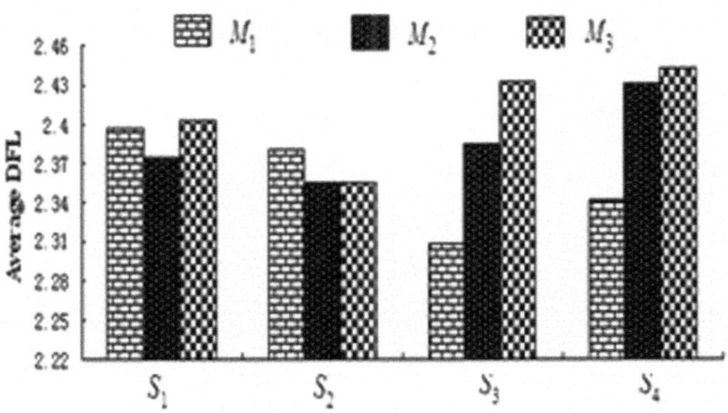

Fig. 10.9 Average DFL value of providers

10.3.3 Provider Evaluation Method Based on FAHP Weighting Model

The previous section proposed a service provider selection model based on DSSI theory, which has the following characteristics. First, the selected evaluation indicators are universal and unique, simplifying the decision-making process; second, the model takes into account the randomness of user needs and the fluctuation of provider service capabilities; finally, the decision results are objective and dynamic, and the most suitable provider can be dynamically selected at different stages. Despite these advantages, the model also has certain limitations. For example, although it quantifies the fluctuation of the provider's service capabilities, it does not give the reasons for the fluctuation. In other words, for those providers who cannot get outsourcing service contracts due to low DFL values, they may not know where (price, technology, etc.) they are not doing well and how to improve to enhance their competitiveness.

Based on the above analysis, this section will give a multi-dimensional service quality (Quality of Service, QoS) evaluation method based on the FAHP weighting model. First, define a multi-dimensional QoS vector containing truthfulness attributes, and establish a fuzzy hierarchy with various attributes of QoS and their corresponding truthfulness; second, use the FAHP weighting model to calculate the fuzzy weight vector and combined fuzzy weight vector of each hierarchy; finally, use the combined fuzzy weight vector and ordinal utility function value to get the evaluation result.

(1) Multi-dimensional QoS vector and ordinal utility function on the candidate provider set

Assume the set of service providers is $S = \{s_1, s_2, \ldots, s_m\}$, where s_i, $i = 1,2,\ldots, m$ is the i th provider, and m is the number of providers. The service quality attributes considered in the model are represented by q_i, $i = 1, 2, \ldots, n$ indicates that, without causing confusion, q_i is also used to represent the value of this attribute. The truthfulness of QoSq_{fid} (s) is defined as the user's evaluation of the service provider's corresponding QoS attribute, i.e., q_{fid} (s) $=$ $(fid_1(s), fid_2(s), \ldots, fid_n(s))$, where $fid_i(s) = \sum_{i=1}^{m_i} r_{ij}(s)/m_i$ represents the user's evaluation of the ith QoS attribute, $r_{ij}(s)$ represents the evaluation value given for the ith QoS attribute when the service is provided for the jth time, m_i represents the total number of times the ith QoS attribute has been evaluated.

Definition 10.13 Multi-dimensional QoS vector.

Suppose the multi-dimensional QoS vector M is a 2 n -dimensional expandable vector, in the form of:

$$M = \left(q_1, q_2, \ldots, q_n, fid_{q_1}, fid_{q_2}, \ldots, \text{fi } d_{q_n}\right) \quad (10.38)$$

In the formula, the first n components of M, q_i, $i = 1, 2, \ldots, n$ are the objective values of QoS (the data is published by the service provider or obtained by third-party monitoring), and the last n components $fi\ d_{q_i}$, $i = 1, 2, \ldots, n$ are the corresponding subjective evaluation values of QoS (the data comes from the evaluation of the service user).

Note:

① For some attributes with ambiguity, such as service attitude, credibility, etc., levels are often used to simply describe in practice. Therefore, before using these data for model calculations, these data need to be quantified—defuzzified, which can be processed using the α-cut and membership function in fuzzy set theory. The specific process will not be described in detail here. ② QoS attribute values can be positive attributes, such as service reliability, the larger the value, the higher the quality; it can also be negative attributes, such as service time, the larger the value, the lower the quality. Therefore, it is necessary to normalize the data so that the attribute values operate within the same range. The normalization process can be represented by the following formula:

$$q'_{ik} = \begin{cases} \dfrac{q_{ik} - q_{ik}^{\min}}{q_{ik}^{\max} - q_{ik}^{\min}}, \text{ if } q_{ik}^{\max} - q_{ik}^{\min} \neq 0 \\ 1 \qquad\qquad , \text{ if } q_{ik}^{\max} - q_{ik}^{\min} = 0 \end{cases}, \text{ (Positive attribute)} \tag{10.39}$$

$$q'_{ik} = \begin{cases} \dfrac{q_{ik}^{\max} - q_{ik}}{q_{ik}^{\max} - q_{ik}^{\min}}, \text{ if } q_{ik}^{\max} - q_{ik}^{\min} \neq 0 \\ 1 \qquad\qquad , \text{ if } q_{ik}^{\max} - q_{ik}^{\min} = 0 \end{cases}, \text{ (Negative attribute)} \tag{10.40}$$

In the formula, q_{ik} represents the value of provider s_i on the kth QoS attribute; $q_{ik}^{\min} = \min_{\forall s_i \in S}(q_{ik})$ represents the minimum value of all providers in the set S on the kth QoS attribute; $q_{ik}^{\max} = \max_{\forall s_i \in S}(q_{ik})$ represents the maximum value of all providers in the set S on the kth QoS attribute.

Definition 10.14 Ordinal utility function on set S.

Suppose $f_1(q_1\ (s_i)), f_2\ (q_2\ (s_i)), \ldots, f_n\ (q_n\ (s_i))$ are respectively the real-valued ordinal utility functions satisfying the weak order relation "·" on the set $S = \{s_1, s_2, \ldots, s_m\}$, then the monotonically increasing composite function $f(s) = f(f_1\ (q_1\ (s_i)), f_2\ (q_2\ (s_i)), \ldots, f_n\ (q_n\ (s_i))$ is an ordinal utility function on S.

Note:

① The weak order relation "·" here refers to a binary relation on the set S. For any two s_i and s_j in the set S, it satisfies: a. Connectivity, $\forall\ s_i, s_j \in S$, $s_i \cdot s_j$ or $s_j \cdot s_i$ or both; b. Transitivity, $\forall\ s_i, s_j, s_k \in S$, if $s_i \cdot s_j$ and $s_j \cdot s_k$, then $s_i \cdot s_k$; c. Indifference, $s_i \cong s_j$ if and only if $s_i \cdot s_j$ and $s_j \cdot s_i$.

② When "·" is a weak order relation on S, there exists a real-valued ordinal utility function f, for $\forall s_i, s_j \in S, f(s_i) \geq f(s_j) \Leftrightarrow s_i \cdot s_j$ And a. if t is a strictly increasing function of f, then $t(f(S_i)) \geq t(f(S_j)) \Leftrightarrow S_i \cdot S_j s$, where f_i, $i = 1{,}2$ are ordinal utility functions, then $g(s_i) \geq g\ (s_j) \Leftrightarrow s_i \cdot s_j$.

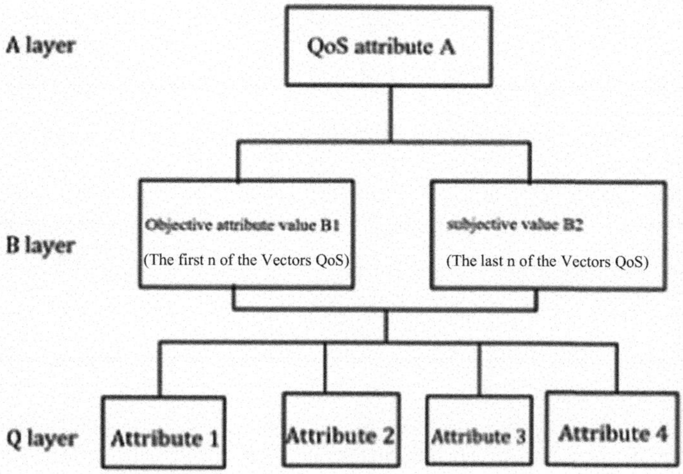

Fig. 10.10 Hierarchical structure of multi-dimensional QoS FAHP

(2) Multi-dimensional QoS FAHP hierarchy and fuzzy weight vector

Based on the previous Definition 10.13, a multi-dimensional QoS FAHP hierarchy can be established, as shown in Fig. 10.10, and the calculation method of the fuzzy weight vector of QoS attributes is given.

① Construct the pairwise comparison matrix for each level. Let B layer's pairwise comparison matrix for A layer be $P = \begin{bmatrix} p_{11} & p_{12} \\ p_{21} & p_{22} \end{bmatrix}_{2\times 2}$ where p_{12} represents the importance of objective attribute values relative to subjective evaluation values, and p_{21} is the opposite; in most cases, $p_{12} = p_{21} = 1$ can be taken, that is, they are equally important in the provider evaluation model. The pairwise comparison matrix of layer q to layer B is

$$Q^1 = \begin{bmatrix} q_{11}^1 & q_{12}^1 & \cdots & q_{1n}^1 \\ q_{21}^1 & q_{22}^1 & \cdots & q_{2n}^1 \\ & & \ddots & \\ q_{n1}^1 & q_{n2}^1 & \cdots & q_{nn}^1 \end{bmatrix}_{n\times n} \text{ and } Q^2 = \begin{bmatrix} q_{11}^2 & q_{12}^2 & \cdots & q_{1n}^2 \\ q_{21}^2 & q_{22}^2 & \cdots & q_{2n}^2 \\ & & \ddots & \\ q_{n1}^2 & q_{n2}^2 & \cdots & q_{nn}^2 \end{bmatrix}_{n\times n} \quad Q^2$$

where $q_{ij}^1 = \frac{q_i'}{q_j'}$, and q'_i and q'_j are the mean values of q_i and q_j after standardization; $q_{ij}^2 = \frac{fidq_i'}{fidq_j'}$, $fi\ d_{q_i'}$ and $fi\ d_{q_j'}$ are respectively the normalized averages of $fi\ d_{q_i}$ and $fi\ d_{q_j}$.

② Judge the consistency of the above pairwise comparison matrices. For the second-order pairwise comparison matrix P, because $p_{12} = \frac{1}{p_{21}}$, $p_{21} = \frac{1}{p_{12}}$, so $p_{12}p_{21} =$

$1 = p_{11}, p_{21}p_{12} = 1 = p_{22}$, according to Definition 10.9, matrix P is a consistent matrix.

For Q^1, Q^2, because $q_{ik}^1 = \frac{q_i^{\cdot}}{q_k^{\cdot}} = \frac{q_i^{\cdot}}{q_j^{\cdot}}\frac{q_j^{\cdot}}{q_k^{\cdot}} = q_{ij}^1 q_{jk}^1, q_{ik}^2 = \frac{fid_{q_i^{\cdot}}}{fid_{q_k^{\cdot}}} = \frac{fid_{q_i^{\cdot}}}{fid_{q_j^{\cdot}}}\frac{fid_{q_j^{\cdot}}}{fid_{q_k^{\cdot}}} = q_{ij}^2 q_{jk}^2$ therefore, according to Definition 10.9, the matrices Q^1 and Q^2 are also consistent matrices.

③ Let's consider P, Q^1, and Q^2 Transformed into a fuzzy consistent matrix. According to Theorem 10.1, P, Q^1, Q^2 The corresponding fuzzy consistent matrices are $R_P\left[\log_\alpha p_{ij} + 0.5\right]_{2\times 2}$, $R_{Q_1} = \left[\log_\alpha q_{ij}^1 + 0.5\right]_{n\times n}$, $R_{Q_2} = \left[\log_\alpha q_{ij}^2 + 0.5\right]_{n\times n}$

④ Calculate the fuzzy weight vector and combined fuzzy weight vector for each level. According to Definition 10.10, the weight vector of the fuzzy consistent matrix $R = \left[r_{ij}\right]_{n\times n}$ is $\omega = (\omega_1, \omega_2, \ldots, \omega_n)^T$ It can be determined by the constrained programming problem P1 (i.e., Eq. 10.28). In order to solve the weight vector $\omega = (\omega_1, \omega_2, \ldots, \omega_n)^T$, the constrained programming problem P1 can be transformed into the following unconstrained programming problem P2 using the Lagrange function:

$$P2{:}\min L(\omega,\lambda) = \sum_{i=1}^{n}\sum_{j=1}^{n}\left(\log_\beta \omega_i - \log_\beta \omega_j + 0.5 - r_{ij}\right)^2 + 2\lambda\left(\sum_{j=1}^{n}\omega_j - 1\right) \tag{10.41}$$

Let $\frac{\partial L(\omega,\lambda)}{\partial \omega_i} = 0$, we can get $4\sum_{j=1}^{n}\frac{(\log_\beta \omega_i - \log_\beta \omega_j + 0.5 - r_{ij})}{\omega_i \ln\beta} + 2\lambda = 0$, after rearranging, we can get $2\sum_{j=1}^{n}(\log_\beta \omega_i - \log_\beta \omega_j + 0.5 - r_{ij}) + \lambda\omega_i \ln\beta = 0, i = 1, 2, \ldots, n$, so we have:

$$\sum_{j=1}^{n}\left[\sum_{j=1}^{n}2\left(\log_\beta \omega_i - \log_\beta \omega_j + 0.5 - r_{ij}\right) + \lambda n\beta^{\frac{1}{n}\sum_{j=1}^{n} r_{kj}}\right] \tag{10.42}$$

The solution is:

$$\omega_i = \frac{\beta^{\frac{1}{n}\sum_{j=1}^{n} r_j}}{\sum_{k=1}^{n}\beta^{\frac{1}{n}\sum_{j=1}^{n} n_j}}, i = 1, 2, \ldots, n \tag{10.43}$$

Note:

① The weight $\omega_i = \frac{\beta^{\frac{1}{n}\sum_{j=1}^{n} r_{ij}}}{\sum_{k=1}^{n} \beta^{\frac{1}{n}\sum_{j=1}^{n}\sum_{j=1}^{n} r_{kj}}}$, $i = 1, 2, \ldots, n$ is a function of the base β, (where $\beta > 1$), thus for $R = \left[r_{ij}\right]_{n\times n}$, there is a family of weight vectors:

$$W = \left\{ \left(\omega_1(\beta), \omega_2(\beta), \ldots, \omega_n(\beta)^T\right) | \omega_i(\beta) = \frac{\beta^{\frac{1}{n}\sum\limits_{j=1}^{n} r_{ij}}}{\sum\limits_{k=1}^{n} \beta^{\frac{1}{n}\sum\limits_{j=1}^{n} r_{kj}}}, \beta > 1, i = 1, 2, \ldots, n \right\} \tag{10.44}$$

In the formula, $\frac{\omega_i(\beta)}{\omega_j(\beta)}$ is a strictly increasing function with respect to β, and $\lim_{\beta\to\infty}\frac{\omega_i(\beta)}{\omega_j(\beta)} = \infty$, $\lim_{\beta\to 1}\frac{\omega_i(\beta)}{\omega_j(\beta)} = 1$, which also indicates that by increasing the value of β, the resolution of the weight sizes can be improved. When β is very large, some weight values will tend to zero, which is not conducive to computer processing. Therefore, the value of β can be determined according to the preference of the decision maker.

② The order of the components of the weight vector is independent of the value of β, but the resolution of the weight sizes can be adjusted by adjusting the value of β. The following is an example to illustrate this point.

Let $R = \begin{bmatrix} 0.5 & 0.5738 & 0.4070 \\ 0.4271 & 0.5 & 0.3332 \\ 0.5933 & 0.6668 & 0.5 \end{bmatrix}$ be a fuzzy consistent matrix, take $\beta = e, \beta = \mathrm{e}^{10}, \beta = \mathrm{e}^{20}$, then the corresponding weight vectors can be obtained as follows:

$$\omega(\mathrm{e}) = (0.3304, 0.3070, 0.3626)$$
$$\omega\left(\mathrm{e}^{10}\right) = (0.2490, 0.1194, 0.6316)$$
$$\omega\left(\mathrm{e}^{20}\right) = (0.1305, 0.030, 0.8395)$$

It can be seen that on the one hand, the order of the components of the weight vector (the third component, the first component, the second component) is independent of the value of β; on the other hand, as the value of β increases, the difference (resolution) between the values of each component becomes larger and larger, the largest component value will approach 1, and the smallest component value will approach 0.

According to Definition 10.11, the correspondence between the weight vectors at each level and the combined weight vector can be obtained. For clarity, Table 10.11 intuitively shows their correspondence.

(3) Provider evaluation method based on fuzzy weight vector and ordinal utility function

Table 10.11 Correspondence between the weight vectors at each level and the combined weight vector

Weight vector	
(B to A)	$\omega_1^{R_P}$ $\omega_2^{R_P}$
Weight vector	
(q to B)	
q layer to B layer	
Objective data	
(B_1)	
Subjective data (B_2)	
Combined weight vector	
(q to A)	
Attribute q_1	$\omega_1^{R_{Q_1}}$ $\omega_1^{R_{Q_2}}$
$\omega_1^* = \omega_1^{R_{Q_1}} \omega_1^{R_P} + \omega_1^{R_{Q_2}} \omega_2^{R_P}$	
Attribute q_2	$\omega_2^{R_{Q_1}}$ $\omega_2^{R_{Q_2}}$
$\omega_n^* = \omega_n^{R_{Q_1}} \omega_1^{R_P} + \omega_n^{R_{Q_2}} \omega_2^{R_P}$	
Attribute q_n	$\omega_n^{R_{Q_1}}$ $\omega_n^{R_{Q_2}}$
$\omega_n^* = \omega_n^{R_{Q_1}} \omega_1^{R_P} + \omega_n^{R_{Q_2}} \omega_2^{R_P}$	

This section will use the combined fuzzy weight vector and ordinal utility function calculated earlier to comprehensively evaluate the service quality of the provider. According to the provider set defined earlier S, multi-dimensional QoS vector M, and a set of ordinal utility functions $f_k\,(qk\,(s_i)), k = 1, 2, \ldots, n,$, then the comprehensive service quality of each provider $S_1, S_2, \cdots, S_m$ is evaluated based on the value of $f\,(S)$. For simplicity, a strictly monotonically increasing linear function can be selected as the ordinal utility function for each QoS attribute, that is $f_k(q_k(s_i)) = \omega_k \cdot q_k(s_i), \quad k = 1, 2, \ldots, n.$

Let $H = \left[h_{ij}\right]_{m\times n}$ represent the QoS attribute matrix, where $h_{ij} = q_i \ \left(s_j\right)$ is the value of provider $s_j \in S$ for QoS attribute $q_i \in \{q_1, q_2, \ldots, q_n\}$. Let L be the standard matrix after H is defuzzified and normalized, and L is divided into rows to get $L = [L_1, L_2, \ldots, L_m]^T$, take $f(s_i) = \sum_{j=1}^{n} \omega_j \cdot q_j(s_i) = L_i \cdot \omega$, then we have:

$$s_i \cdot s_j \Leftrightarrow L_i \cdot \omega \geq L_j \cdot \omega, \omega = (\omega_1, \omega_2, \cdots \omega_n)^T, i, j = 1, 2, \cdots, m \tag{10.45}$$

Equation (10.45) indicates that the necessary and sufficient condition for provider s_i to be superior to s_j is $L_i \cdot \omega \geq L_j \cdot \omega$, where ω is the fuzzy weight vector calculated in the previous section. Therefore, the FAHP weighting mode is the main basis for service quality evaluation.

In addition, in the specific evaluation decision model, since the values of the weight vector components represent the weights of each QoS attribute, the decision maker (service user) can adjust the value of the parameter β according to their own preferences and emphasis to make the decision scheme more in line with their needs. For example, if the weight of service reliability is larger in the QoS attribute weight vector, and the weight of service price is smaller, the decision maker can, according to their own needs (production plan, maintenance difficulty, etc.), if they pay more attention to the reliability of the service, increase the value of β to amplify the weight of the corresponding attribute, while reducing the weight of other attributes.

(4) Case Analysis

In this section, we will verify the feasibility and effectiveness of the proposed method through a case study. Without loss of generality, assume that there are ten candidate service providers, i.e., $S = \{s_1, s_2, \ldots, s_{10}\}$. Considering the industrial characteristics of MRO services, four QoS attributes are used, namely price q_1, time q_2, skill q_3 and reliability q_4, forming an eight-dimensional QoS vector $M = \left(q_1, q_2, q_3, q_4, fid_{q_1}, fid_{q_2}, fid_{q_3}, fid_{q_4}\right)$. The FAHP hierarchy of the eight-dimensional QoS vector is shown in Fig. 10.11.

The randomly generated service quality information data for the ten providers is shown in Table 10.12. The normalized service quality information data is shown in Table 10.13.

For the hierarchical structure in Fig. 10.11, the pairwise comparison matrices of each level and their fuzzy consistency matrices are calculated as follows:

$$P = \begin{bmatrix} 1 & 1 \\ 1 & 1 \end{bmatrix}; \quad R_P = \begin{bmatrix} 0.5 & 0.5 \\ 0.5 & 0.5 \end{bmatrix}$$

$$Q_1 = \begin{bmatrix} 1 & 0.6158 & 0.65 & 0.78 \\ 1.6238 & 1 & 1.0555 & 1.2666 \\ 1.5385 & 0.9474 & 1 & 1.2 \\ 1.2821 & 0.7895 & 0.8333 & 1 \end{bmatrix},$$

$$R_{Q_1} = \begin{bmatrix} 0.5 & 0.4117 & 0.4216 & 0.4548 \\ 0.5883 & 0.5 & 0.5098 & 0.5430 \\ 0.5784 & 0.4902 & 0.5 & 0.5332 \\ 0.5452 & 0.4570 & 0.4668 & 0.5 \end{bmatrix},$$

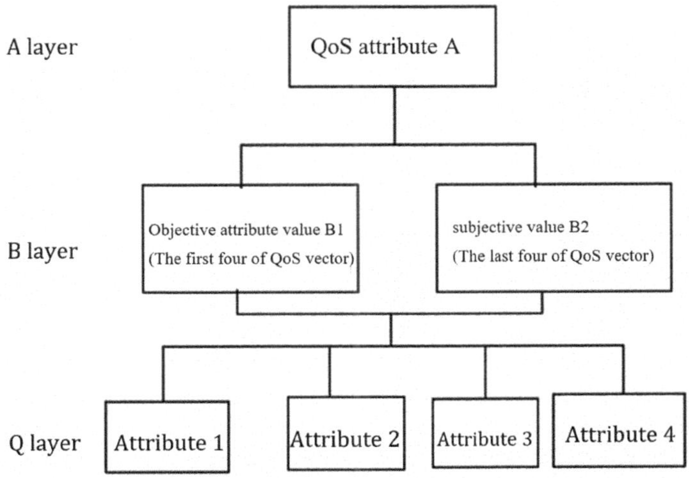

Fig. 10.11 FAHP hierarchy of the eight-dimensional QoS vector

Table 10.12 Service quality information of providers

Project	q_1	q_2	q_3	q_4	q_{fid}
s_1	0.14	175	0.90	0.92	0.99 0.94 0.90 0.95
s_2	0.20	150	0.95	0.98	0.95 0.89 0.93 0.92
s_3	0.10	158	0.92	0.91	0.95 0.85 0.90 0.91
s_4	0.15	152	0.85	0.95	0.88 0.90 0.90 0.88
s_5	0.18	165	0.92	0.98	0.85 0.90 0.95 0.95
s_6	0.20	175	0.90	0.88	0.99 0.94 0.92 0.90
s_7	0.15	190	0.94	0.92	0.95 0.95 0.92 0.95
s_8	0.16	155	0.88	0.96	0.95 0.88 0.95 0.95
s_9	0.18	150	0.92	0.92	0.92 0.92 0.90 0.95
s_{10}	0.15	145	0.92	0.88	0.92 0.90 0.92 0.90

$$Q_2 = \begin{bmatrix} 1 & 1.0651 & 1.5976 & 1.0365 \\ 0.9389 & 1 & 1.5 & 0.9732 \\ 0.6259 & 0.6667 & 1 & 0.6488 \\ 0.9648 & 1.0275 & 1.5413 & 1 \end{bmatrix},$$

$$R_{Q_2} = \begin{bmatrix} 0.5 & 0.5115 & 0.5853 & 0.5065 \\ 0.4885 & 0.5 & 0.5738 & 0.4950 \\ 0.4147 & 0.4262 & 0.5 & 0.4212 \\ 0.4934 & 0.5049 & 0.5788 & 0.5 \end{bmatrix}$$

Table 10.13 Normalized service quality information of providers

Project	q_1	q_2	q_3	q_4	q_{fid}			
s_1	0.6	0.3333	0.5	0.4	1	0.9	0	1
s_2	0	0.8889	1	1	0.7143	0.4	0.6	0.57144
s_3	1	0.7111	0.7	0.3	0.7143	0	0	0.4286
s_4	0.5	0.8444	0	0.7	0.2143	0.5	0	0
s_5	0.2	0.5556	0.7	1	0	0.5	1	1
s_6	0	0.3333	0.5	0	1	0.9	0.4	0.2587
s_7	0.5	0	0.9	0.4	0.7143	1	0.4	1
s_8	0.4	0.7778	0.3	0.8	0.7143	0.3	1	0.2587
s_9	0.2	0.8889	0.7	0.4	0.5	0.7	0	1
s_{10}	0.5	1	0.7	0	0.5	0.5	0.4	0.2587
Average	0.39	0.6333	0.6	0.5	0.6071	0.57	0.38	0.5857

According to formula (10.44), we can calculate the weight vectors and combined weight vectors of R_P, R_{Q_1}, R_{Q_2}, here we take $\beta = \mathrm{e}^{10}$ (the value of β can be adjusted according to the preferences and focus of the decision-maker), then we have:

$$\begin{aligned}\omega_{R_p} &= (0.5, 0.5)\\ \omega_{R_{Q_1}} &= (0.1393, 0.3367, 0.3051, 0.2189)\\ \omega_{R_{Q_2}} &= (0.3082, 0.2717, 0.1314, 0.2887)\\ \omega^* &= (0.2238, 0.3042, 0.2183, 0.2538)\end{aligned}$$

According to formula (10.45), $f(s_i) = L_i\omega$, the utility function values of each service provider can be obtained as:

$$\begin{aligned}&f(s_1) = 0.4463, f(s_2) = 0.7424, f(s_3) = 0.6690,\\ &f(s_4) = 0.5464, f(s_5) = 0.6203,\\ &f(s_6) = 0.2105, f(s_7) = 0.4098, f(s_8) = 0.5946,\\ &f(s_9) = 0.5694, f(s_{10}) = 0.5688\end{aligned}$$

Therefore, the comprehensive ranking of service quality of these ten candidate providers is $S_2 \cdot S_3 \cdot S_5 \cdot S_8 \cdot S_9 \cdot S_{10} \cdot S_4 \cdot S_1 \cdot S_7 \cdot S_6$.

As mentioned earlier, the parameter β has a certain impact on the decision result. To illustrate this intuitively, a sensitivity analysis experiment will be conducted on the parameter β. Now, the hierarchical weight vector, combined weight vector, utility function values of each candidate provider and their rankings under different β values are calculated respectively. The results are listed in Tables 10.14, 10.15 and 10.16. Figure 10.12 intuitively depicts the change rule of the utility function values of the candidate providers with the change of β values.

Table 10.14 Changes in hierarchical weight vector with β values

β	ωR_{Q_1}	ωR_{Q_2}
e	(0.2370, 0.2588, 0.2563, 0.2479)	(0.2565, 0.2532, 0.2355, 0.2548)
e^2	(0.2243, 0.2676, 0.2624, 0.2456)	(0.2628, 0.2562, 0.2216, 0.2594)
e^3	(0.2122, 0.2765, 0.2684, 0.2430)	(0.2690, 0.2590, 0.2083, 0.2638)
e^4	(0.2004, 0.2852, 0.2742, 0.2401)	(0.2750, 0.2615, 0.1955, 0.2679)
e^5	(0.1891, 0.2940, 0.2799, 0.2371)	(0.2809, 0.2638, 0.1834, 0.2719)
e^6	(0.1782, 0.3026, 0.2853, 0.2338)	(0.2867, 0.2658, 0.1719, 0.2757)
e^7	(0.1678, 0.3113, 0.2906, 0.2303)	(0.2923, 0.2676, 0.1609, 0.2792)
e^8	(0.1579, 0.3198, 0.2956, 0.2267)	(0.2978, 0.2692, 0.1505, 0.2826)
e^9	(0.1484, 0.3283, 0.3005, 0.2229)	(0.3031, 0.2705, 0.1407, 0.2857)
e^{10}	(0.1393, 0.3366, 0.3051, 0.2189)	(0.3082, 0.2717, 0.1314, 0.2887)
e^{11}	(0.1307, 0.3339, 0.3096, 0.2149)	(0.3133, 0.2727, 0.1226, 0.2915)
e^{12}	(0.1225, 0.3531, 0.3138, 0.2107)	(0.3181, 0.2734, 0.1143, 0.2941)
e^{13}	(0.1147, 0.3611, 0.3178, 0.2064)	(0.3229, 0.2740, 0.1065, 0.2965)
e^{14}	(0.1073, 0.3690, 0.3216, 0.2021)	(0.3275, 0.2744, 0.0992, 0.2988)
e^{15}	(0.1003, 0.3768, 0.3252, 0.1977)	(0.3320, 0.2747, 0.0924, 0.3009)
e^{16}	(0.0937, 0.3845, 0.3286, 0.1932)	(0.3364, 0.2748, 0.0859, 0.3029)
e^{17}	(0.0875, 0.3921, 0.3318, 0.1887)	(0.3406, 0.2748, 0.0799, 0.3047)
e^{18}	(0.0816, 0.3995, 0.3347, 0.1842)	(0.3447, 0.2746, 0.0743, 0.3064)
e^{19}	(0.0761, 0.4068, 0.3375, 0.1796)	(0.3487, 0.2744, 0.0690, 0.3079)
e^{20}	(0.0709, 0.4139, 0.3401, 0.1751)	(0.3526, 0.2740, 0.0640, 0.3093)

As can be seen from Tables 10.14 and 10.15, as the value of β increases, on the one hand, the numerical difference between the components of the hierarchical weight vector and the combined weight vector becomes larger and larger. In layman's terms, the larger values become larger, and the smaller values become smaller, that is, the "superiority and inferiority" become more obvious; on the other hand, the value of β has a certain impact on the ranking of candidate providers, but the impact is small. This impact is mainly reflected in the ranking of s_9 and s_{10}. As can be seen from Table 10.16, the data of s_9 and s_{10} are very close, and the reliability and corresponding truthfulness of s_{10}, but the time of s_{10} is better than s_9, so when the value of β changes causing the weight of the reliability attribute to increase, s_9 is better than s_{10}; similarly, when the value of β changes causing the weight of the time attribute to increase, s_{10} is better than s_9. It can be seen that the change in the value of parameter β will not affect the ranking of the better and worse providers among the candidate providers, but it will affect the ranking of the intermediate providers, and its utility function value will also fluctuate significantly with the change in the value of β, as shown in Fig. 10.12.

Table 10.15 Changes in the combined weight vector and provider ranking as β changes

β	ω^*	Candidate provider ranking
e	(0.2467, 0.2560, 0.2459, 0.2514)	$s_2 \cdot s_3 \cdot s_5 \cdot s_8 \cdot s_{10} \cdot s_9 \cdot s_4 \cdot s_1 \cdot s_7 \cdot s_6$
e^2	(0.2436, 0.2619, 0.2420, 0.2525)	$s_2 \cdot s_3 \cdot s_5 \cdot s_8 \cdot s_{10} \cdot s_9 \cdot s_4 \cdot s_1 \cdot s_7 \cdot s_6$
e^3	(0.2406, 0.2677, 0.2383, 0.2534)	$s_2 \cdot s_3 \cdot s_5 \cdot s_8 \cdot s_{10} \cdot s_9 \cdot s_4 \cdot s_1 \cdot s_7 \cdot s_6$
e^4	(0.2377, 0.2734, 0.2349, 0.2540)	$s_2 \cdot s_3 \cdot s_5 \cdot s_8 \cdot s_{10} \cdot s_9 \cdot s_4 \cdot s_1 \cdot s_7 \cdot s_6$
e^5	(0.2350, 0.2789, 0.2316, 0.2545)	$s_2 \cdot s_3 \cdot s_5 \cdot s_8 \cdot s_{10} \cdot s_4 \cdot s_1 \cdot s_7 \cdot s_6$
e^6	(0.2325, 0.2842, 0.2286, 0.2547)	$s_2 \cdot s_3 \cdot s_5 \cdot s_8 \cdot s_9 \cdot s_{10} \cdot s_4 \cdot s_1 \cdot s_7 \cdot s_6$
e^7	(0.2300, 0.2894, 0.2257, 0.2548)	$s_2 \cdot s_3 \cdot s_5 \cdot s_8 \cdot s_9 \cdot s_{10} \cdot s_4 \cdot s_1 \cdot s_7 \cdot s_6$
e^8	(0.2278, 0.2945, 0.2231, 0.2546)	$s_2 \cdot s_3 \cdot s_5 \cdot s_8 \cdot s_9 \cdot s_{10} \cdot s_4 \cdot s_1 \cdot s_7 \cdot s_6$
e^9	(0.2257, 0.2994, 0.2208, 0.2543)	$s_2 \cdot s_3 \cdot s_5 \cdot s_8 \cdot s_9 \cdot s_{10} \cdot s_4 \cdot s_1 \cdot s_7 \cdot s_6$
e^{10}	(0.2238, 0.3042, 0.2183, 0.2538)	$s_2 \cdot s_3 \cdot s_5 \cdot s_8 \cdot s_9 \cdot s_{10} \cdot s_4 \cdot s_1 \cdot s_7 \cdot s_6$
e^{11}	(0.2220, 0.3088, 0.2161, 0.2532)	$s_2 \cdot s_3 \cdot s_5 \cdot s_8 \cdot s_9 \cdot s_{10} \cdot s_4 \cdot s_1 \cdot s_7 \cdot s_6$
e^{12}	(0.2203, 0.3132, 0.2141, 0.2534)	$s_2 \cdot s_3 \cdot s_5 \cdot s_8 \cdot s_9 \cdot s_{10} \cdot s_4 \cdot s_1 \cdot s_7 \cdot s_6$
e^{13}	(0.2188, 0.3176, 0.2122, 0.2515)	$s_2 \cdot s_3 \cdot s_5 \cdot s_8 \cdot s_{10} \cdot s_9 \cdot s_4 \cdot s_1 \cdot s_7 \cdot s_6$
e^{14}	(0.2174, 0.3217, 0.2104, 0.2504)	$s_2 \cdot s_3 \cdot s_5 \cdot s_8 \cdot s_{10} \cdot s_9 \cdot s_4 \cdot s_1 \cdot s_7 \cdot s_6$
e^{15}	(0.2161, 0.3258, 0.2088, 0.2493)	$s_2 \cdot s_3 \cdot s_5 \cdot s_8 \cdot s_{10} \cdot s_9 \cdot s_4 \cdot s_1 \cdot s_7 \cdot s_6$
e^{16}	(0.2150, 0.3297, 0.2073, 0.2480)	$s_2 \cdot s_3 \cdot s_5 \cdot s_8 \cdot s_{10} \cdot s_9 \cdot s_4 \cdot s_1 \cdot s_7 \cdot s_6$
e^{17}	(0.2140, 0.3334, 0.2058, 0.2467)	$s_2 \cdot s_3 \cdot s_5 \cdot s_8 \cdot s_{10} \cdot s_9 \cdot s_4 \cdot s_1 \cdot s_7 \cdot s_6$
e^{18}	(0.2132, 0.3371, 0.2045, 0.2453)	$s_2 \cdot s_3 \cdot s_5 \cdot s_8 \cdot s_{10} \cdot s_9 \cdot s_4 \cdot s_1 \cdot s_7 \cdot s_6$
e^{19}	(0.2124, 0.3406, 0.2032, 0.2438)	$s_2 \cdot s_3 \cdot s_5 \cdot s_8 \cdot s_{10} \cdot s_9 \cdot s_4 \cdot s_1 \cdot s_7 \cdot s_6$
e^{20}	(0.2118, 0.3439, 0.2021, 0.2422)	$s_2 \cdot s_3 \cdot s_5 \cdot s_8 \cdot s_{10} \cdot s_9 \cdot s_4 \cdot s_1 \cdot s_7 \cdot s_6$

10.4 Service Spare Parts Forecasting and Management Model Based on Improved Stochastic Programming

In the MRO service chain, scientific and reasonable spare parts management (production, storage and sales supply) is a necessary prerequisite for the efficient operation of the service chain. Its management goal is to effectively coordinate various activities in the spare parts supply process, while ensuring the availability of spare parts, minimize the downtime of faulty equipment and the overall cost. Using limited resources to achieve the maximum degree of demand response is the common demand of all members in the service chain.

The related research results of service spare parts management can be roughly divided into two categories: the first is the demand forecasting and inventory control of spare parts; the second is the production and supply plan of spare parts. From the existing research results, it can be seen that because the decision-making goal of spare parts management is to maximize the satisfaction level while minimizing costs, mathematical programming models are the most commonly used optimization methods. Since they are mostly aimed at the management optimization problems of general service spare parts, these programming models are deterministic, that is, their

Table 10.16 Changes in the utility function values of candidate providers with the value of β

β	Utility function values of candidate providers									
	s_1	s_2	s_3	s_4	s_5	s_6	s_7	s_8	s_9	s_{10}
e	0.46	0.72	0.68	0.52	0.62	0.21	0.45	0.57	0.54	0.55
e^2	0.46	0.73	0.67	0.52	0.62	0.21	0.44	0.58	0.54	0.55
e^3	0.45	0.73	0.67	0.52	0.62	0.21	0.44	0.58	0.55	0.56
e^4	0.45	0.73	0.67	0.53	0.62	0.21	0.43	0.58	0.55	0.56
e^5	0.45	0.73	0.67	0.53	0.62	0.21	0.43	0.58	0.56	0.55
e^6	0.45	0.74	0.67	0.53	0.62	0.21	0.42	0.59	0.56	0.55
e^7	0.45	0.74	0.67	0.54	0.62	0.21	0.42	0.59	0.56	0.55
e^8	0.45	0.74	0.67	0.54	0.62	0.21	0.42	0.59	0.56	0.55
e^9	0.45	0.74	0.67	0.54	0.62	0.21	0.41	0.59	0.57	0.56
e^{10}	0.45	0.74	0.67	0.55	0.62	0.21	0.41	0.59	0.57	0.56
e^{11}	0.45	0.74	0.67	0.55	0.62	0.21	0.41	0.60	0.57	0.56
e^{12}	0.44	0.74	0.67	0.55	0.62	0.21	0.40	0.60	0.57	0.56
e^{13}	0.44	0.75	0.67	0.55	0.62	0.21	0.40	0.60	0.57	0.58
e^{14}	0.44	0.75	0.67	0.56	0.62	0.21	0.40	0.60	0.57	0.58
e^{15}	0.44	0.75	0.67	0.56	0.62	0.21	0.40	0.60	0.57	0.58
e^{16}	0.44	0.75	0.67	0.56	0.62	0.21	0.39	0.60	0.58	0.59
e^{17}	0.44	0.75	0.67	0.56	0.62	0.21	0.39	0.60	0.58	0.59
e^{18}	0.44	0.75	0.67	0.56	0.62	0.21	0.39	0.61	0.58	0.59
e^{19}	0.44	0.75	0.67	0.56	0.62	0.22	0.39	0.61	0.58	0.59
e^{20}	0.44	0.75	0.67	0.57	0.62	0.22	0.38	0.61	0.58	0.59

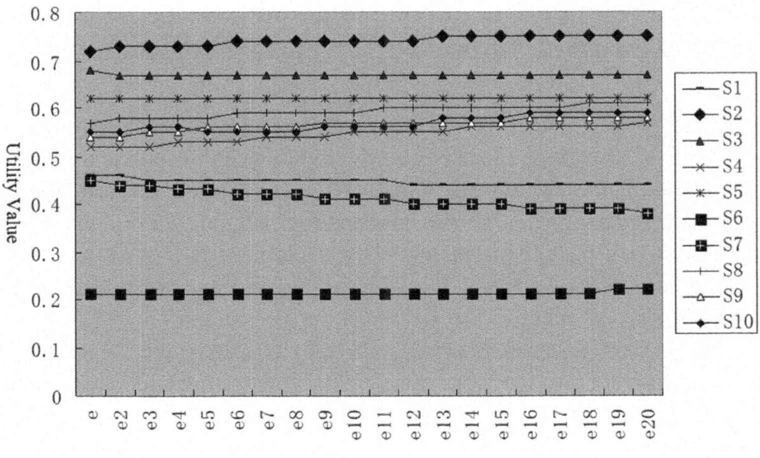

Fig. 10.12 The change rule of the utility function value of the candidate provider with the value of β

input parameters and constraint parameters are all deterministic variables. However, MRO spare parts are different from ordinary spare parts. Their working environment is extremely harsh (such as high temperature, high pressure, high load), which will inevitably lead to their special and uncertain production and storage [10]. Therefore, in the optimization management problem of MRO service spare parts, this speciality and uncertainty must be considered.

For this reason, based on the industrial characteristics of MRO service spare parts, an improved stochastic programming model will be given in this section, which integrates the main activities of spare parts management (production, storage and supply) into a unified modeling framework, while fully considering the randomness and uncertainty in the process of spare parts production and storage, and quantifying them in the mathematical model. The model combines stochastic programming and multi-choice parameter programming. In order to depict the randomness in the production process of spare parts, the production time capability is quantified as a "known" probability distribution random variable; in addition, in order to depict the uncertainty in the storage process of spare parts, the storage cost is quantified as a multi-choice parameter variable, that is, the constraint parameter is set as a set of numbers, not a fixed number, and the elements of this set can be adjusted by the decision maker according to actual needs. For the improved stochastic programming model, since it cannot be directly solved by conventional mathematical software, it is first transformed into an equivalent nonlinear mixed integer programming model through the Lagrange interpolation polynomial method, and then the optimal spare parts management plan is obtained by using LINGO software.

10.4.1 Spare Parts Management Model Based on Improved Stochastic Programming

For the optimization problem of MRO spare parts management, a novel stochastic programming model containing multi-choice parameters is established. First, the model description and model assumptions are given, and on this basis, the objective function and constraint conditions of the model are studied and discussed, and then the corresponding mathematical model is given.

(1) Model description and model assumptions

Consider the multi-product, multi-level, multi-period spare parts management in the MRO service chain, Fig. 10.13 depicts its operating framework.

Users collect and manage equipment status data, estimate their spare parts demand using statistical analysis tools, and demand arrives at the distribution center and spare parts manufacturers respectively. Products are supplied from manufacturers to distribution centers, and then from distribution centers to spare parts users. The randomness and uncertainty in spare parts production and storage are considered, and they are quantified as random variables and multi-choice parameter variables

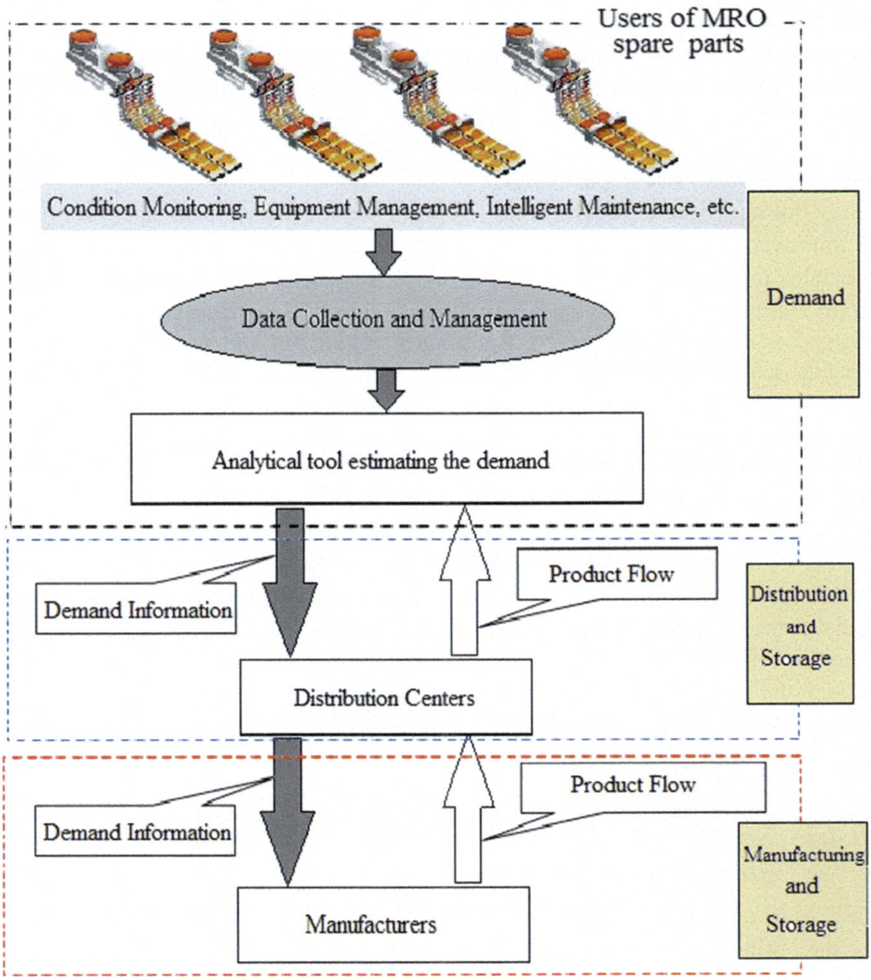

Fig. 10.13 MRO spare parts management operating framework

in the corresponding mathematical programming model. The optimization goal of the model is to minimize the total cost of spare parts management, and the decision variables include production, storage, and product flow. It is worth mentioning that the constraint parameter in the model is a multi-choice variable, and the variable set available for selection can be adjusted according to actual needs, which is an improvement on the existing stochastic programming model. In addition, because the model quantifies the uncertainty in the process of spare parts production and storage at the same time, its decision results are more scientific and in line with reality.

For the need to establish a model, the following model assumptions are given.

Assumption 10.10 Spare parts manufacturers and distribution centers can both store spare parts.

Assumption 10.11 Due to the random factors in spare parts production, the production time capability is quantified as a "known" probability distribution random variable.

Assumption 10.12 In order to deal with the uncertainty in the storage process of spare parts, the storage cost of the manufacturer is quantified as a multi-choice parameter variable.

Table 10.17 lists the symbols and explanations used in the model.

(2) Stochastic planning model with multiple choice parameters

The decision goal is to minimize the total cost, which includes production costs (production and assembly of spare parts), storage costs, logistics costs, etc. In addition, the difference between the actual production time and the production time capacity is also considered in the objective function. There are two situations for this difference: the first is when the former is greater than the latter, and the second is the opposite. In practice, both situations will hinder the maximization of production efficiency. Therefore, it is hoped that the expected value of the difference between the actual production time and the production time capacity will be minimized, that is, $E\left(\sum_{t=1}^{T} p_t \left| C_t^{prod\ time} - \sum_{i=1}^{I} (T_i^{setup} \alpha_{it} + T_i^{prod} \beta_{it}) \right|\right)$ reaches the minimum, where p_t is the penalty cost for the difference between $C_t^{prod\ time}$ and $\sum_{i=1}^{I} \left(T_i^{setup} \alpha_{it} + T_i^{prod} \beta_{it}\right)$. Based on the above analysis, the minimization objective function is:

$$\sum_{i=1}^{I}\sum_{t=1}^{T}\left(C_{it}^{\text{setup}}\alpha_{it} + C_{it}^{prod}\beta_{it}\right) + \sum_{i=1}^{I}\sum_{t=1}^{T}\sum_{j=1}^{J} C_{ij}^{\text{stor}}\Gamma_{ijt} + \sum_{i=1}^{I}\sum_{t=1}^{T}\sum_{n=1}^{N} C_{in}^{\text{stor}}\mu_{int}$$
$$+\sum_{i=1}^{I}\sum_{j=1}^{J}\sum_{n=1}^{N}\sum_{t=1}^{T} C_{jn}^{trans}T_{ijnt}^{1} + \sum_{i=1}^{I}\sum_{n=1}^{N}\sum_{m=1}^{M}\sum_{t=1}^{T} C_{nm}^{trans}T_{inmt}^{2}$$
$$+E\left(\sum_{t=1}^{T} p_t \left| C_t^{\text{prod time}} - \sum_{i=1}^{I}\left(T_i^{\text{setup}}\alpha_{it} + T_i^{\text{prod}}\beta_{it}\right)\right|\right) \tag{10.46}$$

From expression (10.46), it can be seen that due to C_t^{prod}.
The following are the corresponding model constraints.

Constraint 10.1 The upper bound constraint of storage cost for spare parts manufacturer j.

$$\sum_{i=1}^{I} C_{ij}^{stor} \le C_j^{stor} = \{C_j^{stor(1)}, C_j^{stor(2)}, C_j^{stor(3)}, C_j^{stor(4)}\} \tag{10.47}$$

Table 10.17 Symbols and explanations used in the model

Subscript	Explanation
i	MRO parts
j	Parts manufacturer
n	Distribution center
m	User
t	Stage
Decision variables	Description
α_{it}	Whether part i is assembled at stage t, it is a 0–1 variable
β_{it}	The output of part i at stage t
γ_{ijt}	The inventory of part i at parts manufacturer j at stage t
μ_{int}	The inventory of part i at distribution center n at stage t
T^1_{ijnt}	The product flow of part i from parts manufacturer j to distribution center n at stage t
T^2_{inmt}	The product flow of part i from distribution center n to user m at stage t Input variables
C^{setup}_{it}	The assembly cost of part i at stage t
C^{prod}_{it}	The production cost of part i at stage t
C^{stor}_{ij}	The storage cost of part i at parts manufacturer j
T^{prod}_{i}	The production time of part i
T^{setup}_{i}	The setup time of part i
M^{demand}_{imt}	The requirements matrix of user m to part i at stage t
$C^{prod\ time}_{t}$	The production time capacity at stage t, which is a random variable with a probability density of $f\ (C^{prod\ time}_{t})$
C^{trans}_{jn}	The logistics cost from spare parts manufacturer j to distribution center n
C^{trans}_{nm}	The logistics cost from distribution center n to user m
L	A very large number
I	The number of types of spare parts
T	The number of stages
J	The number of manufacturers
N	The number of distribution centers
M	The number of users

Note: The parameter on the right side of Constraint 10.1 C^{stor}_{j} is a set of multi-choice variables, not a fixed number. Setting the constraint parameter in the model to a multi-choice variable is another improvement to the existing linear programming model for spare parts management.

Constraint 10.2 If parts of the spare parts are being produced, the assembly parameter is limited to 1.

$$\beta_{it} - L\alpha_{it} \leq 0 \tag{10.48}$$

Constraint 10.3 The assembly variable is defined as a 0–1 variable.

$$\alpha_{it} \in \{0, 1\} \tag{10.49}$$

Constraint 10.4 The decision variable β_{it} is a non-negative integer.

$$\beta_{it} \geq 0, \beta_{it} \in Z \tag{10.50}$$

Constraint 10.5 The product balance constraint of the spare parts manufacturer.

$$\Gamma_{ijt} = \Gamma_{i,j,t-1} + \beta_{it} - \sum_{n=1}^{N} T_{ijnt}^{1} \tag{10.51}$$

Constraint 10.6 The product balance constraint of the distribution center.

$$\mu_{int} = \mu_{i,n,t-1} + \sum_{j=1}^{J} T_{ijnt}^{1} - \sum_{m=1}^{M} T_{ijmt}^{2} \tag{10.52}$$

Constraint 10.7 The requirements matrix of user m to part i at stage t.

$$M_{imt}^{demand} = \sum_{n=1}^{N} T_{inmt}^{2} \tag{10.53}$$

Constraint 10.8 All of parts will be distributed.

$$\sum_{t=1}^{T} \beta_{it} = \sum_{n=1}^{N} \sum_{m=1}^{M} \sum_{t=1}^{T} T_{inmt}^{2} \tag{10.54}$$

Constraint 10.9 All products supplied to the distribution center will be delivered to users.

$$\sum_{j=1}^{J} \sum_{n=1}^{N} \sum_{t=1}^{T} T_{inmt}^{1} = \sum_{n=1}^{N} \sum_{m=1}^{M} \sum_{t=1}^{T} T_{inmt}^{2} \tag{10.55}$$

Constraint 10.10 The distribution decision variables are all non-negative integers.

$$\Gamma_{ijt}, \mu_{int}, T^1_{ijnt}, T^2_{inmt} \geq 0, \Gamma_{ijt}, \mu_{int}, T^1_{ijnt}, T^2_{inmt} \in Z \tag{10.56}$$

By combining the above objective function and constraints, we can obtain the improved stochastic programming model with multiple choice parameters as follows:

$$\begin{aligned} \min z = & \sum_{i=1}^{I}\sum_{t=1}^{T}\left(C_{it}^{\text{setup}}\alpha_{it} + C_{it}^{\text{prod}}\beta_{it}\right) + \sum_{i=1}^{I}\sum_{t=1}^{T}\sum_{j=1}^{J} C_{ij}^{stor}\gamma_{ijt} + \sum_{i=1}^{I}\sum_{t=1}^{T}\sum_{n=1}^{N} C_{in}^{stor}\mu_{int} \\ & + \sum_{i=1}^{I}\sum_{j=1}^{J}\sum_{n=1}^{N}\sum_{t=1}^{T} C_{jn}^{trans} T^1_{ijnt} + \sum_{i=1}^{I}\sum_{n=1}^{I} C_{it}^{\text{setup}}\alpha_{it} \\ & + E\left(\sum_{t=1}^{T} p_t \left| C_t^{\text{prod time}} - \sum_{i=1}^{I}\left(T_i^{\text{setup}}\alpha_{it} + T_i^{\text{prod}}\beta_{it}\right)\right|\right) \end{aligned} \tag{10.57}$$

s.t.

$$\sum_{i=1}^{I} C_{ij}^{stor} \leq C_j^{stor} = \left\{C_j^{stor(1)}, C_j^{stor(2)}, \ldots, C_j^{stor(h_j)}\right\} \tag{10.58}$$

$$\beta_{it} - L\alpha_{it} \leq 0 \tag{10.59}$$

$$\alpha_{it} \in \{0, 1\} \tag{10.60}$$

$$\beta_{it} \geq 0 \tag{10.61}$$

$$\beta_{it} \in Z \tag{10.62}$$

$$\Gamma_{ijt} = \Gamma_{i,j,t-1} + \beta_{it} - \sum_{n=1}^{N} T^1_{ijnt} \tag{10.63}$$

$$\mu_{int} = \mu_{i,n,t-1} + \sum_{j=1}^{J} T^1_{ijnt} - \sum_{m=1}^{M} T^2_{inmt} \tag{10.64}$$

$$M_{imt}^{demand} = \sum_{n=1}^{N} T^2_{inmt} \tag{10.65}$$

$$\sum_{t=1}^{T} \beta_{it} = \sum_{n=1}^{N}\sum_{m=1}^{M}\sum_{t=1}^{T} T^2_{inmt} \tag{10.66}$$

$$\sum_{j=1}^{J}\sum_{n=1}^{N}\sum_{t=1} TT^1_{ijnt} = \sum_{n=1}^{N}\sum_{m=1}^{M}\sum_{t=1}^{T} T^2_{inmt} \tag{10.67}$$

$$\Gamma_{jt}, \mu_{int}, T^1_{ijnt}, T^2_{inmt} \geq 0, \forall i; \forall j; \forall n; \forall m; \forall t \tag{10.68}$$

$$\Gamma_{jt}, \mu_{int}, T^1_{ijnt}, T^2_{inmt} \in Z, \forall i; \forall j; \forall n; \forall m; \forall t \tag{10.69}$$

10.4.2 Model Solution

In this section, first analyze the mathematical model established in the previous section and discuss how to solve it. Then, based on model analysis and discussion, provide specific steps for solving the model.

1. **Model analysis**

In the objective function (10.57) of the model, $C_t^{\text{prod time}}$ it is a random variable with a probability distribution; In addition, in formula (10.58), C_j^{stor} is a multi choice parameter variable. Therefore, models (10.57)–(10.69) are stochastic programming models that contain multiple choice parameters and cannot be directly solved using conventional mathematical software. To achieve this, it is necessary to first convert it into an equivalent deterministic model, and then solve it using LINGO Version 11.0 or MAPLE 12.0 to obtain the optimal solution of the model.

In order to obtain a deterministic model equivalent to the proposed stochastic programming model, it is necessary to first derive the analytical expression in the objective function $E = (\sum_{t=1}^{T} p_t \cdot \left| C_t^{\text{prod time}} - \sum_{i=1}^{I} T_i^{\text{setup}} \cdot \alpha_{it} + T_i^{\text{prod}} \cdot \beta_{it} \right|)$;

In addition, it is necessary to handle the multiple selection parameters C_j^{stor} in the constraint conditions. In this process, the properties of expected random variables and the Lagrange interpolation polynomial method are used.

2. **Model solving**

Based on the analysis and discussion of the above model, provide the following specific steps for solving the model.

Derived analytical expression $E = (\sum_{t=1}^{T} p_t \cdot \left| C_t^{\text{prod time}} - \sum_{i=1}^{I} T_i^{\text{setup}} \cdot \alpha_{it} + T_i^{\text{prod}} \cdot \beta_{it} \right|)$.

The assumed $C_t^{\text{prod time}}$ probability density function is $f\left(C_t^{\text{prod time}}\right)$, where $0 \leq C_t^{\text{prod time}} < +\infty$. For simplicity, it will be $\sum_{i=1}^{I} (T_i^{\text{setup}} \cdot \alpha_{it} + T_i^{\text{prod}} \cdot \beta_{it})$ written as g_t, $g_t = \sum_{i=1}^{I} \left(T_i^{\text{setup}} \cdot \alpha_{it} + T_i^{\text{prod}} \cdot \beta_{it}\right)$ Using the properties of expectations, there are

$$E\left(\sum_{t=1}^{T} p_t \cdot \left| C_t^{\text{prod time}} - g_t \right)\right|\right) = \sum_{t=1}^{T} p_t \cdot E\left(\left| C_t^{\text{prod time}} - g_t \right|\right) \tag{10.70}$$

To further derive the analytical expression of (10.25), the following is first derived $E\left(\left| C_t^{\text{prod time}} - g_t \right|\right) E\left(\left| C_t^{\text{prod time}} - g_t \right|\right)$

$$\begin{aligned}
E\left(\left| C_t^{\text{prod time}} - g_t \right|\right) &= \int_0^{+\infty} \left| C_t^{\text{prod time}} - g_t \right| \cdot f\left(C_t^{\text{prod time}}\right) dC_t^{\text{prod time}} \\
&= \int_0^{g_t} \left(g_t - C_t^{\text{prod time}}\right) \cdot f\left(C_t^{\text{prod time}}\right) dC_t^{\text{prod time}} \\
&\quad + \int_{g_t}^{+\infty} \left(C_t^{\text{prod time}} - g_t\right) \cdot f\left(C_t^{\text{prod time}}\right) dC_t^{\text{prod time}} \\
&= \int_0^{+\infty} g_t \cdot f\left(C_t^{\text{prod time}}\right) dC_t^{\text{prod time}} \\
&\quad - \int_0^{+\infty} C_t^{\text{prod time}} \cdot f\left(C_t^{\text{prod time}}\right) dC_t^{\text{prod time}} \\
&\quad + 2\left[\int_{g_t}^{+\infty} C_t^{\text{prod time}} \cdot f\left(C_t^{\text{prod time}}\right) dC_t^{\text{prod time}} - \int_{g_t}^{+\infty} g_t \right. \\
&\quad \left. \cdot f\left(C_t^{\text{prod time}}\right) dC_t^{\text{prod time}}\right] = g_t - E\left(C_t^{\text{prod time}}\right) \\
&\quad + 2\left[\int_{g_t}^{+\infty} C_t^{\text{prod time}} \cdot f\left(C_t^{\text{prod time}}\right) dC_t^{\text{prod time}}\right. \\
&\quad \left. - \int_{g_t}^{+\infty} g_t \cdot f\left(C_t^{\text{prod time}}\right) dC_t^{\text{prod time}}\right]
\end{aligned} \tag{10.71}$$

Therefore, it can be concluded that

$$\begin{aligned}
E\left(\sum_{t=1}^{T} p_t \cdot \left| C_t^{\text{prod time}} - g_t \right)\right|\right) &= \sum_{t=1}^{T} p_t \cdot \left\{ g_t - E\left(C_t^{\text{prod time}}\right) \right. \\
&\quad + 2\left[\int_{g_t}^{+\infty} C_t^{\text{prod time}} \cdot f\left(C_t^{\text{prod time}}\right) dC_t^{\text{prod time}} \right. \\
&\quad \left.\left. - \int_{g_t}^{+\infty} g_t \cdot f\left(C_t^{\text{prod time}}\right) dC_t^{\text{prod time}}\right]\right\}
\end{aligned} \tag{10.72}$$

Transform constraints containing multiple selection parameters.

Next, we will use the Lagrange interpolation polynomial method to handle multiple selection parameters.

By Constraint (10.13), there are

$$\sum_{i=1}^{I} C_{ij}^{\text{stor}} \leq C_j^{\text{stor}} = \left\{ C_j^{\text{stor}(1)}, C_j^{\text{stor}(2)}, \ldots, C_j^{\text{stor}(h_j)} \right\} \tag{10.73}$$

Let $f(t_j)$ be a real valued function and $0, 1, 2, \cdots, h_j - 1$ be h_j nodes. Let $C_j^{\text{stor}(1)}, C_j^{\text{stor}(2)}, \cdots, C_j^{\text{stor}(h_j)}$ be the corresponding function values of interpolation polynomials at different h_j nodes, that is

$$f(t_j) = C_j^{\text{stor}(t_j+1)}, t_j = 0, 1, 2, \ldots, h_j - 1 \tag{10.74}$$

Let $P_{h_j-1}(t_j)$ it be $(h_j - 1)$ interpolation polynomial of degree, and insert it at a given node, that is

$$P_{h_j-1}(v) = C_j^{\text{stor}(v+1)}, v = 0, 1, 2, \ldots, h_j - 1, j = 1, 2, \ldots, J \tag{10.75}$$

Therefore, according to the Lagrange interpolation formula, an interpolation polynomial with the j th multiple choice parameter can be constructed

$$P_{h_j-1}(t_j) = \sum_{r=0}^{h_j-1} C_j^{\text{stor}(r+1)} \cdot \prod_{\substack{k=0 \\ k \neq r}}^{h_j-1} \frac{t_j - k}{r - k}, j = 1, 2, \ldots, J \tag{10.76}$$

So, an equivalent deterministic constraint was obtained

$$\sum_{i=1}^{I} C_{ij}^{\text{stor}} \leq \sum_{r=0}^{h_j-1} C_j^{\text{stor}(r+1)} \cdot \prod_{\substack{k=0 \\ k \neq r}}^{h_j-1} \frac{t_j - k}{r - k}, j = 1, 2, \ldots, J \tag{10.77}$$

Obtain an equivalent deterministic mathematical programming model based on the above two steps.

Substitute expressions (10.72) and (10.77) into models (10.57)–(10.69) to obtain a deterministic model

$$\begin{aligned} z = &\sum\nolimits_{i=1}^{I} \sum\nolimits_{t=1}^{T} \left(C_{it}^{\text{setup}} \cdot \alpha_{it} + C_{it}^{\text{prod}} \cdot \beta_{it} \right) \\ &+ \sum\nolimits_{i=1}^{I} \sum\nolimits_{t=1}^{T} \sum\nolimits_{j=1}^{J} C_{ij}^{\text{stor}} \cdot \Gamma_{ijt} + \sum\nolimits_{i=1}^{I} \sum\nolimits_{t=1}^{T} \sum\nolimits_{n=1}^{n} C_{ij}^{\text{stor}} \cdot \mu_{int} \end{aligned}$$

$$
\begin{aligned}
&+\sum_{i=1}^{I}\sum_{j=1}^{J}\sum_{n=1}^{N}\sum_{t=1}^{T} C_{jn}^{\text{trans}} \cdot T_{ijnt}^{1} + \sum_{i=1}^{I}\sum_{n=1}^{N}\sum_{m=1}^{M}\sum_{t=1}^{T} C_{nm}^{\text{trans}} \cdot T_{inmt}^{2} \\
&+\sum_{t=1}^{T} p_t \cdot \left[g_t - E\left(C_t^{\text{prod time}}\right)\right. \\
&+2\left(\int_{g_t}^{+\infty} C_t^{\text{prod time}} \cdot f\left(C_t^{\text{prod time}}\right) dC_t^{\text{prod time}}\right. \\
&\left.\left.-\int_{g_t}^{+\infty} g_t \cdot f\left(C_t^{\text{prod time}}\right) dC_t^{\text{prod time}}\right)\right]
\end{aligned}
\tag{10.78}
$$

Subject to:

$$
\sum_{i=1}^{I} C_{ij}^{\text{stor}} \le \sum_{r=0}^{h_j-1} C_j^{\text{stor}(r+1)} \cdot \prod_{\substack{k=0 \\ k \ne r}}^{h_j-1} \frac{t_j - k}{r - k}, \forall j \tag{10.79}
$$

$$
\beta_{it} - L \cdot \alpha_{it} \le 0 \tag{10.80}
$$

$$
\alpha_{it} \in \{0, 1\} \tag{10.81}
$$

$$
\beta_{it} \ge 0 \tag{10.82}
$$

$$
\beta_{it} \in \mathbb{Z} \tag{10.83}
$$

$$
\Gamma_{ijt} = \Gamma_{i,j,t-1} + \beta_{it} - \sum_{n=1}^{N} T_{ijnt}^{1} \tag{10.84}
$$

$$
\mu_{int} = \mu_{i,n,t-1} + \sum_{j=1}^{J} T_{ijnt}^{1} - \sum_{m=1}^{M} T_{inmt}^{2} \tag{10.85}
$$

$$
M_{imt}^{\text{demand}} = \sum_{n=1}^{N} T_{inmt}^{2} \tag{10.86}
$$

$$
\sum_{t=1}^{T} \beta_{it} = \sum_{n=1}^{N}\sum_{m=1}^{M}\sum_{t=1}^{T} T_{inmt}^{2} \tag{10.87}
$$

$$\sum_{j=1}^{J}\sum_{n=1}^{N}\sum_{t=1}^{T} T^{1}_{ijnt} = \sum_{n=1}^{N}\sum_{m=1}^{M}\sum_{t=1}^{T} T^{2}_{inmt} \tag{10.88}$$

$$\Gamma_{ijt}, \mu_{int}, T^{1}_{ijnt}, T^{2}_{inmt} \geq 0, \forall i; \forall j; \forall n; \forall m; \forall t \tag{10.89}$$

$$\Gamma_{ijt}, \mu_{int}, T^{1}_{ijnt}, T^{2}_{inmt} \in \mathbb{Z}, \forall i; \forall j; \forall n; \forall m; \forall t \tag{10.90}$$

Among $g_t = \sum_{i=1}^{I}\left(T_i^{\text{setup}} \cdot \alpha_{it} + T_i^{\text{prod}} \cdot \beta_{it},\right)\forall t$

Using LINGO Version 11.0 or MAPLE12.0 to solve the above nonlinear mixed integer programming models (10.78)–(10.90).

10.4.3 Case Analysis

In this section, a numerical example will be used to verify the feasibility and effectiveness of the proposed model. Firstly, provide relevant parameter settings; Then, based on the proposed model and solution method, provide decision results and conduct sensitivity analysis of parameters; Finally, the effectiveness of the improvement on traditional planning models was demonstrated through comparative experiments.

1. **Parameter settings**

Assuming two MRO spare parts manufacturers ($J = 2$) produce two different types of spare parts, spare parts 1and spare parts 2($I = 2$), the produced spare parts can be stored in the manufacturer and distribution center. Furthermore, assuming there are three distribution centers ($N = 3$) and four users ($M = 4$), consider six months of spare parts management, with each month corresponding to a stage ($T = 6$), and the user's spare parts demand information is shown in Table 10.18.

At different stages, the production costs of spare part 1 are \$150, \$120, \$110, \$130, \$120, and \$140, respectively. The assembly costs are \$40, \$50, \$50, \$70, \$60, and \$70 respectively. Correspondingly, the production and assembly costs for spare part 2 at different stages are \$80, \$60, \$70, \$60, \$50, \$80, as well as \$50, \$40, \$50, \$60, \$40, \$50, respectively. The production time for spare parts 1 and 2 is 8 h and 6 h respectively, and the assembly time for both types of spare parts is 6 h.

The storage cost of two types of spare parts in the manufacturer is \$80, and the upper bound set of adjustable storage costs for manufacturer 1 and manufacturer 2 (multiple choice parameter variables) is {\$600, 800} and {\$800, \$1000}, respectively. The storage costs for spare part 1 in the three distribution centers are \$80, \$60, and \$70, respectively, while the storage costs for spare part 2 are \$90, \$70, and \$80, respectively.

Without loss of generality, assuming that the production time t capacity $C_t^{\text{prod time}}$ at a stage follows a truncated normal distribution (decision-makers

Table 10.18 User spare parts demand information

	First month	Second month	Third month	Fourth month	Fifth month	Sixth month
User1/Spare Part 1	38	27	2	174	22	78
User2/Spare Part 1	11	23	116	184	168	17
User3/Spare Part 1	192	120	90	147	118	7
User4/Spare Part 1	192	17	164	71	114	179
User1/Spare Part 2	4	67	153	19	10	68
User2/Spare Parts 2	72	97	166	178	76	125
User3/Spare Parts 2	29	88	179	20	85	26
User4/Spare Parts 2	26	90	130	70	86	155

can choose other random distributions as needed), the probability density function of $C_t^{\text{prod time}}$ is

$$f\left(C_t^{\text{prod time}}\right) = \frac{1}{K_t \sigma_t \sqrt{2\pi}} \exp\left[-\frac{1}{2}\left(\frac{C_t^{\text{prod time}} - \mu_t}{\sigma_t}\right)^2\right], 0 \le C_t^{\text{prod time}} < +\infty \tag{10.91}$$

where σ_t and μ_t are distribution parameters, K_t are normalization constants $K_t > 0$, and satisfy $\int_0^{+\infty} f(C_t^{\text{prod time}}) dC_t^{\text{prod time}} = 1$. It is worth noting that the setting of probability distribution will have a certain impact on the numerical results of the example, but it will not affect the validation of the model, as the proposed model is independent of the probability distribution setting of the variable.

According to $\int_0^{+\infty} f(C_t^{\text{prod time}}) dC_t^{\text{prod time}} = 1$, it can be concluded that

$$\begin{aligned} K_t &= \frac{1}{\sigma_t \sqrt{2\pi}} \int_0^{+\infty} \exp\left[-\frac{1}{2}\left(\frac{C_t^{\text{prod time}} - \mu_t}{\sigma_t}\right)^2\right] dC_t^{\text{prod time}} \\ &= 1 - \Phi\left(\frac{-\mu_t}{\sigma_t}\right) = \Phi\left(\frac{\mu_t}{\sigma_t}\right) \end{aligned} \tag{10.92}$$

Among $\Phi(z)$ is the distribution function of the standard normal distribution. Therefore, based on formulas (10.72) and (10.92), it can be calculated that

$$E\left(\sum_{t=1}^{T} p_t \cdot \left| C_t^{\text{prod time}} - g_t \right|\right) = \sum_{t=1}^{T} p_t \cdot \left\{ g_t \cdot \left\{ 1 - \frac{1 - \sqrt{1 - \exp\left[-\left(\frac{g_t - \mu_t}{\sigma_t}\right)^2\right]}}{\sqrt{2} K_t} \right\} \right.$$

$$+ \frac{\sigma_t}{\sqrt{2\pi} K_t} \cdot \left\{ 2\exp\left[-\frac{1}{2}\left(\frac{g_t - \mu_t}{\sigma_t}\right)^2\right] - \exp\left[-\frac{1}{2}\left(\frac{\mu_t}{\sigma_t}\right)\right]^2 \right\}$$

$$\left. + \frac{\mu_t}{2\sqrt{2} K_t} \cdot \left\{ 1 + \sqrt{1 - \exp\left[-\left(\frac{\mu_t}{\sigma_t}\right)^2\right]} - 2\sqrt{1 - \exp\left[-\left(\frac{g_t - \mu_t}{\sigma_t}\right)^2\right]} \right\} \right\} \quad (10.93)$$

Assuming $\mu_t = 27.86$ and $\sigma_t = 1.15$ (decision-makers can p_t adjust their values as needed), then $K_t = \Phi\left(\frac{\mu_t}{\sigma_t}\right) = 1$. Further assume that the penalty cost is 1, which is the simple unit resource cost.

The logistics cost from each manufacturer to each distribution center is \$60, and the logistics cost from the distribution center to the user is listed in Table 10.19.

2. **Result and parameter analysis**

Since the equivalent deterministic models (10.78)–(10.90) are nonlinear mixed integer programming models, they can be directly solved using LINGO Version 11.0 or MAPLE 12.0. Although different parameter settings can affect the complexity of the model, they do not change the type and properties of the model, so the optimal solution of the model can also be obtained in a short period of time. In view of this, the parameter analysis experiment below did not consider the solution time of the model.

Here, using LINGO Version 11.0 to solve the model in the example, the optimal solution is \$881,635. The non-zero decision variables are listed in Table 10.20. Therefore, the optimal strategy for spare parts management (production, assembly, storage, etc.) is obtained through this model. Taking the values of decision variables α_{25}, β_{25}, Γ_{2j5}, μ_{2n5}, T^1_{2jn5} and T^2_{2nm5} as an example, we will specifically explain how to obtain the optimal management strategy.

From Table 10.20, it can be seen that $\alpha_{25} = 1$ and $\beta_{25} = 267$, indicating that in the fifth stage, spare part 2should be assembled and the production volume is 267. $\Gamma_{215} = 10$ and $= \Gamma_{225}1665$ indicate that in the fifth stage, the storage capacity

Table 10.19 Logistics cost information from distribution center to users

	User 1(\$)	User 2(\$)	User 3(\$)	User 4(\$)
Distribution center 1	30	50	70	45
Distribution center 2	60	35	48	55
Distribution center 3	65	52	46	40

Table 10.20 Non zero decision variable values for truncated normal distribution examples

α_{11}	α_{12}	α_{13}	α_{14}	α_{15}	α_{16}	α_{21}	α_{22}	α_{23}	α_{24}	α_{25}	α_{26}	β_{11}	β_{12}	β_{13}	β_{14}
1	1	1	1	1	1	1	1	1	1	1	1	433	187	372	576
β_{15}	β_{16}	β_{21}	β_{22}	β_{23}	β_{24}	β_{25}	β_{26}	Γ_{215}	Γ_{121}	Γ_{221}	Γ_{122}	Γ_{222}	Γ_{123}	Γ_{223}	Γ_{124}
422	281	131	342	628	287	267	364	10	433	131	620	473	992	1101	1568
Γ_{224}	Γ_{125}	Γ_{225}	Γ_{126}	Γ_{226}	T^1_{1111}	T^1_{1121}	T^1_{1131}	T^1_{2111}	T^1_{2121}	T^1_{2131}	T^1_{1112}	T^1_{1122}	T^1_{1132}	T^1_{2112}	T^1_{2122}
1388	1990	1655	2271	2019	38	11	384	4	72	55	27	23	137	67	97
T^1_{2132}	T^1_{1113}	T^1_{1123}	T^1_{1133}	T^1_{2113}	T^1_{2123}	T^1_{2133}	T^1_{1114}	T^1_{1124}	T^1_{1134}	T^1_{2114}	T^1_{2124}	T^1_{2134}	T^1_{2134}	T^1_{1125}	T^1_{1135}
178	2	116	254	153	166	309	174	184	218	19	178	90	22	168	232
T^1_{2115}	T^1_{2125}	T^1_{2135}	T^1_{1116}	T^1_{1126}	T^1_{1136}	T^1_{2116}	T^1_{2126}	T^1_{2136}	T^2_{1111}	T^2_{1221}	T^2_{1331}	T^2_{1341}	T^2_{1112}	T^2_{1222}	T^2_{1332}
10	76	171	78	17	186	68	125	181	38	11	192	192	27	23	120
T^2_{1342}	T^2_{1342}	T^2_{1223}	T^2_{1333}	T^2_{1343}	T^2_{1114}	T^2_{1224}	T^2_{1334}	T^2_{1344}	T^2_{1115}	T^2_{1225}	T^2_{1335}	T^2_{1345}	T^2_{1116}	T^2_{1226}	T^2_{1336}
17	2	116	90	164	174	184	147	71	22	168	118	114	78	17	7
T^2_{1346}	T^2_{2111}	T^2_{2221}	T^2_{2331}	T^2_{2341}	T^2_{2112}	T^2_{2222}	T^2_{2222}	T^2_{2342}	T^2_{2113}	T^2_{2223}	T^2_{2333}	T^2_{2343}	T^2_{2114}	T^2_{2224}	T^2_{2334}
179	4	72	29	26	67	97	88	90	153	166	179	130	19	178	20
T^2_{2344}	T^2_{2115}	T^2_{2225}	T^2_{2335}	T^2_{2345}	T^2_{2116}	T^2_{2226}	T^2_{2336}	T^2_{2346}	t_1						
70	10	76	85	86	68	125	26	155	1						

of spare part2in manufacturers 1 and 2 is 10 and 1665, respectively $T^1_{2115} = 10$, $T^1_{2125} = 76$, and $T^1_{2135} = 171$ indicate that in the fifth stage, the product flows of spare parts 2 from manufacturer 1 to distribution centers 1, 2, and 3 are 10, 76, and 171, respectively $T^2_{2225} = 76$, $T^2_{2335} = 85$, and $T^2_{2345} = 86$ indicate that in the fifth stage, the product flow of spare parts 2 from distribution center 2 to user 2 is 76, and the product flow of spare parts 2 from distribution center 3 to user 3 and user 4 is 85 and 86, respectively. For the zero decision variables not listed in Table 10.20, a value of zero indicates that in the corresponding stage, the corresponding spare parts do not need to be assembled, and there is no production, inventory, or product flow. In this way, the optimal spare parts management strategy can be obtained based on the values of all decision variables.

It is worth mentioning that decision-makers can generate different random constraints and expand the set of multiple selection parameters based on actual situations. In the case analysis, there are three parameters related to random variables, namely p_t, μ_t and σ_t. In order to analyze the impact of these parameters on the optimal value of the objective function, the following parameter analysis experiments will be conducted.

1. Set p_t to different values

The recalculated optimal values are listed in Table 10.21, and it can be seen that the optimal value of the objective function increases with increasing p_t. Figure 10.14 more intuitively depicts this trend and provides an approximate analytical expression for the trend curve.

2. Set to $C_t^{\text{prod time}}$ follow different probability distributions, that is, have different probability density function

Table 10.21 Trend of optimal values with changes p_t

p_t	1	2	3	4	5	6	7	8	9	10
Optimal value (ten thousand)	88.1635	91.1989	94.2343	97.2697	100.305	103.341	106.376	109.411	112.4467	115.4821

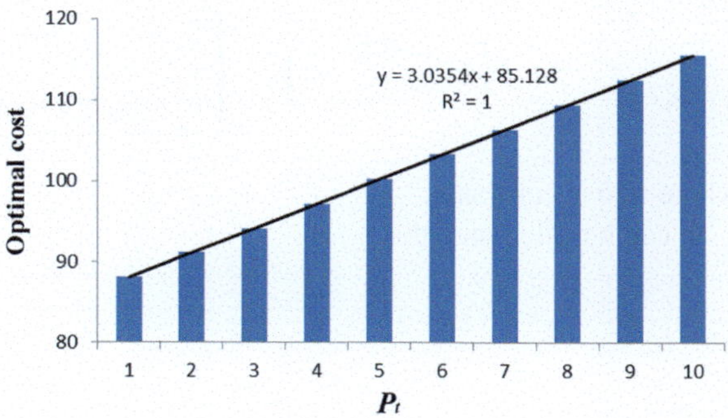

Fig. 10.14 T p_t rend of optimal value with parameter variation

Exponential distribution

Assuming the probability density function of a random variable $C_t^{\text{prod time}}$ is

$$f\left(C_t^{\text{prod time}}\right) = \begin{cases} \lambda_t \cdot \exp\left(-\lambda_t C_t^{\text{prod time}}\right) & C_t^{\text{prod time}} \geq 0 \\ 0 & C_t^{\text{prod time}} < 0 \end{cases} \tag{10.94}$$

Among them λ_t is the distribution parameter, also $\lambda_t > 0$ known as the rate parameter, and.

Similar to the previous derivation, it can be calculated $E\left(\sum_{t=1}^{T} p_t \cdot \left|C_t^{\text{prod time}} - g_t\right|\right)$ as follows

$$\begin{aligned} E\left(\sum_{t=1}^{T} p_t \cdot \left|C_t^{\text{prod time}} - g_t\right|\right) &= \sum_{t=1}^{T} p_t \cdot E\left(\left|C_t^{\text{prod time}} - g_t\right|\right) \\ &= \sum_{t=1}^{T} p_t \cdot \left[\left(\frac{2}{\lambda_t}\right) e^{-\lambda_t g_t} + g_t - \frac{1}{\lambda_t}\right] \end{aligned} \tag{10.95}$$

The optimal value of the objective function recalculated under exponential distribution is \$881,037, with penalty cost $p_t = 1$ and rate parameter $\lambda_t = 0.01$. Similarly, analyze the impact of these two parameters on the optimal value and present the results in Table 10.22.

As shown in Table 10.22, the optimal value increases with the increase of p_t, but decreases with the increase of λ_t. In addition, the rate at which the optimal value p_t increases is greater than the rate at which λ_t decreases, indicating that the influence of parameters p_t on the model's optimal value is greater than that of parameters λ_t. The Figs. 10.15 and 10.16 below illustrate this more intuitively.

Uniform distribution

Assuming the probability density function of a random variable $C_t^{\text{prod time}}$ is

$$f\left(C_t^{\text{prod time}}\right) = \begin{cases} \frac{1}{b-a} & a \leq C_t^{\text{prod time}} \leq b \\ 0 & \text{otherwise} \end{cases} \tag{10.96}$$

where a, b is constants.

It can be calculated that

Table 10.22 Trend of optimal values with p_t and λ_t

	p_t									
λ_t	1	2	3	4	5	6	7	8	9	10
0.01	88.1037	91.1391	94.1745	97.2099	100.2453	103.2807	106.3161	109.3515	112.3869	115.4243
0.02	88.0439	91.0792	94.1146	97.1503	100.1857	103.2208	106.2563	109.2917	112.3272	115.3626
0.03	87.9841	91.0194	94.0548	97.0905	100.1259	103.1610	106.1964	109.2318	112.2674	115.3028
0.04	87.9243	90.9597	93.9951	97.0306	100.0660	103.1013	106.1367	109.1721	112.2075	115.2429
0.05	87.8644	90.8999	93.9353	96.9709	100.0061	103.0415	106.0769	109.1123	112.1479	115.1833
0.06	87.8045	90.8401	93.8754	96.9111	99.9465	102.9819	106.0170	109.0524	112.0882	115.1236
0.07	87.7488	90.7803	93.8157	96.8512	99.8866	102.9221	105.9574	108.9928	112.0284	115.0638
0.08	87.6850	90.7205	93.7559	97.7914	99.8268	102.8622	105.8976	108.9330	111.9687	115.0041
0.09	87.6251	90.6609	93.6963	96.7317	99.7671	102.8025	105.8379	108.8733	111.9091	114.9495
0.1	87.5652	90.6011	93.6365	96.6718	99.7072	102.7426	105.7780	108.8134	111.8493	114.8847

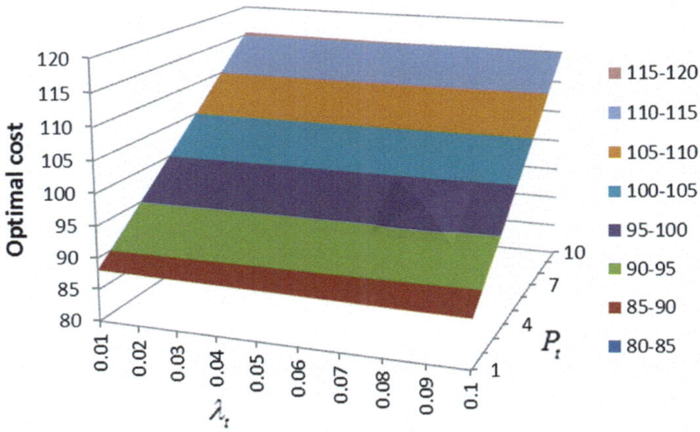

Fig. 10.15 Trend of optimal value with parameter p_t and λ_t (3D)

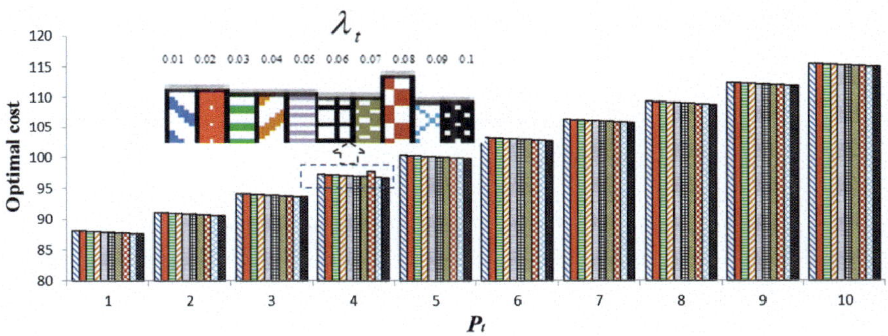

Fig. 10.16 Trend of optimal value with parameter p_t and λ_t (two-dimensional)

$$\sum_{t=1}^{T} E\left(p_t \cdot \left|C_t^{\text{prod time}} - g_t\right|\right) = \sum_{t=1}^{T} \frac{p_t}{b-a} \cdot \left[\frac{1}{2}\left(a^2+b^2\right) - g_t \cdot (a+b) + g_t^2\right] \tag{10.97}$$

The optimal value of the recalculated objective function is \$12,754,960, with penalty cost $p_t = 1$ and two distribution parameters $a = 26$ and (10.1m_{it}09), respectively. Table 10.23 lists the non-zero decision variable values under uniform distribution.

3. **Comparative experiment**

As mentioned earlier, compared with existing mathematical programming models for spare parts management, the stochastic programming model with multiple selection parameters proposed in this chapter mainly has two improvements. Firstly, in response to the uncertainty of MRO spare parts production, the production time

Table 10.23 Non zero decision variable values for uniformly distributed examples

α_{11}	α_{14}	α_{15}	α_{21}	α_{22}	α_{23}	α_{26}	β_{11}	β_{14}	β_{15}	β_{21}	β_{22}	β_{23}	β_{26}	Γ_{216}	Γ_{121}
1	1	1	1	1	1	1	992	576	703	121	342	1172	384	10	992
Γ_{221}	Γ_{122}	Γ_{222}	Γ_{123}	Γ_{223}	Γ_{124}	Γ_{224}	Γ_{125}	Γ_{225}	Γ_{126}	Γ_{226}	μ_{111}	μ_{121}	μ_{131}	μ_{122}	μ_{213}
111	992	453	992	1625	1568	1625	2271	1625	2271	2009	27	515	17	372	19
μ_{223}	μ_{233}	μ_{224}	μ_{115}	μ_{125}	μ_{135}	T^1_{1111}	T^1_{1121}	T^1_{1131}	T^1_{2111}	T^1_{2121}	T^1_{2131}	T^1_{2231}	T^1_{2112}	T^1_{2122}	T^1_{2132}
455	70	257	78	24	179	65	526	401	4	72	45	10	67	97	178
T^1_{2113}	T^1_{2123}	T^1_{2133}	T^1_{1132}	T^1_{1114}	T^1_{1124}	T^1_{1134}	T^1_{1115}	T^1_{1125}	T^1_{1135}	T^1_{2116}	T^1_{2126}	T^1_{2136}	T^2_{1111}	T^2_{1221}	T^2_{1331}
172	621	379	137	174	184	218	100	192	411	68	125	181	38	11	192
T^2_{1341}	T^2_{1112}	T^2_{1222}	T^2_{1332}	T^2_{1342}	T^2_{1213}	T^2_{1223}	T^2_{1233}	T^2_{1243}	T^2_{1114}	T^2_{1224}	T^2_{1334}	T^2_{1344}	T^2_{1115}	T^2_{1225}	T^2_{1335}
192	27	23	120	17	2	116	90	164	174	184	147	71	22	168	118
T^2_{1345}	T^2_{1116}	T^2_{1226}	T^2_{1236}	T^2_{1346}	T^2_{2111}	T^2_{2221}	T^2_{2331}	T^2_{2341}	T^2_{2112}	T^2_{2222}	T^2_{2332}	T^2_{2342}	T^2_{2113}	T^2_{2223}	T^2_{2333}
114	78	17	7	179	4	72	29	26	67	97	88	90	153	166	179
T^2_{2343}	T^2_{2114}	T^2_{2224}	T^2_{2334}	T^2_{2344}	T^2_{2215}	T^2_{2225}	T^2_{2235}	T^2_{2245}	T^2_{2116}	T^2_{2226}	T^2_{2336}	T^2_{2346}	t_1		
130	19	178	20	70	10	76	85	86	68	125	26	155	1		

capacity is modeled as a probability distribution "known" random variable; The second is to model the manufacturer's storage cost as a multi choice parameter variable in response to the uncertainty of MRO spare parts storage. To demonstrate the effectiveness of the above two improvements, the following model comparison experiments were conducted.

1. Set production time capacity as a regular variable and a random variable, respectively

If the production time capability in the objective function (10.46) is a common variable $C{*}_t^{\text{prod time}}$, then a new model without random variables in the objective function can be written as

$$
\begin{aligned}
\text{Min}: z = & \sum\nolimits_{i=1}^{I}\sum\nolimits_{t=1}^{T}\left(C_{it}^{\text{setup}}\cdot\alpha_{it}+C_{it}^{\text{prod}}\cdot\beta_{it}\right)+\sum\nolimits_{i=1}^{I}\sum\nolimits_{t=1}^{T}\sum\nolimits_{j=1}^{J}C_{ij}^{\text{stor}}\cdot\Gamma_{ijt}+\\
&\sum\nolimits_{i=1}^{I}\sum\nolimits_{t=1}^{T}\sum\nolimits_{j=1}^{J}C_{in}^{\text{stor}}\cdot\mu_{int}\\
&+\sum_{i=1}^{I}\sum_{j=1}^{J}\sum_{n=1}^{N}\sum_{t=1}^{T}C_{jn}^{\text{trans}}\cdot T_{ijnt}^{1}+\sum_{i=1}^{I}\sum_{n=1}^{N}\sum_{m=1}^{M}\sum_{t=1}^{T}C_{nm}^{\text{trans}}\cdot T_{inmt}^{2}\\
&+\sum_{t=1}^{T}p_t\cdot\left|C{*}_t^{\text{prod time}}-\sum_{i=1}^{I}\left(T_i^{\text{setup}}\cdot\alpha_{it}+T_i^{\text{prod}}\cdot\beta_{it}\right)\right|
\end{aligned}
\tag{10.98}
$$

Subject to (10.79)–(10.90).

For simplicity, the proposed model with random variables is denoted as Ω_o, and the new model without random variables is denoted as Ω_r. Without loss of generality, it is assumed that $C_t^{\text{prod time}}$ in the Ω_r follows a truncated normal distribution, as shown in (10.44), where σ_t and μ_t are distribution parameters, $K_t = \Phi\left(\frac{\mu_t}{\sigma_t}\right)$. In order to compare the fairness of the experiment, both the model Ω_r and Ω_o the calculation are based on the parameter settings in Sect. 10.4.1. In addition, when the distribution $C_t^{\text{prod time}}$ is given, take $C_t^{\text{prod time}}$ become the expected value $\frac{\sigma_t}{K_t}\varphi\left(\frac{\mu_t}{\sigma_t}\right)+\mu_t$ of $C_t^{\text{prod time}}$. Table 10.24 presents the comparison results of the two models. The difference in optimal values (O_1 and O_2) under different models can evaluate the value of stochastic programming and measure the cost of ignoring uncertainty in decision-making [11, 12].

From Table 10.24, it can be seen that, although the optimal values O_1 and O_2 both increase with the increase of production time capacity, O_2 are significantly better than O_1 others. Their improvement rate is approximately from 3 to 20%, indicating the effectiveness of modeling production time capacity as a random variable to characterize the random factors of spare parts production. In addition, the advantage of the optimal value O_2 is also influenced by the production time capability (reflected in the parameters μ_t) and its fluctuations (reflected in the parameters σ_t). The larger the fluctuations, the more it can reflect the advantages of the model Ω_r. Figure 10.17

Table 10.24 Comparison of models Ω_o and Ω_r

Production time capability				
In Ω_o: Modeling as a regular variable$C*_t^{\text{prod time}}$		In Ω_r: Modeling as a random variable $C_t^{\text{prod time}}$		Improvement rate
$C_t^{\text{prod time}}$	O_1 (Ten thousand)	Distribution parameters of $C_t^{\text{prod time}}$	O_2 (Ten thousand)	$\frac{O_1-O_2}{O_1} \times 100\%$
24.00	87.66	$\mu_t = 24, = \sigma_t 1$	84.54	3.56%
25.00	89.13	$\mu_t = 25, = \sigma_t 3$	85.08	4.54%
26.00	92.47	$\mu_t = 26, = \sigma_t 5$	86.96	5.96%
27.00	93.98	$\mu_t = 27, = \sigma_t 7$	88.83	5.48%
28.03	98.82	$\mu_t = 28, = \sigma_t 9$	91.52	7.39%
29.14	104.34	$\mu_t = 29, = \sigma_t 11$	95.15	8.81%
30.39	115.05	$\mu_t = 30, = \sigma_t 13$	101.94	11.40%
31.75	122.79	$\mu_t = 31, = \sigma_t 15$	109.61	10.73%
33.28	144.22	$\mu_t = 32, = \sigma_t 17$	120.38	16.53%
34.76	168.54	$\mu_t = 33, = \sigma_t 19$	135.20	19.78%

shows the advantages of the model Ω_r, which is particularly suitable for spare parts production environments with high randomness.

2. Set the upper bound of storage cost as a regular variable and a multiple choice parameter variable, respectively

If the upper bound of storage cost is modeled as a regular variable $C_j^{*\text{stor}}$ in Constraint (10.2), the constraint can be rewritten as

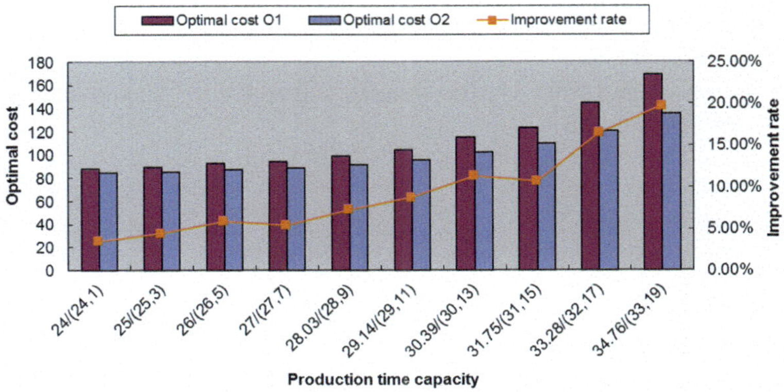

Fig. 10.17 Optimal values and corresponding improvement rates of model Ω_o and Ω_r

$$\sum_{i=1}^{I} C_{ij}^{\text{stor}} \le C_j^{*\text{stor}}, \forall j \tag{10.99}$$

Therefore, a new model with no multiple parameter variables to choose from in the constraint is.

Min: (10.78).

Subject to (10.99), (10.80)–(10.90).

For simplicity, the model with multiple choice parameter variables is Ω_M denoted as, and the new model without multiple choice parameter variables is denoted as Ω_o'. For the sake of fairness, the calculations of the model Ω_M and Ω_o' are based on the parameter settings in Sect. 10.4.1. In addition, when multiple parameter variables C_j^{stor} are given (the elements $C_j^{\text{stor}(1)}, C_j^{\text{stor}(2)}, \ldots, C_j^{\text{stor}(h_j)}$ of the set C_j^{stor} are given), they will be set $C_j^{*\text{stor}}$ to the average value of these elements, that is $\frac{1}{h_j}\sum_{\xi=1}^{h_j} C_j^{\text{stor}(\xi)}$ The comparison results of the two models Ω_M and Ω_o' are listed in Table 10.25.

From Table 10.25, it can be seen that in both models, the optimal values O_3 and O_4 both increase with the increase of the upper bound on storage costs, and O_4 are significantly better than O_3. More importantly, the improvement rate also increases with the increase of the upper bound of storage costs, approximately from 2 to 17%, which verifies the effectiveness of modeling the upper bound of storage costs as a multi choice parameter variable to characterize storage uncertainty factors. In addition, the advantage of the optimal value O_4 is also influenced by the degree of difference between the elements of the set C_j^{stor}. The larger the difference between the elements, the more it can reflect the advantage of the model Ω_M. Figure 10.18 intuitively illustrates the advantages of the model Ω_M, which means that models Ω_M with multiple parameter variables are particularly suitable for spare parts storage environments with high uncertainty.

Table 10.25 Comparison of Models Ω_M and Ω'_o

Upper bound of storage cost				
In Ω'_o: Modeling as a regular variable C_j^{*stor}		In Ω_M: Modeling as a multiple choice parameter variable $C_j^{stor} = \left\{C_j^{stor(1)}, C_j^{stor(2)}, \cdots, C_j^{stor(h_j)}\right\}$		Improvement rate
C_j^{*stor}	O_3 (Ten thousand)	The elements of a set C_j^{stor}	O_4 (Ten thousand)	$\frac{O_3-O_4}{O_3} \times 100\%$
410	86.07	{400, 420}	83.94	2.47%
480	86.82	{460, 500}	84.17	3.05%
540	89.93	{510, 570}	84.66	5.86%
620	90.45	{580, 660}	85.33	5.66%
730	93.22	{680, 780}	86.14	7.59%
860	93.94	{800, 920}	87.03	7.36%
1010	100.68	{940, 1080}	89.75	10.86%
1180	104.51	{1100, 1260}	91.26	12.68%
1390	108.06	{1300, 1480}	94.58	12.47%
1600	120.17	{1500, 1700}	99.49	17.21%

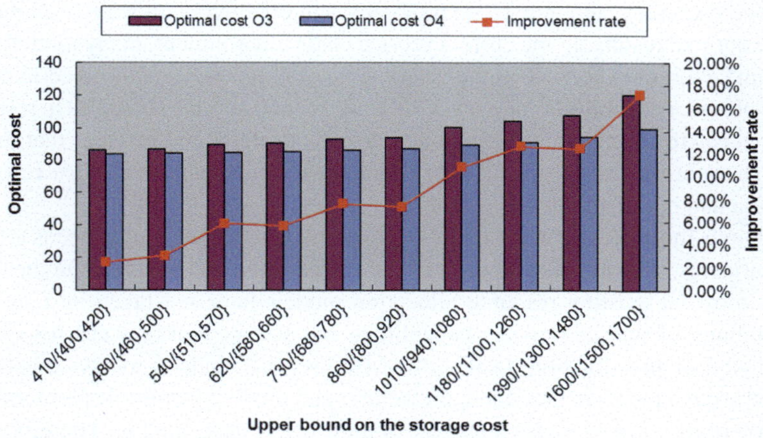

Fig. 10.18 Optimal values and corresponding improvement rates of model Ω'_o and Ω_M

10.5 Service Resource Configuration Based on Fuzzy Stochastic Programming and Profit Sharing Model

After experiencing various maintenance management modes such as post maintenance, preventive maintenance, and production maintenance, modern production equipment management has entered a new stage, and correspondingly, equipment maintenance has also entered a new period of development. Due to the complex structure, multiple components, wide coverage, and long operating cycle of large-scale industrial equipment, MRO services for large-scale industrial equipment are bound to become the focus and difficulty of high-end development in the manufacturing industry and manufacturing service field, and gradually become a hot research topic in the academic community at home and abroad. Early research on MRO services mainly focused on the aviation industry, and now it has gradually expanded to various industries. Scientific and reasonable MRO services are a necessary guarantee for process manufacturing enterprises (such as metallurgical plants, chemical plants, energy industries, etc.) to ensure uninterrupted production. On the other hand, for some typical service industries (such as transportation, telecommunications, etc.), ensuring the normal operation of operational equipment through MRO services is a key factor in their normal operation, reducing costs, and improving service quality. Therefore, more and more enterprises are paying attention to the procurement and usage management of MRO services.

The allocation and optimization of MRO service resources are the key to achieving value co creation among all members of the service chain, and the core issue is how to maximize user satisfaction by utilizing limited service resources. The resources here include not only tangible products (such as spare parts), but also intangible services (such as expert knowledge). MRO service resource allocation mainly solves the problem of how to obtain service delivery capability, how to ensure the availability of service resources, and how to scientifically and effectively allocate resources to achieve optimal results. In the current service chain environment, many random and uncertain factors coexist, such as the randomness of user needs and the uncertainty of service provider capabilities, making it difficult for their service strategies to respond to user needs to the maximum extent and provide professional services that match them. Therefore, there is still a lot of room for improvement in service resource allocation and optimization management.

Although most service resource allocation models consider user needs as key input variables, they mostly assume that their needs are deterministic. On the other hand, almost all existing resource allocation models have not taken into account the ambiguity of service provider capabilities, but assume that they can respond to any demand in any situation. In fact, the MRO service chain is a service network centered around the main product manufacturer and involving multiple partners. The provision of its services is also a process of multi-party participation. Therefore, the uneven capabilities of partners will inevitably lead to ambiguity in service levels. It can be seen that quantifying the uncertainty of demand and the ambiguity of provider capabilities is the key to achieving flexible resource allocation.

A service resource allocation scheme based on fuzzy stochastic programming and profit sharing mode will be proposed to address these issues. Firstly, based on the actual MRO service chain operation model, an MRO service management framework is abstracted, consisting of MRO service users, Service Management Platform (SMP), and Service Allocation System (SAS);Then, model the user requirements as a fuzzy random variable to quantify the randomness of service requirements; On the other hand, resource allocation costs and provider cost budgets are modeled as fuzzy random variables and multi choice parameter variables respectively to quantify the fuzziness and uncertainty of their service capabilities. Furthermore, with the expected profits of SMP and SAS as the optimization objective, a multi-objective fuzzy stochastic programming model containing multiple selection parameters is established with the corresponding period, corresponding MRO service prices, and the number of resources matched with them as decision variables, and a profit sharing mode is introduced into the model. Finally, the Lagrange interpolation polynomial method and global criterion method are used to solve the model and obtain the optimal service resource allocation scheme.

10.5.1 Service Resource Configuration Based on Fuzzy Stochastic Programming and Profit Sharing

In this section, a multi-objective fuzzy stochastic programming model with multiple selection parameters will be proposed for the MRO service resource allocation problem. On this basis, research and discuss the objective function and constraint conditions of the model, and then provide corresponding mathematical models for model analysis and comparison.

1. **Model description and assumptions**

Figure 10.19 depicts the MRO service management framework considered in the model, and the modeling methods used at the corresponding level are shown on the right. At the user level, the demand for MRO services arises when they need to perform preventive maintenance on equipment or when random equipment malfunctions require timely repair. The service management platform SMP is a core member of the service chain, managing random demands from users, including collecting and responding to demand information, and providing paid MRO services to users based on their needs. The Service Configuration System (SAS) is the service resource guarantee organization for this service chain, serving as a reserve repository for MRO services, spare parts, and expert knowledge. It manages and configures the resources within it, providing SMP with paid MRO services that match user needs.

It can be seen that SMP plays a leading role in the MRO service chain, responsible for collecting and managing requirements, while SAS is SMP's service reserve library (including MRO service provision, service spare parts supply, expert technical support, etc.), used to complete the configuration of requirements. SMP and SAS division of labor cooperation, mutual benefit and win–win situation; In a competitive market environment, in order to ensure a stable service configuration to respond to user needs, SMP is willing to share a certain proportion of profits with SAS to achieve long-term cooperation and win–win.

Consider the operation of an MRO service chain for one cycle (including multiple periods), with the expected profits of SMP and SAS as optimization objectives, the corresponding period, the price of corresponding MRO services, and the number of

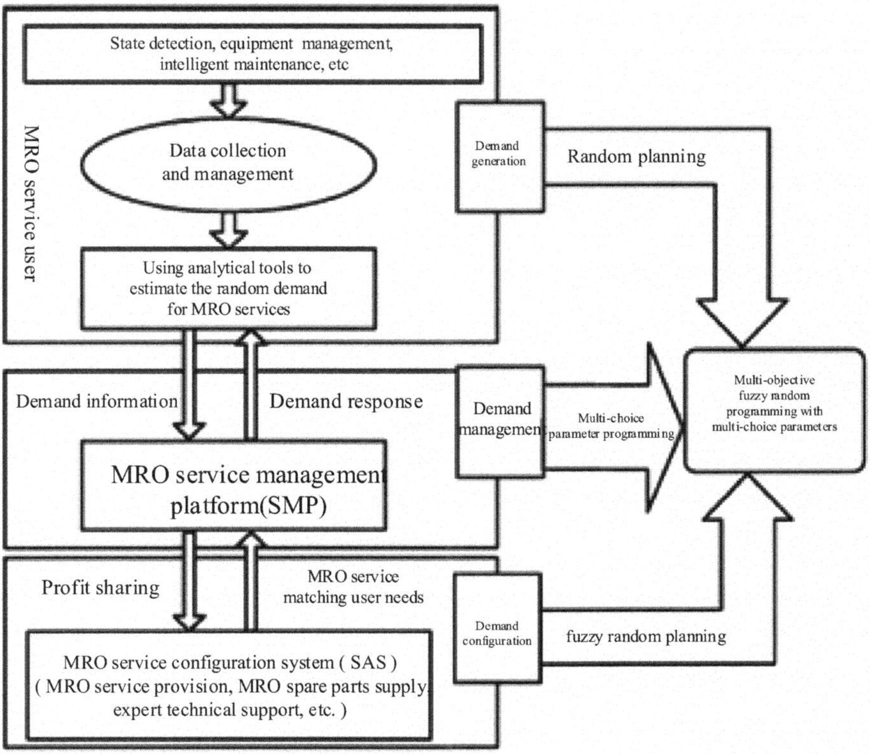

Fig. 10.19 MRO service management framework

matching resources as decision variables, and the standard deviation of shared profits between SMP and SAS and SMP's cost budget as constraints. Establish a multi-objective fuzzy stochastic programming model with multiple selection parameters, The model includes random variables representing random MRO service demand, multiple choice parameter variables representing SMP fuzzy cost budgeting, and multivariate fuzzy random variables representing SAS fuzzy resource allocation costs. By solving the model, the optimal resource allocation plan for both parties is ultimately obtained, which can achieve a win–win service price and corresponding resource matching quantity for SMP and SAS.

Before establishing a mathematical planning model, make the following model assumptions.

Assuming 10.13 At different times, SMP can provide multiple types of MRO services according to user needs and charge related service fees.

Assuming 10.14 SAS provides SMP with the necessary resources for MRO services (such as service spare parts, expert knowledge, etc.) for a fee, to quantify the impact of random factors on resource prices, it is assumed that the resource price per unit is a non negative, normally distributed random variable.

Assuming 10.15 SAS needs to manage and allocate the resources required for related MRO services, in order to quantify the uncertainty of its service capabilities, it is assumed that the related costs are a multivariate fuzzy random variable.

Assuming 10.16 Quantifies the randomness of MRO service user demand, it is assumed to be a random variable with a known probability density function.

Assuming that on October17th, in order to achieve a win–win situation for long-term cooperation, SMP agrees to share profits with SAS and both parties share certain risks.

The symbols and meanings used in the model are shown in Table 10.26.

2. **A multi-objective fuzzy stochastic programming model with multiple selection parameters**

For the members SMP and SAS in the MRO service chain, the optimization objective of the model is to maximize their total profit expectations while satisfying the

Table 10.26 Symbols and meanings used in the model

subscript	
i	MRO services
t	Period
Decision variables	
d_{it}	The price of MRO services i during the period t
q_{it}	Number of resources matched with MRO services i during the period t
Input variables	
m_{it}	There source price that matches the MRO service i during the period t is a random variable, and $m_{it} \sim N\left(\mu_{it}, \sigma_{it}^2\right)$
$\widetilde{f}_{it}$	Th t e resource allocation cost applied to MRO services i during the period is a fuzzy random variable
α_{it}	The proportion of profit sharing generated by MRO services i during the period t,$(0 < \alpha_{it} < 1)$
r_{it}	Th t e random i demand of users for MRO services during the period, with a probability density function of $f(r_{it})$
RT_M	The maximum risk tolerance of service management platform SMP under profit sharing mode
RT_A	The maximum risk tolerance of resource allocation system SAS under profit sharing mode
B_t	Th t e cost budget of SMP during the period is a multi choice parameter,$B_t \in \left\{B_t^1, B_t^2, \cdots, B_t^{k_t}\right\}$
L_t	Th t e probability of SAS meeting budget constraints during the period,$(0 < L_t < 1)$
N	The number of types of MRO services
T	Number of periods

constraint of shared profit standard deviation, that is the maximum risk tolerance value. Below are the expressions for their total expected profit and standard deviation of shared profit:

$$P_M = \sum_{i=1}^{N}\sum_{t=1}^{T}\{(1-\alpha_{it})\cdot[d_{it}\cdot E(r_{it}) - E(m_{it})\cdot q_{it}]\} \quad (10.100)$$

$$P_A = \sum_{i=1}^{N}\sum_{t=1}^{T}\{[E(m_{it}) - E(\tilde{f}_{it})]\cdot q_{it} + \alpha_{it}\cdot[d_{it}\cdot E(r_{it}) - E(m_{it})\cdot q_{it}]\} \quad (10.101)$$

$$S_M = \sum_{i=1}^{N}\sum_{t=1}^{T}\left[(1-\alpha_{it})\cdot\sqrt{d_{it}\cdot Var(r_{it}) - q_{it}\cdot Var(m_{it})}\right] \quad (10.102)$$

$$S_A = \sum_{i=1}^{N}\sum_{t=1}^{T}\left[\alpha_{it}\cdot\sqrt{d_{it}\cdot Var(r_{it}) - q_{it}\cdot Var(m_{it})}\right] \quad (10.103)$$

Among them, $E(\cdot)$ and $Var(\cdot)$ respectively represent the mathematical expectation and variance of random variables, P_M and P_A are the total expected profits of SMP and SAS, S_M and S_A are the standard deviations of shared profits of SMP and SAS, respectively. Thus, the following multi-objective fuzzy stochastic programming model with multiple selection parameters can be obtained:

$$\text{Max}\left(P_M = \sum_{i=1}^{N}\sum_{t=1}^{T}\{(1-\alpha_{it})\cdot[d_{it}\cdot E(r_{it}) - E(m_{it})\cdot q_{it}]\}\right. \quad (10.104)$$

$$\text{Max}\left(P_A = \sum_{i=1}^{N}\sum_{t=1}^{T}\{[E(m_{it}) - E(\tilde{f}_{it})]\cdot q_{it} + \alpha_{it}\cdot[d_{it}\cdot E(r_{it}) - E(m_{it})\cdot q_{it}]\}\right. \quad (10.105)$$

s.T

$$\sum_{i=1}^{N}\sum_{t=1}^{T}\left[(1-\alpha_{it})\cdot\sqrt{d_{it}\cdot Var(r_{it}) - q_{it}\cdot Var(m_{it})}\right] \le RT_M \quad (10.106)$$

$$\sum_{i=1}^{N}\sum_{t=1}^{T}\left[\alpha_{it}\cdot\sqrt{d_{it}\cdot Var(r_{it}) - q_{it}\cdot Var(m_{it})}\right] \le RT_A \quad (10.107)$$

$$\Pr\left[\sum_{i=1}^{N} m_{it}\cdot q_{it}\right] \le B_t \ge L_t, B_t \in \left\{B_t^1, B_t^2, \ldots, B_t^{k_t}\right\}, t = 1, 2, \ldots, T \quad (10.108)$$

$$q_{it}, d_{it} \ge 0, i = 1, 2, \ldots, N; t = 1, 2, \ldots, T \quad (10.109)$$

$$q_{it}, d_{it} \in \mathbb{Z}, i = 1, 2, \ldots, N; t = 1, 2, \ldots, T \tag{10.110}$$

Among them, Eqs. (10.104) and (10.105) indicates that the optimization objective of the model is to maximize the expected profit of SMP and SAS; Eqs. (10.106) and (10.107) indicates the constraint of meeting the standard deviation of shared profits; Eq. (10.108) represents the cost constraint of SMP, Pr[·] represents the probability of event occurrence, and $B_t \in \left\{B_t^1, B_t^2, \cdots, B_t^{k_t}\right\}$ represents that B_t is a multi choice parameter variable; Eqs. (10.109) and (10.110) indicate that the decision variable is non negative and an integer.

3. **Model comparison and analysis**

It can be seen that models (10.104)–(10.110) belong to mathematical programming models in form, and compared with traditional mathematical programming models for solving such problems, there are mainly the following improvements:

1. The objective function (10.105) contains fuzzy random variables, which can quantify the uncertainty of resource allocation costs

In the actual operation of the MRO service chain, resource allocation costs will vary with changes in uncertain factors such as personnel, materials, and environment. If it is set as a constant in a mathematical planning model, it is difficult to quantify its uncertainty, so the decision results obtained lack rationality to a certain extent. In the model of this chapter, resource allocation cost $\widetilde{f}_{it}$ is not a constant, but a multidimensional variable in the form of $\left(f_{it}^1, f_{it}^2, f_{it}^3, f_{it}^4\right)$. Decision makers can set corresponding values for each variable f_{it}^k, $k = 1, 2, 3, 4$ based on the actual situation to reflect the uncertainty of resource allocation costs. In the subsequent case analysis, the influence of fuzzy random variables $\widetilde{f}_{it}$ on the optimal solution of the model will be discussed through sensitivity analysis experiments of parameters.

2. The constraint (10.1m_{it}09) contains random parameters multiple choice parameters B_t, which can quantify the uncertainty of SMP service capability

Constraint (10.108) is a random constraint that restricts the probability of a random event occurring within a certain range. Compared with traditional deterministic constraints, random constraints can better quantify the impact of random factors on decision variables, making decision results more realistic. It is worth mentioning that in this random constraint, the cost budget B_t of SMP is set $\left\{B_t^1, B_t^2, \cdots, B_t^{k_t}\right\}$ as a multi choice parameter, which means that decision-makers can set a set of selectable values for it when they are uncertain about the exact value of the cost budget. This allows the uncertainty of SMP cost budget to be reflected in the model, thereby quantifying the uncertainty of its service capability.

From the above analysis, it can be seen that setting multiple selection parameters in the random constraints of the model can better quantify the uncertain factors in real life, which is also an improvement of the existing mathematical programming models for solving such problems in this paper. In most mathematical programming models, the constraint conditions are deterministic (if the constraint conditions (10.108) is set as a deterministic constraint $\sum_{i=1}^{N} m_{it} \cdot q_{it} \leq B_t, t = 1, 2, \cdots, T$, where m_{it} and B_t are general parameters, although deterministic constraints can simplify the solving process, they cannot quantify the uncertainty of the input parameters of the model. Therefore, the optimality of the decision results needs to be further improved.

10.5.2 Model Solution

As analyzed earlier, the proposed model can effectively quantify the uncertain factors in the operation of the actual MRO service chain, such as resource allocation costs, SMP service capabilities, etc. However, to solve this multi-objective stochastic programming model with multiple selection parameters, traditional solving methods cannot be directly applied, but it needs to be transformed into an equivalent deterministic single objective programming model. Therefore, it is necessary to:

1. Derive the analytical expression in the objective function $E\left(\widetilde{f_{it}}\right)$;
2. Convert random constraints containing multiple selection parameters (10.108) is a deterministic constraint;
3. Transform multi-objective optimization problems into single objective optimization problems.

Based on the above analysis, the following first provides the definitions and propositions used in the model solving process, and then provides the specific steps for model solving.

1. **Definition and Proposition**

Based on references [11, 12], the following basic definitions and propositions are given.

Definition 10.14 Expectations of Fuzzy Random Variables ζ.

For fuzzy random $\mu : \Re \to [0, 1]$ variables ζ, if their membership function is, then their expected value is

$$E(\zeta) = \int_{0}^{+\infty} \mathrm{Cr}\{\zeta \geq r\}dr - \int_{-\infty}^{0} \mathrm{Cr}\{\zeta \geq r\}dr \tag{10.111}$$

where

$$Cr\{\zeta \geq r\} = \frac{1}{2}(\mathrm{Pos}\{\zeta \geq r\} + Nec\{\zeta \geq r\}), \quad Cr\{\zeta \geq r\} = 1 - Cr\{\zeta \geq r\} \tag{10.112}$$

The definition of $\mathrm{Pos}\{\zeta \geq r\}$ and $Nec\{\zeta \geq r\}$ in the above equation is as follows

$$\mathrm{Pos}\{\zeta \geq r\} = \sup_{\zeta \geq r} \mu(\zeta), \quad Nec\{\zeta \geq r\} = 1 - \mathrm{Pos}\{\zeta \geq r\} \tag{10.113}$$

where represents the supremum $\sup(\cdot)$ and r is a real number.

Definition 10.15 Basis Functions in Lagrange Interpolation.

There is a triangular array on the interval $[a, b]$ as follows:

$$\begin{array}{l} x_1^{(1)} \\ x_1^{(2)}, x_2^{(2)} \\ \vdots \\ x_1^{(n)}, x_2^{(n)}, \cdots, x_n^{(n)} \\ \vdots \end{array} \tag{10.114}$$

A polynomial can be constructed from Eq. (10.114)

$$l_k^{(n)}(x) = \prod_{\substack{l=1 \\ l \neq k}}^{n} \left(\frac{x - x_l^{(n)}}{x_k^{(n)} - x_l^{(n)}}\right) = \frac{\omega_n(x)}{\omega_n'\left(x_k^{(n)}\right) \cdot \left(x - x_k^{(n)}\right)}, 1 \leq k \leq n \tag{10.115}$$

Amon $l_k^{(n)}(x)$ g them, $\omega_n(x) = \prod_{l=1}^{n} \left(x - x_l^{(n)}\right)$ it is called the basis function in Lagrange interpolation.

Proposition 10.1 The basis function $l_k^{(n)}(x)$ in Lagrange interpolation is a degree polynomial $n - 1$ and satisfies.

$$l_k^{(n)}(x_l^{(n)}) = \delta_{kl} = \begin{cases} 1, \ l = k \\ 0, \ l \neq k \end{cases}, l, k = 1, 2, \ldots, n \tag{10.116}$$

Definition 10.16 Lagrange interpolation polynomial.

Construct a polynomial of the following degree $\leq n - 1$ for any n value y_k, $1 \leq k \leq n$

$$L_n(x) = \sum_{k=1}^{n} y_k \cdot l_k^{(n)}(x) \tag{10.117}$$

If the basis function in the equation $l_k^{(n)}(x)$ is Lagrange interpolation, $L_n(x)$ is called Lagrange interpolation polynomial.

Proposition 10.2 Global criterion method

For the following multi-objective planning problems:

$$\begin{aligned} &\max\ Z = [Z_1(x), Z_2(x), \ldots, Z_K(x)]^T \\ &\text{s.t. } x \in X \end{aligned}$$

Among them, $x \in R^n$ represents the n dimensional decision variable. X is a set of feasible solutions that satisfy the constraint conditions. It can be transformed into an equivalent single objective programming problem according to the following steps:

1. For each objective x^i, $i = 1, 2, \cdots$ function $Z_k(x)$, $k = 1, 2, \cdots, K$, calculate the corresponding $Z_k^{\max}$, $Z_k^{\max} = \text{Max}\{Z_k(x^1), Z_k(x^2), \cdots, \}$, where is the feasible solution;
2. The transformed single objective planning problem is:

$$\begin{aligned} &\min\ G(x) = \left\{ \sum_{k=1}^{K} \left[\frac{Z_k^{\max} - Z_k(x)}{Z_k^{\max}} \right]^P \right\}^{\frac{1}{P}} \\ &\text{s.t. (10.1 } x \in X \text{ 19)} \end{aligned}$$

Among them $1 \le P \le +\infty$, generally take P as 2.

2. **Model solving steps**

As mentioned earlier, for model (10.104)–(10.110) cannot be directly solved using conventional methods of mathematical programming models. Therefore, in the process of solving the model, definitions or propositions such as the expectation of fuzzy random variables, Lagrange interpolation polynomial method, and global criterion method introduced in 6.3.1 will be used respectively. Below are the specific steps for solving the model:

1. An analytical expression for deriving the expectation $E(\tilde{f}_{it})$ fuzzy random variables $\tilde{f}_{it}$

Assuming $\tilde{f}_{it}$ is a four-dimensional fuzzy random variable, that is $\tilde{f}_{it} = (f_{it}^1, f_{it}^2, f_{it}^3, f_{it}^4)$ and its trapezoidal membership function is $\mu_{\tilde{f}_{it}}(x)$, that is

$$\mu_{f_{it}}(x) = \begin{cases} 0 \ , & x \le f_{it}^1 \\ \frac{x - f_{it}^1}{f_{it}^2 - f_{it}^1} \ , & f_{it}^1 \le x \le f_{it}^2 \\ 1 \ , & f_{it}^2 \le x \le f_{it}^3 \\ \frac{f_{it}^4 - x}{f_{it}^4 - f_{it}^3} \ , & f_{it}^3 \le x \le f_{it}^4 \\ 0 \ , & x \ge f_{it}^4 \end{cases} \tag{10.118}$$

According to Eq. (10.111), the analytical expression of $E(\tilde{f}_{it})$ to be obtained needs to be calculated $\text{Cr}\{\tilde{f}_{it} \ge r\}$ separately for and $\text{Cr}\{\tilde{f}_{it} \le r\}$. From Eq. (10.112), it can be inferred that the first step is to calculate $\mathrm{Cr}\{\tilde{f}_{it} \geq r\}$. Therefore, by first calculating according to Eq. (10.113):

$$\mathrm{Pos}\{\tilde{f}_{it} \geq r\} = \sup_{x \geq r} \mu_{f_{it}}(x) = \begin{cases} 1 \quad , & r \leq f_{it}^3 \\ \frac{f_{it}^4 - r}{f_{it}^4 - f_{it}^3} \, , & f_{it}^3 \leq r \leq f_{it}^4 \\ 0 \quad , & r \geq f_{it}^4 \end{cases} \tag{10.119}$$

$$Nec\{\tilde{f}_{it} \geq r\} = 1 - \mathrm{Pos}\{\tilde{f}_{it} < r\} = 1 - \begin{cases} 0, & r \leq f_{it}^1 \\ \frac{r - f_{it}^1}{f_{it}^2 - f_{it}^1}, & f_{it}^1 \leq r \leq f_{it}^2 \\ 1, & r \geq f_{it}^2 \end{cases}$$

$$= \begin{cases} 1, & r \leq f_{it}^1 \\ \frac{f_{it}^2 - r}{f_{it}^2 - f_{it}^1}, & f_{it}^1 \leq r \leq f_{it}^2 \\ 0, & r \geq f_{it}^2 \end{cases} \tag{10.120}$$

Calculate from Eq. (10.112) again

$$\mathrm{Cr}\{\tilde{f}_{it} \geq r\} = \frac{1}{2} \cdot \left(Pos\{\tilde{f}_{it} \geq r\} + \mathrm{Nec}\{\tilde{f}_{it} \geq r\}\right)$$

$$= \begin{cases} 1, & r \leq f_{it}^1 \\ \frac{2f_{it}^2 - f_{it}^1 - r}{2(f_{it}^2 - f_{it}^1)}, & f_{it}^1 \leq r \leq f_{it}^2 \\ \frac{1}{2}, & f_{it}^2 \leq r \leq f_{it}^3 \\ \frac{f_{it}^4 - r}{2(f_{it}^4 - f_{it}^3)}, & f_{it}^3 \leq r \leq f_{it}^4 \\ 0, & r \geq f_{it}^4 \end{cases} \tag{10.121}$$

$$\mathrm{Cr}\{\tilde{f}_{it} \leq r\} = 1 - \mathrm{Cr}\{\tilde{f}_{it} \geq r\} = \begin{cases} 0, & r \leq f_{it}^1 \\ \frac{r - f_{it}^1}{2(f_{it}^2 - f_{it}^1)}, & f_{it}^1 \leq r \leq f_{it}^2 \\ \frac{1}{2}, & f_{it}^2 \leq r \leq f_{it}^3 \\ \frac{f_{it}^4 - 2f_{it}^3 + r}{2(f_{it}^4 - f_{it}^3)}, & f_{it}^3 \leq r \leq f_{it}^4 \\ 1, & r \geq f_{it}^4 \end{cases} \tag{10.122}$$

Therefore, according to Eqs. (10.111), (10.121), and (10.122), it can be obtained that:

$$E(\tilde{f}_{it}) = \int_0^{+\infty} \mathrm{Cr}\{\tilde{f}_{it} \geq r\} dr - \int_{-\infty}^{0} \mathrm{Cr}\{\tilde{f}_{it} \geq r\} dr$$

$$= \int_0^{f_{it}^1} dr + \int_{f_{it}^1}^{f_{it}^2} \frac{2f_{it}^2 - f_{it}^1 - r}{2(f_{it}^2 - f_{it}^1)} dr + \int_{f_{it}^2}^{f_{it}^3} \frac{1}{2} dr$$

$$+\int_{f_{it}^3}^{f_{it}^4} \frac{f_{it}^4 - r}{2\left(f_{it}^4 - f_{it}^3\right)} dr - \int_{f_{it}^1}^{f_{it}^2} \frac{r - f_{it}^1}{2\left(f_{it}^2 - f_{it}^1\right)} dr$$

$$-\int_{f_{it}^2}^{f_{it}^3} \frac{1}{2} dr - \int_{f_{it}^3}^{f_{it}^4} \frac{f_{it}^3 - r}{2\left(f_{it}^4 - f_{it}^3\right)} dr - \int_{f_{it}^4}^{f_{it}^0} dr$$

$$= f_{it}^1 + \int_{f_{it}^1}^{f_{it}^2} \frac{f_{it}^2 - r}{f_{it}^2 - f_{it}^1} dr + \int_{f_{it}^3}^{f_{it}^4} \frac{f_{it}^3 - r}{f_{it}^4 - f_{it}^3} dr$$

$$+ f_{it}^4 = \frac{1}{2}\left(f_{it}^1 + f_{it}^2 + f_{it}^3 + f_{it}^4\right) \tag{10.123}$$

From Eq. (10.123), it can be seen that there is a linear relationship between the expected value $E\left(\tilde{f}_{it}\right)$ of a four-dimensional fuzzy random variable $\tilde{f}_{it}$ and its components, and the positions of each component are the same in the analytical expression of the expected value. In the following example analysis, we will further discuss the impact of changes in each component of $\tilde{f}_{it}$ on their expectations, and then analyze their impact on the optimal solution of the model.

2. Convert Random Constraint $\Pr\left[\sum_{i=1}^{N} m_{it} \cdot q_{it} \leq B_t\right] \geq L_t, B_t \in \left\{B_t^1, B_t^2, \ldots, B_t^{k_t}\right\}, t = 1, 2, \ldots, T$

One of the characteristics of the proposed model is the inclusion of random variables m_{it} and multiple choice parameters B_t in the constraint conditions, which m_{it} reflects the randomness of service resource prices and B_t reflects the fuzziness of SMP cost budgets. Therefore, this constraint effectively quantifies the uncertainty factors of MRO service chain operation in the network environment. However, to solve mathematical programming problems with such constraints, it is necessary to first convert the constraint into a deterministic constraint. The specific process is as follows:

3. Convert random constraints to deterministic constraints

Let $A_t = \sum_{i=1}^{N} m_{it} \cdot q_{it}, t = 1, 2 \cdots, T$, among them $m_{it} \sim N\left(\mu_{it}, \sigma_{it}^2\right)$ and t is independent of each other, A_t is also a random variable that follows a normal distribution, so there is

$$E(A_t) = E\left(\sum_{i=1}^{N} m_{it} \cdot q_{it}\right) = \sum_{i=1}^{N} E(m_{it} \cdot q_{it}) = \sum_{i=1}^{N} q_{it} \cdot E(m_{it}) = \sum_{i=1}^{N} q_{it} \cdot \mu_{it} \tag{10.124}$$

$$Var(A_t) = Var\left(\sum_{i=1}^{N} m_{it} \cdot q_{it}\right) = \sum_{i=1}^{N} Var(m_{it} \cdot q_{it}) = \sum_{i=1}^{N} q_{it}^2 \cdot Var(m_{it}) = \sum_{i=1}^{N} q_{it}^2 \cdot \sigma_{it}^2 \tag{10.125}$$

By utilizing Eqs. (10.26) and (10.27), as well as the monotonic property of the distribution function of the standard normal distribution $\Phi(z)$, the constraint $\Pr\left[\sum_{i=1}^{N} m_{it} \cdot q_{it}\right] \le B_t \ge L_t, B_t \in \left\{B_t^1, B_t^2, \ldots, B_t^{k_t}\right\}, t = 1, 2, \ldots, T$ can be simplified as:

$$\sum_{i=1}^{N} q_{it} \cdot \mu_{it} + u_{L_t} \cdot \sqrt{\sum_{i=1}^{N} q_{it}^2 \cdot \sigma_{it}^2} \le B_t, B_t \in \left\{B_t^1, B_t^2, \ldots, B_t^{k_t}\right\}, t = 1, 2, \ldots, T \tag{10.126}$$

Among them u_{L_t} is the upper quantile $(1 - L_t)$ of the standard normal distribution.

4. Using Lagrange interpolation polynomials to handle multiple selection parameters $B_t \in \left\{B_t^1, B_t^2, \ldots, B_t^{k_t}\right\}$

Let $g(w_t)$ be a real-valued function, $0, 1, 2, \ldots, k_t - 1$ be k_t nodes, and let $B_t^1, B_t^2, \ldots, B_t^{k_t}$ e the corresponding function value of the interpolation polynomial at k_t different nodes, that is, $g(w_t) = B_t^{w_t+1}, w_t = 0, 1, 2, \ldots, k_t - 1$.

Let $P_{k_t-1}(w_t)$ be the $k_t - 1$ order interpolation polynomial, and $P_{k_t-1}(v) = B_t^{v+1}, v = 0, 1, 2, \ldots, k_t - 1; t = 1, 2, \ldots, T$, then according to the Lagrange interpolation formula (10.117), we can construct the t th multi-parameter interpolation polynomial:

$$P_{k_t-1}(w_t) = \sum_{h=0}^{k_t-1} B_t^{h+1} \cdot \prod_{\substack{\theta=0 \\ \theta \ne h}}^{k_t-1} \frac{w_t - \theta}{h - \theta}, t = 1, 2, \ldots, T, w_t = 0, 1, 2, \ldots, k_t - 1 \tag{10.127}$$

1. Obtain deterministic constraints after transformation

Based n (a) and (b), the transformed constraints can be obtained as

$$\sum_{i=1}^{N} q_{it} \cdot \mu_{it} + u_{L_t} \cdot \sqrt{\sum_{i=1}^{N} q_{it}^2 \cdot \sigma_{it}^2} \le \sum_{h=0}^{k_t-1} B_t^{h+1} \cdot \prod_{\substack{\theta=0 \\ \theta \ne h}}^{k_t-1} \frac{w_t - \theta}{h - \theta}, t = 1, 2, \ldots, T, w_t = 0, 1, 2, \ldots, k_t - 1 \tag{10.128}$$

2. Obtain an equivalent deterministic model from (1) and (2)

If known the probability density function of r_{it} is $f(r_{it})$, then

$$E(r_{it}) = \int_{0}^{+\infty} r_{it} \cdot f(r_{it}) dr_{it} \tag{10.129}$$

$$Var(r_{it}) = E\left(r_{it}^2\right) - [E(r_{it})]^2 = \int_0^{+\infty} r_{it}^2 \cdot f(r_{it})dr_{it} - \left(\int_0^{+\infty} r_{it} \cdot f(r_{it})dr_{it}\right)^2 \quad (10.130)$$

Substituting Eqs. (10.123), (10.129)–(10.130) into the original models (10.104)–(10.110), an equivalent multi-objective deterministic model can be obtained as follows:

$$\max\ P_M = \sum_{i=1}^{N}\sum_{t=1}^{T}\left\{(1-\alpha_{it}) \cdot \left[d_{it} \cdot \int_0^{+\infty} r_{it} \cdot f(r_{it})dr_{it} - \mu_{it} \cdot q_{it}\right]\right\} \quad (10.131)$$

$$\max\left(P_A = \sum_{i=1}^{N}\sum_{t=1}^{T}\left\{\left[\mu_{it} - \frac{1}{2}\left(f_{it}^1 + f_{it}^2 + f_{it}^3 + f_{it}^4\right)\right] \cdot q_{it}\right\} + \alpha_{it} \cdot \left[d_{it} \cdot \int_0^{+\infty} r_{it} \cdot f(r_{it})dr_{it} - \mu_{it} \cdot q_{it}\right]\right) \quad (10.132)$$

s.t

$$\sum_{i=1}^{N}\sum_{t=1}^{T}\left\{(1-\alpha_{it}) \cdot \sqrt{d_{it} \cdot \left[\int_0^{+\infty} r_{it}^2 \cdot f(r_{it})dr_{it} - \left(\int_0^{+\infty} r_{it} \cdot f(r_{it})dr_{it}\right)^2\right] - q_{it} \cdot \sigma_{it}^2}\right\} \le RT_M \quad (10.133)$$

$$\sum_{i=1}^{N}\sum_{t=1}^{T}\left\{\alpha_{it} \cdot \sqrt{d_{it} \cdot \left[\int_0^{+\infty} r_{it}^2 \cdot f(r_{it})dr_{it} - \left(\int_0^{+\infty} r_{it} \cdot f(r_{it})dr_{it}\right)^2\right] - q_{it} \cdot \sigma_{it}^2}\right\} \le RT_A \quad (10.134)$$

$$\sum_{i=1}^{N} q_{it} \cdot \mu_{it} + u_{L_t} \cdot \sqrt{\sum_{i=1}^{N} q_{it}^2 \cdot \sigma_{it}^2} \le \sum_{h=0}^{k_t-1} B_t^{h+1} \cdot \prod_{\substack{\theta=0 \\ \theta \ne h}}^{k_t-1} \frac{w_t - \theta}{h - \theta}, t = 1, 2, \ldots, T, w_t = 0, 1, 2, \ldots, k_t - 1 \quad (10.135)$$

$$q_{it}, d_{it} \ge 0, i = 1, 2, \ldots, N; t = 1, 2, \ldots, T \quad (10.136)$$

$$q_{it}, d_{it} \in \mathbb{Z}, i = 1, 2, \ldots, N; t = 1, 2, \ldots, T \quad (10.137)$$

It can be seen that in the converted equivalent model, the objective function Eq. (10.132) no longer contains fuzzy random variables, and the constraint condition Eq. (10.135) no longer contains random constraints and multiple choice parameter variables. At this time, the models (10.131)–(10.137) are multi-objective deterministic models, and the next step is to convert them into a single objective model.

3. Using the global criterion method to transform multiple objectives into single objectives

For each objective function P_M and P_A, calculate the optimal function values $P_M^{\max}$ and $P_A^{\max}$ for each as a single objective programming problem. Therefore, the global criterion method can obtain the objective function of the transformed single objective programming problem:

$$\min\ (G = \left[\left(\frac{P_M^{\max} - P_M}{P_M^{\max}}\right)^{\lambda} + \left(\frac{P_A^{\max} - P_A}{P_A^{\max}}\right)^{\lambda}\right]^{\frac{1}{\lambda}} \tag{10.138}$$

s.t.

$$\sum_{i=1}^{N}\sum_{t=1}^{T}\left\{(1-\alpha_{it})\cdot\sqrt{d_{it}\cdot\left[\int_0^{+\infty} r_{it}^2\cdot f(r_{it})dr_{it} - \left(\int_0^{+\infty} r_{it}\cdot f(r_{it})dr_{it}\right)^2\right]} - q_{it}\cdot\sigma_{it}^2\right\} \le RT_M \tag{10.139}$$

$$\sum_{i=1}^{N}\sum_{t=1}^{T}\left\{\alpha_{it}\cdot\sqrt{d_{it}\cdot\left[\int_0^{+\infty} r_{it}^2\cdot f(r_{it})dr_{it} - \left(\int_0^{+\infty} r_{it}\cdot f(r_{it})dr_{it}\right)^2\right]} - q_{it}\cdot\sigma_{it}^2\right\} \le RT_A \tag{10.140}$$

$$\sum_{i=1}^{N} q_{it}\cdot\mu_{it} + u_{L_t}\cdot\sqrt{\sum_{i=1}^{N} q_{it}^2\cdot\sigma_{it}^2} \le \sum_{h=0}^{k_t-1} B_t^{h+1}\cdot\prod_{\substack{\theta=0\\ \theta\ne h}}^{k_t-1}\frac{w_t-\theta}{h-\theta}, t = 1, 2, \ldots, T, w_t = 0, 1, 2, \ldots, k_t - 1 \tag{10.141}$$

$$q_{it}, d_{it} \ge 0, i = 1, 2, \cdots, N; t = 1, 2, \cdots, T \tag{10.142}$$

$$q_{it}, d_{it} \in \mathbb{Z}, i = 1, 2, \cdots, N; t = 1, 2, \cdots, T \tag{10.143}$$

The transformed objective function has only one objective, which is a single objective, and the optimization objective is minimized by minimizing the deviation between the ideal solution and the optimal solution of the function. For deterministic nonlinear mixed integer programming models (10.138)–(10.143), common mathematical optimization software can be used for solving.

4. Using LINGO 11.0 or MAPLE 12.0 to solve the above nonlinear mixed integer programming model

10.5.3 Case Analysis

This section will validate the feasibility and effectiveness of the proposed model through a numerical example. Firstly, provide relevant parameter settings; Then, based on the proposed model and solution method, provide decision results and conduct sensitivity analysis of parameters.

1. **Parameter settings**

Without loss of generality, assuming that the service management platform can provide users with three different MRO services (such as ladle trolley service, continuous casting main equipment service, and shield tunneling service), analyze the service chain operation for six months, with each month corresponding to a period. Assuming that the random user demand r_{it} follows a more common bilateral truncated normal distribution (decision-makers can choose other probability distributions based on actual situations), the probability density function of r_{it} can be expressed as

$$f(r_{it}) = \frac{1}{K_{r_{it}} \cdot \sigma_{r_{it}} \cdot \sqrt{2\pi}} \cdot \exp\left[-\frac{1}{2}\left(\frac{r_{it} - \mu_{r_{it}}}{\sigma_{r_{it}}}\right)^2\right] \tag{10.144}$$

Among them, $\sigma_{r_{it}}$ and $\mu_{r_{it}}$ are the two parameters of the distribution, $K_{r_{it}}\left(K_{r_{it}} > 0\right)$ is known as normalization constants, which satisfy the condition $\int_{a_{it}}^{b_{it}} f(r_{it})dr_{it} = 1$, $a_{it} \leq r_{it} \leq b_{it}$. so we can derive

$$K_{r_{it}} = \frac{1}{\sigma_{r_{it}} \cdot \sqrt{2\pi}} \cdot \int\limits_{a_{it}}^{b_{it}} \exp\left[-\frac{1}{2}\left(\frac{r_{it} - \mu_{r_{it}}}{\sigma_{r_{it}}}\right)^2\right] dr_{it} = \Phi\left(\frac{b_{it} - \mu_{r_{it}}}{\sigma_{r_{it}}}\right) - \Phi\left(\frac{a_{it} - \mu_{r_{it}}}{\sigma_{r_{it}}}\right) \tag{10.145}$$

Among them $\Phi(z)$ is the distribution function of the standard normal distribution.

According to Eq. (10.45) and the properties of the mathematical expectation of random variables, it can be concluded that:

$$E(r_{it}) == \frac{\sigma_{r_{it}}}{K_{r_{it}}} \cdot \left[\varphi\left(\frac{b_{it} - \mu_{r_{it}}}{\sigma_{r_{it}}}\right) - \varphi\left(\frac{a_{it} - \mu_{r_{it}}}{\sigma_{r_{it}}}\right)\right] + \mu_{r_{it}} \tag{10.146}$$

Among them $\varphi(z)$ is the probability density function of the standard normal distribution.

Due to $Var(r_{it}) = E\left(r_{it}^2\right) - [E(r_{it})]^2$ the derived analytical expression of $E\left(r_{it}^2\right)$ is as follows:

$$E\left(r_{it}^2\right) = \sigma_{r_{it}}^2 + \frac{\sigma_{r_{it}}^2}{K_{r_{it}}} \cdot \left[\frac{a_{it} - \mu_{r_{it}}}{\sigma_{r_{it}}} \cdot \varphi\left(\frac{a_{it} - \mu_{r_{it}}}{\sigma_{r_{it}}}\right) - \frac{b_{it} - \mu_{r_{it}}}{\sigma_{r_{it}}} \cdot \varphi\left(\frac{b_{it} - \mu_{r_{it}}}{\sigma_{r_{it}}}\right)\right]$$
$$+ \mu_{r_{it}}^2 + \frac{2\sigma_{r_{it}} \cdot \mu_{r_{it}}}{K_{r_{it}}} \cdot \left[\varphi\left(\frac{a_{it} - \mu_{r_{it}}}{\sigma_{r_{it}}}\right) - \varphi\left(\frac{b_{it} - \mu_{r_{it}}}{\sigma_{r_{it}}}\right)\right] \tag{10.147}$$

From Eqs. (10.146) and (10.147), it can be obtained that

$$Var(r_{it}) = E\left(r_{it}^2\right) - [E(r_{it})]^2$$
$$= \sigma_{r_{it}}^2 + \frac{\sigma_{r_{it}}^2}{K_{r_{it}}} \cdot \left[\frac{a_{it} - \mu_{r_{it}}}{\sigma_{r_{it}}} \cdot \varphi\left(\frac{a_{it} - \mu_{r_{it}}}{\sigma_{r_{it}}}\right) - \frac{b_{it} - \mu_{r_{it}}}{\sigma_{r_{it}}} \cdot \varphi\left(\frac{b_{it} - \mu_{r_{it}}}{\sigma_{r_{it}}}\right)\right]$$
$$+ \mu_{r_{it}}^2 + \frac{2\sigma_{r_{it}} \cdot \mu_{r_{it}}}{K_{r_{it}}} \cdot \left[\varphi\left(\frac{a_{it} - \mu_{r_{it}}}{\sigma_{r_{it}}}\right) - \varphi\left(\frac{b_{it} - \mu_{r_{it}}}{\sigma_{r_{it}}}\right)\right]$$
$$- \left\{\frac{\sigma_{r_{it}}}{K_{r_{it}}} \cdot \left[\varphi\left(\frac{b_{it} - \mu_{r_{it}}}{\sigma_{r_{it}}}\right) - \varphi\left(\frac{a_{it} - \mu_{r_{it}}}{\sigma_{r_{it}}}\right)\right] + \mu_{r_{it}}\right\}^2 \tag{10.148}$$

To obtain the optimal solution of the model, Table 10.27 lists the input parameters of the model. Where μ_{it} and σ_{it} are distribution parameters of random variables m_{it}, representing the mean and standard deviation of resource prices, respectively, used to calculate $E(m_{it})$ of the objective functions (10.104) and (10.105), $Var(r_{it})$ of constraints (10.106) and (10.107), and $\Pr\left[\sum_{i=1}^{N} m_{it} \cdot q_{it} \le B_t\right]$ of constraints (10.108); $\tilde{f}_{it}$ isa four-dimensional fuzzy random variable, $\tilde{f}_{it} = \left(f_{it}^1, f_{it}^2, f_{it}^3, f_{it}^4\right)$ where each component represents the possible value of resource allocation cost, used to calculate $E\left(\tilde{f}_{it}\right)$ of the objective function (10.105); $\alpha_{1t} = 0.2, \alpha_{2t} = 0.3, \alpha_{3t} = 0.25$ is sharing ratios of profits generated by the three MRO services are 0.2, 0.3, and 0.25, respectively, used to constrain conditions (10.106) and (10.107); a_{it}, b_{it}, $\mu_{r_{it}}$ and $\sigma_{r_{it}}$ are the distribution parameters of random variables (as shown in Eqs. (10.144) and (10.145)), representing the lower, upper, mean, and standard deviation of user demand, respectively, used to calculate $E(r_{it})$ in the objective functions (10.104) and (10.105), $Var(m_{it})$ of the constraint conditions (10.106) and (10.107); RT_M and RT_A represent the maximum risk tolerance values of SMP and SAS, respectively, used for constraint conditions (10.106) and (10.107); L_t and B_t is the parameters of constraint condition (10.108) represent the probability of SMP meeting budget constraints and the set of budget cost values, respectively; N is the number of MRO service types; T is the number of operating cycles.

2. **Sensitivity analysis of calculation results and parameters**

Substitute the input parameters from Table 10.27 into the model, solve the corresponding analytical expressions for the first three steps, and obtain a multi-objective nonlinear mixed integer programming model equivalent to the original model. According to the steps in 6.3.2, it is necessary to first solve for $P_M^{\max}$ and $P_A^{\max}$. Specifically, the function Eqs. (10.131) and (10.132) can be used as objective functions for two independent single objective nonlinear mixed integer programming models, with constraints ranging from Eqs. (10.133) to (10.137).

Table 10.27 Input parameters of the model

Distribution parameters of resource prices	$\mu_{it} = 60, \sigma_{it} = 7, i = 1, 2, 3; t = 1, 2, \ldots, 6$
Resource allocation costs	$\tilde{f}_{it} = (10, 12, 13, 15), i = 1, 2, 3; t = 1, 2, \ldots, 6$
Profit sharing ratio	$\alpha_{1t} = 0.2, \alpha_{2t} = 0.3, \alpha_{3t} = 0.25, t = 1, 2, \ldots, 6$
Distribution parameters of demand	$a_{it} = 66, b_{it} = 80, \mu_{r_{it}} = 72, \sigma_{r_{it}} = 4, t = 1, 2, \ldots, 6$
Maximum risk tolerance of SMP	$RT_M = 5000$
The maximum risk tolerance of SAS	$RT_A = 1000$
Probability of SMP meeting budget constraints	$L_t = 0.95$
Global criterion method parameters	$\lambda = 2$
SMP Budget Cost Value Collection	$B_t \in \{80,000,\ 100,000\}, t = 1, 2, \ldots, 6$
The number of MRO service types	$N = 3$
Number of operating cycles	$T = 6$

Due to the deterministic nature of the transformed model, it can be directly solved using Lingo software, result that $P_M^{\max}$ and $P_A^{\max}$ is 520886 and 843,259, respectively. Then, a new objective function is constructed using Eq. (10.40), which converts the multi-objective model into a single objective model. Similarly, Lingo software is used to solve the problem, and the optimal solution G^* of the model is 0.9943. The values of each decision variable are listed in Table 10.28. Among them, d_{it} is the price of i type of MRO service in period t, and q_{it} is the number of resources matched with i type of MRO service in period t. For example, $d_{25} = 423$ represents the price for the second MRO service in the fifth period is 423; $q_{36} = 203$ represents the number of resources that match the 6th period with the 3rd MRO service is 203; $\omega_i = 1, i = 1, 2, \ldots, 6$ is Adjoint decision variables in random constraints (10.43).

From Table 10.28, it can also be seen that the price and configuration quantity of MRO services vary with different periods and types of services, reflecting the dynamism of decision-making. This is because in the process of seeking the optimal service strategy, the uncertainty in the MRO service chain (such as random user needs, fuzzy configuration costs, etc.) is fully considered and quantified, making the model more consistent with reality. Therefore, the obtained service strategy is more scientific and reasonable.

In the multi-objective fuzzy stochastic programming model mentioned above, the parameters related to fuzzy and random variables are $(f_{it}^1, f_{it}^2, f_{it}^3, f_{it}^4)$, μ_{it}, σ_{it}, $\mu_{r_{it}}$, $\sigma_{r_{it}}$, and L_t. To better analyze the impact of these parameters on the optimal solution of the model, the following parameter sensitivity experiments are conducted.

Table 10.28 Values of decision variables

q_{11}	d_{11}	q_{12}	d_{12}	q_{13}	d_{13}	q_{14}	d_{14}	q_{15}	d_{15}
154	615	108	615	137	615	145	615	149	615
q_{16}	d_{16}	q_{21}	d_{21}	q_{22}	d_{22}	q_{23}	d_{23}	q_{24}	d_{24}
173	615	142	484	208	490	195	571	89	727
q_{25}	d_{25}	q_{26}	d_{26}	q_{31}	d_{31}	q_{32}	d_{32}	q_{33}	d_{33}
113	423	168	466	224	545	190	545	182	545
q_{34}	d_{34}	q_{35}	d_{35}	q_{36}	d_{36}	w_1	w_2	w_3	w_4
177	545	118	545	203	545	1	1	1	1
w_5	w_6								
1	1								

1. Let $\left(f_{it}^1, f_{it}^2, f_{it}^3, f_{it}^4\right)$ take different values, that is changing the node values of the trapezoidal membership function of fuzzy random variables $\tilde{f}_{it}$

Due to $\left(f_{it}^1, f_{it}^2, f_{it}^3, f_{it}^4\right)$ can reflect the cost of fuzzy allocation of MRO resources, in order to study its impact on the MRO service chain win–win model, the four-dimensional fuzzy vector $\left(f_{it}^1, f_{it}^2, f_{it}^3, f_{it}^4\right)$ was taken from multiple sets of different values, recalculated, $P_A^{\max}$ and the corresponding model optimal solution G^* was obtained. The results are listed in Table 10.29, and Fig. 10.20 more intuitively illustrates the trend of $P_A^{\max}$ and G^* with $\tilde{f}_{it}$.

From Table 10.29 and Fig. 10.20, it can be seen that as the cost of MRO resource allocation increases, the maximum profit $P_A^{\max}$ of SAS decreases correspondingly and the optimal solution G^* of the model increases. This indicates that in real life, if uncertain factors such as personnel and environment lead to an increase in the allocation cost of MRO resources, it will not only reduce the maximum profit of

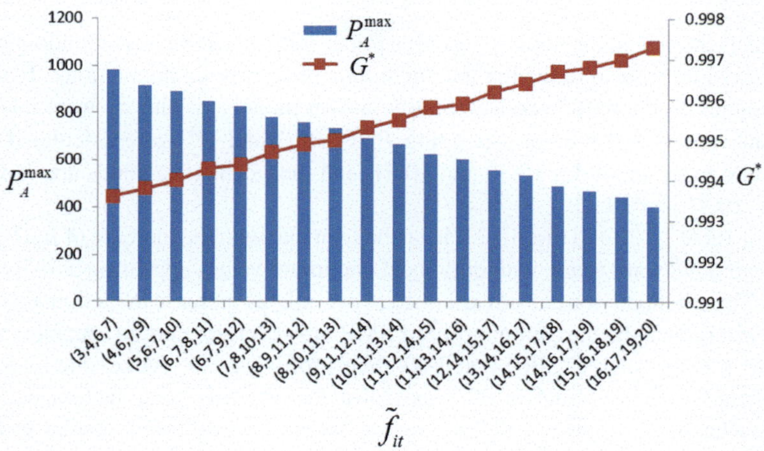

Fig. 10.20 $P_A^{\max}$ and G^* its variation with values $\tilde{f}_{it}$

Table 10.29 The values of $\tilde{f}_{it}$, $E(\tilde{f}_{it})$, P_A^{max} and G^*

$\tilde{f}_{it}$	$E(\tilde{f}_{it})$	P_A^{max} (Thousand)	G^*
(3,4,6,7)	10	982.148	0.9936
(4,6,7,9)	13	915.481	0.9938
(5,6,7,10)	14	893.259	0.9940
(6,7,8,11)	16	848.815	0.9943
(6,7,9,12)	17	826.593	0.9944
(7,8,10,13)	19	782.148	0.9947
(8,9,11,12)	20	759.926	0.9949
(8,10,11,13)	21	737.704	0.9950
(9,11,12,14)	23	693.259	0.9953
(10,11,13,14)	24	671.037	0.9955
(11,12,14,15)	26	626.593	0.9958
(11,13,14,16)	27	604.370	0.9959
(12,14,15,17)	29	559.926	0.9962
(13,14,16,17)	30	537.704	0.9964
(14,15,17,18)	32	493.259	0.9967
(14,16,17,19)	33	471.037	0.9968
(15,16,18,19)	34	448.815	0.9970
(16,17,19,20)	36	404.370	0.9973

SAS, but also hinder the two members of the service chain, SMP and SAS, from achieving a win–win situation. To avoid this situation, appropriate measures can be taken for SAS to stabilize resource allocation costs and promote mutual benefit.

2. Let μ_{it} and σ_{it} take different values, that is, change the mean and mean square deviation of random resource prices m_{it}

μ_{it} reflects the size of the random resource prices that match the corresponding MRO services, while σ_{it} also reflecting the magnitude of their price fluctuations. To study their impact on the MRO service chain win–win model, μ_{it} and σ_{it} were taken as different values, and the corresponding P_M^{max}, P_A^{max} and G^* were calculated. The results are listed in Tables 10.30 and 10.31, and their changing trends are depicted in Figs. 10.21 and 10.22.

From Table 10.30 and Fig. 10.21, it can be seen that with the increase of μ_{it},P_M^{max} is an overall downward trend, with only local maximum values appearing at individual points. This phenomenon can be explained as overall, the increase in random resource prices will reduce the maximum profit of SMP. However, for certain special random resource pricing, such as 50, 54, and 60, their profits may skyrocket due to certain special random factors (such as off peak season factors, directional factors, etc.). On the other hand, as μ_{it} grows, P_A^{max} shows an overall upward trend, similar to P_M^{max}, which also occurs of local minima at certain points. When the growth of μ_{it}, SAS exceeds a certain threshold, P_A^{max} also begins to decrease, indicating that a certain

Table 10.30 The values of μ_{it}, $P_M^{\max}$, $P_A^{\max}$ and G^*

μ_{it}	$P_M^{\max}$(Thousand)	$P_A^{\max}$(Thousand)	G^*
49	378.296	659.926	0.9939
50	520.904	676.593	0.9954
51	344.963	301.845	0.9969
52	328.296	709.926	0.9924
53	311.630	726.593	0.9918
54	520.897	404.198	0.9973
55	278.296	759.926	0.9904
56	261.630	776.593	0.9895
57	244.963	498.451	0.9928
58	228.296	809.926	0.9875
59	211.630	826.593	0.9862
60	520.886	843.259	0.9943
61	178.296	859.926	0.9830
62	161.630	876.593	0.9808
63	144.963	446.121	1.4151
64	128.296	409.925	1.4873

Table 10.31 The values of σ_{it}, $P_M^{\max}$, $P_A^{\max}$ and G^*

σ_{it}	$P_M^{\max}$(Thousand)	$P_A^{\max}$(Thousand)	G^*
1	683.345	967.050	0.9840
2	655.828	944.465	0.9852
3	635.173	929.576	0.9877
4	604.752	909.004	0.9885
5	589.369	883.708	0.9929
6	556.060	860.142	0.9940
7	520.886	843.259	0.9943
8	504.745	774.231	0.9954
9	494.863	725.281	0.9971
10	465.219	699.467	0.9986

range of resource price growth will increase the maximum profit of SAS. However, when the price growth exceeds a certain upper bound, its profit will actually decrease. It is worth mentioning that in the initial stage of growth of μ_{it}, G^* is in a stable state. When the growth of μ_{it} exceeds a certain upper bound, G^* starts to increase sharply, indicating that the increase in resource prices within a certain range will not significantly affect the achievement of a win–win situation between SMP and SAS. However, when the price increases to a certain extent, measures must be taken to control it, otherwise it will not only reduce the profits of both parties, but also hinder them from achieving a win–win situation.

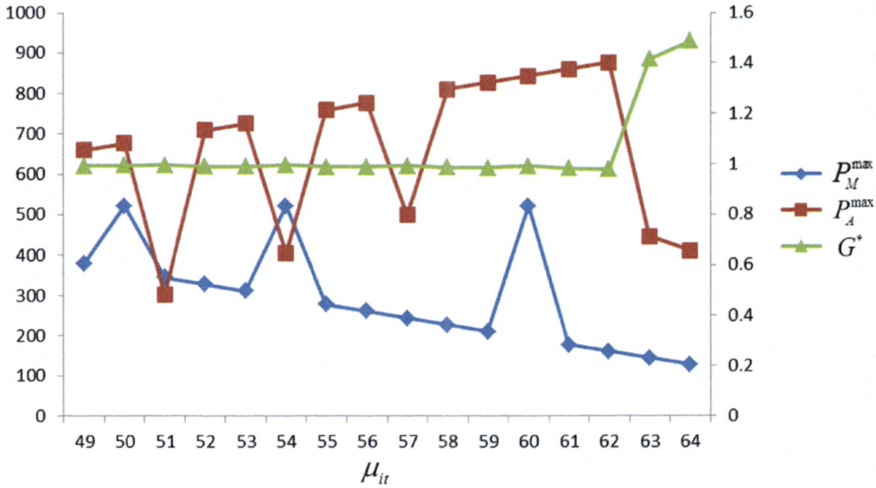

Fig. 10.21 $P_M^{\max}$, $P_A^{\max}$ and G^* its variation with values μ_{it}

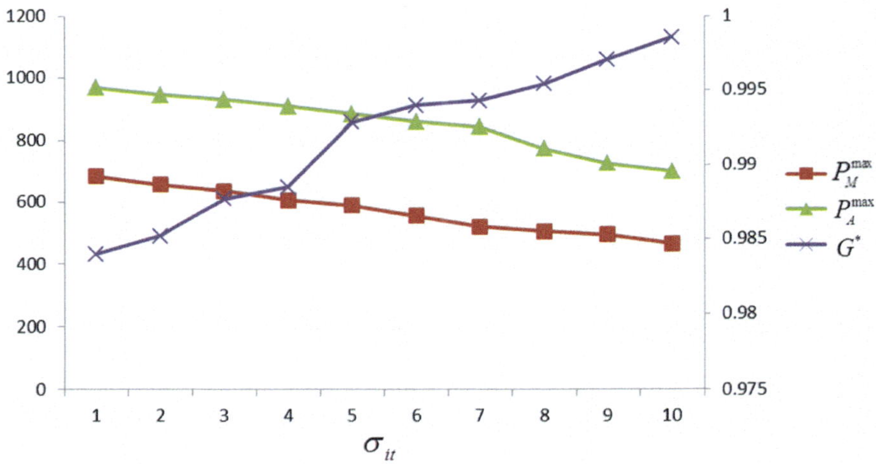

Fig. 10.22 $P_M^{\max}$, $P_A^{\max}$ and G^* its variation with values σ_{it}

From Table 10.31 and Fig. 10.22, it can be seen that with the increase of σ_{it}, $P_M^{\max}$ and $P_A^{\max}$ both show a downward trend, while G^* increasing at the same time. This indicates that as the fluctuation of random resource prices increases, on the one hand, the maximum profits of SMP and SAS will decrease, and on the other hand, it is difficult to achieve a win–win situation for both parties.

3. Take $\left(\mu_{r_{it}}, \sigma_{r_{it}}\right)$ different values, that is, change the parameters that reflect the random demand r_{it} of MRO services

$\mu_{r_{it}}$ reflects the size of random demand, while $\sigma_{r_{it}}$ also reflecting the amplitude of demand fluctuations. To study their impact on the MRO service chain win–win model, take $\left(\mu_{r_{it}}, \sigma_{r_{it}}\right)$ be multiple sets of different values were taken and the corresponding $P_M^{\max}$, $P_A^{\max}$ and G^* were calculated. The results are listed in Table 10.32, and Fig. 10.23 depicts the trend of $P_M^{\max}$, $P_A^{\max}$ and G^* with values $\left(\mu_{r_{it}}, \sigma_{r_{it}}\right)$.

From Fig. 10.23, it can be seen that with the increase of $\mu_{r_{it}}$,the overall trend of $P_M^{\max}$ is upward; When the value of $\mu_{r_{it}}$ is small, $P_M^{\max}$ remains stable as $\sigma_{r_{it}}$ increases, while when the value of $\mu_{r_{it}}$ is large, $P_A^{\max}$ fluctuates significantly with the increase of $\sigma_{r_{it}}$. This indicates that on the one hand, as user demand increases, the maximum profit of SMP increases; On the other hand, when the demand value is small, the magnitude of its fluctuation has little impact on the maximum profit of SMP. However, when the demand value is large, the magnitude of its fluctuation will also have a significant

Table 10.32 The values of $\left(\mu_{r_{it}}, \sigma_{r_{it}}\right)$, $P_M^{\max}$, $P_A^{\max}$ and G^*

$\left(\mu_{r_{it}}, \sigma_{r_{it}}\right)$	$P_M^{\max}$(Thousand)	$P_A^{\max}$(Thousand)	G^*
(68, 2)	312.997	816.667	0.9895
(69, 2)	133.333	822.222	0.9750
(70,2)	150.000	827.778	0.9777
(71,2)	574.647	833.333	0.9939
(72,2)	655.716	838.889	0.9949
(68,3)	274.746	816.667	0.9880
(69,3)	404.797	822.222	0.9918
(70,3)	613.639	827.778	0.9946
(71,3)	166.667	833.333	0.9798
(72,3)	183.333	838.889	0.9815
(68,4)	116.667	816.667	0.9716
(69,4)	133.333	822.222	0.9746
(70,4)	316.335	827.778	0.9895
(71,4)	478.421	833.333	0.9930
(72,4)	520.886	843.259	0.9943
(68,5)	116.667	816.667	0.9716
(69,5)	133.333	180.655	0.9946
(70,5)	150.000	827.778	0.9777
(71,5)	331.205	833.333	0.9899
(72,5)	616.580	838.889	0.9945
(68,6)	116.667	816.667	0.9716
(69,6)	133.333	822.222	0.9750
(70,6)	150.000	827.778	0.9777
(71,6)	166.667	278.621	0.9933
(72,6)	518.878	838.889	0.9935

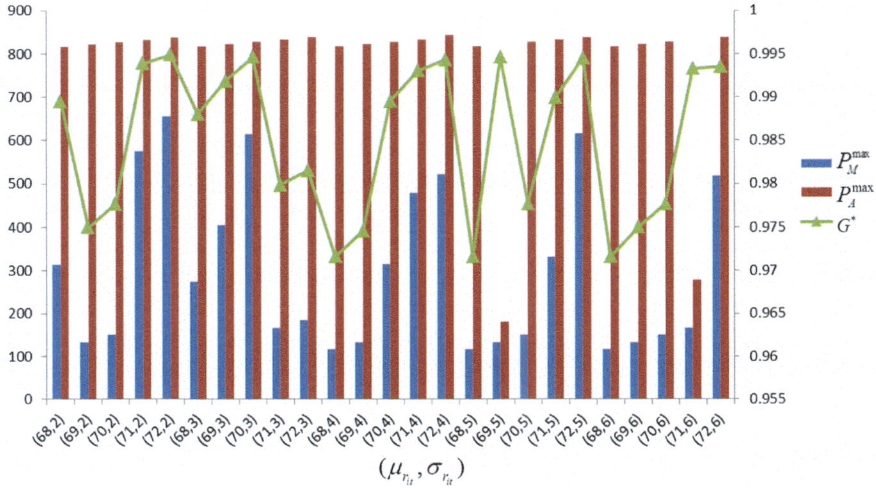

Fig. 10.23 $P_M^{\max}$, $P_A^{\max}$ and G^* its variation with values $(\mu_{r_{it}}, \sigma_{r_{it}})$

impact on the maximum profit of SMP. At the same time, it can also be seen that the changes in $\mu_{r_{it}}$ and $\sigma_{r_{it}}$ have almost no impact on $P_A^{\max}$, only a sudden $P_M^{\max}$ decrease at individual points. In addition, G^* has remained stable with only slight fluctuations at a few points. This indicates that the magnitude of user demand and fluctuations will not directly affect SAS's maximum profit, nor will it significantly hinder SMP and SAS from achieving a win–win situation. It is worth mentioning that both (69, 5) and (71, 6) $P_M^{\max}$ and $P_A^{\max}$ show a significant decrease and G^* increase, indicating that these two situations are not conducive to achieving maximum profits for both parties and hinder their achievement of a win–win situation.

4. Taking L_t different values is changing the parameters that reflect the level of SMP risk preference

The parameter L_t can reflect the risk preference level of SMP, and the smaller its value, the greater the risk preference level of SMP. Table 10.33 lists the corresponding $P_M^{\max}$, $P_A^{\max}$ and G^* values when different values L_t are taken.

From Table 10.33, it can be seen that as the value L_t increases, $P_M^{\max}$ shows an upward trend, G^* show a downward trend, and $P_A^{\max}$ remains unchanged. Figrue 10.23 more intuitively reflects this trend. This indicates that the smaller the risk preference of SMP, the more conservative SMP is, and the greater its maximum profit it can obtain; Although the risk preference level of SMP will not affect the maximum profit of SAS, it will affect the formation of a win–win situation for both parties. The smaller the risk preference level of SMP, the more conducive it is to the formation of a win–win situation.

Table 10.33 The values of L_t, $P_M^{\max}$, $P_A^{\max}$ and G^*

L_t	$P_M^{\max}$(Thousand)	$P_A^{\max}$ (Thousand)	G^*
0.78	398.873	843.259	0.9998
0.80	405.542	843.259	0.9989
0.84	422.430	843.259	0.9975
0.87	461.455	843.259	0.9968
0.91	488.003	843.259	0.9959
0.94	504.237	843.259	0.9950
0.95	520.886	843.259	0.9943
0.97	534.709	843.259	0.9937
0.98	540.916	843.259	0.9935
0.99	589.477	843.259	0.9926
0.994	628.878	843.259	0.9918
0.998	645.182	843.259	0.9887

References

1. Ye Z, Xie M. Stochastic modeling and analysis of degradation for highly correlated products [J] Applied Stochastic Models in Business and Industry, 2015, 31: 16–32
2. Ross S. Stochastic processes [M] New York, NY, USA: Wiley, 1996
3. Talluri S, Sarkis J. A model for performance monitoring of suppliers [J] International Journal of Production Research, 2002, 40 (16): 4257–4269
4. Hong G, Park S, Jang D, Rho H. An effective supplier selection method for constructing a competitive supply relationship [J] Expert Systems with Applications, 2005, 28 (4): 629–639
5. Hu F, Zong Q. Optimal production run time for a refining production system under an extended inspection policy [J] European Journal of Operational Research, 2009, 196 (3): 979–986
6. Mohammadi B, Taleizadeh A, Noorossana R, Samimi H. Optimizing integrated manufacturing and products inspection policy for refining manufacturing system with perfect inspection [J] Journal of Manufacturing Systems, 2014, 37: 299–315
7. Levitin G. The universal generating function in reliability analysis and optimization [M] London: Springer, 2005
8. Satty T. The Analytical Hierarchy Process [M] New York: McGraw Hill, 1980
9. (In Chinese) Xie Jijian, Liu Chengping Fuzzy mathematical methods and their applications [M] Wuhan: Huazhong University of Science and Technology Press, 2006
10. Huiskonen J. Maintenance spare parts logistics: Special characteristics and strategic choices [J] International Journal of Production Economics, 2001, 71 (1): 125–133
11. Zhen L, Wang k. A stochastic programming model for multi-product oriented multi-channel component replenishment [J] Computers & Operations Research, 2015, 60: 79–90
12. Avriel M, Williams A C. The Value of Information and Stochastic Programming [J] Operations Research, 1970, 18 (5): 947–954

Chapter 11
Operation Process Control Based on Cyber-Physical Systems

The Information Physical System (CPS) refers to an embedded system, logistics, coordination and management process, and online service that uses sensors to directly obtain physical data and actuators to act on physical processes. Digital twin is a digital image that contains all information and knowledge, and is a virtual representation of physical products. Data is transmitted through the Internet to achieve intelligent control of the remote operation process of physical entities.

11.1 CPS System

In 2006, the National Science Foundation (NSF) of the United States organized the first international seminar on information physical systems, and the concept of Cyber Physical Systems (CPS) was first proposed, attracting widespread attention from academia and industry. In August 2007, the Presidents Committee of Advisors on Science and Technology (PCAST) elevated CPS as a technological concept to the level of national strategy, rapidly sparking a CPS research boom in the United States. The European Union also launched projects such as ARTEMIS (Advanced Research and Technology for Embedded Intelligence and Systems) in 2007, with plans to invest over 7 billion US dollars in research related to CPS, and listed CPS as an important development direction for intelligent systems. In 2009, Germany clearly stated in the "National Embedded System Technology Roadmap" that CPS will be the fundamental technology for Germany to continue to stay at the forefront of manufacturing. Subsequently, in 2013, CPS was identified as the core technology for Industry 4.0 in the "Industry 4.0 Implementation Recommendations". In 2015, "Made in China 2025" made "promoting the deep integration of informatization and industrialization" one of the important tasks, and proposed that "intelligent manufacturing based on information physical systems, such as intelligent equipment and intelligent factories, is leading the transformation of manufacturing methods". In

M. Liu et al., *Intelligent Predictive Maintenance*, Advanced and Intelligent Manufacturing in China, https://doi.org/10.1007/978-981-97-2677-6_11

May 2016, the State Council issued the Guiding Opinions on Deepening the Integration and Development of Manufacturing Industry and the Internet, which regards information physical systems as an important foundation for the integration and development of manufacturing industry and the Internet, further clarifying the future development of China's CPS.

11.1.1 Definition of CPS

Since the concept of CPS was first proposed in 2006, many institutions and scholars at home and abroad have conducted in-depth research and exploration on CPS, with different concepts and definitions of information physical systems. The definition of CPS by the National Natural Science Foundation (NSF) in the United States focuses on the organic integration and deep collaboration of 3C (control communication computer) technology. CPS is a physical, biological, and engineering system that perceives, controls, and integrates through a computing core (embedded system); The National Institute of Standards and Technology (NIST) in the United States interprets CPS as focusing on the functions of independent units of CPS and their interactions. CPS combines computing, communication, perception, and driving with physical systems, and interacts with the environment (including humans) to varying degrees to achieve time required functionality; The German Academy of Sciences and Engineering (Deutsche Akademie der Technikwissenschaft, Acatech) emphasizes the behavior of data acquisition, processing, and feedback in CPS. Its definition revolves around data acquisition, interpretation, and feedback control. CPS refers to embedded systems, logistics, coordination and management processes, and online services that use sensors to directly obtain physical data and actuators to act on physical processes. Many scholars have defined CPS, and Baheti et al. [1] defined CPS as a highly reliable system in which computing and physical units are tightly integrated and can interact to cope with dynamic uncertain events. Fitzgerald et al. [2] believe that CPS is composed of the physical world and numerous collaborative computational elements that interact with the physical world. In China, Academician He Jifeng was the first to provide the definition of CPS: CPS is a scalable networked system with computing, communication, and control capabilities, and its establishment is based on reliable perception of the environment. According to the definition in the White Paper on Information Physical Systems [4], CPS integrates advanced information technologies such as perception, computing, communication, and control, as well as automatic control technologies, to construct a complex system in which human, machine, object, environment, and information elements in physical space and information space are mapped, timely interacted, and efficiently coordinated. This enables on-demand response, rapid iteration, and dynamic optimization of resource allocation and operation within the system.

11.1.2 Current Research of CPS

Scholars have conducted extensive research on CPS, mainly focusing on CPS modeling, data transmission management technology, and CPS applications. CPS modeling is not only the modeling of independent physical and digital environments, but also the integration and interaction between physical and digital models. Lee et al. proposed a periodic and high error tolerant CPS task model, which achieves real-time task tracking and optimization while ensuring efficiency and stability through the collaborative effect between different subsystems. At the same time, it makes the CPS environment highly fault-tolerant, Able to operate effectively even if some information is not updated in a timely manner. Liu Chun et al. proposed a goal based approach to analyze and establish an event model for information physical fusion systems based on user needs. Wang Yun et al. constructed a preliminary modeling and simulation verification platform for power grid information physical systems based on the fusion model and information model of hybrid systems. This platform can be used to establish continuous and discrete models of the power grid, simulate the dynamic operation of the power grid, simulate the control process and information exchange, and verify the correctness and consistency of information exchange.

The data communication network is also one of the focuses of research on CPS environment construction. Ahmadu et al. established a new transmission protocol for sensor networks in CPS, which can dynamically change the reliability of transmission based on its importance during data collection, and aggregate data from nodes to base stations, effectively reducing energy consumption and bandwidth. Li et al. proposed a distributed joint blocking control and channel management algorithm, which uses Lagrange bidirectional decomposition to solve the problem of maximizing network efficiency. Persson [10] proposed a communication protocol based on CPS environment, which ensures effective information exchange between different individuals and objects through various wireless communication technologies such as Bluetooth, near-field communication, Wi Fi, etc., facilitating data fusion between different sources.

Compared to other fields, CPS's research in the fields of smart grids, biological and medical systems, smart highways, and autonomous driving is slightly more mature, and typical cases have appeared. The application results in these fields all reflect the effectiveness and universal applicability of CPS.

In the field of intelligent manufacturing, its products and intelligent production processes face enormous challenges, and CPS and services are important technological paths to solve this problem.

In terms of production organization, Yao [11] applied the Social Cyber Physical System (SCPS) to 3D printing technology to achieve large-scale personalized customized production. Yanjing et al. [12] proposed technical challenges such as multi-source information perception, spatiotemporal information fusion and interaction, heterogeneous network unified transmission, collaborative control, and engineering standardization, in response to the lack of systematic design for safety production related operation modes in existing Mine Cyber Physical Systems. They elevated the operation status control of mining production activities from a single device and a single system to full system process collaboration. Potente et al. [13] proposed the Cyber Physical Production System (CPPS), which improves the collaborative productivity between people, between people and intelligent devices, and between intelligent devices.

In termsof production planning, Monostori et al. [14] applied CPS to the machining process of mechanical parts, which can automatically generate machining plans and achieve adaptive production. Richard et al. [15] applied a complete CPS system to the practical reuse, refurbishment, and reuse of electrical and electronic equipment, aiming to increase traceability and support decision-making in the refurbishment process. CPS can help the growth of the electrical and electronic equipment recycling market through correct and traceable refurbishment, not only benefiting commercial companies but also benefiting the environment. Berger [16] proposed a reactive production planning and control based on CPPS, which saved production costs.

In terms of production control, Xuefeng [17] proposed a task scheduling algorithm for coal preparation plants based on improved particle swarm optimization algorithm by establishing a CPS model for analysis, and dynamically scheduled the coal preparation process; Pirvu [18] proposed a new CPS model in a distributed production control system, which realizes the intelligence of product manufacturing and assembly. Cuntao [19] designed an automated manufacturing system for automotive aluminum alloy casting wheel hubs based on CPS. Using industrial robots as the platform, a processing system was constructed using cloud computing, big data, and the Internet of Things technologies to achieve remote, reliable, real-time, and precise control of automotive wheel hub production lines, as well as automation and intelligent manufacturing of wheel hubs. Gang et al. [20] considered that existing transformer voltage regulation methods are limited by voltage regulation levels, and it is difficult to achieve efficient stable output voltage regulation based on the voltage curve determined by the scheduling department. They proposed a voltage stabilization dynamic regulation method based on an information physical fusion system model, established an information physical fusion system model for transformers, and implemented a real-time feedback mechanism between information and physical quantities, Control the automatic adjustment of the turns ratio of the primary and secondary coils of the transformer to achieve stable output voltage of the transformer. Liu et al. proposed the concept of Cyber Physical Machine Tool (CPMT), which is a CPS supported machine tool that integrates physical machine tools, machining processes, and computing capabilities. Augmented Reality (AR) technology is used to achieve intuitive and efficient human–machine interaction between humans and CPMT, and auxiliary intelligent windows provide users with intuitive interaction with CPMT to improve the intelligence and autonomy of machine tools.

The production management of manufacturing systems based on CPS achieves mutual influence and real-time feedback between information and physical processes through the analysis, fusion, and mining of environmental perception data through CPS, and autonomously coordinates the production process to achieve efficient production management.

11.1.3 CPS Architecture

(1) CPS operation mode

CPS is a distributed system connected to a network, which, in terms of physical composition, includes components such as sensors, data collection devices, physical devices, computing devices, storage devices, actuators, and communication networks. In terms of logical composition, CPS can be divided into three parts: computing, communication, and control. The computing part includes data collection, data storage, and control computing functions. The communication part includes sensor networks and control networks, which are abstracted as communication functions. The control part includes sensors and control devices, providing both perception and control capabilities.

The operation mode of CPS is shown in Fig. 11.1. The physical layer of CPS perceives the device to obtain environmental information, processes the data and sends it to the information layer, and executes control instructions transmitted by the information layer. After obtaining perceptual data, the information layer calculates the output execution instructions through semantic rules, transmits them to the physical system through actuators, controls the operations of the physical system, and achieves the fusion between the actual system and the virtual system.

In the CPS of manufacturing systems, the physical layer composed of sensors, controllers, etc., and the information layer composed of production line models, computing equipment, etc. are integrated with each other. The information layer perceives the status data of the physical equipment and generates control instructions for the physical layer to execute through data processing and calculation, thereby completing the feedback loop.

(2) CPS architecture

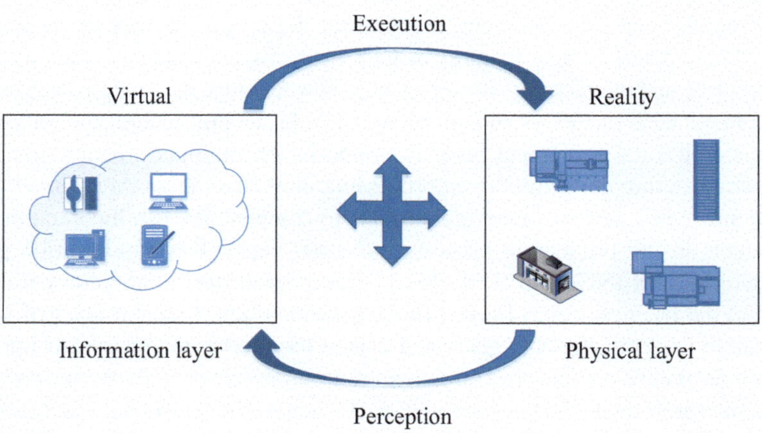

Fig. 11.1 CPS operation mode

Currently, research on the reference architecture of information physical systems is still in the exploratory stage in various countries and organizations. Due to the wide range of fields involved in CPS, academic research on the CPS system is often limited to specific requirements and backgrounds. Based on the composition and operation mode of CPS, expansion and improvement are carried out to design the required system framework.

Fei et al. proposed an architecture for a digital twin workshop (Fig. 11.2), which mainly includes 5 layers: (1) the physical layer, which mainly refers to the physical workshop entities such as workshops, people, machinery and materials, as well as the collection of corresponding production activities in the physical workshop. It is responsible for the physical space realization of product production and processing, and has functions such as self perception, self decision-making, and low-level data collection and transmission; (2) The model layer mainly refers to the virtual workshop and its corresponding virtual production activities in the digital twin workshop, including various models, rule knowledge, etc. of the virtual workshop, responsible for the simulation, analysis, optimization, decision-making, etc. of production activities in the virtual space; (3) The data layer refers to the workshop twin data service platform, which is responsible for providing data support services for the physical workshop, virtual workshop, and workshop service system operation of the digital twin workshop, and has data lifecycle management and processing functions such as generation, processing, integration, and fusion of workshop twin data; (4) The service layer is responsible for providing various workshop production services such as intelligent scheduling, collaborative process planning, product quality management, production process control, equipment health management, and energy efficiency optimization analysis for workshop production; (5) The application layer mainly refers to the application requirements of intelligent manufacturing tasks such as intelligent production, precise control, and reliable operation and maintenance involved in carrying out specific product processing and production.

Ming et al. proposed multiple requirements for CPS, including communication, computing, network, environment, and control, based on the new manufacturing model. They identified new features of CPS in intelligent factories, such as openness, security, and interactivity. Based on these new features, they constructed a cloud service based CPS model, as shown in Fig. 11.3. In the physical execution system, the on-site physical equipment layer is composed of intelligent recognition units, intelligent shooting units, intelligent processing units, intelligent three-dimensional storage units, etc., and uses intelligent robots to complete the handling and picking of products in the production process. Industrial network communication protocols such as PROFINET and IEEE 802.11 can connect the on-site physical device layer and the business control layer. In the integrated management layer of virtual information systems, the data generated during the production process is operated by relational databases deployed on synchronous interactive communication servers. For the enterprise layer, the large amount of data generated in processes such as simulation and the Internet of Things is stored in a medium-sized structured database. Managers can remotely control and manage the visual dynamic production process of product customization through a dynamic simulation platform; Basic information

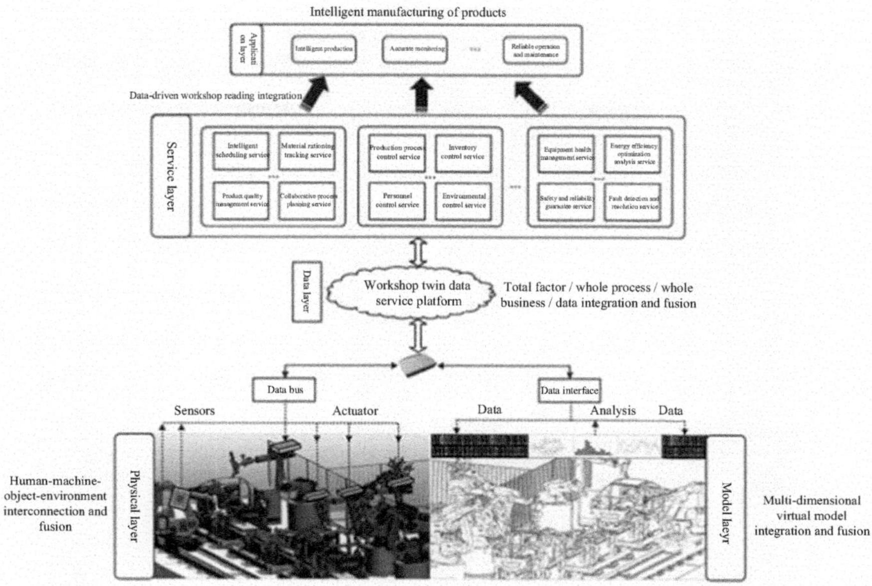

Fig. 11.2 Reference system architecture of digital twin workshop

of product components can be traced through industrial IoT platforms; At the same time, network servers and structured databases can provide users with some data analysis services and can be optimized according to their needs. At the user interface layer, users can track the production process of orders by operating applications such as computer terminals and mobile terminals, and can participate in the entire product production process in real time.

Liu et al. proposed the concept of Cyber Physical Machine Tool (CPMT) based on the concepts of CPS and the Internet of Things, and presented the structure of a three-layer information physical production system based on CPMT, as shown in Fig. 11.4 for the physical layer, information layer, and cloud service layer.

The physical layer contains all the physical elements involved in the manufacturing system, including machine tools and their components, cutting tools, workpieces, industrial robots, Automatic Guided Vehicles (AGVs), and various data acquisition devices. At the same time, key objects that may generate valuable data during the manufacturing process, such as machine tool spindles, tools, workpieces, etc., are equipped with sensors and actuators, so that real-time data can be collected and assigned tasks can be executed.

The information layer is a cyberspace composed of Digital Twin corresponding to the physical layer, including MTCT (Machine Tool Cyber Twin), each of which represents the digital abstraction of its physical counterpart. On the one hand, utilizing embedded algorithms to analyze real-time data collected from the physical layer enables Digital Twin to monitor and control physical peers. On the other hand, the M2M interface allows Digital Twin to communicate with each other, thereby

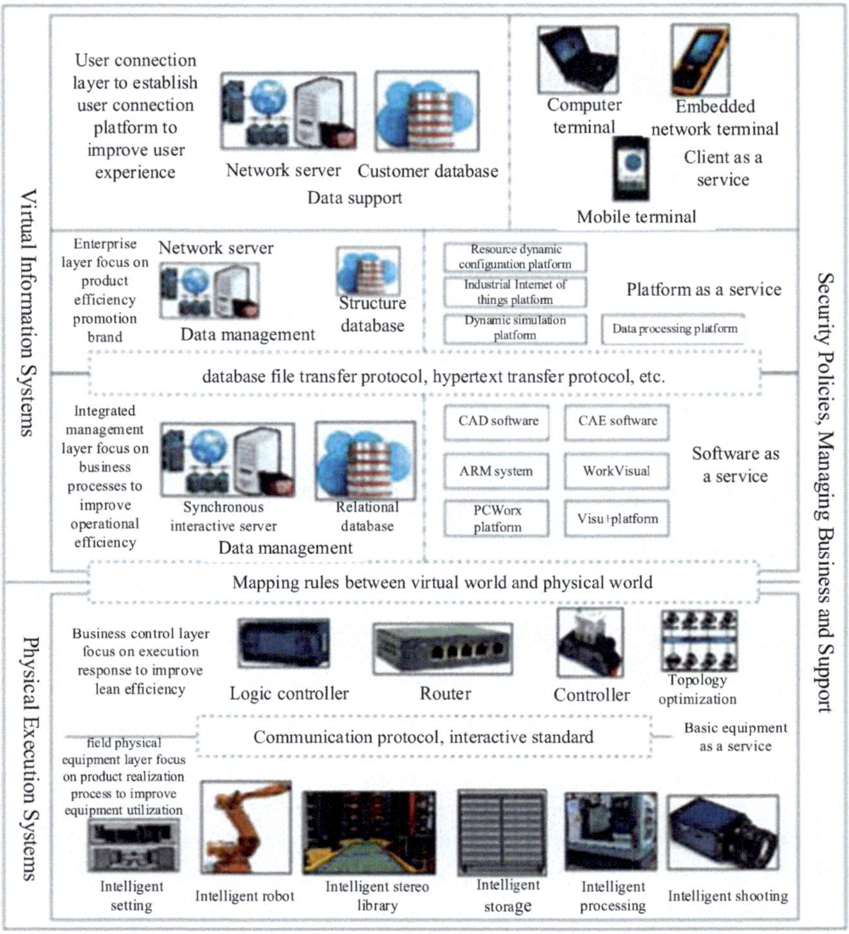

Fig. 11.3 Deployment strategy model based on CPS new features

achieving autonomous cooperation between field level manufacturing equipment. In addition, Digital Twin record the historical information of their physical counterparts and provide it to the cloud through various networks for further analysis and value-added services to directly access on-site manufacturing data.

The service cloud contains various software applications provided by different device manufacturers and third-party service providers, which reside in the service cloud and can access information in Cyberspace through the network.

The typical architecture of the information physics fusion manufacturing system proposed by Zhixia et al. is shown in Fig. 11.5. The entire system can be divided into three levels: sensing layer, communication layer, and decision layer. The sensing layer is the physical layer of sensing, composed of several sensing nodes. On the one hand, it is responsible for perceiving certain physical properties of the concerned

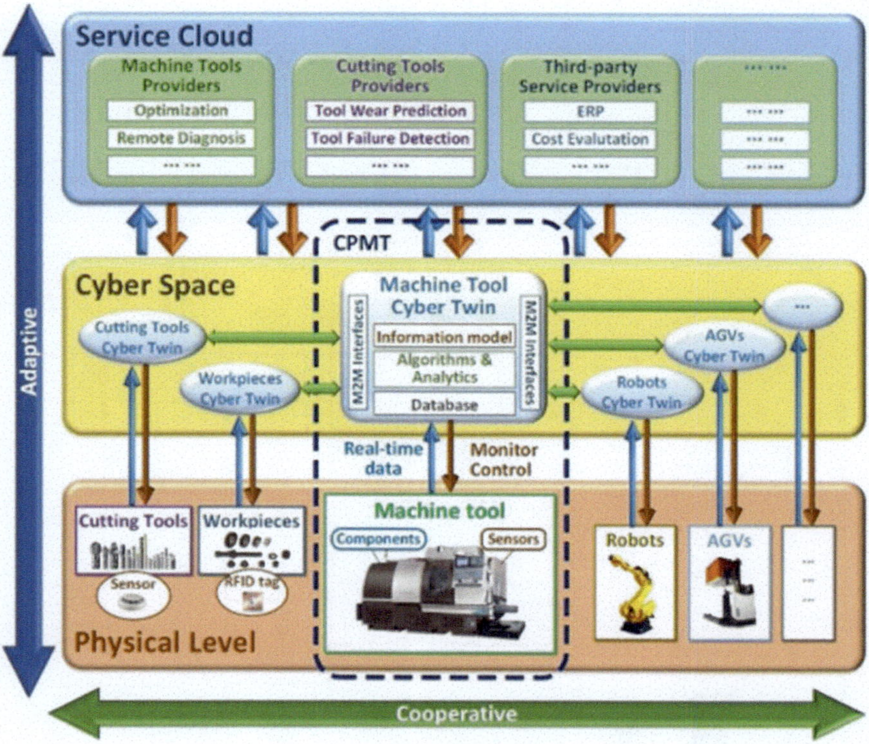

Fig. 11.4 Three layer information physical production system based on CPMT

physical equipment/facility, such as the pose of the positioning fixture, the size, force, and deformation of the workpiece Temperature and other state parameters, or a specific event that occurs, such as the need to replace tools during the machining process; On the other hand, based on the received monitoring commands or control instructions, corresponding operations are executed, such as initiating a processing task or starting to measure a physical property. The collected raw information data is fused and transmitted to the decision-making layer.

Communication refers to the network computing layer, which includes various networks established according to actual needs, such as wired broadband, WiFi, ZigBee, 3G/4G, several communication base stations and network nodes, as well as distributed related databases, knowledge base servers, and information processing servers. It is responsible for the fusion processing and network storage of multi-sensor data, as well as the network transmission and exchange of related data.

The decision-making layer directly interacts with the system by end users, including two functional modules: simulation control center and decision control unit. The simulation control center establishes geometric entity models, integrated information models, and sensing behavior models for each manufacturing element with the support of a virtual manufacturing environment. Based on a complete digital

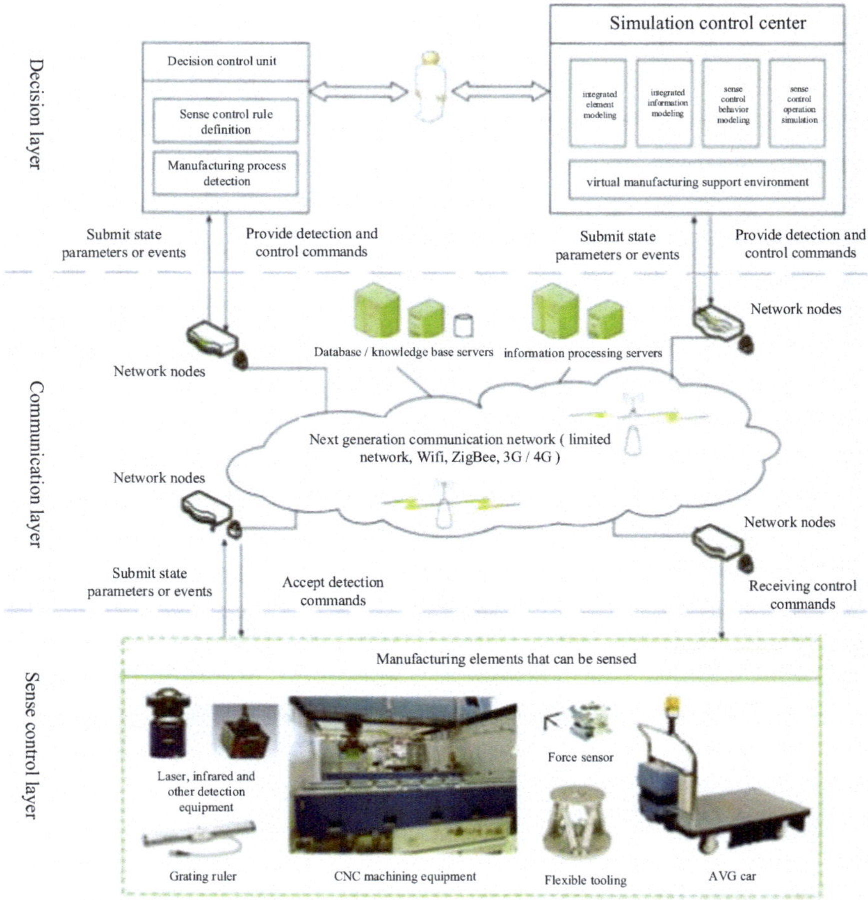

Fig. 11.5 Typical architecture of information physical fusion manufacturing

simulation model and theoretical data of physical properties, it realizes the simulation of sensing operation processes between manufacturing elements, generates initial control schemes, and verifies and improves the digital simulation model through actual data obtained from physical perception, The control scheme obtained through the combination of offline and online simulation is a reliable scheme verified through simulation experiments. The decision control unit provides users with interactive operation functions for defining the sensing rules of manufacturing elements and monitoring the manufacturing process. This activity can also be completed based on the simulation of the sensing operation process, greatly improving the efficiency and safety of manufacturing system operation and control.

From the CPS architecture of the manufacturing systems listed above, it can be seen that generally speaking, the CPS framework mainly covers two functions: (1) Integration: integrating information space with physical space to ensure real-time collection of data information from physical processes, with advanced connectivity capabilities for feedback from control modules to physical processes; (2) Intelligence: With intelligent information processing modules, it can achieve the ability of intelligent data management, analysis, and calculation in the information and control space. Therefore, the functional architecture of information physical systems proposed by the National Institute of Standards and Technology (NIST) in the Framework for Cyber Physical Systems Release 1.0 [24] report is introduced here, including the information physical domain and interconnect domain shown in Fig. 11.6.

Cyber Physical Domain includes physical entities, sensors, actuators that perform functions in the physical world, and media that interact between information entities, exerting control over physical entities through perception, execution, and communication. The main function of the interconnected domain is to connect CPS, collect data from these systems, convert data into information, and analyze information globally to obtain results of operational status monitoring or interactive environment status.

Green arrow: Information flow; Grey arrow: Decision flow; Red arrow: execution flow; Orange arrow: Energy and material flow; Purple arrow: Manage flow.

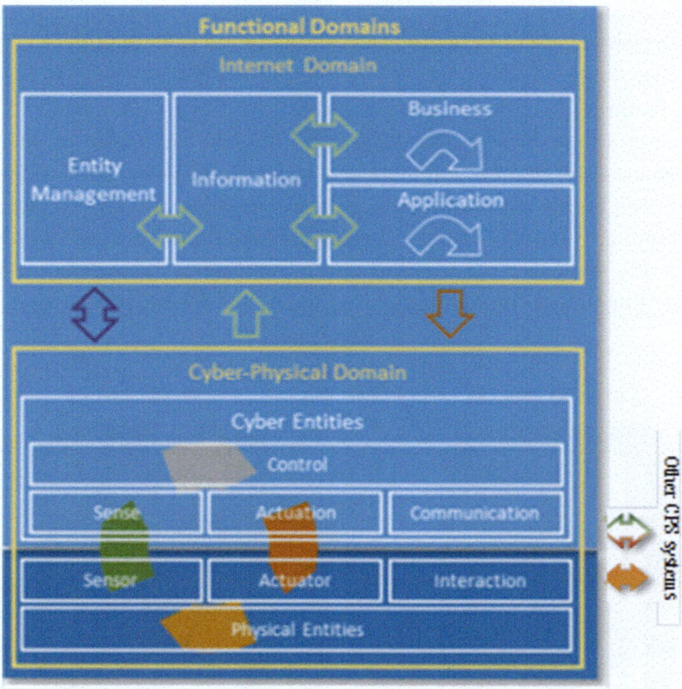

Fig. 11.6 Functional architecture of NIST information physical system

11.2 Intelligent Control Model for Operation Process Based on Digital Twin

The concept of Digital Twin was first proposed by Professor Grieves of the University of Michigan in 2003 in his course. In 2011, Professor Michael Grieves cited the term "Digital Twin" used by his co-author John Vickers to describe the conceptual model in his book "Almost Perfect: Driving Innovation and Lean Products through PLM", which has been used ever since. The conceptual model is shown in Fig. 11.7, which includes three main parts: ① physical products in physical space; ② Virtual products in virtual spaces; ③ The data and information exchange interface between physical space and virtual space [25].

The elements of physical space include all actual and virtual resources in the entire physical world. The actual resources mainly include people, equipment, materials, environment, and other objects that can be collected using information perception devices for data collection; Virtual resources include software resources, knowledge, information, and other resources that cannot be directly collected by intelligent devices. Most of them are existing software systems and some unstructured information, such as human experience, scheduling rules, etc. The physical space encompasses all production factors and is a huge, heterogeneous dynamic system. Virtual space corresponds to physical space. Actual resources have unique corresponding mirrors in virtual space, and their states and behaviors are synchronized with physical spatial features. Specifically, virtual space is a digital description of physical space, which is a dynamically evolving digital model. The model in virtual space constantly adjusts and changes with changes in real-world data. Real time synchronization between virtual space and physical space elements is achieved through data, and the comprehensive data of physical world elements can be reflected in real-time in the digital description model of virtual space; Both can achieve dynamic evolution and real-time interaction through various services, and virtual space can change the

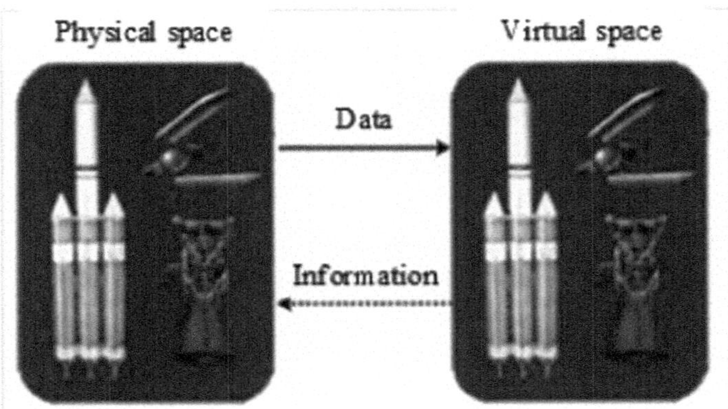

Fig. 11.7 Conceptual model of digital twin [26]

state and behavior of actual spatial elements through services, as well as drive the evolution of virtual models.

In the field of manufacturing, the earliest use of Digital Twin was in the Apollo program of the National Aeronautics and Space Administration (NASA) in the United States. Digital Twin were introduced into the health maintenance and decision-making of spacecraft, and through simulation experiments, they accurately reflect and predict the status of the spacecraft during mission execution, thereby helping pilots make correct decisions. In 2012, NASA proposed the definition of digital twin for aircraft and flight systems: "a comprehensive multi physics, multi-scale, probabilistic simulation model for aircraft or systems that uses the optimal physical model, updated sensor data, and fleet historical data to reflect the state of the twin corresponding to the model". In the same year, in the "Modeling, Simulation, Information Technology, and Processing" roadmap released by NASA, Digital Twin officially entered the public's view, highlighting characteristics such as integration, multi material rationality, multi-scale, and probability.

In the context of mass customization, flexible production, part tracking, product self-awareness, and communication between machines and other products in Industry 4.0, on-site equipment, machines, factories, and individual products will all be connected to the network. As a key concept of Industry 4.0, the Information Physical System (CPS) can be described as a set of physical devices, objects, and facilities that interact with virtual information spaces through communication networks. Among them, each physical device uses its virtual information part as the digital representation of its real device, ultimately forming a "digital twin". In information physical systems, Digital Twin are like virtual representations of physical products, a digital image that contains all information and knowledge. They are connected to the physical part through data transmission from the physical part to the network part, allowing for monitoring and control of physical entities, which can send data to update their virtual models.

11.2.1 Application of Digital Twin Models

At present, Digital Twin have been applied in many industries such as healthcare, education, construction, automotive, and aviation. In the manufacturing field, Digital Twin are considered a key technology for manufacturing enterprises to move towards Industry 4.0 strategic goals. With the full transparency of the production process, decision-makers can easily identify areas that need improvement in product design and related manufacturing processes, and make corresponding operational adjustments. Production management personnel can remotely monitor the factory through the virtual image platform, enabling them to obtain timely information on production, quality, orders, etc. anytime and anywhere, improving management response speed and transparency. By utilizing digital services to improve equipment utilization, improve equipment maintenance quality, optimize energy efficiency, and improve the speed and quality of information services, potential value can be discovered from data, enabling data to services and turning big data in digital factories into meaningful information, enabling intelligent decision-making.

Production scheduling is the neural center for decision optimization, process control, and performance improvement in production workshops. The scheduling driven by Digital Twin is supported by the digital twin system. Through the virtual real mapping and interaction fusion of all elements, all data, all models, and all spaces, a new scheduling mechanism is formed, which includes virtual real response, virtual real interaction, virtual control, and iterative optimization, Realize collaborative matching and continuous optimization of the scheduling elements of "workpiece machine constraint goal". Under the scheduling driven by Digital Twin, scheduling elements are mapped between physical and virtual workshops to form a collaborative optimization network of virtual and real symbiosis. The physical workshop actively perceives the production status, while the virtual workshop analyzes the scheduling status, adjusts scheduling plans, and evaluates scheduling decisions through self-organization, self-learning, and self simulation methods. It quickly determines the range of anomalies, responds quickly, and makes intelligent decisions, with better adaptability to changes, disturbance response, and anomaly resolution capabilities.

Manufacturing energy consumption management refers to the monitoring, analysis, control, and optimization of energy consumption such as water, electricity, gas, heat, and raw materials during the manufacturing process, while effectively ensuring the performance of the manufacturing system and the economic benefits of the enterprise. This enables precise management of energy consumption, achieving energy conservation and emission reduction, reducing manufacturing costs, and maintaining competitiveness. Manufacturing energy consumption management based on Digital Twin refers to the perception of energy consumption information, production factor information, and production behavior status information in the physical workshop through various sensing technologies. In the virtual workshop, production factors and behavior in the physical workshop are truly reflected and simulated. Through continuous interaction between the physical workshop and the virtual workshop in the actual production process, real-time control and iterative optimization of manufacturing energy consumption in the physical workshop are achieved.

The above applications mainly focus on the system level applications of Digital Twin, and there are also related applications in the field of device management for "device Digital Twin". To improve the processing capability of machine tools, Cai et al. proposed the information physical manufacturing of digital twin virtual machine tools, integrating manufacturing data and sensory data onto digital twin virtual machine tools to improve their reliability and capability in information physical manufacturing. Simultaneously utilizing sensory data to extract the machining characteristics of CNC machine tools, enabling them to better reflect their physical corresponding actual states in various applications. It is also proposed to use sensors to capture machine specific characteristics and data and information fusion analysis techniques for modeling and developing digital twin virtual machine tools.

Figure 11.8 shows a schematic diagram of the developed digital twin virtual machine tool integrated with sensing data and machining information. Manufacturing data includes key machining parameters that describe the manufacturing process, such as workpiece material and size, spindle speed, feed rate, cutting depth, G-code, tool position, etc. Sensor data is obtained from various sensors used for processing monitoring and analysis, such as current sensors for power measurement, acceleration sensors for vibration measurement, force meters for force measurement, and acoustic emission sensors. The overall architecture of data collection and management is shown in Fig. 11.9. Manufacturing data is obtained from machine controllers, while sensor data comes from sensor data collection devices. They are organized on a gateway and uploaded to the database via the internet (see Fig. 11.9). Users can access data through the internet using devices such as computers and tablets.

Merge these two types of data sources into one virtual machine tool to build their feature configuration files. For example, in some cases, a machine tool may experience more significant vibration in a specific cutting direction under the same cutting parameters, or when using the same tool and G-code to cut the same part, the machine tool will consume more power than another machine tool of the same type. These characteristics make virtual machine tools more in line with the true state of their physical objects. After each machining task, the characteristic curve of the machine tool is updated, which helps to track and predict the performance of the machine. Therefore, when using virtual machine tools to simulate the manufacturing of new

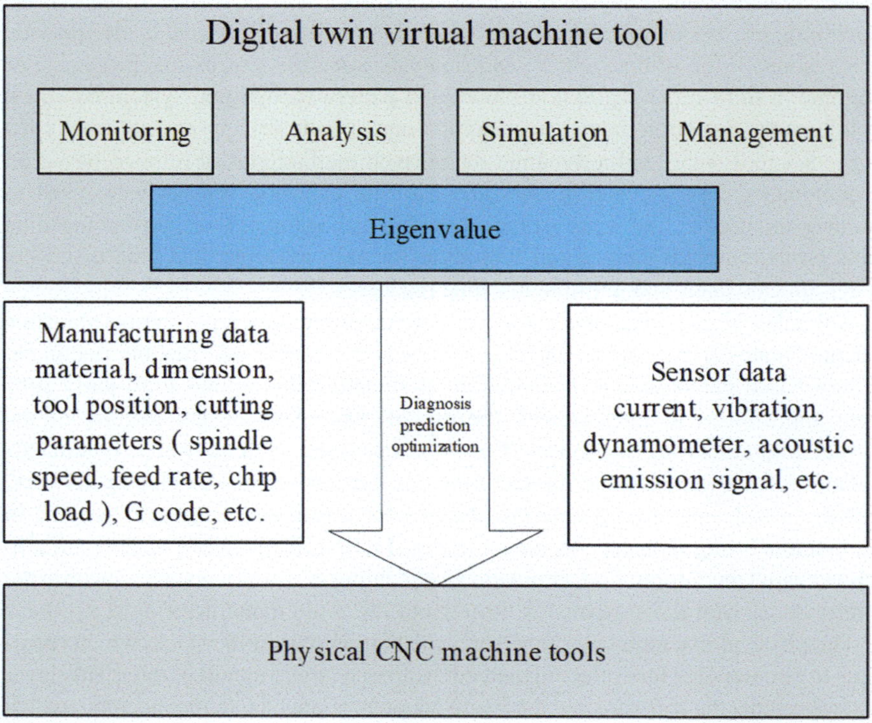

Fig. 11.8 Construction of a digital twin virtual machine tool by integrating sensory data and manufacturing information

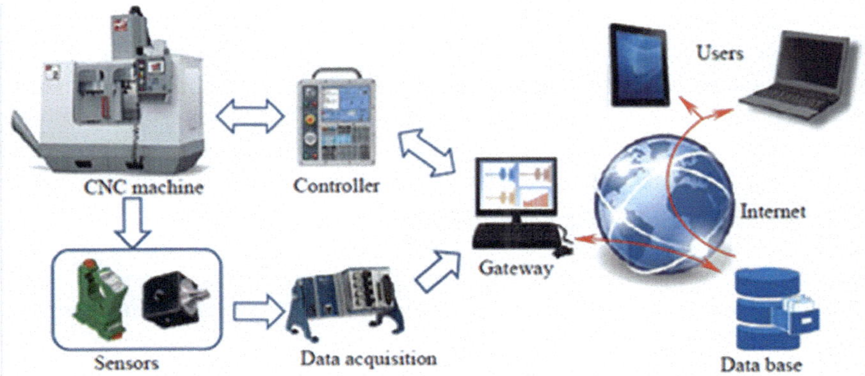

Fig. 11.9 System architecture for data collection and management

parts, the simulation results can reflect the latest state of the machine tool and have higher accuracy. The ultimate goal of building Digital Twin is to provide more accurate diagnostic, predictive, and optimization functions to improve the performance and utilization of physical CNC machine tools.

By collecting and integrating two types of data, it is possible to monitor the machining process more comprehensively and in real-time. These data also provide rich resources for offline review and analysis after the processing process. For example, a third-order regression model and pattern recognition system based on simultaneous detection data of acceleration and spindle driving current were used to predict tool wear, and a dynamic data-driven method was developed based on dynamometer data. Accelerometers and acoustic emission sensors were used to monitor and predict tool wear. At the workshop management level, virtual machine tools provide the advantages and limitations of each machine, and help managers better allocate tasks and multiple machine resources.

Angrishet al. [48] described how to store streaming data representing real-time machine states from physical machine tools in a scalable and flexible document pattern based database, and proposed an architecture that allows third-party software applications to interact with the digital twin. And the architecture of the Virtual Manufacturing Machines (CyMOS) operating system, which commands and controls various virtual manufacturing machines in the manufacturing network, was discussed, opening up new possibilities for information physical systems in the manufacturing industry. As shown in Fig. 11.10, this operating system can also achieve rapid reconfiguration, simulation, scheduling of internal resources within the enterprise, as well as its expansion throughout the entire manufacturing ecosystem.

The physical machine layer represents the bottom part of the stack, which represents the boundaries for collecting and controlling a large amount of data. This layer also represents the software and hardware adapters connected to the machine, which are necessary for communication with the upper layer.

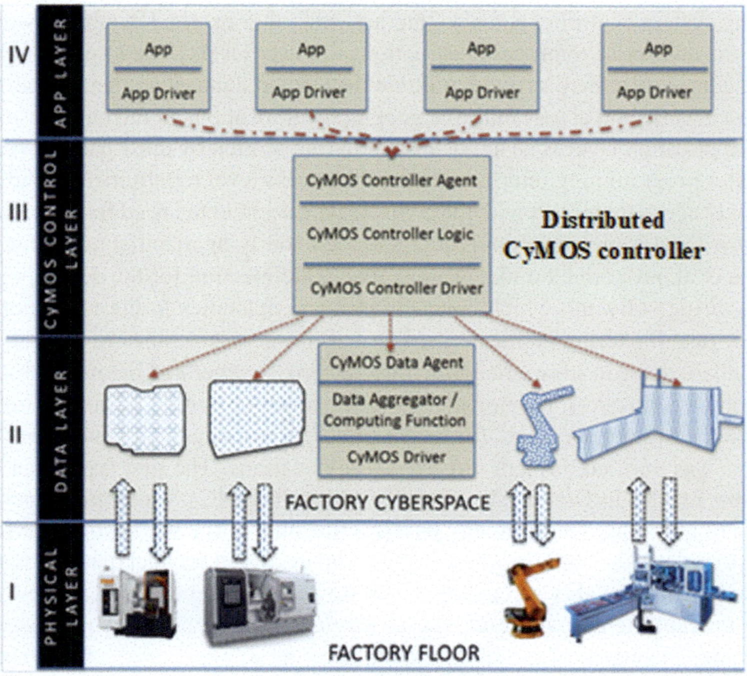

Fig. 11.10 Operating system control architecture based on digital twin

The data layer represents the data communication between the physical machine layer. Each machine in the workshop will have a separate virtualization state, which is supported by a database instance that is constructed to receive real-time streaming data from the machine. Drivers written on this data layer can communicate with the proprietary interfaces of individual machines. The data analysis algorithm embedded in the data layering algorithm is used to analyze and summarize information for higher-level system calls. At the upper level of this layer, each virtualization state of the machine has an agent that communicates with the operating system control layer. This agent is responsible for translating instructions from the control layer and generating the results of requests related to the physical machine, which can be used by any software application or device that requests it.

The operating system control layer is essentially a management program that controls the virtualization status of each machine in the production workshop. It contains information about virtual machine tools in its network, the current state of the machines, and other information collected from external data sources such as ERP or MES systems. At the bottom of this layer, there are necessary drivers for communication with individual Digital Twin. In the middle of the stack is a higher-level computing algorithm used to aggregate data from virtual states, which can transform input requests from applications located on top of the operating system architecture, and then obtain valid data requests from the corresponding virtualization states. The operating system controller can also interact with other external databases within or outside the enterprise to process instructions to lower levels. Due to the complexity at the factory level, there may be multiple distributed controllers distributed within the factory workshop across multiple geographical locations in the enterprise.

The application layer is an application program written by third-party developers who prefer programming interfaces to meet business level requirements. The application only needs to interact with the control layer, without the need for developers to know how to pass specific input requests to the underlying manufacturing machine.

Singh et al. proposed a middleware software architecture for the development of manufacturing software, which connects physical machines in the workshop with customer manufacturing applications. It connects smart machines to a highly scalable database that can store streaming time series data generated by onboard sensors and machine controllers. Meanwhile, the mechanism of writing through third-party applications without the need for direct physical communication with workshop machines was verified through three client applications. The first type is an application that resides in Digital Manufacturing Shared (DMC), which implements the function of querying data from any physical machine in the workshop; The second application is a Python application used to compare digital product data and machine generated data for quality screening; The third application builds a LabView application that can interact with middleware services. The architecture proposed here enables the ecosystem of intelligent manufacturing applications to be established and deployed in the workshop through open-source software and hardware devices, thereby reducing the cost of manufacturing software development.

The high-level logical structure of middleware is shown in Fig. 11.11. As mentioned earlier, each physical machine with a corresponding controller (such as SIEMENS, FANUC, MAZAK) has a corresponding MTCONNECT agent that continuously transmits data from the machine to any service that requests them. Then store the basic flow data from the controller and any sensors related to the machine in a structured database. To make this data useful, the script divides it into three hierarchical structures. The bottom layer is the machine information layer, which contains the raw data of all machines in the workshop; The second layer contains process related information, especially the processes executed by the machine. If allowed, this layer can obtain input from other workshop ERP or MES systems. The third layer is the summary information layer, which is calculated by the first, second, and second layers and provided through a universal access library. Any third-party application will retrieve data through the General Access Library (GAL). The hierarchy below GAL provides hardware abstraction, which means that any third-party software does not need to directly access any hardware equipment in the workshop to achieve appropriate functionality. This middleware architecture is very similar to the hardware abstraction provided by the Android operating system, allowing any software developer to build applications without directly accessing any specific hardware included in the smartphone.

By establishing a digital twin of machine tools, effective control of physical machine tools can be achieved. At the same time, as a digital twin, virtual machine tools can analyze and calculate the production status and other data of physical

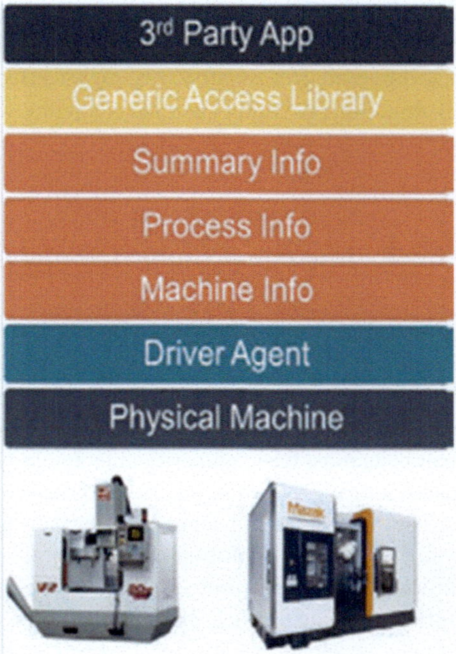

Fig. 11.11 An advanced middleware architecture

machine tools, achieving corresponding functions such as prediction, maintenance, and decision-making.

In addition, Siemens has proposed the concept of Digital Twin, dedicated to helping manufacturing enterprises build production system models that integrate manufacturing processes in the information space, achieving full digitalization from physical space to manufacturing execution. Dassault Company, based on a digital twin 3D experience platform, continuously improves the product design model of the information world through user feedback to optimize the product model of the physical world for complex product user interaction needs. Tesla aims to develop a digital twin for each vehicle, enabling synchronous data transmission between the car and the factory. In the development process of new passenger aircraft represented by the Boeing 787, Boeing fully adopts digital twin technology, defining three-dimensional product manufacturing information and three-dimensional design information into the three-dimensional model of the product, achieving high integration and collaboration of product design, tooling design, part processing, component assembly, and component inspection, and establishing an integrated application system for three-dimensional digital design and manufacturing, Created a new model for aircraft design and manufacturing, ensuring the development cycle and quality of Boeing aircraft. GE has developed the Predix digital twin platform to collect various information throughout the entire lifecycle of individual assets, analyze and predict past and present performance as well as future information, and conduct scenario testing and future optimization. Through this platform, abnormal bearing detection of jet engines can detect potential faults 15–30 days in advance. In addition, Predicting the lifespan of blades and arranging maintenance schedules reasonably can also save huge costs. Digital Twin are an effective means of integrating physics and information, providing clear ideas, methods, and implementation approaches for the implementation of CPS.

11.2.2 Intelligent Control Model for Operation Process

Information systems have added both human computer systems (HCS) and information physical systems (CPS) to manufacturing systems. The United States proposed the CPS theory in the early twenty-first century, which aims to achieve a perfect mapping and deep integration of information systems and physical systems. Digital twin is its most basic and critical technology, and the performance and efficiency of manufacturing systems can also be greatly improved.

The most essential feature of the new generation of intelligent manufacturing systems is that their information systems have added cognitive and learning functions. Information systems not only have powerful perception, computational analysis, and control capabilities, but also have the ability to learn, improve, and generate knowledge as shown in Fig. 11.12. The new generation of artificial intelligence technology will form a new generation of human information physical systems: (1) People will transfer some cognitive and learning based mental labor to information systems, which have the ability to recognize and learn; (2) By enhancing intelligence through human–machine hybrid technology, deep human–machine integration will fundamentally improve the ability of manufacturing systems to handle uncertainty problems, greatly optimizing manufacturing system performance [29].

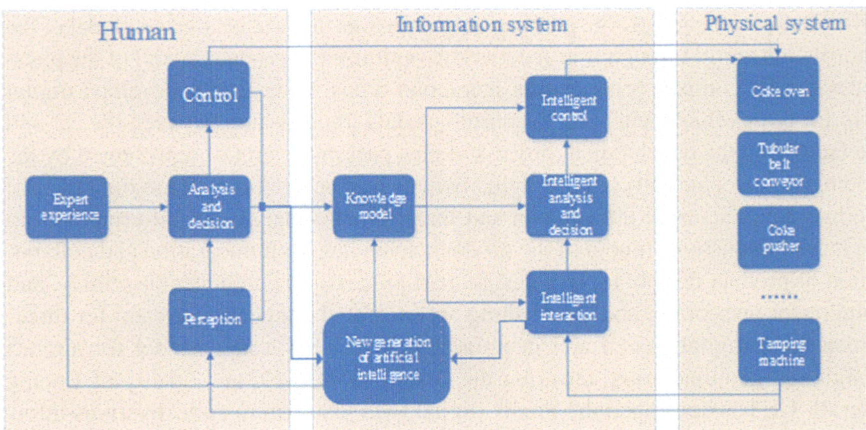

Fig. 11.12 Basic mechanism of the new generation intelligent manufacturing system

The new generation of intelligent manufacturing further highlights the central position of human beings, which is a comprehensive integrated large system that coordinates "human", "information system", and "physical system"; Will elevate the quality and efficiency of the manufacturing industry to a new level, laying a better material foundation for people's better lives; It will liberate humans from more physical and mental labor, enabling them to engage in more meaningful creative work, and human society will truly enter the "intelligent era".

The new generation of intelligent manufacturing isa large system composed of three functional systems: intelligent products, intelligent production, and intelligent services, as well as two supporting systems: industrial intelligent networking and intelligent manufacturing cloud. The transformation of the industrial model with intelligent services as the core is the theme of the new generation of intelligent manufacturing. In the era of intelligence, the full lifecycle services of products such as market, sales, supply, operation and maintenance are all endowed with new content due to new technologies such as the Internet of Things, big data, and artificial intelligence.

In the field of intelligent operation and maintenance of equipment, the digital twin based operation process control shown in Fig. 11.13 can achieve remote error compensation, regulation, and intelligent control of equipment operation process.

The operational status of sensor measurement equipment, as well as the dust level in the workshop, environmental factors such as temperature and humidity, which are known to affect equipment functionality, and the scrap rate of products. In the data detection stage, a Complex Event Processing (CEP) engine in the workshop digital twin model detects a complex pattern that indicates the onset of abnormal equipment behavior and degradation processes, or a warning of changes in production process status. CEP sends an event to the Prediction stage, triggering an online predictive analysis or optimization analysis service that uses big data-driven artificial intelligence methods such as statistical learning, machine learning, or deep learning to provide

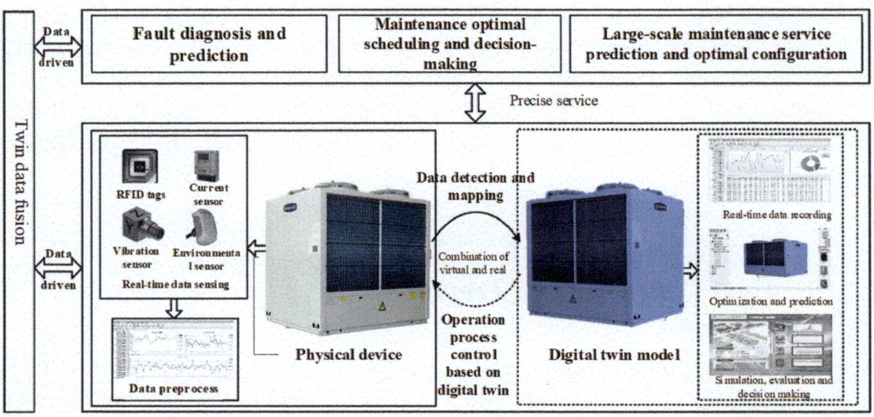

Fig. 11.13 Intelligent control model based on digital twin

RUL prediction or optimization of device operation status parameters. This prediction event triggers simulation, evaluation, and decision-making functions, which are simulated and evaluated through simulation and evaluation models. In the decision-making stage, a human–machine joint optimization decision-making strategy (such as domain expert decision-making, machine autonomous decision-making, human–machine enhanced hybrid decision-making, etc.) is adopted to formulate and provide online recommendations for the best parameters of equipment operation, the best time for equipment maintenance, and the best time to order relevant spare parts, And develop equipment operation control strategies and maintenance scheduling strategies. During the execution phase, by configuring and continuously monitoring key performance indicators through actuators, the process of data detection, optimization and prediction, simulation, evaluation and decision-making, and execution is completed to achieve continuous improvement of enterprise business performance.

11.3 Data-Driven Production Process Parameter Optimization Case Analysis

By conducting field research on the production process of a chemical plant shown in Fig. 11.14 and analyzing relevant production data, it is determined that the crude benzene production process is the analysis object. Based on the crude benzene production process, the key nodes affecting the crude benzene yield are preliminarily determined, and the correlation analysis of its operating data is carried out to determine the main influencing factors and control strategies that affect the crude benzene yield, assisting in optimizing the production process of the chemical plant. In the intelligent decision-making stage, the first step is to use Principal Component Analysis (PCA) method to analyze the key operational parameters that affect crude benzene yield and determine the main influencing factors on crude benzene yield; Then, a radial basis neural network (RBNN) was used to model the yield and quality of crude benzene, and the prediction model between the main influencing factors and crude benzene yield and quality was determined. By solving the prediction model, the optimal production control parameters for crude benzene were obtained, which provided control strategies for intelligent optimization control of the production process.

11.3.1 Data Collection and Preprocessing

We have collected production data related to each section of the factory in June 2018, as shown in Table 11.1. 744 sets of effective data were compiled, including various parameter indicators of coal blending in the drum cooling section, coking process

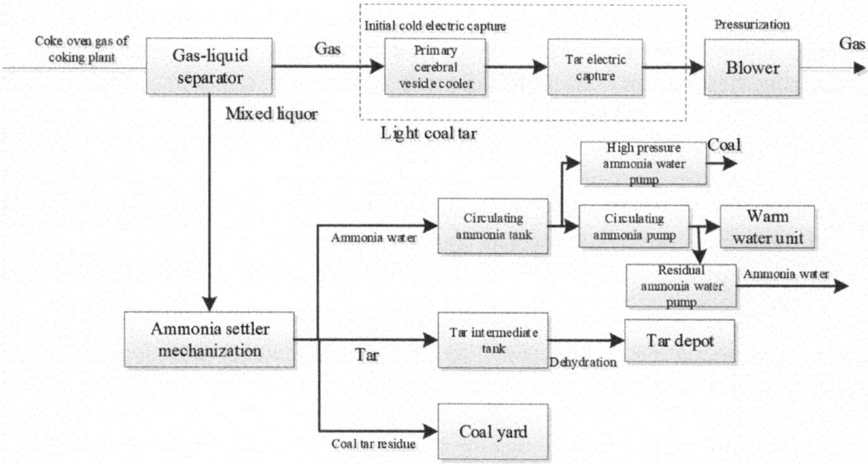

Fig. 11.14 Process flow diagram of the drum cooling section of a chemical plant

parameter indicators, and corresponding daily coke quality, gas yield, and energy consumption parameter indicators.

Unprocessed on-site report data may have the following issues: inconsistency, noise content, high latitude, etc. There are various factors that can cause deviation, such as human errors, data degradation, intentional errors, etc. The recorded production data cannot truly reflect normal production conditions, increasing the difficulty of data analysis. By grasping data trends and identifying anomalies, noise, outliers, and examining unusual values can be discovered. It is necessary to preprocess the original production data to improve data quality and enhance the efficiency, accuracy, and performance of the data analysis and mining process. Process the data as follows:

Table 11.1 List of production reports

Workshop section	Report name
Drum cooling section	Daily report of electric capture operation record for cross cut tube primary cooler
	Daily report of blower operation
Ammonium sulfate section	On site operation record of ammonium sulfate
	Daily report of central control operation record for ammonium sulfate system
Crude benzene section	Daily report of operation records for crude benzene washing station
	Daily report of operation records for crude benzene distillation tube furnace position

1. Missing value handling: Due to the potential risk of data errors caused by manual report entry, if one or more records in the dataset are missing, the method of ignoring these records should be adopted;
2. Cleaning up outliers: Due to unexpected production conditions or human factors, the recorded production data cannot truly reflect the normal production situation. It is necessary to clean up these inconsistent data before analyzing them. If there are errors in data input, they can be corrected by comparing with the previous and subsequent production data, or the record can be directly removed;
3. Remove duplicate data: Remove records in the dataset that have the same features and labels.

Due to the inconsistent dimensionality of different features in the dataset, there may be significant differences in numerical values, which may affect the results of data analysis if not processed. Therefore, it is necessary to normalize the data in a certain proportion (within a specific interval) for comprehensive analysis. The methods of normalization mainly include the following:

1. Min max standardization: Linear transformation of the original data to map the resulting values between [0, 1], with a transformation function of [0, 1],

$$x^* = \frac{x - min}{max - min}$$

Among them, max is the maximum value of the sample data, and min is the minimum value of the sample data. This method can enhance the stability of attributes with very small variances and also maintain entries with 0 in sparse matrices.

2. Z-score standardization: Standardize the data by giving the mean and standard deviation of the original data, and the processed data conforms to the standard normal distribution, that is, the mean is 0 and the standard deviation is 1. The conversion function is,

$$x^* = \frac{x - \mu}{\sigma}$$

Among them, μ is the mean of the sample data and σ is the standard deviation of the sample data.

Figure 11.15 shows the data after min max normalization, eliminating the influence of dimensionality and order of magnitude.

11.4 System Modeling and Computational Analysis

11.4.1 Mechanism Analysis

(1) Factors affecting the benzene washing section

	V1	V2	V3	V4	V5	V6	V7	V8	V9	V10
1	.375	.166666666667	.052631578947	.002424242424	.5	.75	.538461538462	.789473684211	.025147137507	.024038461538
2	.375	.217948717949	.052631578947	.001212121212	.5	.75	.538461538462	.684210526316	.025949705725	.024[illegible]
3	.250	.153846153846	.052631578947	.003636363636	.4	.75	.615384615385	.526315789474	.024612092028	.023237179487
4	.250	.128205128205	.052631578947	.004848484848	.5	.75	.615384615385	.684210526316	.026484751204	.023237179487
5	.375	.153846153846	.052631578947	.003636363636	.5	.75	.615384615385	.684210526316	.027019796683	.024038461538
6	.313	.153846153846	.052631578947	.004848484848	.5	.81	.615384615385	.736842105263	.028089887640	.024839743590
7	.313	.589743589744	.052631578947	.003636363636	.5	.81	.615384615385	.631578947368	.027287319422	.024839743590
8	.313	.474358974359	.052631578947	.001212121212	.5	.75	.615384615385	.631578947368	.027019796683	.024038461538
9	.313	.461538461538	.052631578947	.001212121212	.4	.81	.538461538462	.578947368421	.027822364901	.022435897436
10	.250	.679487179487	.052631578947	.002424242424	.5	.81	.538461538462	.736842105263	.029159978598	.023237179487
11	.313	.525641025641	.014354066986	.001212121212	.4	.75	.615384615385	.578947368421	.026752273943	.024038461538
12	.313	.538461538462	.052631578947	.001212121212	.4	.81	.615384615385	.526315789474	.022204387373	.024839743590
13	.313	.564102564103	.052631578947	.006060606061	.5	.94	.538461538462	.631578947368	.023274478331	.024839743590
14	.313	.294871794872	.052631578947	.002424242424	.5	.81	.538461538462	.684210526316	.021669341894	.024038461538
15	.500	.410256410256	.052631578947	.001212121212	.5	.88	.692307692308	.631578947368	.022471910112	.024038461538
16	.438	.282051282051	.052631578947	.001212121212	.5	.81	.615384615385	.631578947368	.022471910112	.024038461538
17	.438	.230769230769	.052631578947	.001212121212	.5	.88	.615384615385	.736842105263	.023006955591	.024038461538
18	.438	.333333333333	.052631578947	.001212121212	.5	.88	.615384615385	.578947368421	.021936864633	.024038461538
19	.500	.153846153846	.052631578947	.001212121212	.6	.88	.615384615385	.526315789474	.021669341894	.024038461538
20	.625	.269230769231	.052631578947	.002424242424	.6	.94	.615384615385	.631578947368	.025949705725	.023237179487
21	.563	.423076923077	.052631578947	.001212121212	.6	.88	.615384615385	.578947368421	.024879614767	.024839743590

Fig. 11.15 Schematic diagram of standardized dataset

The lean oil containing 0.2–0.4% crude benzene (washed oil after benzene removal) is pumped from the lean woil tank to the top of the 2 # benzene washing tower, pressurized by a semi rich oil pump at the bottom, and then sent to the top of the 1 # benzene washing tower. The rich oil flowing out of the 1 # benzene washing tower (washed oil after benzene removal) contains about 2.5% benzene, which is then sent to the benzene removal process. The lean oil after benzene removal is cooled and recycled.

In the process of benzene washing production, the factors that affect the yield of crude benzene mainly include the following:

1) Absorption temperature: The absorption temperature refers to the average temperature of the contact surface between the gas and washing oil in the benzene washing tower, which depends on the temperature of the gas and washing oil, as well as the influence of atmospheric temperature. The absorption temperature affects the crude benzene recovery rate by affecting the sparsity of absorption mass transfer and the change of mass transfer driving force. Increasing the absorption temperature can increase the molecular diffusion rate, thereby increasing the mass transfer coefficient between the gas film and the liquid film, but not significantly, The increase in crude benzene pressure on the washing oil surface is significant;
2) Washing oil molecular weight: When other conditions are constant, the smaller the washing oil molecular weight, the higher the concentration of free crude benzene and the better its absorption;
3) Washing oil circulation volume: Increasing the circulating washing oil volume can reduce the crude benzene content in the washing oil, increase the absorption driving force, and increase the mass transfer sparsity of the liquid film, thereby improving the crude benzene recovery rate and reducing the post tower loss of crude benzene. However, the circulating washing oil volume should not be too large, otherwise the load and various losses of the benzene removal equipment will increase;
4) Wash oil quality: Conduct a full distillation analysis of the circulating wash oil, determine the degree of deterioration of the wash oil quality based on the

analysis results, adjust the regeneration amount and temperature, and timely supplement fresh wash oil;

5) Benzene content in lean oil: The benzene content in lean oil is one of the main factors determining the content of benzene group hydrocarbons in the gas after the tower. When other conditions are constant, the higher the crude benzene content in the lean oil entering the tower, the greater the post tower loss;
6) Absorption surface area: The absorption surface area of the benzene washing tower is the surface area of the packing inside the tower. The larger the surface area of the seasoning, the longer the contact time between the gas and the washing oil, and the more complete the recovery process.
7) The pressure and flow rD_g ate of gas: When the gas pressure increases, the diffusion coefficient will decrease, and the sparsity of gas film mass transfer will be reduced. But as the gas pressure increases, the partial pressure of benzene hydrocarbons in the gas will increase proportionally, resulting in a significant increase in mass transfer kinetics and an increase in absorption rate. Increasing the gas flow rate can improve the sparsity of gas film mass transfer, thereby strengthening absorption mass transfer. However, the gas flow rate should not be too high to avoid increasing the entrainment of mist and increasing the resistance inside the tower.

(2) The principle of benzene removal from rich oil in the benzene removal section

When heating a liquid mixture, when the sum of vapor partial pressures of each component reaches a certain total system pressure, direct water vapor is introduced into the benzene removal and flow control system, which obviously reduces the distillation temperature. That is, at a lower temperature, crude benzene is completely evaporated from the wash oil. Therefore, the amount of direct steam used has an extremely important impact on the operation of benzene removal distillation. The direct steam introduced during the benzene removal distillation process is superheated steam to prevent water vapor from condensing inside the tower and entering the lean oil at the bottom of the tower. When the temperature of the direct superheated steam entering the benzene removal tower is higher than the washing oil temperature, the amount of direct steam will be proportionally reduced with its degree of overheating. In the benzene removal section, the factory uses a tube furnace heating method for stripping and removing benzene. The evaporation rate of each component from the rich oil in the benzene removal tower depends on the following factors: steam pressure of each component at the bottom oil temperature of the tower, operating pressure inside the tower, direct steam amount, and circulating washing oil amount.

2. Correlation analysis of data

The crude benzene recovery process is a high-order object with multiple parameter inputs and outputs, with many and long control channels, slow dynamic response, complex internal mechanisms, interrelated parameters, high nonlinearity, and high control requirements for production objects. Therefore, it is far from enough to establish a crude benzene recovery yield model solely based on the linear structure of variables. Firstly, based on the process mechanism and relevant analysis methods, identify the variable forms that are most closely related to the objective function. Then, introduce these variable forms into the model, and find a suitable optimization model for the problem of crude benzene recovery yield for modeling.

In the face of numerous factors that may affect the yield of crude benzene, in order to accurately and effectively determine the key factors affecting the yield of crude benzene, it is necessary to conduct correlation analysis on the factors that may affect the yield of crude benzene. By analyzing the relationships between different features or data, key influencing and driving factors are identified to optimize the crude benzene recovery process. Currently, there are five mainstream correlation analysis methods.

1) Icon related analysis: By visualizing data, it is difficult to identify trends and connections solely from a data perspective. However, plotting data points into charts can clearly reflect their trends and connections;
2) Covariance and covariance matrix: covariance is used to measure the overall error of two variables. If the trends of the two variables are consistent, the covariance is positive, indicating a positive correlation between the two variables. If the covariance is negative, indicating a negative correlation between the two variables. If the two variables are independent of each other, the covariance is 0;
3) Correlation coefficient: a statistical indicator that reflects the degree of closeness between variables. The value range of the correlation coefficient is $[-1, 1]$, where 1 indicates complete correlation between two variables, -1 indicates complete negative correlation between two variables, 0 indicates no correlation between two variables, and the closer the data approaches 0, the weaker its correlation;
4) Univariate regression and multiple regression: Regression analysis can establish statistical methods between two or fewer groups of variables;
5) Information entropy or mutual information: The method of measuring the prior relationship between features is called mutual information, which can determine which type of feature is closely related to the final result.

Based on the production operation data from May 26, 2018 to June25, 2018, establish a data file and perform min max normalization processing. After removing abnormal data, perform univariate regression on the factors to determine whether there is correlation between variables, and model and merge the variables with correlation.

(1) Correlation analysis between control variables and crude benzene yield

The correlation between the control variables and the crude benzene yield is shown in Fig. 11.16, with the crude benzene yield calculated based on the crude benzene yield per unit of raw gas.

The corresponding correlation coefficients are shown in Table 11.2. For the correlation coefficient of the heat map, variable 25 (crude benzene moisture content) has no correlation with other related factors, and observation of the dataset shows that it is basically a fixed value (0.1). Therefore, it is considered to delete this variable.

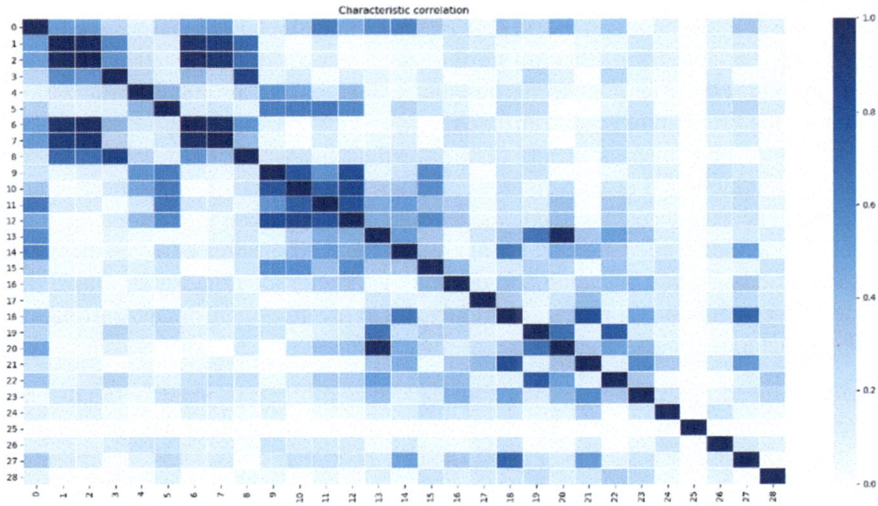

Fig. 11.16 Heat map of the correlation between control variables and crude benzene yield

(2) Correlation analysis between control variables and crude benzene yield

The correlation between various variables, including control variables and crude benzene mass, is shown in Fig. 11.17.

The corresponding correlation coefficient table is shown in Table 11.3. As above, for the correlation coefficient of the heat map, variable 25 (crude benzene moisture content) has no correlation with other related factors, and observation of the dataset shows that it is basically a fixed value (0.1). This variable has no correlation with crude benzene production and quality, so it is considered to be deleted.

The correlation between control variables 1–15 is shown in Fig. 11.18. To identify which correlated variables can be integrated, the next step is to perform variable dimensionality reduction. Principal component analysis is a data dimensionality reduction technique that can transform a large number of correlated variables into a small set of unrelated variables, which become principal components (Figs. 11.19 and 11.20).

The calculation steps of principal component analysis are as follows:

Step 1: Take the data matrix $X \in R^{m*n}$ under normal operating conditions, where m is the number of samples and n is the number of features:

$$(\mathrm{X}_1, \mathrm{X}_2, \ldots, \mathrm{X}_n) = \begin{bmatrix} x_1(1) & x_2(1) & \ldots x_n(1) \\ x_2(2) & x_2(2) & \ldots x_n(2) \\ \ldots & \ldots & \ldots\ldots \\ x_1(m) & x_2(m) & \ldots x_n(m) \end{bmatrix}$$

Among them: $X_1 = (x_1(1), x_1(2), \ldots, x_1(m),)^T, \quad i = 1, 2, \ldots, n$ n is the number of features.

Table 11.2 Corresponding table of correlation coefficients

	471.25000	8.37500	7.87500	0.51375	44.25000	21.63750	6.56250	5.05000	2.78750	23.27500	24.83750	24.76250	23.85000	89.37500
471.25000	1.00000	0.49436	0.48723	0.26070	0.08725	0.28757	0.49767	0.50648	0.17278	0.18811	0.33764	0.64131	0.45740	0.57586
8.37500	0.49436	1.00000	0.97732	0.55690	0.13873	0.09985	0.94279	0.88740	0.69115	0.06540	0.01349	0.14220	0.00973	0.02406
7.87500	0.48723	0.97732	1.00000	0.51911	0.16017	0.08062	0.97532	0.92339	0.66780	0.10069	0.03646	0.13043	0.01379	0.02309
0.51375	0.26070	0.55690	0.51911	1.00000	0.22790	0.08629	0.39785	0.25238	0.85755	0.08205	0.15153	0.02380	0.12704	0.08697
44.25000	0.08725	0.13873	0.16017	0.22790	1.00000	0.40192	0.12746	0.06576	0.25887	0.53220	0.45735	0.15683	0.39307	0.04764
21.63750	0.28757	0.09985	0.08062	0.08629	0.40192	1.00000	0.12807	0.14900	0.08574	0.62429	0.63211	0.64271	0.58214	0.03221
6.56250	0.49767	0.94279	0.97532	0.39785	0.12746	0.12807	1.00000	0.97182	0.53412	0.07975	0.00317	0.16283	0.01391	0.02846
5.05000	0.50648	0.88740	0.92339	0.25238	0.06576	0.14900	0.97182	1.00000	0.38096	0.05256	0.01122	0.15926	0.04591	0.06851
2.78750	0.17278	0.69115	0.66780	0.85755	0.25887	0.08574	0.53412	0.38096	1.00000	0.13716	0.17215	0.05538	0.17052	0.15375
23.27500	0.18811	0.06540	0.10069	0.08205	0.53220	0.62429	0.07975	0.05256	0.13716	1.00000	0.77460	0.53077	0.81478	0.06619
24.83750	0.33764	0.01349	0.03646	0.15153	0.45735	0.63211	0.00317	0.01122	0.17215	0.77460	1.00000	0.72793	0.84407	0.28235
24.76250	0.64131	0.14220	0.13043	0.02380	0.15683	0.64271	0.16283	0.15926	0.05538	0.53077	0.72793	1.00000	0.77886	0.43105
23.85000	0.45740	0.00973	0.01379	0.12704	0.39307	0.58214	0.01391	0.04591	0.17052	0.81478	0.84407	0.77886	1.00000	0.41779
89.37500	0.57586	0.02406	0.02309	0.08697	0.04764	0.03221	0.02846	0.06851	0.15375	0.06619	0.28235	0.43105	0.41779	1.00000
6.33750	0.61468	0.04178	0.04317	0.07156	0.04621	0.28174	0.09066	0.13742	0.20141	0.17868	0.33441	0.50308	0.42039	0.48886
1.38750	0.32364	0.04196	0.03261	0.15283	0.11742	0.21233	0.00332	0.01863	0.17397	0.57022	0.55547	0.36701	0.56874	0.30942
20.62500	0.23767	0.18217	0.19590	0.02843	0.09329	0.09821	0.23401	0.20731	0.06141	0.19008	0.17733	0.29792	0.32104	0.03832
91.37500	0.04308	0.14288	0.18097	0.03896	0.04121	0.02574	0.16162	0.16739	0.11646	0.01508	0.02806	0.04344	0.03304	0.16120
185.25000	0.37897	0.05621	0.05497	0.08150	0.19109	0.25940	0.07764	0.15635	0.15728	0.19710	0.22514	0.21763	0.18657	0.34025
3.35000	0.27471	0.04098	0.06404	0.28480	0.17110	0.24636	0.11477	0.13014	0.21140	0.19714	0.06813	0.13597	0.13624	0.65375
41.25000	0.47601	0.03053	0.03684	0.11589	0.05475	0.01061	0.03508	0.00013	0.16772	0.01663	0.24302	0.35604	0.36083	0.98098
182.00000	0.19041	0.05864	0.06473	0.01207	0.04958	0.01331	0.05566	0.09988	0.03870	0.01462	0.00458	0.01357	0.01625	0.32048
402.25000	0.35057	0.08093	0.12653	0.28371	0.13308	0.09549	0.21993	0.24232	0.22682	0.06452	0.17978	0.32386	0.29472	0.50034
130.00000	0.05287	0.17500	0.19490	0.08765	0.05825	0.20314	0.23566	0.19091	0.12506	0.16529	0.10216	0.17549	0.13526	0.35311

(continued)

Table 11.2 (continued)

0.36000	0.11474	0.08161	0.10276	0.05001	0.02653	0.02863	0.11106	0.11164	0.05213	0.11305	0.01584	0.05878	0.05673	0.11312
0.10000	4.01E-16	4.42E-15	7.94E-15	2.24E-15	2.95E-15	5.95E-15	1.67E-15	7.98E-16	1.59E-15	1.16E-14	2.87E-15	1.45E-14	1.10E-14	0
22.00000	0.10613	0.10383	0.10036	0.15164	0.19452	0.20871	0.11025	0.11385	0.03365	0.20937	0.16366	0.17238	0.16277	0.04540
4382.0000	0.32908	0.13614	0.12420	0.02263	0.11050	0.26385	0.12177	0.15771	0.01545	0.16516	0.21826	0.22607	0.17027	0.14903
0.25929	0.10362	0.00597	0.00290	0.01882	0.02495	0.11809	0.01173	0.01138	0.06627	0.07387	0.02152	0.08593	0.03425	0.12378
6.33750	1.38750	20.62500	91.37500	185.25000	3.35000	41.25000	182.00000	402.25000	130.00000	0.36000	0.10000	22.00000	4382.0000	0.25929
0.61468	0.32364	0.23767	0.04308	0.37897	0.27471	0.47601	0.19041	0.35057	0.05287	0.11474	4.01E-16	0.10613	0.32908	0.10362
0.04178	0.04196	0.18217	0.14288	0.05621	0.04098	0.03053	0.05864	0.08093	0.17500	0.08161	4.42E-15	0.10383	0.13614	0.00597
0.04317	0.03261	0.19590	0.18097	0.05497	0.06404	0.03684	0.06473	0.12653	0.19490	0.10276	7.94E-15	0.10036	0.12420	0.00290
0.07156	0.15283	0.02843	0.03896	0.08150	0.28480	0.11589	0.01207	0.28371	0.08765	0.05001	2.24E-15	0.15164	0.02263	0.01882
0.04621	0.11742	0.09329	0.04121	0.19109	0.17110	0.05475	0.04958	0.13308	0.05825	0.02653	2.95E-15	0.19452	0.11050	0.02495
0.28174	0.21233	0.09821	0.02574	0.25940	0.24636	0.01061	0.01331	0.09549	0.20314	0.02863	5.95E-15	0.20871	0.26385	0.11809
0.09066	0.00332	0.23401	0.16162	0.07764	0.11477	0.03508	0.05566	0.21993	0.23566	0.11106	1.67E-15	0.11025	0.12177	0.01173
0.13742	0.01863	0.20731	0.16739	0.15635	0.13014	0.00013	0.09988	0.24232	0.19091	0.11164	7.98E-16	0.11385	0.15771	0.01138
0.20141	0.17397	0.06141	0.11646	0.15728	0.21140	0.16772	0.03870	0.22682	0.12506	0.05213	1.59E-15	0.03365	0.01545	0.06627
0.17868	0.57022	0.19008	0.01508	0.19710	0.19714	0.01663	0.01462	0.06452	0.16529	0.11305	1.16E-14	0.20937	0.16516	0.07387
0.33441	0.55547	0.17733	0.02806	0.22514	0.06813	0.24302	0.00458	0.17978	0.10216	0.01584	2.87E-15	0.16366	0.21826	0.02152
0.50308	0.36701	0.29792	0.04344	0.21763	0.13597	0.35604	0.01357	0.32386	0.17549	0.05878	1.45E-14	0.17238	0.22607	0.08593
0.42039	0.56874	0.32104	0.03304	0.18657	0.13624	0.36083	0.01625	0.29472	0.13526	0.05673	1.10E-14	0.16277	0.17027	0.03425
0.48886	0.30942	0.03832	0.16120	0.34025	0.65375	0.98098	0.32048	0.50034	0.35311	0.11312	0	0.04540	0.14903	0.12378
1.00000	0.32510	0.03062	0.08648	0.63654	0.21227	0.44135	0.43092	0.32391	0.09907	0.13313	2.75E-15	0.16233	0.49235	0.04207
0.32510	1.00000	0.38761	0.07003	0.14004	0.28493	0.26046	0.03193	0.34862	0.13267	0.20834	2.47E-16	0.05737	0.08446	0.19502
0.03062	0.38761	1.00000	0.08342	0.34895	0.25908	0.10699	0.31266	0.37977	0.42577	0.16924	1.01E-15	0.04532	0.31381	0.12886
0.08648	0.07003	0.08342	1.00000	0.29215	0.07672	0.17632	0.38349	0.08796	0.16065	0.14742	9.72E-15	0.07622	0.24226	0.15920
0.63654	0.14004	0.34895	0.29215	1.00000	0.06233	0.31556	0.79904	0.08546	0.47938	0.17612	5.92E-15	0.20022	0.71255	0.18652

(continued)

Table 11.2 (continued)

0.21227	0.28493	0.25908	0.07672	0.06233	1.00000	0.65842	0.07617	0.75382	0.12422	0.07520	2.21E-15	0.02058	0.20448	0.23976
0.44135	0.26046	0.10699	0.17632	0.31556	0.65842	1.00000	0.31459	0.48073	0.39572	0.08614	0	0.03023	0.14267	0.13035
0.43092	0.03193	0.31266	0.38349	0.79904	0.07617	0.31459	1.00000	0.08274	0.57189	0.29886	4.23E-15	0.15174	0.52682	0.18834
0.32391	0.34862	0.37977	0.08796	0.08546	0.75382	0.48073	0.08274	1.00000	0.31809	0.01463	1.16E-15	0.00710	0.10059	0.31818
0.09907	0.13267	0.42577	0.16065	0.47938	0.12422	0.39572	0.57189	0.31809	1.00000	0.17404	4.11E-15	0.06975	0.18947	0.26315
0.13313	0.20834	0.16924	0.14742	0.17612	0.07520	0.08614	0.29886	0.01463	0.17404	1.00000	1.02E-15	0.19304	0.06459	0.04340
2.75E-15	2.47E-16	1.01E-15	9.72E-15	5.92E-15	2.21E-15	0	4.23E-15	1.16E-15	4.11E-15	1.02E-15	1.00000	7.23E-16	0	1.74E-15
0.16233	0.05737	0.04532	0.07622	0.20022	0.02058	0.03023	0.15174	0.00710	0.06975	0.19304	7.23E-16	1.00000	0.14150	0.10414
0.49235	0.08446	0.31381	0.24226	0.71255	0.20448	0.14267	0.52682	0.10059	0.18947	0.06459	0	0.14150	1.00000	0.01132
0.04207	0.19502	0.12886	0.15920	0.18652	0.23976	0.13035	0.18834	0.31818	0.26315	0.04340	1.74E-15	0.10414	0.01132	1.00000

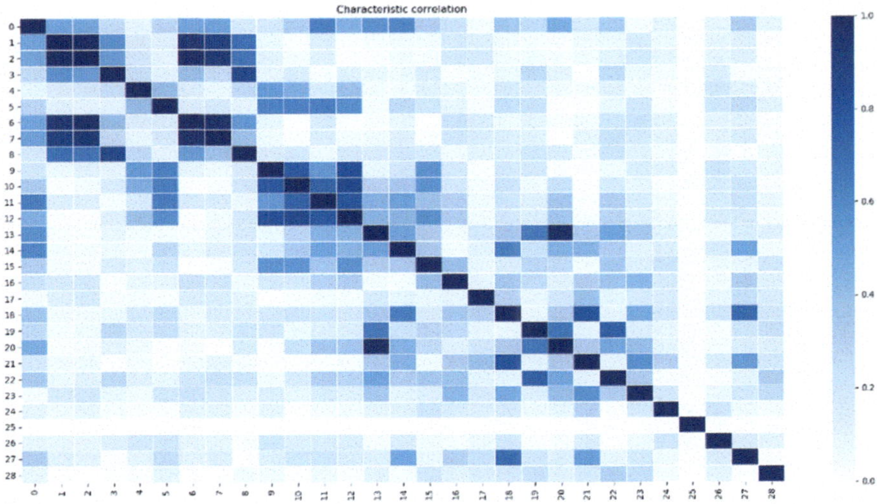

Fig. 11.17 Heat map of the correlation between control variables and crude benzene mass

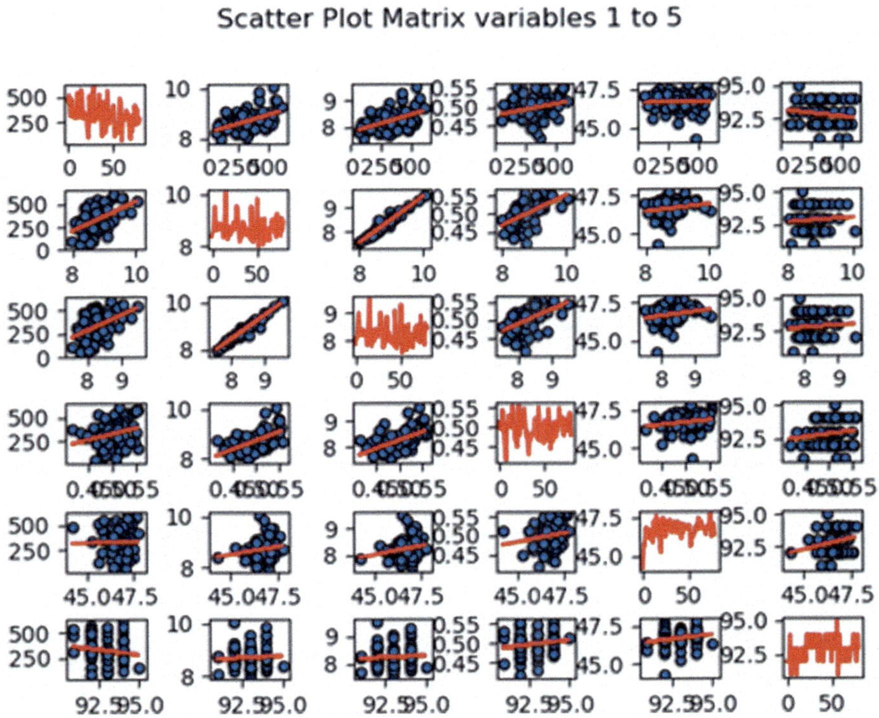

Fig. 11.18 Schematic diagram of correlation between control variables 1–5

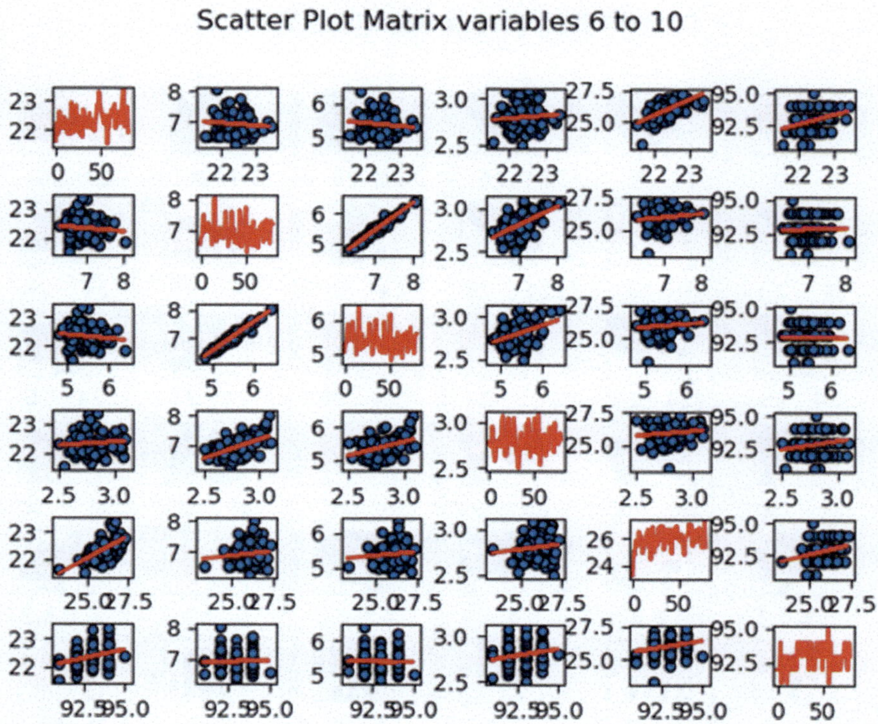

Fig. 11.19 Schematic diagram of correlation between control variables 6–10

Step 2: Eliminate the differences in dimensionality and order of magnitude among the feature variables, standardize the data, and obtain a normalization matrix;

Step 3: Establish a covariance matrix R based on the standardized data matrix, which is a statistical indicator reflecting the degree of correlation between the standardized data. The larger the value, the more necessary principal component analysis is to be performed on the data:

$$R = \begin{bmatrix} r_{11} & r_{12} & \cdots & r_{1n} \\ r_{21} & r_{22} & \cdots & r_{2n} \\ \vdots & \vdots & \ddots & \vdots \\ r_{m1} & r_{m2} & \cdots & r_{11} \end{bmatrix}$$

r_{ij} is the correlation coefficient between characteristic x_i and characteristic x_j, and the calculation formula is $r_{ij} = \frac{\sum_{k=1}^{n}(x_i(k)-\bar{x_i})(x_j(k)-\bar{x_j})}{\sqrt{\sum_{k=1}^{n}(x_i(k)-\bar{x_i})^2 \sum_{k=1}^{n}(x_j(k)-\bar{x_j})^2}}$

Step 4: Calculate the eigenvalues and eigenvectors of the covariance matrix R: $|\lambda I - R| = 0$.

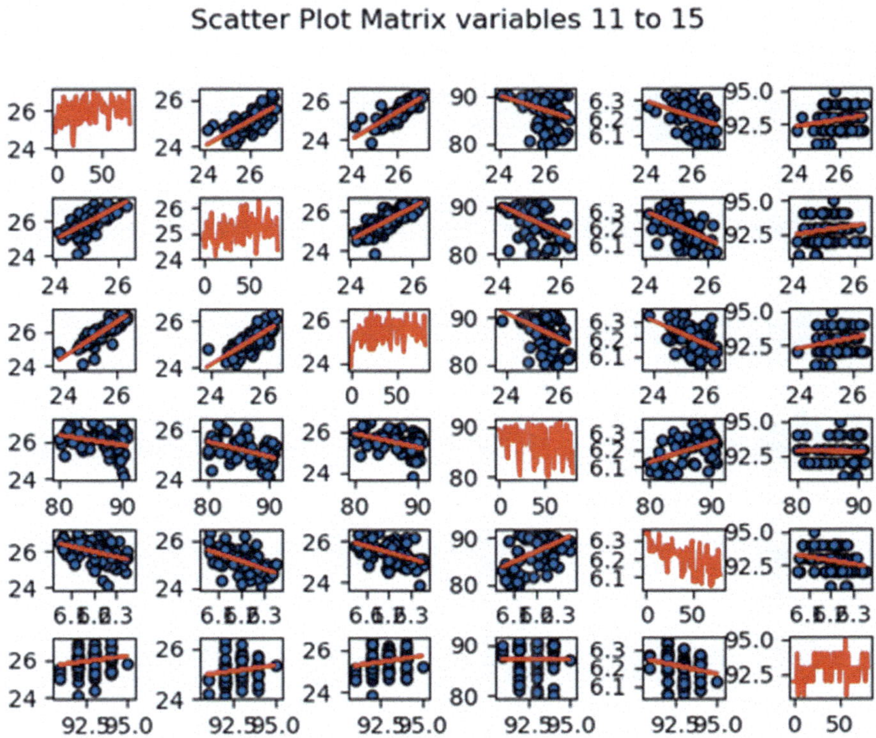

Fig. 11.20 Schematic diagram of correlation between control variables 11–15

Step 5: Arrange the feature values in order of their size. The eigenvalues are the variances of each principal component, and their magnitude reflects the calculation of principal component contribution rate and cumulative contribution rate:

Contribution rate: $\frac{\lambda_i}{\sum_{k=1}^{n}\lambda_k}(i = 1, 2, \cdots, p)$.

Accumulated contribution rate: $\frac{\sum_{k=1}^{i}\lambda_k}{\sum_{k=1}^{n}\lambda_k}(i = 1, 2, \cdots, n)$.

(5) Calculate the eigenvalues with a contribution rate of 80–90%, with p as the main component.
(6) Establish an initial fact Z_i or x_i loading matrix and explain the principal components. Factor loading is the correlation coefficient R between principal components and original features, which reveals the degree of correlation between principal components and can better explain the meaning between them.

$$Z = \begin{bmatrix} z_{11} & z_{12} & \cdots & z_{1n} \\ z_{21} & z_{22} & \cdots & z_{2n} \\ \vdots & \vdots & \ddots & \vdots \\ z_{m1} & z_{m2} & \cdots & z_{mn} \end{bmatrix}$$

The principal component analysis and variance contribution rate obtained through controlling variables are shown in Table 11.3.

The cumulative variance contribution rate is 80.501%, and the component score coefficients are shown in Table 11.4.

The principal components integrated through principal component analysis are shown in Fig. 11.21.

3. Radial basis function neural network modeling

From the perspective of optimizing operation and control, the prerequisite for achieving crude benzene recovery yield control and process optimization is to obtain feedback information on crude benzene yield. The complexity of the crude benzene recovery process and the lag in yield increase the difficulty of achieving closed-loop control of direct product yield. After such a large event, it is impossible to know the adjustment of operators in a timely manner, let alone achieve optimized control of crude benzene yield. Using certain technical means to estimate product quality is of great significance for process monitoring and optimization operations.

Yield estimation is essentially a modeling problem, which involves constructing a mathematical model that describes the functional relationship between measurable key operating variables, controlled variables, disturbance variables, and product yield. Based on process operation data, the estimated value of product yield is obtained. In the process of crude benzene recovery, due to the complex properties of crude benzene and the essential nonlinearity of the crude benzene process, the above relationship exhibits strong nonlinearity. According to different modeling methods, there are two solutions to achieve crude benzene yield estimation: (a) Establish a strict mechanism model, calculate the process material balance, and use a different method to trigger from the process mechanism, On the basis of reasonable simplification assumptions, the impact of various aspects of the process on product yield has been comprehensively considered, and yield estimation has high accuracy. However, due to the large size and high computational complexity of the model, as well as the lack of prior process mechanism analysis, significant deviations may occur, which limits the yield estimation of the mechanism model; (b) Establish a statistical model for product yield and identify the model based on existing process operation data and corresponding product quality data. This method is relatively simple and easy to implement on site, especially since the emergence of neural network technology represented by BP network, its strong non-linear processing ability and self-learning function have provided new favorable tools for product estimation.

Radial basis neural networks can naturally partition the input space, allowing the input vector to be extended to the high hidden unit space, thereby greatly reducing local minima. If the center selection of the RBNN network is appropriate, the hidden layer only requires a few neurons to achieve good approximation effect, and the learning speed is fast, with the characteristic of approaching the global optimum. Compared with traditional mechanism modeling methods, using RBNN neural networks to establish nonlinear system models not only has a few bit gradient, but also has higher accuracy and self use ability. The non-linear function of the hidden layer function in this article is selected as the Gaussian function.

Table 11.3 Explanation of total variance correspondence table (Extraction method: principal component analysis)

Ingredients	Initial eigenvalue			Extract sum of squares loading		
	Amount to	% of variance	Accumulated%	Amount to	% of variance	Accumulated%
1	6.261	23.188	23.188	6.261	23.188	23.188
2	5.056	18.724	41.912	5.056	18.724	41.912
3	3.468	12.844	54.756	3.468	12.844	54.756
4	3.286	12.170	66.926	3.286	12.170	66.926
5	1.418	5.251	72.177	1.418	5.251	72.177
6	1.192	4.413	76.591	1.192	4.413	76.591
7	1.056	3.911	80.501	1.056	3.911	80.501
8	0.958	3.548	84.049			
9	0.885	3.278	87.327			
10	0.612	2.268	89.595			
11	0.539	1.998	91.593			
12	0.464	1.719	93.312			
13	0.313	1.160	94.471			
14	0.277	1.025	95.496			
15	0.251	0.928	96.424			
16	0.227	0.840	97.264			
17	0.174	0.644	97.908			
18	0.124	0.459	98.367			
19	0.113	0.419	98.786			
20	0.100	0.371	99.157			
21	0.082	0.303	99.459			
22	0.054	0.199	99.659			
23	0.035	0.128	99.786			

(continued)

Table 11.3 (continued)

Ingredients	Initial eigenvalue			Extract sum of squares loading		
	Amount to	% of variance	Accumulated%	Amount to	% of variance	Accumulated%
24	0.032	0.119	99.905			
25	0.012	0.045	99.950			
26	0.009	0.035	99.985			
27	0.004	0.015	100.000			

Table 11.4 Component score coefficient matrix (Extraction method: principal component analysis)

	Ingredients						
	1	2	3	4	5	6	7
V1	−0. 119	0.079	0.039	−0. 031	0.113	0.018	−0. 152
V2	−0. 031	0.187	−0. 013	0.018	0.001	−0. 007	−0. 049
V3	−0. 030	0.190	−0. 008	0.010	−0. 049	−0. 015	−0. 040
V4	0.008	0.123	−0. 031	0.084	0.395	0.194	0.203
V5	0.059	0.070	0.105	−0. 096	0.059	0.106	−0. 065
V6	0.095	0.018	0.131	−0. 090	−0. 012	0.246	0.157
V7	−0. 036	0.183	−0. 010	−0. 008	−0. 135	−0. 077	−0. 082
V8	−0. 043	0.168	0.003	−0. 007	−0. 208	−0. 115	−0. 133
V9	0.019	0.148	−0. 038	0.063	0.301	0.185	0.168
V10	0.102	0.060	0.155	−0. 071	−0. 050	−0. 088	−0. 120
V11	0.122	0.048	0.109	−0. 010	−0. 065	0.056	−0. 040
V12	0.127	−0. 001	0.066	0.034	−0. 114	0.154	0.067
V13	0.133	0.045	0.103	0.016	−0. 080	−0. 035	−0. 082
V14	−0. 102	−0. 015	0.159	−0. 108	0.215	0.001	0.022
V15	−0. 116	−0. 013	0.084	0.029	−0. 075	0.034	−0. 127
V16	0.097	0.037	0.063	0.072	0.040	−0. 333	−0. 092
V17	0.031	−0. 042	0.126	0.154	0.160	−0. 226	−0. 215
V18	−0. 022	0.020	0.019	0.127	−0. 345	0.298	0.402
V19	−0. 093	−0. 022	0.121	0.178	−0. 118	0.084	−0. 087
V20	0.046	−0. 006	−0. 129	0.221	0.090	−0. 007	−0. 198
V21	−0. 090	−0. 025	0.170	−0. 108	0.227	0.002	0.018
V22	−0. 052	−0. 001	0.186	0.148	−0. 114	0.111	0.131
V23	−0. 066	0.016	0.043	−0. 224	−0. 205	−0. 021	0.033
V24	0.003	−0. 042	0.203	0.086	0.127	0.088	0.079
V25	0.002	−0. 032	−0. 097	−0. 039	0.021	0.516	−0. 345
V27	−0. 045	0.011	−0. 020	0.076	0.001	−0. 297	0.515
V28	0.074	−0. 002	−0. 070	−0. 176	0.098	−0. 079	0.285

RBNN neural network is a feedforward neural network, generally divided into a 3-layer structure, as shown in Fig. 11.22.

Radial basis neural network function model: $y = \sum_{i=1}^{h} w_i . \exp(-\frac{1}{2\sigma_i^2}(\|x - c_i\|^2))$ $i = 1, 2, \ldots, h$. Among them, h is the number of hidden layer nodes; c_i is the center vector of the i-th node in the hidden layer; σ_i is the shape parameter corresponding to the i-th center vector; w_i is the weight from the i-th node in the hidden layer to the output layer.

FAC1_1	FAC2_1	FAC3_1	FAC4_1	FAC5_1	FAC6_1	FAC7_1
-2.59306	-1.56248	-1.35014	1.84584	-.28410	2.10342	1.12262
-1.50452	-.63716	.28834	.97137	-.99224	.84225	.64001
-1.86331	-.65707	-.58725	.92147	.20966	.55254	.56658
-.80748	.33378	.13072	1.01128	-1.10511	-.73712	-.54291
-.17596	1.81041	.37271	1.64619	-.13220	.50611	.14606
-.99241	-.01004	.19814	1.26033	.12824	-.04302	.43181
-1.08086	-.35585	.26281	.23232	-1.15984	-.57175	-.25968
-.74019	-.47059	1.46523	-.07795	-1.29645	-1.55189	-.81259
-.66827	-.55841	1.03815	-.20066	-1.39471	-1.55612	-1.00442
.23674	.77527	.48326	1.23588	-1.22712	-.60241	-.18032
-.25649	-.08514	-.46638	1.69045	.02488	-1.44325	.13801
-1.34913	-.10602	-.45416	1.02976	.74078	.71274	-.36071
-.38873	.36330	.27276	.67809	-.49285	.32997	-.92578
-1.16129	-1.00773	.27657	.31278	.95804	-.35212	1.03211
-1.17642	-.37851	.35671	.55170	.73612	-1.12375	-.37279
-.68215	1.54413	.42339	.68991	1.20969	-.38317	.94682
-.96269	3.15320	.07213	.92604	-.38532	-1.69681	.74665
.45117	-.17008	1.68998	-.62895	.17667	1.33034	-.04081
-1.21024	-1.48555	.60094	.23847	.10447	-.60638	.83177
-1.64307	-.94049	.08450	-.07253	-.27789	-1.06796	-.25128
-1.43643	-.75389	-.35124	.09580	-.48353	-1.23061	-.31989
.35504	-.42699	.54028	.75858	-1.03219	.39351	.83076

Fig. 11.21 Principal component sample table

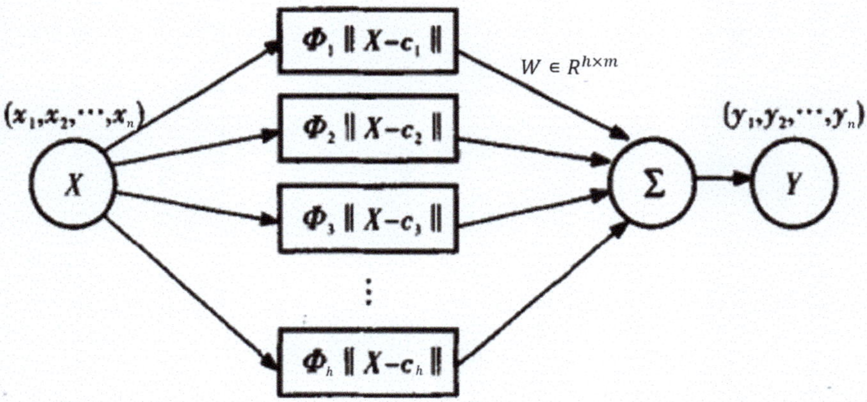

Fig. 11.22 Schematic diagram of RBNN model

After conducting principal component analysis on the control variables, the 28 control variables in the original dataset were integrated into 7 principal components. Establish a neural network model for crude benzene yield using the RBNN model. Input 7 variables and output the crude benzene yield. Establish a crude benzene yield prediction model as shown in Fig. 11.23 using the parameters obtained from the RBNN network learning process.

Prediction model for crude benzene yield:

$$y_{qual}(z_1, z_2, z_3, z_4, z_5, z_6, z_7) = \sum_{i=1}^{10} W_{1i} * G_{1i}(Z)$$

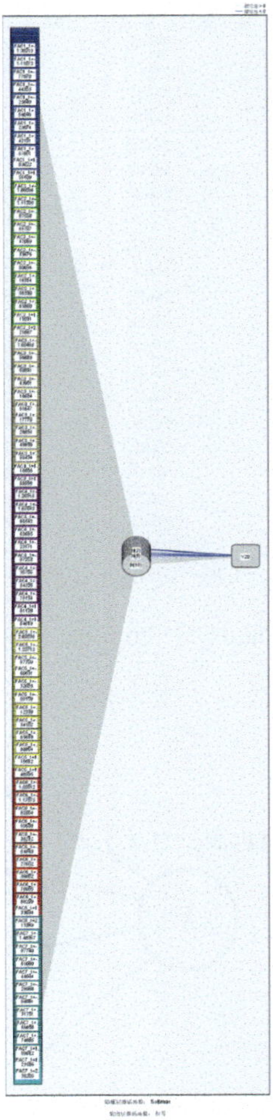

Fig. 11.23 Schematic diagram of crude benzene yield prediction model

Simulation experiments were conducted on the actual crude benzene yield using trained crude benzene yield models, and the results in Fig. 11.24 were obtained.

By conducting simulation experiments on actual crude benzene mass using trained crude benzene mass models, a prediction model for crude benzene recovery mass can be obtained, as shown in Fig. 11.25.

Quality prediction model for crude benzene:

$$y_{quan}(z_1, z_2, z_3, z_4, z_5, z_6, z_7) = \sum_{i=1}^{10} W_{1i} * G_{1i}(Z)$$

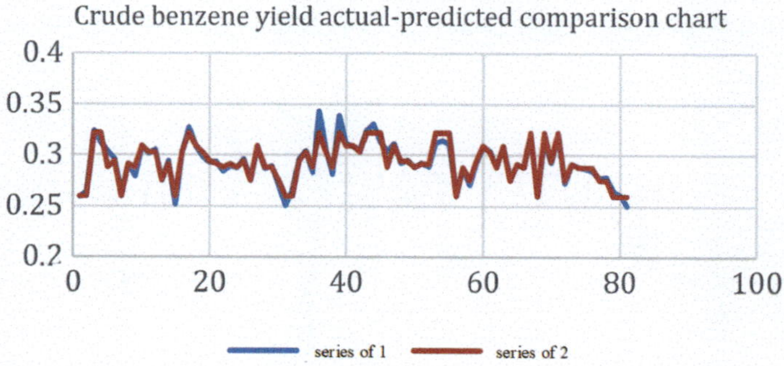

Fig. 11.24 Comparison chart of actual and predicted crude benzene yield (with an accuracy rate of 98.7%)

Fig. 11.25 Schematic diagram of crude benzene quality prediction model

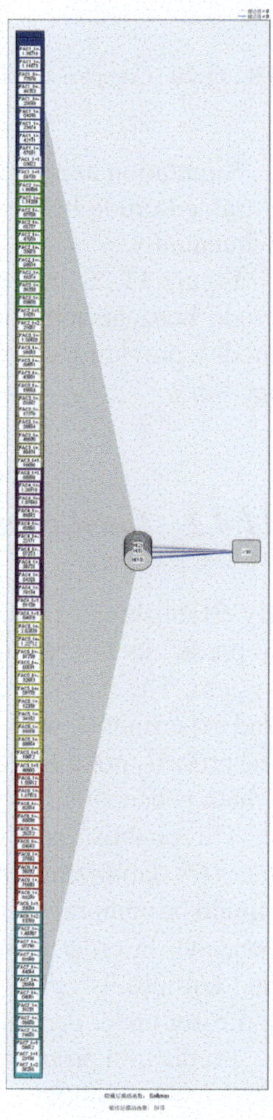

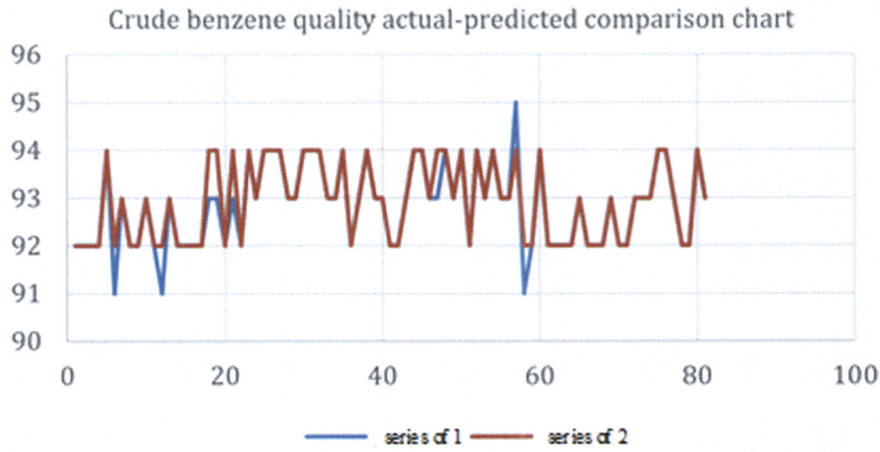

Fig. 11.26 Comparison chart of actual and predicted crude benzene quality (accuracy 99.99%)

Simulation experiments were conducted on the actual crude benzene yield using a trained crude benzene yield model, and the results in Fig. 11.26 were obtained sequentially.

Figure 11.26 and relative error data indicate that the prediction accuracy of the crude benzene mass model is significantly improved by establishing the RBNN model, providing strong support for achieving optimized control of crude benzene recovery.

11.4.2 Intelligent Production Optimization Strategy

By establishing a prediction model for crude benzene yield and quality, it is possible to predict its future output value based on the current status and future inputs of the system. The predictive model can be obtained through actual production numbers and experiments, without the need for a deep understanding of the internal stimuli of the process, nor the need for complex system identification and modeling operations, which is beneficial for improving the robustness of the system.

The establishment of a mathematical model for crude benzene yield provides a basis for optimizing calculations. The control objectives of the crude benzene section should be comprehensively considered in terms of both the yield and quality of crude benzene. In order to fully improve the control effect of the crude benzene section and tap into its operational potential, it is obviously necessary to establish a clear indicator under certain control parameters and achieve optimal results.

Production process optimization is the study of the system characteristics (structure, mechanism, process parameters) of a process, and based on this, by adjusting process parameters to achieve coordinated connections between systems and achieve the maximum effectiveness of the system.

The yield and quality of crude benzene are two important indicators for measuring the economic benefits of the crude benzene section in chemical plants. Therefore, using the yield and quality of crude benzene as the objective functions for intelligent optimization problems has practical significance.

Using Genetic Algorithm (GA) to solve the objective function. GA is a random, iterative, and purifying search algorithm based on the genetic mechanism of natural selection kernel population, with wide applicability. It has been widely applied in function optimization, combinatorial optimization, pattern recognition, and neural networks, and has become a new approach and method for solving highly complex problems.

Genetic algorithm is a universal optimization algorithm based on the principle of biological evolution, which introduces the evolutionary ideas of biological aggregation of genetic inheritance and "natural selection, survival of the fittest". The solution of optimization problems is regarded as the optimization process of feasible solutions, and its process is shown in Fig. 11.27. Each solution is represented as an individual using genetic coding, and an optimization objective function is used to determine the fitness of the individual for evaluation. By using genetic operators such as crossover and compilation, the population is combined to generate the next generation of individuals, gradually evolving towards an optimized population.

The basic steps of genetic algorithm:

Step 1: Randomly generate the initial population in the search space and encode it into a binary bit string form;

Step 2: Calculate the objective function for each individual, adjust it to obtain the corresponding fitness value, and calculate the individual's selection rate;

Step 3: Generate the next generation of individuals through genetic operator operation:

Copying refers to individuals copying according to their fitness values. The larger the adaptation value of the string, the more opportunities it will have to provide one or more descendants in the next generation;

Crossover: Randomly select a pair of parents from natural selection, determine the crossover position, and exchange the Weichuan defined by the crossover position of the two parents to generate a new pair of offspring loci;

Mutation: Randomly changing a bit string value with the minimum probability, which is relatively secondary to replication and crossover operations. Its purpose is to lose some useful genetic factors and restore the role of airway string diversity.

Prediction model for crude benzene yield:

$$y_{qual}(z_1, z_2, z_3, z_4, z_5, z_6, z_7) = \sum_{i=1}^{10} W_{1i} * G_{1i}(Z)$$

Quality prediction model for crude benzene:

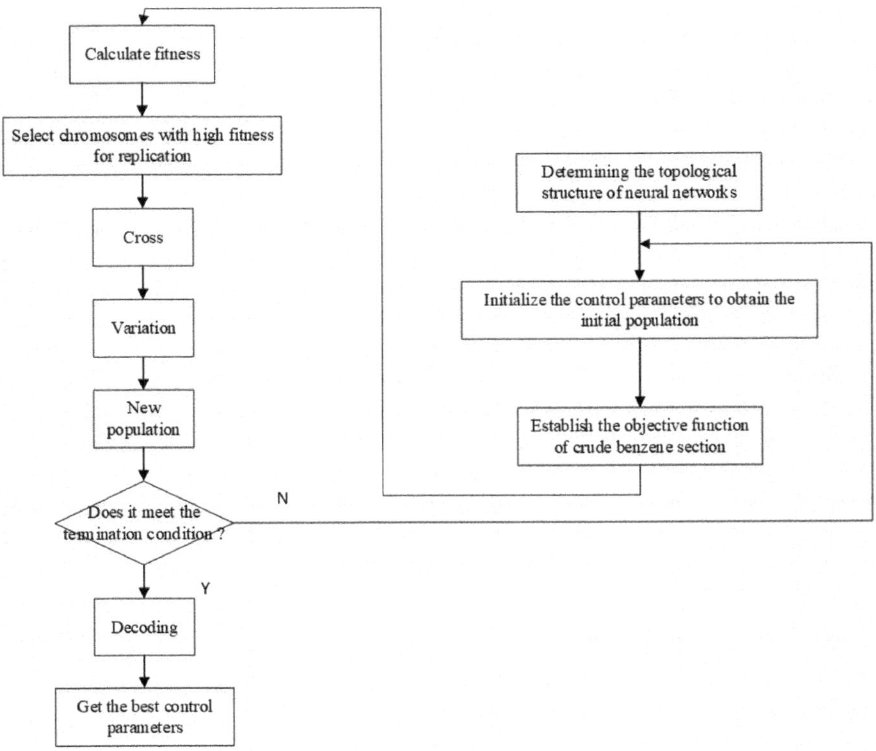

Fig. 11.27 Genetic algorithm flowchart

$$y_{quan}(z_1, z_2, z_3, z_4, z_5, z_6, z_7) = \sum_{i=1}^{10} W_{1i} * G_{1i}(Z)$$

$$G_{ki}(X) = \exp(-\frac{1}{2\sigma_{ki}^2}(\|x - c_{ki}\|^2) \quad k = 1, 2, \cdots, 10$$

Objective function: $f(.) = y_{qual} + y_{ayan}$

By qualitatively analyzing the process mechanism of crude benzene recovery, and comprehensively using correlation analysis and principal component analysis to determine the key factors affecting crude benzene yield and quality, a radial basis function neural network prediction model for crude benzene yield and quality was established, providing a basis for subsequent intelligent predictive control.

Introducing intelligent algorithms into the crude benzene recovery process to establish an objective function, and solving the objective function to determine the optimization control parameters of the crude benzene section, provides the possibility for enterprises to achieve intelligent production.

References

1. Baheti R, Gill H. Cyber physical systems The Impact of Control Technology [C] Control Systems Society Washington D. C., USA: IEEE, 2011:161–166
2. Fitzgerald J, Larsen P G, Verhoef M. From Embedded to Cyber Physical Systems: Challenges and Future Directions [M] Collaborative Design for Embedded Systems Springer Berlin Heidelberg, 2014
3. (In Chinese) He Jifeng Cyber Physical Systems [J]. Communication of the Chinese Computer Society, 2010,6 (1): 25–29
4. (In Chinese) China Forum onthe Development of Information Physical Systems. White Paper on Information Physical Systems [R]. Beijing, 2017,3
5. Lee J., Kang G. S. Development and use of a new task model for cyber physical systems: A real-time scheduling perspective [J] The Journal of Systems and Software, 2017, 126:45–56
6. (In Chinese) LiuChun, Huang Ranran, Han Daojun Analysis of Event Models for Target Based Information Physical Fusion Systems [J]. Computer Science, 2017,44 (4)
7. (In Chinese) Wang Yun, Liu Dong, Weng Jiaming, etc Research on Modeling and Simulation Verification Platform for Power Grid Information Physical Systems [J] Chinese Journal of Electrical Engineering, 2018,38 (1)
8. Ahmadi H, Abdelzaher T. An Adaptive Reliability Cyber Physical Transport Protocol for Spatial temporary Data [C] Real Time Systems Symposium, 2009: 238–247
9. Li K, Liu Q, Wang F. Joint Optimal Congestion Control and Channel Assignment for Multi radio Multi Channel Wireless Networks in Cyber Physical System [C] 2009 symposia and workshops on Ubiquitous, Autonomous and Trusted Computing in Junction with the UIC'09 and ATC'09 Conferences 2009: 456–460
10. Persson M, Hakansson A. A Communication Protocol for Different Communication Technologies in Cyber Physical Systems [J] Procedia Computer Science, 2015, 60:1697–1706
11. Yao X, Lin Y. Emerging manufacturing parade shifts for the incoming industrial revolution [J] The International Journal of Advanced Manufacturing Technology, 2016,85 (5–8): 1665–1676
12. (In Chinese) SunYanjing, Hua Gang, Dou Linming,etc Research and Challenges of Information Physical Systems in Mining Engineering [J]. Coal Science and Technology. 2018, 46 (2)
13. Potente T, Jasinski T, Wolff B. Productivity Increase through Industrialization of Overload in Cyber Physical Production Systems [J] Advanced Materials Research, 2013, 769 (14): 359–366
14. Monostori L, K Á D Á R B, Bauernhansl T. Cyber physical systems in manufacturing [J] CIRP Annals Manufacturing Technology, 2016,65 (2): 621–641
15. Richard G. Sharpe, Paul A. Goodall, Aaron D. Neal, et al. Cyber Physical Systems in the re use, refurbishment and recycling of used Electrical and Electronic Equipment Journal of Cleaner Production 170 (2018) 351–361
16. Berger C, Berlak J, Reinhart G. Service based Production Planning and Control of Cyber Physical Production Systems [C] Bled eConference, Bled, Slovenia, 2016
17. (In Chinese) Lei Xuefeng, Liu Weidong, Zhang Shen Research on Task Scheduling of Information Physical Systems in Coal Preparation Plants [J] Industrial and Mining Automation, 2015, 41 (3): 61–65
18. Pirvu B, Zamfirescu C, Gorecky D. Engineering insights from an anthropocentric cyber physical system: A case study for an assembly station [J] Mechatronics, 2016, 34:147–159
19. (In Chinese) Yin Cuntao Design of Automotive Hub Manufacturing System Based on CPS [J]. Automotive Manufacturing Technology, 2017,10
20. (In Chinese) XuGang, Zhao Miaoying Research on the Application of Information Physical Systems in Transformer Voltage Regulation Control [J]. Computer Application Research, 2019, 36 (2)
21. Liu C, Cao S, Tse W, et al. Augmented Reality assisted Intelligent Window for Cyber Physical Machine Tools [J] Journal of Manufacturing Systems, 44 (2017) 280–286
22. (In Chinese) TaoFei, Cheng Ying, et al. Theory andTechnology of Information Network Fusion in Digital Twin Workshop [J]. Computer Integrated Manufacturing System, 2017,23 (8)

23. (In Chinese) Chen Ming, Liu Jiangshan,Li Jielin,etcResearch on Intelligent Factory Deployment Strategy Based on New Characteristics of Information Physical Systems [J]. Chinese Journal of Engineering Machinery, 2017,15 (5)
24. Cyber Physical Systems Public Working Group Framework for Cyber Physical Systems Release 1.0 [R]. 2016.5
25. (In Chinese) Zhuang Cunbo, Liu Jianhua, Xiong Hui, etc The connotation, architecture, and development trend of product Digital Twin [J]. Computer Integrated Manufacturing Systems, 2017,23 (4)
26. Stavropoulos P., Papachalampopoulos A., Vasiliadis E., Tool wear predictability estimation in milling based on multi-sensor data International Journal of Advanced Manufacturing Technology, 2016,82 (1-4): 509-521
27. Cai Y, Starly B, Cohen P, et al. Sensor data and information fusion to construct digital wins virtual machine tools for cyber physical manufacturing [J] Procedia Manufacturing 10 (2017) 1031–1042
28. Singh S, Angrish A, Barkley J. et al. Streaming Machine Generated Data to Enable a Third Party Ecosystem of Digital Manufacturing Apps [J] Procedia Manufacturing 10 (2017) 1020–1030
29. (In Chinese) Zhou Ji, Li Peigen,Zhou Yanhong, Wang Baicun, Zang Jiyuan, Meng Liu Edward New Generation Intelligent Manufacturing Engineering, 4 (2018): 11–20

The manufacturer's authorised representative in the EU is Springer Nature Customer Service Centre GmbH, Europaplatz 3, 69115 Heidelberg, Germany. If you have any concerns regarding our products, please contact ProductSafety@springernature.com

Printed and bound by CPI Group (UK) Ltd, Croydon, CR0 4YY
30/09/2026
02225253-0016